DOCTRINES OF DEVELOPMENT

Doctrines of Development examines development as idea and practice from its early nineteenth-century invention to the late twentieth century, where the authors question whether development has ceased to have purpose.

The authors trace the idea of development from its origin by distinguishing it from the idea of progress and the dissatisfaction with progress in utilitarian, positivist, idealist and critical moments of western thought.

Extended case histories from Australia, Canada and Kenya illustrate why and how development doctrine accompanied early capitalism. Shorter case studies of India, Latin America and Austria are used to examine how development theory has been part of the history of development doctrine.

The authors argue that trusteeship – the intent of one to act on behalf of another – has been a powerful force in the formation of development doctrine and that despite claims for a 'new orthodoxy', trusteeship remains central to increasingly perverse theories of 'alternative' development.

M.P. Cowen is Reader in Economics at London Guildhall University.
R.W. Shenton is Associate Professor in the Department of History, Queen's University, Canada.

DOCTRINES OF DEVELOPMENT

M.P. COWEN and R.W. SHENTON

LONDON AND NEW YORK

First published 1996
by Routledge
11 New Fetter Lane, London EC4P 4EE

Simultaneously published in the USA and Canada
by Routledge
29 West 35th Street, New York, NY 10001

Typeset in Perpetua by Solidus (Bristol) Limited
Printed and bound in Great Britain by
Biddles Ltd, Guildford and King's Lynn

British Library Cataloguing in Publication Data
A catalogue record for this book is available from the British Library

Library of Congress Cataloguing in Publication Data
A catalogue record for this book has been requested

ISBN 0-415-12515-4
0-415-12516-2 (pbk)

CONTENTS

Part III

PREFACE

The doctrine of development, as it appears in this book, embodies the intent to develop. It is the question 'What is development?' that makes the existence of intentions to develop obvious. This is so if only because responses to the question of development usually present an image of something created anew, or improved, or renewed, or of the unfolding of potential which has the capacity to exist but which presently does not do so. Yet, to intend to develop does not necessarily mean that development will result from any particular action undertaken in the name of development. However, the existence of an intent to develop does mean that it is believed that it is possible to act in the name of development and that it is believed that development will follow from actions deemed desirable to realise an intention of development. An intention to develop becomes a doctrine of development when it is attached, or when it is pleaded that it be attached, to the agency of the state to become an expression of state policy.

When, as is often the case, the question 'What is intended by development?' is confused with the question 'What is development', an intention to develop is routinely confused with an immanent process of development. In its classical origin, and not only in ancient Greece, development was understood as a natural process in which phases of renewal, expansion, contraction and decomposition followed each other sequentially according to a perpetually recurrent cycle. In the modern world, a world in which it is artifice rather than nature that provides the analogue for the understanding of movement, development has increasingly come to refer to a discontinuous process in which destruction and renewal are simultaneous, as much as sequential. However, the essential unity of creation and destruction contained within the process of development has not changed; it still involves destruction. If we take the development of real-estate as a commonplace example of a process of development, we intuitively know that the property developer causes the old to be destroyed in order to create the new. Here we see how a process of development works through its negative dimension of decay, decomposition and destruction as much as through its positive dimension embodying an image of the new. One result of the confusion between the intentions to develop and an idea of a process of development is that the negative dimension of the latter remains hidden.

Any doctrine of development faces the inherent difficulty of bringing an intention to develop to bear upon a process of development. Indeed, most intentions to develop are themselves responses to what are deemed to be the undesirable effects – unemployment, impoverishment – of processes of development. Thus, in posing the question of 'What is intended by development?', a fundamental objection to destruction, and usually to the destruction of the old, is time and again given voice on the very grounds of the intention of development itself. Logically, however, it is difficult to understand how it is possible for development to be intended without the belief that destruction will create improvement, the purpose of the intent to develop. Belief in making development happen can only be grounded in a process of development, in that it is the process of development, and not the intention to develop, which makes destruction a necessary part of development. The difficulty of development is such that modern development doctrine is based upon a reversal in the order of the positive and negative dimensions of development as a process. It was the apprehension of the destructive dimension of a process of development which, as we shall show, was the starting point for the modern intention to develop. Intention, here, was to give order to a particular process of development, the development of capitalism, which, it was believed, embodied no developmental purpose and whose destructive dimension was poverty and the unemployment of the potential of productive power.

This book, like Caesar's Gaul, is divided into three parts. Part I is devoted to the invention of the modern doctrine of development in the first half of the last century and to the elaboration and critique of this doctrine. We begin, in Chapter 1, with the origin of modern development doctrine as espoused by the Saint-Simonians in early nineteenth-century France. It was the Saint-Simonians, and Auguste Comte in particular, who formulated development doctrine in response to what they regarded as the faults of the received idea of progress. Indeed, the idea of development that was embedded in the older idea of progress, that there was an unfolding of potential and that the new was a purposeful improvement on the old, was subsumed by development doctrine. Development doctrine, as we show, rested upon the intent to develop through the exercise of trusteeship over society. Trusteeship is the intent which is expressed, by one source of agency, to develop the capacities of another. It is what binds the process of development to the intent of development. Once development doctrine was 'in the air', no longer would it be possible to treat progress and development as if they were synonyms for one another. This is why evolution, which is also often conflated with development, came to appear in Darwin as 'progressive development'. Here progress was what qualified development but it was not its substitute.

There is a genealogy of the Saint-Simonian doctrine which runs from and through the nineteenth century to the present. One genealogical line, from Comte to John Stuart Mill and then, in the late nineteenth century, to the Fabian socialists, domesticated the doctrine for Britain and served to renew colonial trusteeship for the administration of development in the twentieth-century British imperial arena, especially in Africa. This book stemmed from our work on Fabian colonialism in Africa during the course of which we realised that the conceptual roots of the intention to develop had to be understood before we could give an account of the historical practice of development in Africa. Development practice in Africa could not be separated from what was purported to be development in Britain itself. A conceptual understanding requires concepts. Concepts are not jargon and the conceptual roots of development are not the jargon of development. Jargon, as Theodor Adorno makes clear, arises when the conceptual basis of phenomena is eclipsed by treating phenomena as the administrative matter and means of policy.[1] The jargon of development, therefore, makes 'development' serve as the administrative means of policy.

We are well aware, in starting with Saint-Simonian doctrine, that the modern intent to develop had a pre-nineteenth-century heritage and that one immediate conceptual source of the doctrine was the philosophy of Immanuel Kant. Kant's 1784 essay, 'Idea for a universal history with cosmopolitan intent', has been used to show the extent to which eighteenth-century thinking continues to find its way into official and quasi-official development doctrine. In particular, the emphasis which Kant put upon 'the destitution into which human beings plunge each other' has been emphasised as the positive source of development, as the means of what Kant hoped would be the 'inevitable escape' from destitution.[2] However significant Kant may be, we start with the positivist doctrine of development because it was what entered into state practice of the mid-nineteenth century as the positive means to confront the urban destitution and unemployment of labour which the development of capitalism had left in its wake. It was, we argue, from then that the intent to develop was given constructive purpose and that development doctrine was perpetuated as the practice to deal with the surplus population of the process of development.

The counterpoint to development, what is generally called underdevelopment, arose at the very moment when the positive doctrine of development was finding an official voice. This is why the seemingly strange theology of Cardinal Newman is taken up in Chapter 2. Newman's 1845 *Essay on the Development of Christian Doctrine* dealt with the criteria of development, asking the question of whether it was possible to evaluate the phenomena of development. However much his focus was on Christian doctrine, the way in which Newman established the difference between 'corrupt' and 'true development' bears a remarkable resemblance to the contemporary meanings of 'underdevelopment' and 'development'. The resemblance between corruption and underdevelopment is not fortuitous. Newman raised the distinction between the internal and external of development which, when transposed to the time and place of capitalist development, lies at the heart of underdevelopment theory. Newman's 'tests' of development, designed to evaluate the course of the development, were initially couched within the framework of the positivist method. Later, however, in his *Grammar of Assent* and in the revised edition of the *Essay* (1878), Newman appreciated how the evaluation of development restricted its compass and expanded the meaning of development to make the interior of development that which gave personal belief to doctrine, rather than simply being the belief of doctrine subject to the external authority of conscience.

For Newman, as for what follows about the idea of development in Hegel and Marx, in Chapter 3, destruction, decay and decomposition are of the essence of development. We argue that while it is difficult to find a doctrine of development in Hegel, Hegel did formulate a principle of development which Marx used for the purpose of explaining why the internal or interior of development resided in the human capacity and desire, without censor, to produce, create and imagine a future different from the present. For Marx, the doctrine of development throttled what he understood to be true development. We concentrate on his 1840s writings in which he scathingly attacked the German variants of Saint-Simonian doctrine as contained in both 'true socialism' and the writings of Friedrich List. Marx later developed his critique of capital which represented productive force as if it were the internal of development because it possessed the source of external authority to direct the work of labour power. However, what we emphasise is Marx's understanding of why the socialist version of trusteeship only reproduced the way in which the possibility of true development had been alienated by making the intent to develop reside in an authority which was external and prior to the process of development itself.

For Marx, the economic of development was constrained within a restricted domain of development. Capitalism developed progressively because it was through an internal process of development, devoid of constructive intent, that the potential of productive force expanded. This potential, however rich it made the class of capital, did not realise the potential embodied in the capacities to create and imagine if freed from the dictates of production. It was the expanded domain of development which encompassed the potential for universal freedom, a domain of the future which hung over the present restricted domain of development. What had been virtue for Newman, the subjection of the internal to an external authority of development, was horrific for Marx. Thus, Chapter 3 ends with a discussion of Friedrich List precisely because this apostle of national development asserted intent for the internal of development. As Marx's invective against him stressed, List's case of special pleading was for the state to become the external authority of development with the bourgeoisie as its trustee. For Marx no such authority was possible. Trusteeship made an external authority dominate over the internal process of development and thereby cemented the doctrine of development over the idea of development whose principle could only be found in the expanded domain. It was, for Marx, where the intent to develop subsumed the process of development within such a restricted domain that the possibility of freedom contained within the process of development was negated.

Part II of this book offers a series of case studies which serve two purposes. Chapters 4 and 5 contain extended studies of mid-century Victoria, Australia, and Quebec, Canada, and of twentieth-century Kenya. In one sense, our choice of these cases is random in that if we had chosen either proximate cases, such as New South Wales, Ontario, and Tanzania, or, indeed, different countries on other continents, we are confident that we would still have been able to provide forceful evidence for our argument. In another sense, the choice was determined by the particularities of the argument itself. Thus, Victoria was chosen because Hans Arndt, writing on the historical origin of modern development from his base at the Australian National University, teasingly argued that this origin was to be found in 1850s Victoria where the new self-governing colony intended development to create new, and renew existing, infrastructure to attract new populations to do the work of development.[3] Now as we all know, it may be dangerous to tease. The evidence which Arndt presented, in the form of an essay on 'economical development' by the influential engineer Charles Mayes, demonstrated exactly the opposite of what Arndt intended. Mayes argued for economic development in order to lock a surplus population into Victoria, whence, he feared, labour power would otherwise emigrate. Exactly the same intent informed development doctrine in Canada where the francophone part of the new self-governing alliance between Upper and Lower Canada attempted to prevent the emigration of the francophone population across the border to the United States.

The case of Kenya shows that development practice, in the passages from colonial to post-colonial government, continues to be what it had been in the nineteenth century. Development doctrine intends to do the opposite of what the idea of development, conveying a fluidity of movement from the old to the new, implies. Its intention in practice is to confront, compensate, and pre-empt this fluidity of movement in order to renew the agrarian conditions of development by locking up population in the countryside. Such is the agrarian doctrine of development.

Shorter case studies, of India, Latin America, Austria and, even more fleetingly, South Korea, serve a different purpose but one which is also partly that of the Canadian case in Chapter 4. The purpose of these studies is to show both how theory takes a hold of official practice and how practice becomes, in turn, a part of the theory for the doctrine of development. Thus, in the short case study of India, found in Chapter 1, we show how the differences between progress and development, the idea of development, and development doctrine were reiterated cyclically through official policy and practice. As such the case of India provides, as it were, a parallel history of nineteenth-century development. Our discussion of nineteenth-century positivism in Latin America, located in Chapter 2, can be seen as the now largely forgotten backdrop to more recent issues of dependency and underdevelopment, issues which have played such a large part in theorising about development. Austria, discussed in the early part of Chapter 7, arises both as an immediate backdrop to the discussion of Schumpeter's theory and the arena in which that theory has been put to work in the historiography of economic development. Alice Amsden's study on South Korea is used, towards the end of Chapter 7, to show why Schumpeter has become so significant for an understanding of East Asian industrialisation. Last, but not least, the extended case study of Britain, in Chapter 6, serves both purposes. Development doctrine, we argue, came to turn-of-century Britain as an interplay of socialist positivism, idealism and, through Joseph Chamberlain, of the constructivist German doctrine of industrial development. The intention of a British doctrine, deriving from List, was to provide the basis for industry in Britain to compete with that of Germany. It was to act against British de-industrialisation in the same way that List's doctrine a half-century earlier was meant to compensate for a loss of productive power which, List assumed, was due to British industrial development. If development doctrine failed to take root in Britain, as indeed did much of what was intended by idealist attempts to construct the common good of community through development, then both were to find an antagonistic place in the British administration of colonial policy. It is our intention in future work on Fabian colonialism to return to the case of Britain and its, especially African, empire. We therefore present the British case in an especially truncated form.

Part III of this volume explicitly takes up development as a twentieth-century phenomenon. Schumpeter's development theory, in Chapter 7, follows the positivist doctrine of development in that it offers a theory of trusteeship based upon an investment bank for society. In so far as the destructive dimension plays a necessary part in the internal process of capitalist development, Schumpeter's theory is akin to those of Newman, Hegel and Marx. However, Schumpeter differed from both earlier doctrines and ideas of development in that, amid the turn-of-century modernist currents of thought, he made self-development an end of development, expressed by the will of the entrepreneur; and the bank, the ephor of development which supervises the internal process of capitalist development, the external authority of development. In so far as Schumpeter's theory has been emended to temper his rejection of the state's source of external authority over development, it has proved to be more amenable for later twentieth-century industrialisation. The way in which Schumpeter contained the restricted domain within the economic system, and made the ephor of development the bridge between the 'system' and what he called 'the social order' of capitalism, allowed List's bourgeois trustees to reappear in a new light.

We conclude, with the end of Chapter 7 and in our *Conclusion*, Chapter 8, by bringing the modern history of development to bear on what has increasingly come to be a new orthodoxy of development. From the 1980s on, and thus during a period in which the constructivist purpose of state development has been ever more rejected as a basis for policy, the increasingly pervasive orthodoxy variously named 'another development', 'alternative development' or 'non-development' has formally rejected the totality of development theory as we have accounted for it in this book. In the face of the perpetual negative dimension of the capitalist process of development, it is not surprising that development doctrine should be dismissed as an intent to develop which has failed. As we argue in Chapter 2, and as has been confirmed by the case studies, the reaction against development doctrine has repeatedly occurred at the very moment in which the constructivist intent to develop was enunciated.

What is significant about the current new orthodoxy of this 'other development' is that, while trusteeship is the condition which makes the intent of development practicable, the assumption of trusteeship is refused by the adherents of 'other development'. This refusal to assume trusteeship opens a chasm between the internal and external of development and, from the vantage-point of development doctrine, unbinds the intent to develop from the process of development itself. Proponents and practitioners of 'other development' find their space to work in this chasm which barely could have been envisaged in the gamut of development theory from Comte to Schumpeter. Two consequences follow.

The first is that, whether or not the state is denuded of the constructive intent to develop, the external authority of development continues to be capital. Moreover, the ephor of development, the bank (or The Bank), continues to work on an international scale of reckoning. Second, because the practice of non-development is dissociated from the progressive of development and thus is confined to the intent to conserve, it reproduces what was always regressive about the process of development, to wit, the decay and decomposition of the old. As such, what was to be fulfilled by the internal of development, a future different from the present, is eclipsed by a recourse to the past. However, because of the process of capitalist development itself, this past is one which can only be imagined by the proponents of 'another development'. 'Another development' is, without intent, merely another doctrine of development.

If there is purpose in this book, it is to show how the idea of development has been entrapped by doctrines of development and to express the hope that the idea of development can be recovered from its entrapment in trusteeship.

Part I

1

THE INVENTION OF DEVELOPMENT

> The inorganic has one final comprehensive law, GRAVITATION. The organic, the other great department of mundane things, rests in like manner on one law and that is, DEVELOPMENT.
>
> (Chambers, 1844)[1]

> We should . . . design a civil service which is development-oriented.
>
> (Z. Cindi of AZAPO, a South African political movement, 1992)[2]

Development seems to defy definition, although not for a want of definitions on offer. One recent development studies text, notably entitled *Managing Development*, offers 'seven of the hundreds of definitions of development'. Here, and typically, the well-taken distinction between development as the means of transitive action and that of an intransitive end of action is conflated with a distinction between the state policy of development and the attempt to empower people, independently of the state, in the name of development. Thus, development is construed as 'a process of enlarging people's choices'; of enhancing 'participatory democratic processes' and the 'ability of people to have a say in the decisions that shape their lives'; of providing 'human beings with the opportunity to develop their fullest potential'; of enabling the poor, women, and 'free independent peasants' to organise for themselves and work together. Simultaneously, however, development is defined as the means to 'carry out a nation's development goals' and of promoting 'economic growth', 'equity' and 'national self-reliance'.[3] Given that there is scarcely a 'Third World' dictatorship which does not at least in part attempt to legitimise its mandate to rule in the name of development nor a development agency which does not espouse the rhetoric of popular empowerment, it is little wonder that we are thoroughly confused by development studies texts as to what development means.

A major source of confusion arises out of the way that the emblem of development is attached to a source of subjective action which is deemed to make development possible. When it is said that 'capitalism develops' we take development to be an immanent and objective process. But when we hear, for instance, that it is desirable for state policy to achieve the goal of 'sustainable development', then we understand it to be that there is a subjective source of action that can be undertaken in the name of development. To will the means of development in the name of a purposive end of development is to presuppose that it is possible for development to happen as the result of decision and choice. Development thus comes to be defined in a multiplicity of ways because there are a multitude of 'developers' who are entrusted with the task of development.

When attempts are made to establish how the means of action are related to a purposive end in the name of development, as is done in the introductory chapters of most textbooks on development, this confusion is compounded. Logically, the confusion arises out of an old utilitarian tautology. Because development, whatever definition is used, appears as both means and goal, the goal is most often unwittingly assumed to be present at the onset of the process of development itself. Thus, for example, we are told in Staudt's text that the goal of development is to enlarge choice. For choice to be exercised, let alone enlarged, it is assumed that there be the desire and capacity to choose as well as knowledge of possible choices. Yet, these three components of choice are routinely assumed to be as much preconditions for the development process as the goals in which the process results. If any one of these conditions is regarded as missing then it is that gap which development is invoked to bridge. To seek recourse in the rhetoric of 'empowerment' is not to solve the development problem but to replicate it. Either people have power to exercise choice, in which case there is no cause for empowerment, or they do not and the task of empowerment is that of the logical problem of development.

The nineteenth-century resolution of the development problem was to invoke trusteeship. Those who took themselves to be developed could act to determine the process of development for those who were deemed to be less-developed. Now, in the last part of the twentieth century, the logical sleight of hand which justified entrusting the means of development to 'developers' is no longer convincing. As doctrine, trusteeship stands condemned as Eurocentrism, an imperial vestige of the post-1945 attempt to improve living standards of poor colonies and poor nations through state administration. Development, when interpreted through the screen of trusteeship, is taken to have no meaning for 'Third World' countries and continents of mass poverty; it has had its time and has failed as an idea and a practice. Yet, it is still contended that 'development' is the means whereby the goal of universal human improvement can be attained. However much the surrogates of 'transformation', 'progress' and even 'revolution' are used to distance the purpose from 'development', we are still drawn back to believe that a surplus population is to be made productive and mass poverty eliminated through some form of empowerment, both through and against the state.

The confusion surrounding development is compounded yet again because attempts to define the word embrace the task of conveying some essential meaning of development. Thus, by way of development studies, students are asked to understand the purpose of what they are studying. By way of development agencies, experts are asked to reflect on the purpose of what they are appraising and managing. However worthwhile the purpose of these tasks, an essential meaning of development cannot be arrived at when its scope is so partial and when it is taken to have no past, or so little that its history can be ignored.

Development, since the 1970s, has come to inform official practice of the 'advanced' capitalist world, but is still more closely associated with practice of confronting poverty and unemployment in Africa, Asia and the southern Americas. It is here that we are led to believe that the domain of development lies. Equally, the period of development is invariably assumed to be a span of imperial and post-colonial history since 1945. The subject of development is that of the imperial state, before and after political dismemberment, while its object is taken to be colonial and Third World peoples. Even when the subject and object of development are inverted and the state is confronted in the name of development, the domain of development remains restricted and its essential meaning is lost.

Our intention here is to unravel some of the confusion about development by opening its scope. We do not add to its list of definitions, nor is it our task to find an essential meaning for a word in everyday use. What we do instead is to take development back to when it was first advanced as a 'hypothesis' to make us conscious of history in the modern world and to a time when it was hoped that an evaluation could be made as to whether 'development' had occurred in the course of history. We take the modern idea of development back to where it was invented, amidst the throes of early industrial capitalism in Europe. The idea of development is necessarily Eurocentric because it was in Europe that it was hoped to provide the constructivist means to compensate for results of the development of capitalism. It was here that development was meant to construct order out the social disorders of rapid urban migration, poverty and unemployment. If, as is now often contended, development has failed, it did so before it was debated as a source of state action at the turn of the twentieth-century beginning of the period of colonial history in Africa and again at the beginning of the end of the colonial period little more than fifty years later.

There are texts, especially in the academic discipline of the sociology of development, that locate the modern roots of 'development' in the early nineteenth century as coterminous with the origin of sociology itself. Bernstein, for example, begins the introduction to his widely used selection of readings *Underdevelopment and Development* by stating that: 'The idea of development as the progressive transformation of society begins to assume a modern form in the writings of the "founding fathers" of social science.'[4] More recent introductory texts, such as Harris's *The Sociology of Development* and Barnett's *Sociology and Development*, trace the genesis of development through the formulations of Comte, Spencer, Durkheim and Weber. Barnett, in particular, begins his text with the experience of the labour migrant, during the period of rapid migration at the turn of the century in Europe, to give focus to the Weberian idea that 'migrants have had to find new ways of solving the problems of order and morality as a result of the disruption resulting from changes affecting their lives'. Sociology developed thus, and created 'development', as 'a science which could bring about order in this suddenly changing and confusing world'.[5] Aspects of the historical consciousness of development certainly appear to be more developed in the academic study of sociology than in economics but blindness remains.

First, these texts fail to root the constructivist idea of development in Europe of the first half of the nineteenth century. The problem of order, as in Barnett's opening to his text, is that of the individual migrant worker facing the fact or prospect of unemployment. There is no witness to the prospect of how official state policy, constructed by means of development, was to be encouraged to manage social disorder and, foremost, the disorder of unemployment. Taking this to be 'common knowledge', Bernstein confirms a 'strategic contrast' between the post-1945 idea of development, attached to state policy, and the 'many earlier conceptions' of development. As Bernstein puts it, it is only 'in the period since the end of the Second World War' that 'development has become a slogan of global aspiration and effort'.[6] By 'global aspiration and effort', Bernstein implies that there was a constructive effort, especially on the part of the United States, to impose social and economic order upon the world.

Second, but not separately, development and progress are seamlessly stitched together. 'There can be no doubt', Aseniero has written in his history of developmentalism, 'that development has become the central organizing concept in terms of which the historical movement and direction of social systems are analysed, evaluated and acted upon.' However, Aseniero then insists that development, as the 'dominant organizing myth of our epoch', has taken over 'the role played by the concepts "progress" in the Enlightenment and "growth" in classical economics'.[7] Development, as in Harris' text, appears as an appreciation of the movement of history through the progress from agrarian to commercial and industrial society.[8] Or, for instance, in *Poverty and Development in the 1990s*, the most recent Open University text, Thomas refers to the two meanings of development: '(1) as an *historical process of social change* in which societies are transformed over long periods; and (2) as consisting of *deliberate efforts aimed at progress* on the part of various agencies, including governments, all kinds of organisations and social movements'.[9] There is little or no suggestion of development as the counterpoint to progress, little recognition that the early development theory of Comte, for instance, was based upon the idea that 'development' may be used to ameliorate the disordered faults of progress.

We likewise find that underdevelopment, as the failure of previously 'attested' capitalist development to reproduce itself in the Third World,[10] has a shallow history. Indeed, it is claimed in a recent *Development Dictionary* that 'Under-development began' on 20 January 1949,[11] the day on which Truman inaugurated his presidency of the United States with the following words:

> We must embark on a bold new program for making the benefits of our scientific advances and industrial progress available for the improvement and growth of underdeveloped areas. The old imperialism is dead – exploitation for foreign profit – has no place in our plans. What we envisage is a program of development based on the concepts of democratic fair dealing.[12]

While it is mentioned that Truman was not the first to use 'underdeveloped', we are informed that the word was probably 'invented' in 1942 to provide an economic basis for post-war peace.[13] In fact, Truman's idea was not new. Bourdillon, the governor of Nigeria, when addressing the Royal Empire Society in London in 1937, said: 'The exploitation theory . . . is dead and the development theory has taken its place.' In 1905, the British Liberal Prime Minister, Campbell Bannerman, responded to Joseph Chamberlain's injunction to develop 'our' estates in Africa with a development design of his own: 'We desire to develop our *undeveloped* estates in this country; to colonise our own country.'[14] The word may be different but the sense is the same. As for development, so for 'underdevelopment': neither was invented during or after the Second World War and neither was originally construed as part of a new imperial project for the colonial and post-colonial 'Third World'.

It was the global scope of 'aspiration and effort', according to Bernstein, which made development and underdevelopment new after 1945. Development was the name, as a slogan, for the attempt to confront world poverty which again was given the name, again as a slogan, of underdevelopment. When Truman pronounced on development, it was an expression of the world power of the United States and not a programme of policy that aspired to a singular national source of state sovereignty. Thus, when Bannerman baited Chamberlain at the turn of the century over a doctrine of development, which Chamberlain espoused, it was to emphasise a narrower scope of British national economy; Chamberlain aspired to construct the wider scope of an integrated and protected British imperial economy.

Time and again, we are impressed by the way that the idea of development is tied to that of state sovereignty. Geoffrey Kay puts it succinctly:

> National sovereignty can have no real meaning unless it is joined to the idea of development as progress towards a social and economic equality from which no nation is debarred for natural reasons. National sovereignty and development defined in this way adhere to each other as closely as the principle of equal rights adheres to that of the freedom of the individual.

But 'development', in its adhesion to nationalism, 'no longer carries conviction':

> Just as the principles of freedom and equality were emasculated in the nineteenth century and reduced to hollow, formal and reactionary incantations, so nationalist cries and the demand for development have been drained of whatever revolutionary content they may once have possessed.[15]

However drained of its meaning as an idea of progress, development came to be embedded in the predicate of social and political order. Progress, Comte had incanted, is the *development* of order. As such, development was the means towards which progress might be ordered but it was not the idea of progress itself. When development acquired a constructivist purpose, as the means by which the state could impose order upon society, it was attached to an idea of development whose origin was different from that of the idea of progress. Whereas the original idea of progress had seen the past as a series of inferior stages which formed a prelude to the future, the idea of development as an immanent process did not necessarily rest, as did the idea of progress, upon a conviction that the future would be an improvement upon the past; nor did it convey the impression that the future was to be projected from a present stage of history which represented the culmination of the past. Again, in the immediate post-1945 period, when 'development' had become so accentuated that many are led to believe that it was then discovered, we find a recognition that the origin of the idea of progress was different from that of development. Indeed, the recognition that progress was different from development was inspired by a theological belief that the idea of progress had been a perversion of Christian doctrine. In 1948, Brunner wrote that progressivism 'is the bastard offspring of an optimistic anthropology and Christian eschatology'; Baillie that 'the doctrine of progress is a Christian heresy. Like all heresies it is essentially a lopsided growth. It is the development of one aspect of the received truth to the neglect of other aspects.'[16] That progress should be understood to be 'lopsided growth' is no surprise to anyone familiar with the political economy of development and underdevelopment.

When underdevelopment appeared in the 1960s, it did so as a concept which is part of a theory of capitalist development. In this sense underdevelopment, like the idea of development itself, is more than a name. Underdevelopment theory (e.g. the work of Gunder Frank) does not, as the interpretation of Truman implies, take underdevelopment to be the name for Third World poverty. It is not simply an object upon which development is designed to act constructively. Rather, underdevelopment is a simultaneous part of a process of development. Underdevelopment is a destructive result of development, but the criteria for why a condition of mass poverty should arise out of development is to be found as part of a theory of development. In this case, the theory is one of capitalist development, but the method that gives rise to a meaning of underdevelopment as the distorted part of some true idea or model of development is to be found in the nineteenth century.

We find a modern meaning of underdevelopment, as part of a theory of development, before 1850. The word was '*corruption*' and the ostensible subject was the development of Christian doctrine. Acton, the historian, claimed that John Henry (Cardinal) Newman's 1845 *Essay on the Development of Christian Doctrine* 'did more than any other book of the time to make the English "think historically, to watch the process as well as the result"'. And Pattison, a cleric, wrote to Newman in 1878:

> Is it not a remarkable thing that you should have first started the idea – and the word – development, as the key to the history of church doctrine, and since then it has gradually become the dominant idea of all history, biology, physics, and in short has metamorphosed our view of every science, and of all knowledge.[17]

Largely through the influence of Newman, it was formally appreciated that however desirable it might be for history to move through 'true' developments, there were doctrines and practices that simultaneously failed to reproduce the original idea of the phenomenon in question. The failure of a mode of development to reproduce itself was referred to as corruption but it conveyed all that underdevelopment means to us today: decay and decomposition, disarticulation and disintegration of what was posited to be an integrated body of thought and practice. Newman's difference between true development and corruption was applied to the body of the Church.

Long before Newman, however, corruption had been an essential part of state theory as when, for instance, it was supposed that 'statecraft', sovereignty and policy were the means to forestall the natural decomposition of the 'body politic'. After Newman, his ideas of what lay behind sovereignty, through his account of what made allegiance different from conscience and what, through development, maintained the identity of the Church, were extended to the body of the state. Corruption became underdevelopment when the idea could be further extended to capitalism itself.

Development, by virtue of Newman's critique of the positive science of progress, was the counterpoint to corruption rather than progress. In latter-day theories of underdevelopment, the economy and society disintegrates with the development of capitalism. Where the development of capitalism is accompanied by industrial progress, according to the logical development of the productive force of a proletariat, then this model of true development is counterposed to the case where capitalism is deemed to be incapable of true development. The old idea of corruption becomes underdevelopment because it happens simultaneously with development; decay and decomposition do not necessarily follow a sequence of procreation and growth but are a simultaneous part of development itself.

That the birth of the constructivist idea of development has remained largely hidden from us is largely due to its eclipse by the dominant notion of the 'age of progress' of the last half of the nineteenth century. This period, which has conveyed to us so much about the progress of industrial capitalism in the nineteenth century, has hidden from view the specific origin of the term 'development' in the chaotic years before 1850. Likewise, when the progress of industrial capital was seen to be faltering, in Britain at any rate, towards the end of the century, 'development' reappeared at the heart of the debate over how state policy towards labour and industry in Britain and the new colonial empire in Africa should be conducted. After the 1910s and until the late 1930s, development, again, was largely hidden from view.[18] We should not be surprised, therefore, to find a common assumption during the 1960s and 1970s, the apogee of 'Third World' state development, that 'development' should have been invented at the beginning of the late colonial period in 1945.

It is striking that in the period during which development was invented and in the years of 'the first development debate' from 1890 to 1910, the idea of development carried with it a constructivist belief and a conviction for trusteeship. When we review the Saint-Simonians, Comte, Mill, and Newman as a possible progenitor of underdevelopment, our purpose is to reveal how a theory of trusteeship was built into the construction of development before 1850.

THE EUROPEAN SETTING OF DEVELOPMENT

By 1844, the year in which Robert Chambers anonymously published his widely read anticipation of Darwin, the *Vestiges of the Natural History of Creation*, the idea of development had already become part of British common sense. What was it about Britain of the time that impelled Chambers to make a proclamation about development? The economic historian A.E. Musson gives us a clue:

> In 1850 . . . half the population was still living in rural areas, over a fifth of the occupied population was engaged in agriculture and the next biggest occupational group was in domestic service; the great majority of industrial workers were skilled handicraftsmen or labourers, working in small workplaces or at home; only a small minority of the total labour force was in factories; many industries were still unmechanized, and in many of those where machinery had been introduced, it had as yet made only limited progress; only in a few industries such as cotton, coal and iron had steam power been introduced on a large scale; water-wheels were still numerous and widespread; much of the production and trade of the country was still in limited local markets. Truly, much of the England of 1850 was not very strikingly different from that of 1750.[19]

In what is now accepted as a standard introduction to the economic and social history of nineteenth-century Britain, E.J. Hobsbawm sharply characterises the significance about the specific period in which 'development' was born:

> No period of British history has been as tense, as politically and socially disturbed, as the 1830s and early 1840s, when both the working class and the middle class, separately or in conjunction, demanded what they regarded as fundamental changes. . . . Much of this tension of the period from 1829 to 1846 was due to [the] combination of working classes despairing because they had not enough to eat and manufacturers despairing because they genuinely believed the prevailing political and fiscal arrangements to be slowly throttling the economy.[20]

In France of the 1840s, the same sense of immediate social fear and turmoil was apparent. Although the transformation of the French economy is widely accepted as having been less advanced than in Britain, and industrial capitalism less 'developed', it was plunged, socially and politically, into even greater turmoil. And here, 'development' had become part of language even earlier than in Britain.

Louis Blanc, in 1841, asked

> Is the poor man a member of society or its enemy? . . . He finds the soil everywhere about him occupied. Can he cultivate the land for himself? No, because the right of the first occupier has become the right of property. Can he gather the fruits that the hand of God has caused to ripen along man's way? No, for as the soil has been appropriated so have the fruits. . . . What shall this unfortunate one do?

And then Blanc answered:

> He will tell you 'I have arms, I have intelligence, I have strength, I have youth; take all that and in exchange give me a bit of bread.' This is what the proletarians say and do today. But even here you may respond to the poor man: 'I have no work to give you.'[21]

Seven years later the *Journal des travailleurs* declared: 'Unemployment is the most hideous sore of the current social organisation.'[22]

After mid-century, these troubled times were to give way in Britain, if not in France, to the bright and certainly less disturbed period associated with much of the reign of Queen Victoria, which masks these tribulations for the unwary. Yet they were not to do so, as Hobsbawm reminds us, until the mid-century had given birth to the modern meaning of being Luddite and Radical, trade unionist and utopian socialist, Democrat and Chartist, along with such words as industry, capitalism, socialism, crisis and sociology.[23] Hobsbawm could and should have added the idea of development. It was the turmoil and fear of revolution in the first half of the nineteenth century that gave birth to the idea of development and it is these unexplored origins that are the source of much present-day confusion about development's meaning.

THE IDEA OF PROGRESS

Our particular view that the modern idea of development was created in the crucible of the first half-century of Western European transition to industrial capitalism stands in stark contrast to the claims of those seeking to establish legitimacy for the supposed modern 'sub-discipline' of development economics. Those who seek historical legitimation for development economics find it in the writings of the Scottish Enlightenment, with prominence given to Adam Smith. The eighteenth-century Scottish writers, it is supposed, formulated the first theory of 'development' by postulating a series of stages of human activity commencing with hunting and fishing, progressing through pastoralism and settled agriculture, and culminating with commerce and manufacture. As one commentator has suggested: 'The four stages, at any rate at the outset of its career, usually took the form of a theory of development embodying the idea of some "natural" or "normal" movement through a succession of different modes of subsistence.'[24]

Yet this attempt by modern developers to legitimate their practice through the claim of ancestral origin in the respectable works of the Scottish Enlightenment, rather than in the rough and tumble of early industrialism, is questionable. For rather than attempting to provide some schema of the 'stages' of human progress through history, what later became the early liberal doctrine of progress (*and not development*), associated with Locke and Smith in hindsight, was the result of an attempt to resolve the Hobbesian problem of how social and thus political order might be attained.

Seen in context, the effort to construct state order was the climax of a wider intellectual confrontation, beginning in the late seventeenth century, over received ideas concerning the nature of human history and the destiny of states. For much of the period encompassed by the history of Western thought the understanding of change was powerfully influenced what was believed to be the life-cycle of all living things.[25] By virtue of the classical analogy with the natural organism, change presented the unfolding of potential through natural cyclical sequences. The cyclical metaphor of change was taken from the way that a growing and maturing organism deposits a seed to re-create life after its own decay and destruction. Thus, the deliberate attempt to construct state order was predicted to degenerate into 'disorder' and 'ruin', the prerequisite conditions for a period of renewed political construction. Successful statecraft, and the art of politics, as in the thought of Machiavelli, was limited to the possibilities of prolonging maturity and forestalling degeneration.

This older meaning of development expressed the Janus-faced character of change. Positive, or constructive, change was created out of the negative moments of destruction and decay. Deliberate intervention could do no more than ameliorate and forestall the destruction of the degenerative period in the ordered, naturally determined, recursive, and finitely given sequences of cyclical change.[26]

It was this sense of historical change that remained predominant until it was challenged by the influence of thought in the period extending from Fontenelle in the late seventeenth century to Hegel a century later, and was gradually, but not wholly, supplanted in nearly every area of intellectual endeavour by the proposition of the alternative idea of progress. This idea of progress connoted a linear unfolding of the universal potential for human improvement that need not be recurrent, finite or reversible. Rather, the progress which was made possible, it was initially believed, by the revelation of God through an increasingly scientific understanding of Nature, was potentially limitless. From the idea of Providence, where God's revelation maps out a design in advance of human effort, to that of secular variants where human purpose is established autonomously, the idea of progress spelt out the possibility of directing the potentially unlimited capacity for improvement through the human effort of labour.

We can best understand the works of the Scottish Enlightenment, for instance, through the transformation in the accepted view of human history and capacity that was made in the seventeenth and eighteenth centuries. Smith's mentor, Adam Ferguson, for example, insisted that man 'would always be improving his subject, and he carries this intention where ever he moves, through the streets of the populous city or the wilds of the forest . . . and is "perpetually busied in his reformations" if also "continually wedded to his errors".' Importantly, for Ferguson, major obstacles to man's 'reformations' were, first, that the person 'does not propose to make rapid and hasty transitions; his steps are progressive and slow. . . . It appears, perhaps, equally difficult to retard or quicken his pace. . . . We mistake human nature, if we wish for a termination of labour, or a sense of repose.'[27] And, second, 'those revolutions of state that remove or withhold the objects of every ingenious study or liberal pursuit, that deprive the citizen of occasions to act as a member of a public, that crush his spirit, that debase his sentiments and disqualify his mind for affairs'.[28] To this day, these obstacles survive as the liberal conditions for overcoming the barriers against progress. Human effort has to be directed to the work of labour within a formal constitution of liberty. The obstacles to progress remain. Indeed, it is the ideas and practices of development themselves that have come to be seen by some, in the twentieth century, as means which entrenched the barriers against progress.

Ferguson's fellow Scot, Adam Smith, took up his pen against these debilitating 'revolutions of state'. Smith tells the reader in his introductory 'Plan of Work' that

> Nations tolerably well advanced as to skill, dexterity, and judgement in the application of labour, have followed very different plans in the general conduct and direction of it. . . . Though these plans

were, perhaps, first introduced by the private interests and prejudices of particular orders of men, without any regard to, or foresight of, their consequences upon the general welfare of society; yet they have given occasion to very different theories of political economy. . . . Those theories have had a considerable influence, not only upon the opinions of men of learning, but upon the public conduct of princes and sovereign states.[29]

However, Smith later warns his reader regarding those same 'private interests' that:

The proposal of any new law or regulation of commerce which comes from this order, ought always to be listened to with great precaution. . . . It comes from an order of men whose interest is never exactly the same with that of the public, who generally have an interest to deceive and even to oppress the public, and who accordingly have, upon many occasions, both deceived and oppressed it.[30]

A large portion of the *Wealth of Nations* can be read as a warning against the machinations of 'the men of commerce'. It is precisely their 'interests' which lay at the bottom of Smith's simultaneous attempt to show how the cyclical rise, stagnation and decline of nations might be forestalled and their continued material improvement prolonged.

Given that God, understood as Nature mediated by reason, was beneficent, and that individuals were guaranteed the 'fruits of [their] labour' through the establishment of secure conditions for exchange by the state *prior* to individual participants entering the market, then, for Smith, material improvement would follow spontaneously from exchange as the natural outcome of individual decisions that were made in production and trade. In order to establish these necessary prior conditions the state was obligated to remove those constraints, whether of feudal taxation or mercantilist restrictions, 'that remove or withhold', to reiterate Ferguson, 'the objects of every ingenious study or liberal pursuit'.[31] As the constraints were eliminated from the realisation of the individual's capacity to work, the individual was also to be imbued with the moral restraint of 'social sympathy' which would balance 'self-love' as if by an 'invisible hand'. Thus by all working according to a principle in nature, material improvement might progress through the course of history.[32]

This end was to be achieved by counterposing the human aversion to work and the natural capacity for work, which would in turn permit, as Hirschman tells us, the 'collapse of the "passions" that had formerly ruled human actions into individuals' primarily pecuniary "interests"'.[33] The belief in the possibility of 'spontaneous order' and improvement arising from free exchange followed from a belief in a natural equality of exchange between individuals who were imbued with the capacity to work but who, as individuals, lacked the ability to calculate the social need for their labour or its fruits. Given the assumption that it would be implausible to expect that all individuals would be able to calculate the precise worth of their individual labour efforts, it was concluded that the social need of all would never be met through conscious cooperation. A social interest or 'moral sentiment', by which the needs of all individuals would be satisfied, could only be established through the well-known formula of individuals pursuing their own 'diverse ends' in the name of their own 'self-love' or 'interest'. As Smith put it in an often quoted passage,

> man has almost constant occasion for the help of his brethren, and it is in vain for him to expect it from their benevolence only. He will be more likely to prevail if he can interest their self-love in his favour, and shew them that it is to their own advantage to do for him what he requires of them. . . . It is not from the benevolence of the butcher, the brewer, or the baker that we expect our dinner, but from their regard to their own interest.[34]

Thus, as is well known, it was the market which, for Smith, provided the arena for transforming the pursuit of individual interest or 'self-love' into the fulfilment of social needs through the mechanism of a potentially infinite series of individual agreements which culminated in the market.

The state, here, stood apart from the market arena, but through its prior establishment of definite conditions for voluntary acts of exchange, its guarantee that the agreements of exchange would satisfy the pursuit of individual interest and, finally, its guarantee of the general sanctity of property, it maintained the essential preconditions for the market's continued operation.

The negative or destructive dimension of the pre-modern conception of historical change was largely exiled from this conception of change as progress. For if the objective course of history was determined by the outcomes of voluntary acts of exchange, then it was reasonable to suppose that all outcomes were positive for progress. Where destruction and decay of human capacity and subsistence was experienced, or where any expected increase in labour productivity and purchasing power could not be observed, then the negative outcome was not a result of the objective process of exchange but was due to subjective failing. Some individuals were deficient in that their capacities to make informed decisions, on the basis of the reason to work productively, were lacking in the virtues of progress.

It is now recognised that Smith's emphasis upon moral judgement, and thus the capacity to make decisions, was not confined to his *Theory of Moral Sentiments* but was an integral part of his work as a whole. Moral judgement was the means by which individuals decided to act. Whatever the motive for action, such as meeting conditions for subsistence, Smith put a premium upon 'sympathy' as the means by which an individual person would be able to account, by way of imagining and feeling and not simply by the rational calculation of interest, for why she or he acted in one way rather than another. Sympathy was a quality which denoted the empathy that one individual felt for the self-love of the other; it was the 'social passion', a passion which could and should not be excluded by the interest of self-love.[35] 'All for ourselves', wrote Smith, 'and nothing for other people, seems, in every age of the world, to have been the vile maxim of the masters of mankind.'[36] Deriving from the providential design of God, 'the Superintendent of the universe' whose manifold sentiments were balanced and were not confined to the single-stranded rational pursuit of utility, sympathy was the quality which the individual was to take into the market if the self-love of each, separately, was to be translated into the love of each for the other.

The subjective capacities for action, formed out of a tension between self-interest and sympathy, were the counterpoint to the objective process of the market. Smith, it has been noted, drew from both Aristotle and Newton. The objective account of progress was an analogy of the Aristotelian idea of 'a Purposeful Nature': 'a progressive force exists in nature apart from any conscious human planning'. But Smith adopted Newtonian mechanics in making a distinction between having self-interest serve as the ruling principle of progress for society and sympathy serve as the regulating force of change. 'Self-interest', it has been suggested by Smith, 'implants the motion to society, and sympathy directs the motion within wholesome constraints.'[37] If self-interest arises out of history, sympathy, with its derivation from the qualities of God, must be consciously re-created as an integral part of human subjectivity if change is to be regulated by way of the social order of progress. If individuals were not imbued with sympathy before they entered the market there would be little reason to suppose that the objective possibility of progress would result in accordance with providential design. Thus, what was not progressive was subjective, and was to be treated as such through poor laws and labour reform, attacks on unproductive landlords and commercial prohibitions, state 'corruption' and all that was part of the early liberal panoply of reforming state stratagems.

What was true for individuals was for Smith, and later more emphatically so for Ricardo, true for nations as well. When Smith intervened in the 'rich country–poor country debate' through his *Inquiry into the Nature and Causes of the Wealth of Nations* it has been argued that he rejected the classical view that national wealth and power was subject to cyclical advance, stagnation and decline, and replaced it with a 'new description of rich and poor countries interlocked in a system of free trade reflecting the realities of a changing world. He could see the possibilities of development (i.e. progress), both for rich and poor countries offered by a system of natural liberty in foreign trade'.[38] This forecast of the possibility of realising a 'progressive state' through the 'natural' and 'spontaneous' elaboration of social order was carried forward into the nineteenth century. It was not, however, without its dissenters who, observing the prevalent political disorder in the wake of the French Revolution and the emerging social disorder consequent on the birth of industrial capitalism, questioned the possibility of boundless human improvement.

THE PESSIMISM OF PROGRESS: THOMAS MALTHUS

Although his *Essay on the Principle of Population* is most often presented as an anodyne argument concerning the limits to the growth of population imposed by agriculture, Malthus' argument, published in 1798 in the wake of arguably one of the first capitalist crises, was primarily a moral one which used the issue of population growth to argue against the possibility, as propounded in the utopian writings of Condorcet and Godwin, of unlimited social perfectibility.

Procreation, for Malthus, was the result of 'the passion between the sexes', which could never be overridden by rational self-interest or moral virtue for the 'pleasures of pure love will bear the contemplation of the most improved reason, and the most exalted virtue'.[39] The growth of population was kept in check only by 'vice' and the 'misery' of poverty with the small assistance of 'moral restraint'. Social misery, taken together with the enduring results of 'the passion between the sexes', resulted in an 'oscillation' between 'progressive' and 'retrograde' movements in population growth but could never provide the basis for the perfection of society. Small, incremental movements toward social improvement were possible, but only in the context of the recognition of the physical limits imposed by 'the Creator' to stimulate human virtue and exertion.

If the avowed target of Malthus' *Essay* was the utopian thought of Godwin and Condorcet, it was also the occasion of an explicit attack upon certain of the possibilities of 'improvement' contained in the writings of Adam Smith. Malthus questioned Smith's discussion of 'the causes which affect the happiness of nations or the happiness and comfort of the lowest orders of society, which is the most numerous class in every nation'.[40] In asking the question of whether national wealth was distinct from that of national 'happiness', Malthus argued that 'perhaps Dy. Adam Smith has considered these two inquiries as still more nearly connected than they really are'. Pointedly, Malthus charged that Smith had 'not stopped to take notice of those instances where the wealth of a society may increase without having any tendency to increase the comforts of the labouring part of it'. These 'comforts' were defined as the 'two universally acknowledged ingredients, health and the command of the necessaries and conveniences of life'. He agreed with Smith that: 'Little or no doubt can exist that the comforts of the labouring poor depend upon the increase of the funds destined for the maintenance of labour, and will be very exactly in proportion to the rapidity of this increase.'[41] But then Malthus suggested that 'perhaps Dy. Adam Smith errs in representing every increase of the revenue or stock of society as an increase in these funds'. For,

> it will not be a real and effectual fund for the maintenance of an additional number of workers, unless the whole, or at least a great part of this increase of the stock or the revenue of the society, be convertible into a proportional quantity of provisions; and it will not be so convertible where the increase has arisen merely from the produce of labour, and not from the produce of land. A distinction will in this case occur, between the number of hands which the stock of society could employ, and the number which its territory could maintain.[42]

Any increase in national wealth due solely to the growth of manufacture without a commensurate increase in the produce of the land, although it might lead to a rise in wages, would do little to benefit the labouring poor, for 'this rise would soon turn out to be merely nominal, as the price of provisions must necessarily rise with it'. Indeed, in Malthus' view, an increase in national wealth arising solely from manufacture might well decrease national happiness in that the increase in wages stemming from it would 'entice many from agriculture and thus tend to diminish the annual produce of the land'.[43]

If an increase in national wealth arising from the growth of manufacture could not break the limits imposed by passion on the increase in national happiness, neither could the diversion 'of some additional capital into the channel of agriculture' because the effects of such a diversion would only be realised very slowly. Nor could foreign trade provide a way out of the conundrum for although it might provide some relief in 'A small country with a large navy, and great inland accommodations for carriage, such as Holland . . . the price of provisions must be very high to make such an importation and distribution answer in large countries less advantageously circumstanced in this respect.'[44] There was, then, for Malthus no escape, in the long run, from the most disheartening reflection that the great obstacle in the way of any extraordinary improvement in society is of a nature that we can never hope to overcome. The perpetual tendency in the race of man to increase beyond the means of subsistence is one of the general laws of animated nature which we can have no reason to expect will change. . . . it is evident that no possible good can arise from any endeavour to slur it over or keep it in the background. On the contrary, the most baleful mischiefs may be expected from the unmanly conduct of daring not to face the truth . . .[45]

In the short run, however, the institution of private property could provide some respite. In agriculture, under 'a system of private property no adequate motive to the extension of cultivation' could exist unless there was a 'profit on the capital which has been employed', but that the need for 'profit' necessarily limited production so as to 'make the actual produce of the earth fall very considerably short of the *power* of production' by excluding 'from cultivation a considerable portion of land which might be able to grow corn'. In other words, the motive to produce had to be that of private profit but the means to do so, through the capitalist organisation of production, would limit actual output below any given limit of the potential to produce. Were it 'possible to suppose that man might be adequately stimulated to labour under a system of common property' actual production might be greatly increased but 'according to all past experience' it seemed 'perfectly visionary to suppose that any stimulant' other than private interest 'should operate on the mass of society with sufficient force and constancy to overcome the natural indolence of mankind'.[46]

Similarly, in manufacture, the institution of private property was absolutely essential to continued production and, as in agriculture, the stimulus to produce depended on 'profit', which in turn depended on the existence of consumers with the ability to buy. For Malthus, however, 'the consumption and demand occasioned by the persons employed in productive labour' could 'never alone furnish a motive to the accumulation and employment of capital'. The consumption of an 'unproductive' landlord class, which he championed, might help but the problem would still remain. Having framed the problem, Malthus wondered 'how it is possible to suppose that the increased quantity of commodities, obtained [produced] by the increased number of productive labourers should find purchasers, without such a fall in price as would probably sink their value below the cost of production'.[47]

Such a fall in price would destroy the stimulus of 'profit' and would result in unemployment and immiseration. Resulting poverty would assist with 'vice' and 'moral restraint' in postponing the ultimate reckoning between the growth of population and the means of subsistence.

From Wallace and Darwin in the late 1850s to Keynes a century later, Malthus' formulation of the poverty problem was a source of inspiration for those whose names, through evolution and state management of the economy, came to evoke popular meanings of development. It is striking that a theorist of pessimism, who attempted to strike at the heart of the central idea of progress by denying the objective possibility of immanent social improvement, should have played such a major part in making the idea of development possible. Yet that Malthus should have inspired the evocation of 'development' is neither fortuitous, coincidental nor paradoxical. Malthus' formulation, amidst his reflection on misery, conveyed his understanding of the limits of positive progress and opened up the negative dimension of history as the necessary premise for an idea of development.

THE DESIGN OF DEVELOPMENT

Those who lived through the depressions of the first decades of the nineteenth century must have seen Malthus' grim predictions about the inevitability of economic crises as borne out. The English crises of 1815 and 1818, with those in France of 1816 and 1825, seemed to have disproved Say's dictum on the necessary balance between production and consumption. In the face of the appalling human consequences of these economic and social disasters those, like Sismondi, who accepted emerging industrial capitalism as a given, looked to state intervention to eliminate its worst horrors as well as to revitalise a world of small independent producers in which production and consumption were to be more tightly bound together. Indeed, Sismondi, as in Kitching's text on development,[48] is given pride of place in the pantheon of anti-industrialism. However, the Saint-Simonians and others viewed capitalism, although not industrialism, as a passing social phase which could be and should be transcended. The transience of capitalism could only be recognised by breaking with the early liberal view of progress as a 'natural' process ungoverned by intention and erecting a theory of development, imbued with an overt sense of design, to take its place.

Although much has been made of the writings of Henri de Saint-Simon, much less attention has been paid to those who styled themselves Saint-Simonians.[49] Yet it was not only through these disciples – who included Blanqui, Enfantin and, for a time, Auguste Comte – that the thoughts of Saint-Simon came to be widely known. More importantly, it was they who created much of what was to become commonly known as Saint-Simonian, and later, positivist thought.[50] Nurtured in the rise of industrial capitalism and the French Revolution, during which their spiritual leader nearly lost his head to the guillotine, the Saint-Simonians posed the same problems that had inspired Adam Smith: the creation of order in a society undergoing radical transformation and the nature of that transformation itself. The answers they provided were, however, markedly different.

'Humanity', they argued, was a 'collective entity' which had 'grown from generation to generation as a man grows in the course of years, according to its own physical law, which has been one of progressive development'.[51] The Saint-Simonians' presumed advance of the 'progressive development of social relationships' in history was coeval with 'the progress of the moral conception by which man becomes more conscious of a social destiny' of 'universal association by and for the constantly progressive amelioration of the moral, physical, and intellectual condition of the human race'. It was a history in which each saw its prosperity and growth bound up with the prosperity and growth of all.[52]

The Saint-Simonians divided human history into what they referred to as 'organic' epochs, which presented 'the picture of union among the members of ever widening associations which determine the combination of their efforts toward a common goal'. The middle ages in Europe had been, in their view, such an epoch. Society had been dominated by the idea of religion while economy had been characterised by the order of feudal corporations which regulated production and exchange. By contrast, critical epochs of human history, such as that which culminated in the French Revolution, were 'filled with disorder; they destroy[ed] former social relations, and everywhere tend[ed] towards egoism'. In doing so they were harmful but were always 'necessary and indispensable; for in destroying antiquated forms which had contributed for a long time to the development of mankind but had finally become harmful, they have facilitated the conception and realisation of better forms'.[53]

From a Saint-Simonian perspective, the thinkers of the Enlightenment and the French Revolution had succeeded in destroying the old bases of political and economic order upon which the old organic epoch rested, but had shown themselves incapable of creating the theory and practice upon which a new organic epoch could be built. By mistaking the 'egoism' of the then-present critical epoch for that which was needed to construct the basis of the new organic epoch, ideas which put a premium upon self-interested action of the rational individual, they had compounded the problem of social disorder. The influence of unreconstructed Enlightenment thought was such that it ran the risk of prolonging the social agony characteristic of an epoch of criticism and destruction while delaying the arrival of the new organic epoch in which 'men' would be once again 'associated', albeit now on a new footing. This new basis of human 'association' would be 'industrialism' itself, guided by 'intellect' and governed by 'sympathy' in place of the prevailing 'egoism' of the 'critical period'. It was the then-current state of the application of 'science' to production that the Saint-Simonians offered as evidence of how egoism, founded on the premise of rational action, culminated in the irrational and destructive qualities of the prevailing critical epoch.

The Saint-Simonians did not 'deny any of the progress' that had been made in industrial production, but questioned whether 'the march toward improvement could not have been much more rapid than it has been'. They lamented that:

> There has been no organised effort, however, to free industry from the narrow confines within which it is at present restricted so that industrial practices may be raised to the height of scientific theories. In industry, everything is still left to the uncertain performances of individual luminaries. Evidence from experimentation, often lengthy and prejudicial, is almost the only means used by industrialists to evaluate their processes; this evidence which each one of them must revise, for, thanks to competition, each is interested in surrounding with mystery the discoveries he has attained so that he may keep them as his monopoly. If theory and practice are integrated at all, the process is always incidental, isolated and incomplete.

Could 'one count', they asked, 'how much effort has been wasted, how much capital buried', because in

> industry, as in science, we find only isolated efforts; the only feeling that dominates all thoughts is egoism. The industrialist is little concerned with the interests of society. His family, instruments of production, and the personal fortune he strives to attain, are his mankind, his universe, and his God. He sees nothing but enemies in those who follow the same career; he lies in wait for them; he spies upon them and has made his happiness and glory consist in ruining them.[54]

In propounding a theory of active self-interest the 'economists' had propounded a theory of order but it was this 'theory itself' which was for the Saint-Simonians 'the main source of disorder':

> *Laissez-faire, laissez passer*! This has been the necessary solution, and this has been the only general principle they have proclaimed. One knows well enough under what influences this maxim was formulated; it carries its own date with it. The economists thought that with one stroke of the pen they could solve all the questions relating to the production and distribution of wealth. They entrusted personal interest with the realisation of that great precept, without wondering whether any individual, no matter how keen his insight, could judge within the limitations of his environment, being, so to speak, in a valley, the totality of which can only be seen from the highest mountain peaks. We are the witnesses of the disasters that have already resulted from this principle

of circumstance, and if more striking examples were needed, they would appear in throngs to testify to the impotence of such a theory designed to make industry productive.[55]

Amidst these 'throngs' was the 'example' of the man 'who lives by his hands' who could not 'join his voice to that of society' in applauding the introduction of steam power to his trade. For if, as the Saint-Simonians willingly acknowledged, the introduction of steam was necessary and potentially beneficial for all and would moreover provide employment to ever growing numbers, the 'statistical calculation' which 'proved that in a certain number of years they would have bread' could offer no consolation to the 'thousands of famished men' whose lives were dislocated by technological change.[56]

Moreover, *laissez-faire* dislocated society in more general terms, for

When there is a rumour that a certain branch of production offers good possibilities, all enterprise and capital flow in that direction; everyone hurries there blindly. No one takes the time to trouble himself about the proper size and the necessary limits of the enterprise. The economists cheer at the sight of this overcrowding, for they recognise from the great number of contestants that the principle of competition has been widely applied. Alas! What is the outcome of this death struggle? A few fortunate ones triumph but at the price of the complete ruin of innumerable victims.[57]

The necessary consequence of these 'uncoordinated efforts' was 'overproduction' with the 'balance between production and consumption' being continually threatened. This in turn caused the 'numberless catastrophes and the commercial crises that terrify the speculators and stop the execution of the best projects'. These 'ruined' 'hardworking men' and 'hurt' 'morality' because these crises brought about the conclusion 'that in order to succeed, more seems to be needed than honesty and hard work'.[58]

We make no claim that the Saint-Simonians were alone in pointing to the misery, unemployment and destruction which accompanied early capitalist industrialisation and which was deemed to be the result of a doctrine which promoted self-interest, the market and competition. In England, for instance, there were critical currents which moved on similar lines and which have come to be interpreted as 'physiocratic anti-commercialism', 'communitarian political economy' and 'the community of goods' as examples of early socialist political economy.[59] Rather, the Saint-Simonians were singular in that they did not seek recourse, as was characteristic of English constructivism, in either 'an unhistorical, imaginatively reconstituted agrarian past'[60] nor a millenarian future that was unhinged from the epoch of the present. That they are known as 'scientific socialists' is an indication, but only partially and sometimes misleadingly so, of the extent to which Saint-Simonians attempted to impose constructive order upon what they took to be industrial disorder of the present. In doing so they conferred agency upon development and gave it constructivist purpose. No longer was development that which happened during a period of history. It became the means whereby an epoch of the present was to be transformed into another through the active purpose of those who were *entrusted* with the future of society.

A THEORY OF TRUSTEESHIP

The remedy for disorder of the critical epoch was clear to the Saint-Simonians. Only those who had the 'capacity' to utilise land, labour and capital in the interests of society as a whole should be 'entrusted' with them. However, the Saint-Simonians regarded prevailing legislation about 'inheritance' to be a major obstacle standing in the way of a solution since they took it to be merely another term for property as it was then constituted. 'Today', the Saint-Simonians argued,

> capacity is only a poor title for credit; to acquire credit, one must first possess something. The accident of birth blindly distributes the instruments of production, whatever they may be; and if the heir, the idle owner, entrusts them to the hands of a skilful worker, it is quite evident that the primary increment accrues to the incapable and lazy owner.[61]

Neither property nor inheritance, the Saint-Simonians argued, had ever been historically sacrosanct. Slavery and serfdom, based upon forms of property, had been largely destroyed, along with nearly all other 'privileges of birth', during the preceding critical epoch. One privilege, however, remained: the right to property by birth rather than by 'right of ability'. All previous disputation, no matter how critical, had left the individual right to inherit property intact. The luminaries of the Enlightenment and Revolution had not disturbed it. The 'English economists' Malthus and Ricardo, who had understood the significance of the right to inherit, had simply attempted in their writings to 'justify' the political organisation in which 'one part of the population lives at the expense of the other part' who nourished 'noble owners in their idleness'. Bentham, in their opinion, shrank from the conclusion of his own doctrine of 'utility'. Private property could neither be justified on the grounds of happiness nor on the grounds of the capacity of the possessors of property to 'use' it in the most productive manner.

Sismondi, who saw the importance of property in the generation of economic crises, limited himself to a critique only of landed property and did 'not dare to attack property in its entirety'.[62] Yet it was property 'in its entirety', particularly property in the form of capital, which the Saint-Simonians saw as at the heart of the problem. How was it to be resolved?

The Saint-Simonians feared that 'some people will confuse [our] system with what they know as community of goods'.[63] And indeed they advocated no such scheme. Rather, they argued for property to be placed in the hands of 'trustees' who would be chosen on the basis of their 'capacity' to decide where and how society's resources should be invested. These 'trustees', they argued, already existed in an embryonic state in the form of 'instinctive endeavour whose manifest purpose is to restore order by leading toward an organisation of material work'. These 'instinctive endeavours' were none other than society's bank and the 'trustees' society's bankers.[64]

Bankers served 'as intermediaries between the workers, who are in need of instruments of work, and the owners of these instruments who either cannot or do not want to use them'. They were, moreover, 'because of their knowledge and connections . . . much more in a position to appraise the needs of industry and the ability of industrialists than . . . idle and isolated individuals'. In order to realise their potential usefulness it would, however, be 'necessary to modify bankers and banks'. Banks and bankers were to be made fit for trusteeship.[65]

Banks were to be reformed in such a way as to create a 'general system of banks'. At the head of the system there was to be a 'central bank representing government in the material order'. This central bank would be 'the depository of all of the riches, of the total fund of production, and all the instruments of work'; in brief, 'of that which today composes the entire mass of individual properties':

> On this central bank would depend banks of the second order which would be merely extensions of the first, by means of which the central bank would keep in touch with the principal localities to know their needs and productive power. The second order would command within the territorial area of their jurisdiction increasingly specialised banks which embrace a less extensive field, the weaker branches of the tree of industry. . . .
>
> All needs would converge in the superior banks; from them all endeavour would emanate. The general bank would grant credit to the localities, which is to say, transfer instruments of work to them only after having balanced and combined the various operations. And these credits would then be divided among the workers by the special banks, representing the different branches of industry.[66]

Bankers were also to be necessarily reformed because a critical facet of egoism was that the 'advantage that might result from the intermediary position of bankers between the idle and the workers' was 'often counter-balanced and even destroyed' by 'various forms of fraud and charlatanism'. In order to combat the destructiveness of egoism, the 'general staff' of bankers had to be made aware of their 'function', their 'social destiny'.[67] Bankers, in short, had to be moralised.

Although only an example of the 'unrefined rudiments' of the new society, reforms of the banking system, coupled with making bankers fit the moral function of intermediation, were an essential example, for the Saint-Simonians, of the ways in which progress could be both tamed and indefinitely extended: 'all the great evolutions of the past, even the most legitimate, that is to say, those which have contributed most to the happiness of mankind, all seemed to have the characteristic of common catastrophe or upheavals at their beginning'. Again taking disorder as the key, they continued:

> Today the situation is no longer the same: mankind knows that it has experienced progressive evolutions. It is acquainted with their nature and extent. It knows the law of these crises which have ceaselessly modified mankind and brought man ever nearer to the normal conditions of existence. Now man can, through the progress of the past, verify the future which his feelings show him. Above all, man can pave the way for the realisation of this future through slow and successive transformation of the present. Mankind should therefore foresee and avoid disorders and violence which have been the conditions for all past progress.[68]

All that is associated, through our contemporary minds, with the method of Positivism – a presumed rejection of metaphysics, claims for empirical method and protocols for prediction – is found here, but it starts with the destructive and negative possibilities of progress. The positive is nowadays distinguished from the normative rather than the negative, but it was the desire to avoid what was destructive about change that fired the attempt to create a system of positive thought. A gap between the positive science of how to evaluate a past and a normative intention to manage was to be filled by the idea of development.

It was the most famous of the Saint-Simonians, Auguste Comte, who, after his break with their leader, attempted to provide the science upon which progress should be based. Comte's seemingly perpetually repeated message was that 'Progress is the development of Order under the influence of Love'. Like the other Saint-Simonians, his project was to attempt to reconcile the moral, intellectual and material qualities of progress with social order. In his mind, the logical goal of human endeavour, progress, had to made consistent with what he deemed to be humanity's most pressing need – social stability and equilibrium. And in attempting to provide his framework of how equilibrium conditions for 'social dynamics' might be established upon the basis of 'social statics', Comte completed the invention of development which the Saint-Simonians had begun.

PROGRESS AND DEVELOPMENT: COMTE AND 'ORDER'

Comte broke with his Saint-Simonian *confrères* over the question of whether there was an existing science of society which was adequate to effect the reconciliation between progress and order. In arguing that an adequate science of society had yet to be created, Comte judged the Saint-Simonian project of social reconstruction to be premature and set about creating a science of history or 'sociology', which was his alternative name for it.[69]

Sociology's object was nothing less than the discovery of the 'laws of human social evolution'. Social evolution, in Comte's words, contained 'two aspects': *development*, which brings after it *improvement*:

> The only ground of discussion is whether development and improvement – the theoretical and the practical aspect – are one; whether development is necessarily accompanied by a corresponding amelioration, or progress, properly so-called. To me it appears that the amelioration is as unquestionable as the development from which it proceeds, provided we regard it as subject, like the development itself, to limits, general and special, which science will be found to prescribe.[70]

Human improvement or progress, the 'practical aspect', had been stilted, however, because development, the 'theoretical aspect', had failed to reconcile progress with order. Comte maintained that 'All previous philosophies had regarded Order as stationary, a conception which had rendered it wholly inapplicable to modern politics.' Positivism, however, could represent 'Order in a totally new light'. It could do so by placing 'Order on the firmest possible foundation, that is on the invariability of the laws of nature' with order in 'Social phenomena, as in all others . . . resting necessarily upon Order in nature, in other words, upon the whole series of natural laws.'[71] As to the reconciliation of order with progress, Comte maintained that,

> Necessary as the reconciliation is, no other system has even attempted it. But the facility with which we are now enabled . . . to pass from the simplest mathematical phenomena to the most complicated phenomena of political life, leads at once to the solution of the problem. Viewed scientifically, it is an instance of that necessary correlation of existence and movement, which we find indicated in the inorganic world, and which becomes still more distinct in Biology. Finding it in all the lower sciences, we are prepared for its appearance in a still more definite shape in Sociology. In Sociology the correlation assumes this form: Order is the condition of all Progress; Progress is always the object of Order. Or to penetrate the question still more deeply, Progress may be regarded simply as the development of Order; for the Order of nature necessarily contains within itself the germ of all possible Progress . . .[72]

For Comte then, the natural world was an ordered one within which progress took place according to natural laws which could be discovered and understood. Detached from the providential order of God and proclaimed to be a religion of Humanity, the positivist reckoning was that humankind was biologically part of nature and as such was subject to laws of social order which could be derived from, and made analogous to, laws of nature. But human capacity was singular in that it had the potential to understand and know the possibility of order. Knowing was the active pursuit of improvement. To reconcile social order with human progress social laws had to be comprehended by humanity in such a way as to ensure their applicability but, Comte argued, 'Owing to the mental and moral anarchy' in which humankind lived 'systematic efforts to gain the higher degrees of Progress' were 'as yet impossible'. This explained, though it did 'not justify, the exaggerated importance' which Comte maintained was 'attributed . . . to material improvements'.[73] Material improvement was one facet of what was taken to be the variability of progress. Progress was restless and inconstant, consisting as it did of the movement of disparate events and uneven forces of history. It had to be given the consistency and morality of order. The means by which progress would be subsumed by order was development.

One part of Comte's method was to contrast what he called 'social dynamics' or the uneven forces and disparate events that characterised human progress with 'social statics' or the overwhelming consistency and morality of order which he saw as immanent in each epoch of human history. To make the intellectual tools which were necessary to highlight the contrast between dynamics and statics, Comte employed a method of reasoning which neither solely deduced the laws of history to be necessary consequences from admitted premises nor simply induced historical laws from admitted historical facts. Rather, in contradistinction to both possibilities, Comte established a method whereby from a simple hypothesis, the whole of the ascertained facts of history could be explained by making tendential laws, established from the hypothesis, less general and more complex.[74] This method, which verified a law not through agreement with lower-level factual independence, but high-level laws, was designed to root the logic of order in history.[75] In his *Course of Positive Philosophy*, Comte set out a hierarchy of sciences to fulfil a logical and historical order. Mathematics came first and set out the most general natural laws that 'govern phenomena most removed from human involvement and manipulation'.[76] Social physics or sociology came last after biology. Later, in the *System*, moral order was added to express the science of sociability but this ultimate science remained speculative. Science is anti-metaphysical in that it does not search for an absolute explanation of the origin or purpose of universal essence. Neither does science compile facts but rather coordinates laws which express relations between particular phenomena. Thus, the laws of social order, more complex than those of earlier or more general sciences, could not be derived from the more general and less complex laws. A science for social order assembles its own materials and particular method, that of history, but must nevertheless conform to the general laws of nature, natural change and progress.

Comte founded his distinction between statics and dynamics in order to contrast the restlessness of progress, the social dynamic, with the consistency and morality of order, the statics of a social system which conveyed the desirable impressions of arrangement and command. In this contrast, the movement of disparate events and uneven forces of history is to be disciplined by reference to the static qualities of order. Progress is subsumed by order through development.

Development, or all that improves or makes life better out of the variation of progress, is captured by the law of the three stages. In the first or fictitious stage, all unknown facts are fictitious moral laws. Inanimate objects give the impression of being animate. Knowing is nothing other than the instinct of feeling and opinion, formed in accordance with hopes and desires since human needs determine instinct and prompt activity. The capacity to act is developed by naturally exercising the power to act through the necessary illusion of arbitrary will.

Thus, the tendency of the fictitious or theological stage is towards individualism, the ego becomes the absolute. While altruism is denied, social sentiment, the order of 'love for one another', is created through a community of opinion and feeling.[77] The second stage, called metaphysical, substitutes entities for deities, the attempt to create systematic thought rather than act through 'the spontaneous creation of the mind'. Thus, in a search for the causes of phenomena, the capacity to act is now regulated, rather than developed naturally, according to the criteria of conceptual laws. But metaphysics suppresses imagination in the name of objectivity and tends to 'idiocy' or 'the deficiency of subjectivity' and thus tends to corrupt morals and disorder in politics.[78] It is in the third positive stage, that egoism is eliminated by altruism through the science of systematic thought. There is harmony between sets of conceptions and observations, or suggestions 'from within' and 'impressions from without' in the necessary purpose of regulating the power to act. The subjective is subordinated to the objective but neither through 'idiocy' nor its opposite of 'madness' the 'excess of subjectivity'.[79] In the third stage, humanity culminates in a 'normal state', where activity regulates the power to know, where 'the original imperfections of the order of nature', 'imperfections' or 'variations' of climate and race are corrected by the intervention of human action, now a system of objective knowledge. Thus, 'the law of development' is nothing other than the consequences of the law of inverse deduction: 'the law of inevitable increase in complication in proportion with the decrease of generality'. The consequences are 'the increase in the liability to imperfection' and 'the increase in the capacity for improvement'.[80] Positivism permits the capacity for improvement to be balanced against variations of the normal state of humanity and development is the fulcrum of the balance. Trusteeship is the political means of development, to make progress orderly, and it is of the third stage of Positivism.

Comte transposed the triad of stages into different arrays, the sets of lower-level laws through which history was ordered. Three natural historical stages of infancy, adolescence and maturity were supposed to correspond to the three states of intelligence: fiction, the hypothetical abstractions of metaphysics and the demonstrative qualities of Positivism. Likewise, the historical epochs of antiquity, feudalism and industrial life not only corresponded to states of intelligence but also to sociability or 'feeling in social instinct': the civic life of antiquity, collectivism of the Middle Ages (the historical period of feudalism) and universal sociability of industrialism.[81] Comte was insistent that phenomena 'always preserve the same arrangement' however much they might vary in degree. 'The fundamental dogma of positive religion', he wrote, was 'the invariability of all laws'.[82] If the arrangement of the phenomena varied, if there was variation in their universal rather than natural order, there would be no social progress. The determining array, however, was the feeling of sociability, the principle of order and the end of progress. Development was about the capacities to determine the capacity to act, but sociability determined activity.

In his usual manner, Comte made the 'three modes of activity', that of conquest or aggressive warfare, military defence and then labour, correspond to all the other arrays or tendential laws. However, he then subsumed labour of the positive stage by sociability. The division of labour, created by the progress of science, only made activity more special and particular and reinforced egoism, which Comte characterised as the numbness of will but insatiability of desire. Therefore, to counter the dispersive effects of the industrial division of labour, the moral ethic of sociability had to be inculcated in industrial society.[83] From the vantage point of statics, industrial activity was the source of universal cooperation. It was through the division of labour that each individual subject acquired a social function. In the final stage of Positivism, the only means by which material need would be directly satisfied was by labour, 'action of a useful kind on man's environment'. 'Labour does at last become the only mode by which material wants will be satisfied; but not until the whole population of the globe is taken into account.'[84] Thus stated, we might suppose that it is the sociability of labour in industrial production which expresses the universality of humanity, natural maturity and the final vindication of positive knowing. Not so, however, because industrial life is promoted by the fictitious individualism of infancy; it is neither regulated by abstraction, by say abstract labour, nor can it be the best source for impressing social harmony upon individual subjectivity. It is military life which, despite its purpose of destruction, is demonstrated to 'possess a moral aptitude' and thus fulfils a social purpose. War is the best school of obedience, source of command and collective association.[85]

The industrial problem was centred upon capital and exchange. The original character of labour is selfish because the practical activity of satisfying material need stimulates personal instinct, fosters the egoism of 'individual importance', corrupts moral purpose and then the intellect. Labour becomes unselfish when 'life assumes a social character, though it be only the life of the family'. Progress 'from selfish to unselfish toil' entails the accumulation of *capital* where capital is nothing other than the 'permanent aggregate of material products'. A material product is distinguished from an intellectual product because of 'two essential economic laws': each individual can produce a surplus over current consumption and products can be preserved or stored for a longer time than is necessary for their reproduction.[86] From these two elementary conditions of labour, sociability can be established as a condition of the accumulation of products. The surplus of the individual worker provides consumption for others and the capacity to store products enables material wealth to be transmitted or distributed. Material wealth does not increase significantly until it is concentrated: 'Hence capital is never very largely increased, until by some means of transmission the wealth obtained by several workers is collected in the hands of a single possessor; who then provides for its proper distribution, after duly seeing to its preservation.'[87] Comte sets out three modes of transmission or distribution which, true to form, correspond to the threefold arrays: Conquest, Exchange and Gift. Conquest is a form of forced exchange but since it is correlated with the first stage of illusion and war, the outcome of distribution is uncertain. Gift is also of antiquity, a part of cooperative sentiment and it is the task of Positivism to transform Gift into a system and thus part of deliberate, knowing sociability: 'The most ancient and the most noble of all the modes of transmitting products will do more to effect the reorganisation of industry, than the empty metaphysics of our gross economy can easily conceive.'[88] At the outset, Comte accuses metaphysical abstraction of giving a primacy to Production and by ignoring Preservation and Transmission, succouring the selfishness of individualism by upholding the mode of voluntary exchange.

It is capital which transforms individualism into the possibility of altruism. Capital forms the basis of the division of labour which makes each worker produce for others and be dependent upon other workers for the instruments of production. Voluntary exchange, however, maintains the illusion that each worker is producing products for him- or herself. To regulate exchange is to make sociability a system, to 'fill the labourer with a sense of his social value'.[89] In the final stage of humanity, positivism, labour would be ordered as the latent quality of sociability was made explicit in society. Since capital ensures the accumulation of material wealth it liberates individuals from material labour and permits the pursuit of the intellect, the knowing of sociability. Positivism is therefore about socialising capital. Capitalists should fulfil the function of capital, to preserve and distribute wealth and act as the trustees for the wealth of society.[90]

Comte had held that capitalists, informed by positivism, should act as trustees for the wealth of humanity. According to their personal merit, 'that is according to their fitness to represent humanity', individuals among the 'proletaries' are classified by humanity. A dictatorship of the industrial patriciate, but educated through the social pressure of the proletaries, establishes the real reward of a worker according to 'a just esteem', the extent to which labour works for humanity and esteem may be refused to the powerful who work for themselves as isolated individuals.[91] The industrial patriciate, or captains of industry, expressed, for Comte, the new social force of temporal power and positive thinkers coupled with those who embodied the spiritual power of humanity. This couple would form the universal basis for a new ruling class in France which could then be extended to the rest of the world.

While Comte, for instance, wanted to retain the Enlightenment idea of progress, he rejected the negative idea that a new social order demanded either or both the destruction of all social institutions and the damning of the past. Whereas the metaphysics of the philosophers of the revolution had criticised social order through the absolute essence of concepts such as liberty, freedom and consciousness, positivism was constructive, aiming to establish the worth of social institutions and use science to further a relative conception of social good. The metaphysical spirit, Comte maintained, could not organise; it could only criticise.[92] While secularism and industrialism were the historical givens for Comte, the positivist project was to recast Catholicism, to reconstruct the church of the Middle Ages to fulfil a secular corporate purpose. To impose order on abrupt change, rather than to celebrate in the spontaneity of change itself, was the aim underlying the belief that what had been destroyed had to be replaced by the purposeful knowledge of science. Comte's Religion of Humanity, his disciples believed, was 'the light and salvation of a disordered and suffering world'.[93]

Comtean positivism, of course, was the true path of knowledge. It was the mode of thought which corresponded to the mature adulthood of humanity whose dominant form of social interaction would be altruism. By elevating altruism or 'universal love' to be the guiding force of the positivist age, Comte directly challenged the logic of the Scottish Enlightenment. Comte argued that 'No calculations of self-interest [could] rival this social instinct.'[94]

Thus, rather than subduing human passions by collapsing them into pecuniary interests in the manner of Smith, Comte, along with Fourier, grasped hold of one 'passion', love, and elevated it from the selfish, egoistic individual love of utilitarianism to the heights of selfless altruism, which would itself be the motive force for reconciling progress and order. For Comte, there was no tension, no interplay between self-interest and sympathy. According to Smith, sympathy had to be constantly re-created in all epochs as a condition for progress. In the positivist scheme, however, a one-dimensional subjective attribute of altruism was collapsed into an objective command which belonged to one epoch of history.

Through the particular altruistic epoch both the ultimate in community and, through the capacity to comprehend how to reconcile progress and order, the application of positive science would be realised. Humanity would now enter a golden age of reconciliation with nature in which social sympathy would triumph over self-love. An essential condition for achieving this triumph of 'social sympathy' was the fulfilment, in Comte's words, of the 'social mission of woman', which arose 'as a natural consequence from qualities peculiar to her nature'. From the most important of natural 'qualities', 'the most essential attribute of the human race, the tendency to place social above personal feeling', Comte pronounced that woman 'is undoubtedly superior to man'.[95] Only through 'this essential attribute', itself rooted in maternity, could mankind, hitherto dominated by force and 'self-love', be made moral and its 'progressive development' assured:

> Love, then, is our principle; Order our basis; and Progress our end. Such is the essential character of the system of life which Positivism offers for the definite acceptance of society; a system which regulates the whole course of our private and public existence, by bringing Feeling, Reason and Activity into permanent harmony. . . . Life in all its actions and thoughts is brought under the control and inspiring charm of Social Sympathy.[96]

When this end was achieved, 'knowledge' could then 'guide action' and: 'The governing classes, clearly perceiving the end they are called on to realise, can reach it directly, in place of wasting their forces on tentative and mistaken efforts'.[97] This was the way in which the next stage of the 'collective development of mankind' could be achieved and the improvement of progress follow. With Comte, development now occupied the gap between progress and order. Once progress had been reconciled with order in the positivist epoch of altruism, development would have done its work and met its end. In the Age of Progress after 1850, and when 'development' disappeared from view in a period of rapid capitalist progress, it might have appeared that the positivist prospect of a golden age of social sympathy was attainable. For ordered progress to be continuously reproduced it was only necessary to develop the encyclopedias of knowledge and then to apply their wisdom.

This procedure would guide the 'wise regulation of an important order of political relationships, those that concern the influence of the advanced nations on the development of the inferior civilizations'. Comte warned against 'metaphysical' and 'theological' formulae which 'because of the absolute nature of their principal conceptions' led 'all civilised men to transfer their ideas, customs and institutions, indiscriminately and often very indiscreetly' to the 'less civilised'. Such a practice, he warned – anticipating colonial developers by a century – would very likely lead to the 'gravest political unrest'. The more one meditated on the subject, Comte declared, the more one felt

> that practice, not less than theory, demands exclusive or at least principal consideration of the most advanced social evolution first of all, and the neglect of less complete types of progress. It is only when we have determined what belongs to the elite of humanity that we can regulate our intervention in the development of more or less backward peoples, by reason of the necessary universality of the fundamental evolution, with due appreciation of the characteristic circumstances of each case.[98]

PROGRESS AND DEVELOPMENT: J.S. MILL AND 'KNOWLEDGE'

Unlike the thinkers of the Scottish Enlightenment and Thomas Malthus, whose works have spawned a veritable academic industry of reinterpretation, those of the Saint-Simonians and Auguste Comte have fallen into a strange kind of academic limbo. While ritualistically hailed in introductory texts as works of the founders of sociology, their writings are simultaneously routinely reviled as intellectual buffoonery and, when not deemed incomprehensible, are treated as having little intellectual value. The dismissive treatment of the positivists is all the more strange considering the long line of thinkers who have acknowledged the impact of positivist thought upon their own and who are rarely as cavalierly dismissed. This list ranges from Proudhon, Leslie Stephen and Joseph Schumpeter to G.B. Shaw and Beatrice Webb, Mauss and Evans-Pritchard. Not the least of those so influenced was John Stuart Mill whose work, more than any other, brought the invention of development to the English-speaking world.[99]

Mill owed a great intellectual debt to the Saint-Simonians and, in particular, to Comte. Mill's debt was widely recognised by authors writing in the late nineteenth and early twentieth centuries but has been largely forgotten in more recent interpretations. Part of the reason for this forgetfulness is the fault of Mill himself, who systematically culled many of the more laudatory references to Comte from later editions of his work without changing the ideas which he had acknowledged in earlier versions to have been lifted from Comte. Since, as so often happens, the more standard texts have been reprinted from the later editions, it is now difficult to see Comte's influence on Mill. The amnesia also stems from the attempts, by both latter-day philosophers and economists, to appropriate Mill in such a way as to lend legitimacy to their disciplines. A Comtean connection, if openly admitted, would be clearly embarrassing for those who are intent on demarcating the separate academic areas of philosophy, economics and sociology.

Contact between Mill and the Saint-Simonians commenced early in the former's career when in 1829, in the midst of his well-known 'mental crisis' occasioned by the bleakness of his utilitarian education, Mill fled to France seeking solace. There, in Paris, he met the Saint-Simonians, the 'writers by whom, more than any others, a new mode of political thinking was brought home to me'. Among them was Comte whose early writings 'seemed to give a scientific shape' to Mill's thought. Comte, Mill reflected, had induced him to stop mistaking 'the moral and intellectual characteristics' of his own era (which had much to do with his breakdown) 'for the normal attributes of humanity'.[100] Back in England during 1831, Mill issued a clarion-call through a series of essays entitled the *Spirit of the Age*, which strikingly echoed Saint-Simonian thought in its analysis of the crisis of British society. 'The affairs of mankind', Mill wrote, 'or any of those smaller political societies which we call nations, are always either in one or the other of two states, one of them in its nature durable, the other essentially transitory. The former of these we may term the *natural* state, the latter the *transitional*.'[101] 'Worldly power' and 'moral influence' in the 'natural' state were 'habitually and indisputably exercised by the fittest persons whom the existing state of society affords'. The 'material interests of the community' in the natural state were 'managed by those of its members who had the greatest capacity for such management'. By contrast, the transitional state could be recognised 'when worldly power, and the greatest existing capacity for world affairs are no longer united but severed', when there were 'no established doctrines' and when the world of 'opinions' was a 'mere chaos'. A transitional state would continue 'until a moral and social revolution . . . has replaced worldly power and moral influence in the hands of the most competent' so that society would be 'once more in its natural state' and able to 'resume its normal progress, at the point where it was stopped before by the social system which it has shivered'.[102] Mill's contrast between the natural and transitional states unmistakably corresponded to the fundamental Saint-Simonian distinction between the organic and critical epochs. As such, the meaning of development, which was drawn from the original distinction between two epochs of history, was carried over into Mill's adaptation of positivism.

For Mill, as for the Saint-Simonians and Comte, 'The first of the leading peculiarities of the present age is that it is an age of transition':

> Much might be said . . . of the mode in which the old order of things has become unsuited to the state of Society and of the human mind. But when almost every nation on the continent has achieved, or is in the course of rapidly achieving, a change in its form of government; when our own country, at all times the most attached in Europe to old institutions, proclaims almost with one voice that they are vicious in the outlines and the details, and *shall* be renovated, and purified and made fit for civilized man, we may assume that a part of the effects of the cause just now pointed out, speak sufficiently loudly for themselves. . . . Society demands, and anticipates, not merely a new machine, but a machine *constructed* in another manner.[103]

Yet if he agreed with the Saint-Simonians on the necessity for creating a new social 'machine', Mill disagreed with them over the means by which the machine might be constructed. Along lines of which our own postmodernist thinkers might well take note, Mill claimed:

> There is one very easy, and very pleasant way of accounting for this general departure from the modes of thinking of our ancestors: so easy, indeed, and so pleasant, especially to the hearer, as to be very convenient to such writers for hire or for applause, as address themselves not to the men of the age that has gone by, but to the men of the age which has commenced. This explanation is that which ascribes the altered state of opinion and feeling to the growth of human understanding. According to this doctrine we reject the sophisms and prejudices which misled the uncultivated minds of our ancestors, because we have learnt too much, and have become too wise. . . . We have

now risen to the capacity of perceiving our true interests; and it is no longer in the power of impostors and charlatans to deceive us.[104]

Though a 'firm believer in the improvement of the age', Mill wrote that he was 'unable to adopt this theory' because it was based on a false interpretation of the characteristics of the 'improvement' that had occurred. An improvement of progress was not the result of the attainment of individual genius, such as that of Comte. Rather 'the grand achievement of the age' was 'the diffusion of superficial knowledge'. And thus 'the intellect of the age' could not be the reason why there should be any social demand for a differently constructed 'social machine':

Not an increase of wisdom, but a cause of the reality of which we are better assured, may serve to account for the decay of prejudices; and this is, increase of discussion. Men may not reason better, concerning the great questions in which human nature is interested, but they reason more. Large subjects are discussed more, and longer, and by more minds. Discussion has penetrated deeper into society; and if no greater numbers than before have attained the higher degrees of intelligence, fewer grovel in the state of abject stupidity . . .[105]

By 'discussion' Mill meant a general level of education in the widest sense of the word. It was through more general education that the new 'social machine' was to be generated and, for Mill, a condition for more education was a radical extension of 'liberty'. Education and liberty played a central role in Comte's idea of 'development', but only in the 'critical' epoch when their role was to challenge and help destroy old ideas. There was no such role for education in the 'organic' epoch; indeed education, in its Millian meaning, was antagonistic to the epoch that was to be ushered in by Comtean positivism. However, it must be emphasised that Mill accepted, with his adherence to a doctrine of 'liberty', the positivist reconciliation of progress and order. The main preoccupation of Book VI of Mill's *Logic*, whose publication in 1843 marked his arrival as a generally influential intellectual figure, was the problem of how to reconcile progress with order.

Mill opened Book VI of the *Logic* by condemning the 'backward state of the Moral Sciences'. His remedy was the application 'of the methods of the physical sciences, duly extended and generalised'.[106] A science of human nature or, in 'Comte's convenient barbarism "sociology"', was Mill's prime aim. He dismissed the argument that the 'science' would necessarily be deterministic and discounted a second objection that its necessary inexactness would vitiate its claim to scientific status. Although he admitted that it was implausible to suppose that it would ever be possible both to set up replicable experiments and be able to amass and analyse all relevant data, Mill argued that similar problems applied to meteorology which, despite its handicaps as science, was recognised as useful. The science of human nature would be an 'inexact science' but it could be a 'science' nonetheless.[107]

Mill's version of psychology, which comprised 'the universal or abstract portion of human nature', was to be the foundation of the science of human nature.[108] Psychology was to be supplemented by 'ethology', which he defined as the 'subordinate science which determines the kind of character produced in conformity to those general laws [of psychology], by any set of circumstances, physical and moral'.[109] Here again, Mill differed sharply from Comte, who had placed 'biology' or, as in Mill's rendering of Comte, 'physiology', as the science from which 'sociology' was to be derived. When Mill defined ethology he moved rapidly from the metaphor of the organic body to that of the mind: 'According to this definition, Ethology [was] the science which correspond[ed] to the art of education in the widest sense of the term, including the formation of national or collective character as well as individual.'[110] Yet in his attempt to bring education to the fore as the means to reconcile progress with order, Mill's reinterpretation of Comte was nothing other than a reconciliation with Comte's schema for development.

By inserting 'psychology' and 'ethology' into Comte's schema while retaining Comte's historical method, which he preferred to call 'inverse deduction', Mill attempted to reconcile positivism with utility. Utility was substituted for the command of order to ensure 'liberty'; to carve out a central place for 'discussion' in the creation of the new 'social machine'; to reconcile his own education with the influence of Comte, the man whom he characterised in the first edition of the *Logic* as 'the greatest living authority on scientific methods in general'.[111]

To grasp the significance of this reworking of positivism which was carried out in order to make it consistent with utilitarianism, it is crucial to see just how wide Mill, in keeping with his father's utilitarian teaching, believed 'education in the widest sense' to be. Education, as it appeared throughout his later works, encompassed the diverse issues of electoral and land reform, birth control and equality for women, and the rights of labour.[112] All were deemed simultaneously necessary prerequisites for, and classrooms of, the 'education' and the 'discussion' which would make the construction of the new 'social machine' possible. The social machine for the new natural state could only be constructed through developing the minds of human beings. Minds could only be developed under the condition of liberty, a condition which was one of 'choice'.

Thus, the development problem was posed long before it came to be associated with colonial and Third World trusteeship. 'Customs', wrote Mill – not as a twentieth-century cultural anthropologist of Africa but as a mid-nineteenth-century commentator on England – might be 'good as customs, and suitable',

> yet to conform to custom, merely *as* custom, does not educate or develop . . . any of the qualities which are distinctive endowments of a human being. The human faculties of perception, judgement, discriminative feeling, mental activity, and even moral preference, are exercised only in making choice. . . . The mental and moral, like the muscular powers, are improved only by being used. [emphasis added]

The emphasis upon 'choice' is unmistakable; Mill continued:

> He who lets the world, or his own portion of it, choose his plan of life for him, has no need of any other faculty than the ape-like one of imitation. He who chooses his plan for himself, employs all of his faculties. He must use observation to see, reasoning and judgement to foresee, activity to gather materials for decision, discrimination to decide, and when he has decided, firmness and self-control to hold to his deliberate decision. . . . It is possible that he might be guided in some good path, and kept out of harm's way without any of these things. But what will be his comparative worth as a human being?[113]

Choice was not only the capacity to choose, a condition for development, but a quality of being human as opposed to being 'ape-like'. The individual person, with each human subject different from the other, was the result of choice. 'Individuality', in Mill's words, was 'the same thing as development, and it is only the cultivation of individuality which produces, or can produce, well-developed human beings.'[114] The end of development was for Mill, as it was for Comte, the reconciliation of progress and order.

Progress not only coexisted with order, but each was necessary for the other. Again, in Mill's words, 'a party of order or stability, and a party of progress or reform, are both necessary elements of a healthy state of political life. . . . Each of these modes of thinking derives its utility from the deficiencies of the other.' A contradiction between progress and order, where what was negative for one gave positive purpose to the other, would continue until development had done its work, when 'the one or the other shall have so enlarged its mental grasp as to be a party equally of order and of progress, knowing what is fit to be preserved from what ought to be swept away'.[115] Only then might it be possible to enter into Mill's 'stationary state'.

Mill's stationary state is much vaunted as the negation of progress.[116] But the failure to distinguish the idea of progress from that of development, here as in so many other cases, gives a misleading impression of what is implied in all allusions to a stationary state of economic growth. For Mill the stationary state was a natural state where mere material progress, prompted by necessity in which 'minds' were 'engrossed by the art of getting on', would shed its false character of increasing wealth and produce its 'legitimate effect, that of abridging labour'. Here the art of getting on would be replaced by the 'Art of Living'. When Mill gave a vision of a stationary state he saw it as a possible outcome of a foreshortened period of his transitional state. However astute the perception that Mill was one of the most influential early exponents of modern environmentalism, the 'art of living' is made possible by the development designed to act against the chaos of progress. For instance, the realisation of Mill's stationary state was what today's developers would call 'sustainable development'. Mill expressed his case for a stationary state in words which should make its modern-day exponents blush for their intellectual and literary inadequacy:

> A world from which solitude is extirpated, is a very poor ideal. Solitude, in the sense of being often alone, is essential to any depth or meditation of character; and solitude in the presence of natural beauty and grandeur, is the cradle of thoughts and aspirations which are not only good for the individual, but which society can ill do without. Nor is there much satisfaction in contemplating a world with nothing left to the spontaneous activity of nature; with every rood of land brought into cultivation, which is capable of growing food for human beings; every flowery waste or natural pasture ploughed up, all quadrupeds or birds which are not domesticated for man's use exterminated as his rivals for food, every hedgerow or superfluous tree rooted out, and scarcely a place left where a wild shrub or flower could grow without being eradicated as a weed in the name of improved agriculture.

Mill continued to emphasise that the solitude of natural space could be achieved through a change in mind over time:

> If the earth must lose that great portion of its pleasantness which it owes to things that the unlimited increase in wealth and population would extirpate from it, for the mere purpose of enabling it to support a large, but not a better or happier population, I sincerely hope, for the sake of posterity, that they will be content to be stationary, long before necessity compels them to it.[117]

However, the development necessary to bring about the stationary state could only happen autonomously in societies which were not bound by 'custom' and in which tolerance and rational discussion were permitted. An India, for instance, as argued by Mill following his father, needed to be governed despotically through the exercise of trusteeship in order to create the conditions under which 'education', 'choice', 'individuality' – in a word 'development' – might occur. For Mill, wittingly, as for unwitting modern theorists, development could only occur where the conditions of development were already present. Societies in which the conditions were not present had to be guided by those from societies in which such conditions were already extant.

THE INDIAN REITERATION

J.S. Mill was an employee of the British East India Company for more than three decades, and drafted his *Logic* and *Principles of Political Economy* on the Company's stationery while so employed. In spelling out his necessary conditions of development for Indians, Mill stood amidst a long line of British writers who addressed the question of India as a codicil of the European invention of development.[118]

As early as the writings of Thomas Mun in the seventeenth century, India had appeared as a problem of English development. The question which arose was whether England's trade with India constituted a 'drain' on the wealth of the former and thus a power for England's corruption and decline.[119] Through the works of James Steuart and others, coeval with the process of the East India Company's conquest of the subcontinent, the object of inquiry continued to be that of reciprocal effects of Indian and British development.[120]

By the late nineteenth century, the problem addressed by Mun had been reversed. Now it was the supposed 'drain' of Indian developmental capacity through the British connection which, in the writings of Dadabhai Naoroji and Romesh Dutt, was to claim centre stage.[121] In between, and indeed beyond, India was to be at least a partial preoccupation of an extraordinary number of the luminaries of British political economy. Thomas Malthus was to become, during 1804, a charter member of the faculty of the newly established staff training college of the East India Company at Haileybury. James Mill was a senior company administrator, obtaining his position through Jeremy Bentham's intervention. David Ricardo was a Company director, while James Ramsay McCulloch taught at Haileybury. For each as for John Maynard Keynes, who in the next century was to be a civil servant in the India Office and whose early writing on economics returned to James Steuart's concern with Indian monetary problems, India was to be a formative and prime concern for those who are remembered for having inspired, or reflected upon, signal shifts in policy towards Britain.[122]

It was the 'constitution' of Indian society which lay, from the late eighteenth century onwards, at the heart of much of this inquiry. James Mill put the matter thus:

> The nations of Europe became acquainted, nearly about the same period, with the people of America, and the people of Hindustan. Having contemplated in the one, a people without fixed habitation, without political institutions, and with hardly any other arts than those indispensably necessary for the preservation of existence, they hastily concluded, upon the sight of another people

inhabiting great cities, cultivating the soil, connected together by an artificial system of subordination, exhibiting monuments of great antiquity, cultivating a species of literature, exercising arts, and obeying a monarch whose sway was extensive, and his court magnificent, that they had suddenly passed from the one extreme of civilization to the other.

The elder Mill argued that 'the force of observation, demonstrated the necessity of regarding the actual state of the Hindus' at the point of British conquest 'as little removed from that of half civilized nations'. In the face of this evidence, however, a 'saving hypothesis' was adopted, Mill argued, by those who championed India. This hypothesis stated that the 'situation in which the Hindus [were] now beheld' was 'a state of degradation'. The inhabitants of Hindustan had formerly been in 'a state of high civilization; from which they had fallen through the miseries of foreign conquest and subjugation'.[123] This 'saving hypothesis', despite its modern overtones, echoes a pre-modern cyclical view of development.

Chief amongst those who had invoked the saving hypothesis were the Orientalists – who sought an authentic basis for restoration of Indian greatness in Persian and Sanskrit texts – and Edmund Burke, who, through his speeches during the impeachment of Warren Hastings, had argued for a reconstitution of British rule on the basis of what he took to be the authentic constitution of Indian society.

For Burke, it was English rule itself which bore primary responsibility for the then-present state of Indian regress. 'The Tartar invasion was mischievous, but it is our protection which destroys India', he wrote. 'Young boys' without 'society or sympathy with the natives' ruled on behalf of the company 'animated with all the avarice of age and all the impetuosity of youth.' They rolled into India 'one after another; wave after wave . . . new flights of birds of prey and passage'. Indian nationalists were later to reproduce the sentiments of Burke's words of condemnation when he argued that:

Every rupee of profit made by an Englishmen is lost forever to India. With us there are no retributory superstitions, by which a foundation of charity compensates, through ages, to the poor, for the injustices of a day. With us no pride erects stately monuments which repair the mischiefs which pride had produced, and which adorn a country out of its own spoils.

Having destroyed, 'England' had not re-created in India, so Burke continued:

England has erected no churches, no hospitals, no palaces, no schools; England has built no bridges, made no highroads, cut no navigations, dug out no reservoirs. Every other conqueror of every other description has left some monument, either of state or beneficence, behind him. Were we to be driven out of India this day, nothing would remain, to tell that it had been possessed during the inglorious period of our dominion, by anything better than the ourang-outang or the Tiger.[124]

If nothing had been built by Britons in India, much of importance, in Burke's view, had been destroyed. Most important was the laying of waste to the Indian *ancien régime* and the 'prescriptive constitution' which it had embodied. No mere paper document, this constitution was 'an idea of continuity' which extended 'in time . . . and in space'. The constitution was 'a constitution made by what is ten thousand times better than choice, it is made by the peculiar circumstances, occasions, tempers, dispositions, and moral, civil, and social habitudes of the people which disclose themselves only in a long space in time'.[125] Warren Hastings' rule, on behalf of the East India Company, had corrupted the value of a constitution in India. If it was necessary that Englishmen 'must govern' Indians, then Indians, if the destruction was to stop, had to be governed 'upon their own principles and maxims and not upon ours; that we must not think to force them to our narrow ideas, but extend ours to take in theirs; because to say that people shall change their maxims, lives and opinions, is what cannot be'.[126]

It was both for the sake of Indians and the English that the period of corruption should be halted because Burke believed that the 'breakers of law in India became the breakers of law in England'. Nabobs, returned with their gains from the break-up of the Indian constitution, threatened through their purchase of political place and privilege to corrupt and undermine the English constitution, in Burke's view. Speaking in Parliament during the impeachment of Warren Hastings, Burke spelt out why the House of Commons was not merely voting in the trial of one official but was in the process of laying down 'a set of maxims and principles to be the rule and guide of future governors in India', and in particular for 'Lord Cornwallis who was about to proceed thither as Governor General'.[127]

For Cornwallis, the restoration of the Indian constitution was synonymous with the official recognition of 'a regular gradation of ranks' necessary to preserve the 'order in civil society' of India. Central to the recognition of hierarchical order was a recognition of the rights to landed possession, the rights which Cornwallis took to be the Indian equivalent of rights claimed by the aristocracy of England of which he himself was a member.[128] As such, he followed Burke and others for whom landed property was integral to a prescriptive constitution. Burke had argued that the houses of the aristocracy were the 'public repositories and houses of Record' of the English constitution. These 'repositories' were safeguarded by the 'traditionary politics of certain families' more than by 'anything in Laws and order of the State'.[129]

Cornwallis, however, adduced an additional reason for making a permanent settlement of property and taxation with the Zemindars of Bengal. In assuming that the '*zemindari*' were the Indian analogue of his aristocratic class, Cornwallis argued for their encouragement of improvement of the land and the extension of cultivation. In response to those who argued against a permanent fixing of taxes and their revision on a decennial basis, Cornwallis proposed that:

> In a country where the landlord has a permanent property in the soil it will be worth his while to encourage his tenants . . . to improve that property. . . . But when the lord of the soil himself . . . is only to become the farmer for the lease of ten years, and . . . then . . . exposed to the demands of a new rent, which may perhaps be dictated by ignorance or rapacity, what hopes can there be, I will not say of improvement, but of preventing desolation. Will it not be in his interest, during the early part of his term, to extract from the estate every possible advantage for himself, and . . . to exhibit his lands at the end of it in a state of ruin . . .?[130]

The *zemindari* were to be 'restored to such circumstances as to enable them to support their families with decency'. A 'regular gradation of ranks' thus was to be supported 'which is nowhere more necessary than in this country, for preserving order in civil society'.[131] Under a permanent settlement, Cornwallis promised 'wealth and happiness to the intelligent and industrious part of the individuals of the country'. Such was to be one example of how the case for trusteeship was to be tied to justification for landed privilege, an example which was to recur in different times and places during the course of colonial history.

In planning the permanent settlement of the Bengal *zemindari*, Cornwallis availed himself of the services of the noted Orientalist and friend of Burke, Sir William Jones, precisely the type of thinker whom James Mill was later to inveigh against. Cornwallis went on to argue that 'one third of the company's territory in Hindostan is now a jungle inhabited only by wild beasts'. A ten-year lease would never 'induce any proprietor to clear away the jungle'. Cornwallis added the argument of progressive development to that of the stability of the prescriptive constitution of Burke.[132] His views were to find support in the writings of Thomas Malthus who had justified, as we have seen, the role of landed proprietors in slowing the inevitable decline of the condition of humankind.

Credit has been given to Malthus, for example by Stokes, for having '"discovered" the law of rent', or the category of differential rent, during the course of his 'professional duties at the East India College'.[133] Repeatedly conflated with taxation, arguments about the theory of rent and its application became the important question, as Stokes has shown, of British rule in India. For Malthus, rent, at its simplest, was defined as that portion of the value of the whole produce which remains to the owner of the land, after all the outgoings belonging to its cultivation, of whatever kind, have been paid, including profits to the capital employed, estimated according to the usual and ordinary rate of the profits of agricultural stock at the time being.[134]

Under this theory, all land would generate the same quantity of rent if all land was of the same quality. Land, however, was of varying quality. Initially, only the best land would be brought into production. As population and the demand for foodstuffs grew, land of lower quality would be farmed. In order for this lower quality land to be tilled, the returns from it would have to be at least equal to the 'outgoings' or average costs of production. Given that produce from poorest-quality land would determine the price of agricultural commodities, the owners of land superior to that of poorest quality, and which was needed to satisfy the demand for agricultural produce, would realise a 'bounty of Nature' in the form of a rent associated with the quality of the soil. Thus, Malthus assumed that landlords would have a keen interest in finding those tenants to work the land who would do so most efficiently. It was for this reason that Malthus maintained that 'the interest of the landlord is strictly and necessarily connected with that of the state' in ensuring an increase in agricultural productivity.[135]

In Malthusian terms, it had been the over-extraction of rent, in the form of tax, by the pre-conquest 'Oriental despots' of India, which accounted for Indian poverty. The permanent fixing of taxes to be paid by the Bengal *zemindari* would not only then satisfy the demands for prescriptive constitution of Burke but would also set India on the path to a progressive economic state.[136] However, as Ambirajan has pointed out, this was not to be:

There was no legal way by which landlords could collect the rent from their tenants. Thus many Zemindars found it difficult to pay their taxes regularly. When they failed to pay their taxes, the lands were taken from them and auctioned off. Most of the lands were acquired by the nouveau riche of Calcutta, who ruled their possessions through unscrupulous agents. Poverty was the invariable result of such rack-renting and mismanagement.[137]

In failing to meet the expectations of its progressive economic purpose, the permanent settlement with the Bengali *zemindari* simultaneously undercut its Burkean justification. The 'nouveau riche of Calcutta' were not Burke's 'houses' in which the prescriptive constitution was lodged. Moreover, uncollected revenues placed the East India Company in a state of 'financial stringency'. Financial compulsion opened the door to the expansion of the alternative policy of a '*ryotwari* settlement' with those who were taken to be the true cultivators of the soil. This was the policy begun by Alexander Read and Thomas Monro in parts of the Madras Presidency from 1792 onwards.[138]

By 1815, the economic reasoning behind the *zemindari* settlement was regarded as faulty by the East India Company's Court of Directors, and by Monro, who based himself on a particular reading of Adam Smith. The effect of the permanent settlement was to 'augment the landlord's rent, not the profit of the cultivator'. The Directors now believed, as recounted by Ambirajan, that 'there was no reason to suppose that the rent of the landlord would be automatically spent on capacity-creating investments'. On the other hand, the cultivator, under a *ryotwari* settlement, 'would have every inducement to continue his industry, although there should be no surplus to be paid in the shape of rent to the landlord'.[139] This view meshed with the theoretical position of Malthus' antagonists in the famous debate over the nature and significance of rent as a category of political economy.

For Ricardo, rent paid to a landlord was always a deduction from either the wages of labour or the profits of capital that accrued without exertion of the landowner. For this reason he believed that 'the interest of the landlord' was 'always opposed to every other class in the community.'[140] James Ramsay McCulloch, who lectured on political economy at Haileybury, repeated this particular criticism of Malthus in his correspondence with Ricardo and his edition of Smith's *Wealth of Nations*.[141] It was James Mill, in his *History of British India*, who was to deliver the most devastating critiques of landlordism in India. We have already seen how Mill dissented sharply from the Orientalist view of India as a high civilisation thrown upon hard times. In particular he attacked the work of Cornwallis' mentor, Sir William Jones.

'It was unfortunate', Mill thought of Jones, 'that a mind so pure, so warm in the pursuit of truth, and so devoted to oriental learning . . . should have adopted the hypothesis of a high state of civilization in the principal countries of Asia.' Mill characterised the *zemindari* before the permanent settlement as petty tyrants. He cited East India Company documents brought before the House of Commons to the effect that the *zemindars* had 'plundered all below them' and 'enriched themselves with the spoils of the country': 'The whole system thus resolved itself, on the part of public officers, into habitual extortion and injustice; which produced, on that of the cultivator, the natural consequences – concealment and evasion, by which government was defrauded of a considerable part of its just demands.'[142]

Mill argued that the permanent settlement completely misunderstood the real role and position of the *zemindari*. Because, Mill explained, the Zemindar 'collected the rents of a particular district, lived in comparative splendour, and his son succeeded him when he died . . . it was inferred without delay' that the

Zemindars 'were the proprietors of the soil, the landed nobility and gentry of India'. There had been a failure to understand that Bengal was not Sussex. The *zemindari*, Mill opined, may have governed their ryots with a 'despotic power' but 'they did not govern them as tenants of theirs'. The latter had a 'hereditary possession' rather than a tenancy 'at will or by contract' from which it 'was unlawful to displace him'. Moreover, 'for every farthing the Zemindar drew from the ryot, he was bound to account' to his masters. Unlike the idealised English system of landlord, capitalist farmer/tenant and agricultural wage-labourer in India, Mill maintained that:

> Three parties shared in the produce of the soil. That party to any useful purpose most properly deserves the name proprietor, to whom the principal share of the produce for ever belongs. To him who derives the smallest share of the produce the title owner least of all belongs. In India to the sovereign the profit of the land may be said to have wholly belonged. The ryot obtained a mere subsistence. . . . The Zemindar enjoyed . . . a compensation for his services. To the government belonged more than half the gross produce of the soil.[143]

Thus, while 'the English were actuated not only by an enlightened, but a very generous policy when they resolved to create in favour of individuals a permanent property in the soil . . . conducive . . . to the increase of its produce, and the happiness of the people', they had completely misread the Indian reality. What was worse, for Mill, was that they had misread the English reality on which they had based the theory that lay behind the permanent settlement.

'When our countrymen draw theories from England,' Mill wrote with great sarcasm, 'it would be good if they understood England.' It was not because England had a landed aristocracy that agriculture had improved. Agricultural improvement was the result of the effort of 'the immediate cultivators' who had 'increased so wonderfully the produce of the land in England, not only without assistance' from their landlords, 'but often in spite of them' but who were afforded protection by the 'laws of England' against their lords. Far from being the progressive landowners of Malthus, England's lords demonstrated the 'love of domination over the love of improvement and of wealth'. Long leases on stable terms were recognised, Mill maintained, as 'essential to all spirited and large improvement', but the 'proprietors of the soil in England' complained that such leases 'render[ed] their tenantry too independent of them'.

'If the gentlemen of England' were willing 'to sacrifice improvement to the petty portion of arbitrary power' that they exercised over their tenants-at-will, Mill asked, 'what must we not expect from the Zemindars of Hindustan', with minds nurtured in the 'habits of oppression' when asked to choose between improvement and domination? Mill revealed that the bulwark of Burke's prescriptive constitution was a backward, petty tyranny that attempted to export itself to India.[144]

Even this was done ham-handedly. Mill cited the collector of Mindapore that 'all of the Zemindars with whom I have had any communication . . . say that such a harsh and oppressive system was never before resorted to in this country'. The previous practice of imprisoning *zemindari* for failing to pay their tax dues was 'mild and indulgent to them' in comparison with the English practice of depriving them of their lands and position. English practice had 'in the course of a very few years reduced most of the Zemindars of Bengal to distress and beggary'. Another official, Sir Henry Strachey, recounted how the 'great men formerly were the Mussulman rulers, whose places we have taken, and the Hindu Zemindars. These two classes are now ruined and destroyed'.[145] Moreover, the 'ryots, left without any efficient legal protection, were entrusted to the operation of certain motives, which were expected to arise out of the idea of permanent property' where 'practically that permanence had no existence'. It was no wonder, for James Mill, that India starved.[146]

Pervasive poverty had spawned the evil of social disorder. Strachey had elsewhere argued that 'the vices of the people' arose from their poverty and ignorance but 'especially from poverty'. 'When a man had nothing to lose', James Mill asked, 'and everything to gain by disregarding the laws of society, by what power is he to be restrained?' The Bengal permanent settlement, which for Burke was to create social order and for Malthus was to promote 'improvement', had become through its application, for James Mill, the source of chaos and economic regress.[147]

In Mill's opinion, all this need not have been. 'There was an opportunity in India', Mill wrote, 'to which the history of the world presents not a parallel.' The rights of the *zemindari* could have been bought out and tenure 'bestowed upon those from whom alone, in every country, the principal improvement of agriculture must be derived, the immediate cultivators of the soil'. The ryots would continue to be taxed, but directly by the East India Company. Moreover, the amount of taxation would be arrived at scientifically, in keeping with the Ricardian rendition of rent theory. Only that portion remaining after the deduction of 'all outgoings including profit', or in effect that which had formerly constituted the 'unearned increment' of *zemindari* rents, would be appropriated. This measure, along with an imported code of law to replace 'native superstition', would provide both order and progress. Such a policy would be 'worthy to be ranked among the noblest that ever were taken for the improvement of any country, might have helped to compensate the people of India, for the miseries of that misgovernment which they had so long endured'. The opportunity that hitherto had been lost because 'the legislators were English aristocrats; and aristocratical prejudices prevailed' should now, so Mill argued, be grasped.[148]

The departure signalled by Mill's programme was significant. As Barber has argued, 'Mill sought a regime which offered uninhibited scope for private initiative' but 'immense barriers to the natural play of market forces had been erected'. James Mill's mission 'of British administration', Barber has explained, was to remove 'the barriers against market forces', but the mission was contradictory:

> Through the implementation of the scientific tax the market would be perfected. Simultaneously, the state would be assured of the revenues required to ensure security in person and property and to guarantee incorruptibility in the administrative and judicial systems. The resulting structure, in turn, would activate unused potential in the private sector. Before these gains could be realised, however, the hand of the state would necessarily be very visible. The planners' success, of course, was ultimately to be measured by the achievements of the private sector.[149]

The four decades following the publication of James Mill's history of British India were characterised by a renewed aggressive, territorial acquisition on the part of the East India Company coupled with the building of public works and the reform of revenue collection. It was the task of James Mill's son, John Stuart Mill – who had been found employment with the Company just as Bentham had found a position for his father before him – to evaluate the public works and revenue reform policies.

The occasion for the younger Mill's writing of his *Memorandum of the Improvements in the Administration of India* was that of a defence of the Company against the revocation of its governmental privileges in the aftermath of the 1857 mutiny and revolt. J.S. Mill detailed the prodigious construction of canals for irrigation and transport which had been dug and repaired, the railways and roads built, some by public capital, some constructed privately but with government guarantees.[150] All had been made possible, in the view of the younger Mill, by 'fixing and moderating the demands of the Government on the tax-paying population', which accounted for over two-thirds of state revenue. This was, in his opinion, not only 'the most effectual means which could have been adopted for improving the productive resources of the country' but of providing the resources for the 'direct aid of government to industry', the principal means of which were 'irrigation and the means of communication'.[151]

In reviewing the changes that had taken place in the collection of revenue, J.S.

Mill condemned the 'important mistake' of the permanent settlement committed by Cornwallis in Bengal. While Mill averred that Cornwallis had the 'most generous intentions' of protecting the rights of 'occupying tenants', the 'poverty' and 'passive character' of the cultivators rendered this protection 'illusory'. Mill went on to approvingly detail the institution of *ryotwari* settlements by Monro and others as a correction to the errors of Cornwallis in Bengal. Yet these reforms were insufficient. 'It was now known', wrote Mill,

> that in the greater part of India, and without doubt originally throughout the whole, the property in land (so far as that term is applicable at all to India) resides neither in the individual ryot, nor in the great officers who collected revenue for the former native governments, but in the village communities.[152]

It was the 'village community' that was, in Mill's view, the basis for the best form of taxation or rent collection. Mill clearly derived this view, which was imbibed, with variations, throughout Company rule in India, from the work of Sir Henry Maine who had been Law Member of the Supreme Government of India over much of the same period as Mill's own tenure with the Company. In a later review of Maine's work, *Village-Communities in the East and West*, Mill set out the reasons for his approbation.

Mill judged Maine's work to be an 'important contribution to a branch of knowledge in which the author is unrivalled – the philosophy of historical institutions'. Maine challenged 'Political thinkers, who at one time may have been over-confident in their power of deducing systems of social truth from abstract human nature' but who now showed 'a tendency to the far worse extreme, of postponing the universal exigencies of man as man, to the beliefs and tendencies of particular portions of mankind as manifested in their history'.

By examining the historical roots of social institutions, 'particularly land tenure', Maine had applied, in Mill's opinion, a 'powerful solvent to a large class of conservative prejudices . . . which many believe to be essential elements of the conception of social order'. This struck 'at the tendency to accept the existing order of things as final – as an indefeasible fact, grounded on eternal social necessities'.[153] Especially in regard to the supposed usurpation of common right to form landlord property, it opened the question,

> whether the older or the later ideas are best suited to the future; and if the changes from the one to the other were brought about by circumstances which the world has since outgrown – still more if it appears to have been in great part the result of usurpation – it may well be that the principle, at least of the older institutions, is fitter to be chosen than that of the more modern, as the basis of a better and more advanced constitution of society.[154]

Mill maintained that 'It was not Mr. Maine's business, in a purely historical and jurisprudential work, to deduce practical inferences' from his recounting of the manner in which the rights of village-communities in Asia and Europe had been usurped. There were, however, 'certain truths, of a very important character . . . which Mr. Maine's work seems to us to support and illustrate very impressively'. These were, Mill now argued by referring explicitly to Britain, 'that different systems of property in land have existed, and even co-existed . . . but over our entire history since the Norman conquest, we have been gradually transforming one of these systems into another'. Moreover, there was no good reason to prefer the latter system 'under which nearly the whole soil of Great Britain has come to be appropriated by about thirty thousand families – the greater part of it by a few thousand of these'. Given these truths, Mill maintained that,

> if the nation were to decide, after deliberation, that this transmutation of collective landed ownership into individual shall proceed no further. . . . Nay, further, that if the nation thought proper to reverse the process, and move in the direction of reconverting individual property into some new and better form of the collective, as it has been converting collective property into the individual, it would be making legitimate use of an unquestionable moral right . . .[155]

'The nation', Mill proposed, 'ought to take into serious consideration which among the many footings on which the right of land ownership might be placed' would be that 'most beneficial to the whole community, with a view to adopting' such a system. The 'most beneficial' was clearly, for Mill, some form of community tenure.[156]

Yet, if Maine had set out the justification for a move toward community tenure in Britain as much as in India, the movement in India was on the verge of a second betrayal. 'Since the Mutiny a reaction' had 'set in' against the movement toward the recognition of communal property in India. In Oude, before its pre-Mutiny annexation, there had existed 'the Talookdars – a class of functionaries of very various origin, who collected the Government dues from large districts' through 'bodies of undisciplined mercenaries'. These had kept the country in a state of 'bloodshed and warfare' while the Talookdars' possessions were 'swelled by the dispossession, and sometimes the extermination, of entire families of land-holders'. British reform in limiting their powers had 'exasperated' the Talookdars by depriving them of their 'misused position' and they retaliated by becoming 'the only powerful class or body in all of India that did join with the mutineers'. In response, according to Mill, the British administration, unnerved by the Mutiny, 'declared them proprietors of the soil, and delivered over the cultivating classes into their hands'. There was 'great danger' that this act would be replicated throughout the northern provinces of India. If this were to come to pass, opined Mill, 'the whole of the northern provinces' would be 'possessed . . . by a comparatively small body of absolute owners . . . with a vast body under them as tenants at will'.

And this – one of the greatest social revolutions ever effected in any country, with the evil peculiarity of being a revolution not in favour of a majority of the people, but against them – its supporters defend in the name of civilization and political economy; though if there is a truth emphatically taught by political economy . . . it is that the status of an agricultural tenant-at-will is intrinsically vicious, and in a really civilized community ought not exist.

'The exposition given by Mr. Maine of the real nature and history of agricultural customs in India' was to be studied to put a check upon the 'baneful reaction' of restoring a class of landlords.[157]

Fear of this 'baneful reaction', which Mill judged to be a bolstering of Talookdar power in Oude, led him to defend the East India Company from those who sought to disestablish it in the aftermath of the 1857 Mutiny. For while Mill championed the cause of 'village-community' he could not, as we have seen, countenance that of India's rule by the 'representative democracy' of Britain nor, still less, by Indians themselves.

In hearings before the House of Lords on the government of India in the aftermath of the Mutiny, Mill was asked what he believed would be the result of India being governed by a Secretary of State appointed by the British Government and approved by Parliament. His response was unequivocal. Such a government, Mill stated, 'would be the most complete despotism that could exist in a country like this'. Parliament would have little interest or inclination to avail itself of the information necessary for the intelligent government of India. Moreover, the changes in policy for India resulting from changes in governing party in India would be a 'great evil'.[158] When asked whether the then extant East India Company's Court of Directors represented the 'people of India', Mill was again succinctly unequivocal, responding 'Certainly not.'[159] Yet given his hostility to Parliamentary control and his belief that India had not yet 'attained such a degree of civilization and improvement to be ripe for anything like a representative government', he maintained that Company rule was still the best available alternative. If it were not possible to have, either in Britain or in India, 'the advantages given by a representative Government of discussion by persons of all partialities, prepossessions and interests' then 'some substitute was better than none'.

A substitute government, for Mill, was 'a body unconnected with the general government of the country, and containing many persons who have made that department of public affairs the business of their lives; it was the East India Company's then current Court of Directors'.[160] In the 'Petition of the East India Company', drafted by Mill to Parliament in defence of the Company's rights, Mill maintained that 'in no government known to history' had 'appointments to offices, as especially to high offices, been so rarely bestowed on any other consideration than those of personal fitness' than as had been true under the Company's rule. Patronage was justified in this case because 'the dispensers of patronage' had been 'persons unconnected with party, and under no necessity of conciliating Parliamentary support'. Patrons had been drawn 'primarily from middle classes, irrespective of political considerations, and, in large proportion [from] the relatives of persons who had distinguished themselves by their services in India'. If it had not been for this unencumbered trusteeship, Mill flatly stated, 'in all probability India would have been long since lost to this country'.[161]

Among those families which had distinguished themselves by their service to India were the Stracheys, whose Indian connection was well over a century old by the time Mill wrote the above-quoted lines. By the mid-nineteenth century, two scions of this family, John and Richard Strachey, had risen to high office in India. In 1882 the Stracheys co-authored the retrospective of their lives' work, *The Finances and Public Works of India from 1869 to 1881*, and dedicated the book to 'the public servants of all classes the results of whose labours for the people of India [were] herein recorded . . . by their fellow workers'. In their final chapter, the Stracheys addressed the question of the 'future requirements of public works and finance in India'. Here, they reflected that 'the recent remarkable progress of India, which has been placed beyond every reasonable doubt, may be traced up to the natural productive powers of the country, for the development of which greatly increased facilities have been given by the extension of railways and cheap transport'.[162] Yet, there was no time to cry, 'Rest and be thankful.' The Stracheys argued that, as the famines of the 1870s had shown, much more needed to be done. How was this work to be accomplished? Their answer was that 'experience' had 'established beyond dispute that it is within our power both to construct and work railways economically through state agency'. Moreover, the Stracheys continued, it was in the financial interests of the Government of India to do so:

> for, it may without hesitation be said that in the case of lines yielding a profit . . . the amount which would be carried out of the country by a company of foreign capitalists . . . must be greater than the charge incurred under Government management . . . since the profits, even if smaller, would all remain in India.[163]

For their theoretical justification for state development, the Stracheys turned to the writings of none other than J.S. Mill who in his 'matured opinions' represented one of the 'ablest advocates of rational freedom and the foremost opponents of the undue extension of Government action'. In making Mill a force of authority which must command 'attention and respect', the Stracheys quoted Mill's *Principles of Political Economy* at length:

In the particular circumstances of a given age or nation there is scarcely anything, really important to the general interest, which it may not be desirable, or even necessary, that the Government should take upon itself, not because private individuals cannot effectively perform it, but because they will not. At some towns and places there will be no roads, docks, harbours, canals, works of irrigation, hospitals, schools, printing presses, unless the Government establishes them; the public either being too poor to command the necessary resources, or too little advanced in intelligence to appreciate the ends, or not sufficiently practised in cojoint action to be capable of the means.

Their quotation of Mill continued:

This is true, more or less, of all countries inured to despotism, and particularly to those in which there is a very wide distance in Civilization between the people and the Government: as in those which have been conquered and are retained in subjugation by a more energetic and more cultivated people.

And:

In many parts of the world the people can do nothing for themselves which requires large means and combined action; all such things are left undone, unless done by the State. In these cases, the mode in which the Government can most surely demonstrate the sincerity with which it intends the greatest good of its subjects, is by doing the things which are made incumbent on it by the helplessness of the public in such a manner as shall tend not to increase and perpetuate, but to correct, that helplessness.[164]

Such a clear injunction for the empowerment of their charges through the practice of trusteeship had not been lost on Richard Strachey, who had already taken 'greatly to Positivism and the philosophy of Comte'.[165] Yet, if it was Mill's advocacy of the trustee's role that had appealed most to British civil servants in India, such as the Stracheys, it was to a different, but allied, passage of his *Political Economy* that nascent Indian economic thought, as Ganguli has shown, turned to for guidance.

Mill had written that:

A country which makes regular [non-commercial] payments to foreign countries, besides losing what it pays, loses also something more, by the less advantageous terms on which it is forced to exchange its production for foreign commodities. . . . The paying country will pay a higher price for all that it buys from the receiving country, while the latter, besides receiving the tribute, obtains the exportable produce of the tributary country at a lower price.[166]

In the hands of Pritwish Chandra Roy, Subramania Iyer, and then most tellingly in those of Dadabhai Naoroji and Romesh Dutt, this passage was to become the basis of the celebrated 'drain theory' from which all modern theories of underdevelopment in some measure are descended. This drain constituted a corruption of trusteeship. Thus, Naoroji's text was titled ***UnBritish** Rule in India*. Facets of corruption were to include a relative lack of saving, relative scarcity of capital and unnecessary drudgery of labour due to the lack of mechanisation.

To reverse the drain was the work of development, but the work could only be done through the creation of an alternative developmental process. The field of development activity was to be none other than Maine's village-communities in which an increase in labour productivity and incomes would dam up, it was thought, the effect of the drain while creating an autonomous democratic form of development. As Ganguli has written, 'An entire economic philosophy took shape in the last decades of the nineteenth century clustering around the idea of the priority of rural development, not merely as a strategy, but as a value in itself.'[167]

Through Mohandas Gandhi, and then with the establishment of the Panchayat Raj of the 1950s, trusteeship, once having earlier been betrayed, was to be re-established through rural development if only to become itself the target of accusations of corruption and betrayal.[168]

CONCLUSION

A common sense of 'development' is that an immanent process of development, as of the development of capitalism in India, destroys the social value of community. From Burke at the end of the eighteenth century, to the late twentieth-century critics of capitalist development in India, the constructive purpose of policy has been to assert the value of community in the face of the destruction wrought by an immanent process of development. This purpose has been so well assimilated in the course of social action that the value of community has come to be seen as a natural quality of being human. Thus, particular communities, from the village to the urban neighbourhood, are enveloped within the good of human community – the freedom of self-determination and self-realisation.

An historical account of development doctrine, especially that which stems from the nineteenth century, provides a different perspective on what the social value of community might be and why its natural quality arises from the way in which development has been one possible means to construct the positive alternative to the disorder and underdevelopment of capitalism. Community, as the case of India shows, has been cultivated through an intention to develop and this intention to develop has been framed by trusteeship. In development doctrine, development and trusteeship are a part of each other; without trusteeship there is no development doctrine.

The task of development doctrine was to provide a foundation for state policy and, in so far as the doctrine was developed from Comte to Mill, the developmental ideal of policy came to be that of developing a community. In particular, rural development was nothing other than the development of a community of rural producers. Community derived from the state and state policy was necessary for the development of community. The intention to develop, and the way the intention came to be expressed through policy, was the axle upon which the cyclical development of doctrine turned. Each sequence of the turn of policy in India, from Burke to James Mill and then from father to son, and then again from John Stuart Mill to the Indian critics at the end of century, was a reaction against an earlier period of development. As we shall show, the same cyclical development happened in the twentieth century. The corruption which was adduced to each earlier period of development was that of the intention to develop, an intention which failed to realise what was deemed to be the value of community.

Here we have a paradox. The cyclical, pre-modern image of development was an immanent process of development or one, as we shall explain in Chapters 2 and 3, that came to be referred to as the 'external' of development. From its nineteenth-century vantage point, and in the attempt to transcend the idea of Progress, development had lost its cyclical image and, as in the purpose of progressive development, had come to represent the potential and possibility for a linear movement of human improvement. The paradox is that cyclical movement reappeared, whatever the purpose of progressive development, in the intention to develop. It was the intention to develop which embraced the 'internal' of development, namely the conscious authority of autonomous being to determine and realise its potential.

There were two ways out of this paradox. A first, that of Marx (to whom we return in Chapter 3), was to recognise that it was the 'external' development of material and/or 'real' conditions of human existence which was made unconsciously progressive by the expansion of capital and that the internal of development, the conditions for human community, was made possible by the immanence of the development of capitalism. What Marx denied was that there was any possible progress in the social constitution of community; in particular, the development of the particular village community in India was literally reactionary: 'These small stereotypes of social organisms have been to the greater part dissolved, and are disappearing, not so much through the brutal interference of the British tax-gatherer and the British soldier, as to the working of English steam and English free trade.' And, in the well-quoted words which are usually put down to the endemic failings of his Eurocentricism and anti-environmentalism, Marx in 1853 was unambiguous about what the supposed 'good' of village community represented:

> We must not forget that these little communities were contaminated by distinctions of caste and slavery, that they subjugated man to external circumstances instead of elevating man to the sovereign of circumstances, that they transformed a self-developing social state into never changing natural destiny, and thus brought about a brutalizing worship of nature, exhibiting its degradation in the fact that man, the sovereign of nature, fell down on his knees in adoration of Hanuman, the monkey, and Sabbala, the cow.[169]

Destruction of the village community and of artisanal production in the town would accompany, Marx reflected, the realisation of the dual British mission in India, a mission which could not be based upon development doctrine:

> All the British bourgeoisie may be forced to do will neither emancipate nor materially mend the social condition of the mass of the people, depending not only on the development of their productive powers, but on their appropriation by the people. But what they will not fail to do is lay down the material premises for both. Has the bourgeoisie ever done more?

To emphasise what Marx meant by 'human progress' for a human community, he concluded:

> The bourgeois period of history has to create the material basis of the new world – on the one hand universal intercourse founded upon the mutual dependency of mankind, and the means of that intercourse; on the other hand the development of the productive powers of man and the transformation of material production into a scientific domination of natural agencies. . . . then only will human progress cease to resemble that hideous, pagan idol, who would not drink the nectar but from the skulls of men.[170]

We will show in Chapter 3 how the metaphors of progressive development, evidenced by Marx's references to 'social organism', 'inherent organic laws' and 'social state', were part of the evolutionary tempo of the time but that they fall within what we call a restricted domain of development. It is in the expanded domain, where development does have meaning for human community and where it cannot be reduced to the social community of development doctrine, that Marx's prospect for development has the potential to escape from the limits of the paradoxical conundrum of intentional development.

A second way out of the paradox was provided by Newman after 1845. Like Marx, Newman was much affected by the tempo of development and also endeavoured to escape from the limits which progressive development imposed upon development doctrine. Newman's recourse, unlike that of Marx, was to expunge the idea of Progress from development. Development doctrine, for Newman, was the essence of development. God's providential order, the only external form of development, became the internal of development according to the truth and authenticity of religious practice. Thus, 'true' as opposed to 'false' development was evidenced by the belief in, and assent given to, doctrine as it developed and the identity between various practices and the original principle of the Christian revelation. In the course of advancing criteria for what true development meant, Newman prefigured the modern meaning of under-development. He called false development the 'corruption' of doctrine. Corrupt doctrine could develop but its development was little other than what had been entailed by the positivist attempts to make development intentional. Development merely re-created the original condition – pervasive poverty and unemployment – which the immanent process of development had created. It is to under-development that we now turn.

2

DEVELOPMENT WITHOUT PROGRESS

Underdevelopment and J.H. Newman

In the preceding chapter we have argued that the idea of the intentional practice of development was not an invention of the post-1945 international order. The idea of development had been invented to deal with the problem of social disorder in nineteenth-century Europe through trusteeship. In this chapter we will show that, likewise, the idea of underdevelopment was also part of the general phenomenon of development in the mid-nineteenth century. The idea of underdevelopment was referred to then as 'corruption', but it conveyed in its use much of what underdevelopment was to come to mean for the world after 1945. It is our purpose here to show the affinity between the nineteenth- and twentieth-century discussions. This affinity, as we argue in our concluding chapter, is crucial for understanding a substantial part of the problems with present-day theories of self-described 'alternative development'.

We will proceed here by first examining post-war underdevelopment theory in order to uncover some of its less obvious assumptions. We will then move to a discussion of the developmental thought of nineteenth-century Latin America which has been largely ignored by the theorists of 'dependency'. We then turn to the debate between the propounders of underdevelopment theory and their critics. Finally, we will explore a nineteenth-century debate over development which occurred in what is at first glance the unlikely realm of Christian theology, in order to show the ways in which it foreshadowed the modern underdevelopment debate.

POST-WAR 'UNDERDEVELOPMENT THEORY'

Post-war modern underdevelopment theory flows from Paul Baran's *Political Economy of Growth*. Baran wrote: 'It is in the underdeveloped world that the central, overriding fact of our epoch becomes manifest to the naked eye: the capitalist system, once a mighty engine of economic development, has turned into a no less formidable hurdle to economic advancement.'[1] Later, Gunder Frank spelt out why capitalism, as an agent of development, might become such a 'formidable hurdle' to 'advancement' or progress. He wrote of underdevelopment as an historical process which was causally related to the 'pattern of evolution' of developed, industrial societies. In one part of the world, the North, the process whereby producers were separated from their means of production was matched by one in which they were reabsorbed and reintegrated into the production process as proletarian wage-workers. The coupling of these processes happened through the sufferings and privation of capitalist development.[2]

But it was development nonetheless. First, from one reading of Marx, the reintegration of producers into a capitalist production process created the potential for a proletariat to recombine need with industrial capacity without the intervention of capital. This was development as immanent process. Capital may have originated at the intersection between the twin processes of separation and reintegration, but its development could continue only so long as need continued to be separate from capacity. Second, for the positivists, as we have emphasised here, development was the means by which the state might serve to actively contain the disorder of unemployment and destitution. Development, as an active practice of the state, could only be envisaged after the two processes of separation and reabsorption had begun to run through a faltered course of history. This was intentional development, largely forgotten by most theory, including underdevelopment theory.

For Frank neither development as immanent process nor as intended practice was to be found in the other part of the world: 'In the South, in what is today called the Third World, this process [the reabsorption] did not happen at all.' Producers were separated from their means of production in agriculture, manufacture and handicrafts, 'but what never happened was the real reintegration or reabsorption of these people into the productive process in the same way as the North'.[3]

One process without the other, separation without reintegration, is the mark of underdevelopment. One without the other is unbalanced or disharmonious, incoherent or one-sided. One part of the system, it is implied, does not form an active component part of the other. These are images of underdevelopment. The condition of underdevelopment, moreover, is not transitional in the sense that reintegration will happen once the first process is complete. Development means that the two processes happen simultaneously and not sequentially.

Nor, for Frank, is an evaluation of development to start from the premise of dualism: the two parts of a world are not to be treated as if they belonged to two separate, self-enclosed systems. Frank's argument is that the lack of a second process is part of true capitalist development in the North and part of underdevelopment in the South. Thus, were material proof of wage-labour to be offered for an argument that there was a second process of capitalist development in the South, then this proof would be an effect of underdevelopment and not development. Wage-labour for mines, estate agriculture or manufacturing would not in itself register a component part of capitalist development. Rather, the phenomenon would be evaluated as the 'development of underdevelopment' for the South.[4]

Underdevelopment 'develops' because an incoherent coupling of the twin processes of development makes the lack of reintegration and reabsorption in the South a result of the historical fact that the two processes were coherently coupled in the North. Economic development in the South cannot be autonomous of the North as long as capitalism is the engine of development. Free trade, associated with the 'diffusion' of capitalism, was, in this rendition, the origin of a blocked reintegrative process of development. Consumer goods were imported in exchange for the export of primary products; capital goods were imported for primary production capacity. Through international exchange, the potential surplus which would otherwise be available for domestic investment was drained to the North. Economic development in a 'satellite' South could not be 'self-generating, or self-perpetuating'.[5] Or, as Prebisch put it, the consequence of a dependent status is 'that due to external pressure the country cannot decide autonomously what it should do or cease doing'.[6]

However much Frank criticised the school Prebisch founded and represented, that associated with the Economic Commission of Latin America (ECLA), dependency was kept as the link between underdevelopment and the incapacity for development. Intentional development was negated when there was no capability for the state to decide and, thus, to possess policy.

Any development which could be experienced as economic and social improvement, or envisaged as part of the intention of the state to develop newly fledged nation states was the development of the original condition which blocked the reintegrative process of development. Brazil's national motto, 'Order and Progress', was borrowed from Comte, Frank could have pointed out, with the 'Development' left out. Frank insists for Latin America generally, that nineteenth-century Liberal reforms of agriculture and trade 'served to accelerate the very economic process that had stimulated it in the first place'.[7] This is what Frank meant by a key phrase that characterised underdevelopment as being a process of 'continuity in change'. The development of underdevelopment led to changes in economic and social life whose result was only to keep the condition of life the same as it had always been from the time at which Latin America had been drawn into world trade and exchange. Intentional development, therefore, could only happen outside of capitalism and the engine of development had to be socialism. Many nineteenth-century Latin Americans, however, had thought otherwise.

DEVELOPMENT THOUGHT IN NINETEENTH-CENTURY LATIN AMERICA

It is generally agreed that the nineteenth century was a crucial period for the expansion of much of the economic activity that is now interpreted to be symptomatic of, and whose significance is elaborated almost wholly within debates about, 'dependency'. William Glade, in his standard economic history of Latin America, notes that:

> Between 1825 and 1856 . . . the value of exports leaving Buenos Aires had grown from 5.6 to 16 million pesos. Peruvian exports of wool began on a regular basis with a shipment of 5,700 pounds net weight in 1834; by 1846–56, an annual average of 1.5 million pounds net weight was being exported. In the 1815–1820 period, two or three ships sufficed to handle the annual trade between Chile and Great Britain; by 1847 over 300 ships on this route carried export products such as copper ore, guano, wool, and nitrate of soda . . .[8]

The pace quickened dramatically after mid-century. Argentine exports, which stood at £1.4 million in 1853, had grown to £9.2 million in 1873 and then to £74.5 million in 1910. Brazilian exports soared 'from 150.3 million francs in 1839-1844 to 512.3 million francs in 1869-74 to 990.1 million francs in 1901–1905'. In Mexico, meanwhile, 'exports increased from 40.6 million pesos in 1877–78 to 160.7 million pesos in 1900–1901 to 287.7 million pesos in 1910–11'. Other examples could be recited without difficulty.

As Glade has argued, this 'upward trend in Latin American exports' was largely 'the material outcome of the foreign capital inflows', which amounted to some $8.5 billion by 1913. The important expansion of transportation networks, banking, urban populations, and the scale of commercial activity, were all due to the massive movement of money and people to Latin America in the nineteenth century.[9] The general expansion in economic activity has come to be interpreted as the result of the 'pull' of foreign markets and thus to provide a neatly defined chapter in the history of underdevelopment. It has become such a commonplace to understand this key period of Latin American history as one of the operation of unrestrained market forces and free trade that it now constitutes the locus classicus of 'underdevelopment'.

Yet, as Glade also makes clear, 'much more was involved than mere routine business transactions carried on by free public enterprises in a pure market context'. The foreign investment boom of 1850 to 1914 'got underway with strong assistance from supporting state intervention'. Moreover, this state intervention was characteristic of the period as a whole. According to Glade, 'beyond the brief but inconclusive flirtations with *laissez-faire* during the years immediately following political emancipation from Spain and Portugal . . . there does not appear to have been any radical departure from a *dirigismo* of one sort or another'.[10]

Policies of state intervention and direction, or what Glade refers to as '*fomento*' or intentional development, took many forms. For agriculture, the payment of export bounties along with exemptions from import and export taxes, and the awarding of concessionary contracts were the rule. In infrastructure 'cash subsidies' based upon large public borrowings were prevalent. Companies undertaking port improvement, the creation of municipal utilities and the building of railways were often the recipients of direct public subsidies and/or investment return guarantees. In a number of instances, including that of guano mining in Peru and railway management in Chile, state participation became direct via participation in mixed private/public enterprises.[11] To quote Glade again:

> The fact that Latin American governments were the main borrowers of external funds is perhaps the most revealing single indicator of the degree of state involvement in economic processes during the late nineteenth-century period, for substantial amounts of these funds were employed to modify the economic indicators of the market place and to influence private investment to flow into the channels which would guide national economic life towards an export-oriented pattern of development.

Glade continues:

In other words, during a period in which national private capital was as yet insufficiently developed to assume the leadership in the drive for economic progress, national governments, buttressed by their superior borrowing capacity abroad, stepped in to become the dominant force expressing what purported to be the 'national interest'. Consequently, a very large segment of the foreign capital which came to Latin America between 1850 and 1914 was injected into the local economic processes through the instrumentality of the state, as public investment and as subsidies designed to attract and influence private investment. *Given the scale on which the process operated, it would be entirely misleading to construe the era in terms of a simple laissez-faire version of economic liberalism.*[12]

However, Glade argued, the desire for development should not be taken to mean 'that the governments of the region were necessarily performing their development functions effectively'. According to Glade, expectations of the intention to develop were not realised because:

In the era of positivism, state intervention was not employed to revamp the domestic institutional framework in any radical fashion. Instead, what was involved was a state-fostered development of the external sector to obtain some sort of maximal growth within a context of minimal restructuring of domestic institutions.[13]

Glade's preface to his general comment, namely 'In the era of positivism', must be emphasised.

'Nowhere else', Harold Davis wrote, 'did positivism achieve a stronger hold upon the directing class of society than in Latin America'.[14] Davis was writing in 1961 immediately before the thesis of 'development of underdevelopment' became so pre-eminent in Latin American history. It is remarkable that neither this observation, nor Glade's aside that 'Development economics appears, in many respects, to be the present day counterpart to positivism', has found a significant place in the late-twentieth-century 'development debate'. Indeed, despite the monumental work of Leopold Zea and the later important contributions of Hale, Crawford and others, 'positivism' does not merit even an entry in the index of Frank's foundational *dependentista* work *Capitalism and Underdevelopment in Latin America*.

Why this should be so is at first puzzling. For, as Zea repeatedly shows, an examination of the concerns of nineteenth-century Latin American positivist thought demonstrates a sharp focus on the internal determination of development. The Argentine Saint-Simonian Esteban Echeverría repeatedly evinces this concern.[15] Exiled by the dictator Rosas, whom he accused of having hijacked the Argentine revolution against Spain, Echeverría wrote in 1837 that 'progress' was nothing other than:

the law of development and the necessary end of all free society. . . . But every people, every society, has its laws or conditions peculiar to its existence, which arise from their customs, their history, their social state, their physical, intellectual and moral necessities, from the very nature of the soil which providence willed that they should live on and inhabit perpetually. . . . Normal, true progress means that a people moves toward the development and exercise of its activity with those conditions peculiar to its existence.[16]

It was in an attempt to discover precisely what the 'law' of development was and how it might be followed that Echeverría, along with a host of other Latin American thinkers, turned to positivism. For Echeverría, as for so many of the Latin American '*pensadores*', the revolutions against Spain, although successful in securing independence, had ultimately failed. Made in the name of 'liberty', these movements had brought only despotism. The liberators had failed to establish a new order of liberty.

Although the diagnosis of the specific causes of the failed revolution varied from author to author, the common characteristic in their understanding of the betrayal was that there had been a failure which was internal to Latin American society, as it was assumed, and one which had not permitted Hispanic Americans to break with their colonial past. As such, the failure to break with Spanish colonialism culminated in a continuation of colonialism after independence.

The perpetuation of colonial practice was responsible for the failure of liberation. Echeverría exclaimed: the 'great thought of the revolution' had 'not been carried out'. 'We are independent', he argued, in words that would be echoed throughout sub-Saharan Africa more than a century later, 'but we are not free.' It was no longer the 'arms of Spain' which oppressed Latin America but its 'traditions'. 'The body' of Latin America had been emancipated, 'but not its mind'.

Out of the failure of the revolution to lead to 'political emancipation' and a 'social emancipation' had come the anarchy of the counter-revolution of Rosas and the triumph of the antithesis of progress – 'the stationary idea'.[17] 'The [Latin] American Revolution', insisted Echeverría, 'like all great revolutions of the world, was solely concerned with tearing down the Gothic structure erected by tyranny and force during the centuries of ignorance, and it had neither time nor a period of peace sufficient for building a new structure.'

The failed leaders of independence 'knew, doubtless, that the mind of the people was not prepared to evaluate its importance; they recognised that there were in their sentiments, in their customs, in their way of thinking, certain reactionary instincts against everything that was new and which they did not understand'. It was, however, 'necessary to attract the votes and the strength of the multitudes to the new cause by offering them the bait of omnipotent sovereignty'. But once the people began to exercise their sovereignty, after having overthrown Spanish rule, 'it was difficult to curb them'. So it was that 'the principle of the power of the masses would produce all the disasters that it has produced, and would end in the sanction and establishment of despotism'.[18]

Echeverría's contemporary, Domingo Sarmiento – Argentine *pensadore*, one-time president, and founder of the avowedly positivist Parana School – set out a parallel analysis of what he took to be Latin American, specifically Argentinian, backwardness and chaos. Chaotic stagnation, for Sarmiento in 1845, initially arose from the particular Latin American human inheritance of the 'two different races, the Spanish and the native' which, with the addition of the 'negro race', had formed a 'homogeneous whole'.[19] This new people was characterised by a 'love of idleness and an incapacity for industry'. Sarmiento's racially founded causation created an 'unfortunate result':

> owing to the incorporation of native tribes, effected by the process of colonisation. The American aborigines live in idleness, and show themselves incapable, even under compulsion, of hard and protracted labor. This suggested the idea of introducing negroes into America, which has produced such fatal results.[20]

Neither had 'the Spanish race' shown itself more energetic than the aborigines, when it has been left to itself in the 'wilds of America'. Sarmiento's 'wilds of America itself', in which 'pastoral life' was followed, reminded him

> of the Asiatic plains, which imagination covers with Kalmuck, Cossack or Arab tents. The primitive life of nations – a life essentially barbarous and unprogressive – the life of Abraham, which is that of Bedouin of today, prevails on the Argentine plains, although modified in a peculiar manner by civilization. The Arab tribe which wanders through the wilds of Asia, is united under the rule of one of its elders or of a warrior chief; society exists, although not fixed in any determined locality. Its religious opinions, immemorial traditions, unchanging customs, and its sentiment of respect for the aged, make altogether a code of laws and a form of government which preserves morality, as it is there understood, as well as order and the association of the tribe.[21]

It was this very 'order' which made 'progress impossible' because there could be 'no progress without permanent possession of the soil, or without cities which are the means of developing the capacity of man for the process of industry, and which enable him to extend his acquisitions'. Absence of 'civilization' explained what was for Sarmiento the abortive nature and tyrannical result of the struggle for Argentine independence from Spain. Before 1810, 'two distinct, rival and incompatible forms of society, two different forms of civilization existed in the Argentine Republic: one being Spanish, European and cultivated, the other barbarous, American, and almost wholly of native growth'. The 1810 revolution, according to Sarmiento, 'set these two distinct forms of national existence face to face' in a contest which was to culminate in 'the absorption of one into the other'.[22] To anticipate the trajectory of Sarmiento's positive thought, the dualism which was originally founded upon racial categories ended in the positivist duality between order and progress.

The revolution against Spanish colonialism was the project of 'Buenos Ayres' which, Sarmiento wrote, 'set about making a constitution for itself and the Republic, just as it had undertaken to liberate itself and all South America: that is eagerly, uncompromisingly, and without regard to obstacles'. The 'personification' of a 'utopian' vision, significantly inverted by being wrenched out of its original Latin American arcadian setting and fixed in the European city, was the leader Rivadavia who, Sarmiento argued had

> brought over from Europe men of learning for the press and for the professor's chair, colonies for the deserts, ships for the rivers, freedom for all creeds, credit and the national bank to encourage trade, and all the great social theories of the day for the formation of his government. In a word he brought a second Europe which was to be established in America, and to accomplish in ten years what elsewhere had required centuries.[23]

But this was all for nought. Revolution had unleashed, alongside the 'bravery' and 'daring' of the gaucho and the 'unemployed peasantry', the barbarism which was shown by the opposition of the gaucho and the unemployed towards the law of the city and then towards civilisation itself. Barbarism had triumphed in the overthrow of Rivadavia and his ultimate replacement by the provincial caudillo Rosas, 'the great tyrant of the nineteenth century, who unconsciously revived the spirit of the middle ages, and the doctrine of equality armed with the knife of Danton and Robespierre'. Such was the grand betrayal of the revolution, for Sarmiento, a betrayal which turned the negation of the revolution into the positive prognosis of progress through development.

For Sarmiento and Echeverría, the era of Spanish colonialism was one of order without progress, yet the revolutions against Spain, in offering to carry forward the potential of progress, had only brought chaos and regress. Although they may have differed in their approaches towards the outcome of the revolution, all of the *pensadores* set out dichotomies, such as the opposition of 'civilization and barbarism' or 'republicanism and Catholicism', to explain the failure, chaos and retrogression which was the result of the Spanish/Indian colonial heritage. In order to break free of these dead hands of the past, a 'mental emancipation' was deemed to be necessary.

In the parlance of a later age in another continent, it was necessary to 'decolonise the mind'. However, in nineteenth-century Latin America by way of contrast to twentieth-century Africa, the immediate reference for the attempt at intellectual liberation was antithetical to that of the present. It was to North America, and especially the United States, that a turn was made to find a model for the promise of progress. In accounting for the difference between the two Americas, the Chilean Bilbao argued that:

In non-Catholic and free countries man is sovereign and respects the sovereignty of his fellow man. There are no infallibles who come to power, and all men have faith in the law which guarantees their rights and faith in the vote of all men, which cannot contravene their rights. If there is error, there is no imposition, and they expect the infallible progress of conviction. Such is the politics of a people whose votes cannot be forced or ridiculed. Law is religion, and the religion of *free examination* produces the religion of the law. [Emphasis added]

Bilbao then turned to Latin America:

But in Catholic countries there is a fantastic and real terror when the opposition party wins, because we know and we believe, or foresee rightly, that it is a defeat without hope; it is the enthronement of something infallible and incapable of committing a wrong, something that establishes itself with the inflexibility of vengeance. This is why there are so many revolutions and so much servility.

Catholicism denied liberal principles and institutions

with the infallible word of councils and popes; but the progress of the period has consisted of using the same terms, of occupying position, in accepting the language and terminology of liberty, and in using suffrage, the press, education, and the schools to the discredit of suffrage . . . and in educating slaves of the church and not citizens of the state.[24]

Bilbao's view was that development had occurred in the United States, where Catholicism had been repudiated.

Much, though certainly not all, of the general tone of Sarmiento's analysis of American backwardness was echoed in the work of his contemporary, Juan Bautista Alberdi. In his 1856 *Bases and Points of Departure for the Political Organisation of the Argentine Republic*, Alberdi argued that 'from the sixteenth century to the present Europe has not for a single day ceased to be the source and origin of the civilization of this continent'. For Alberdi 'to try to anglicize the Spanish race', was 'to disregard nature'. The Latin American revolution was 'the product of the development of the human spirit, and has as its end this very development'. This was 'a fact derived from other facts, and it must produce other new facts'. Each nation had 'and must have its own civilization, because each civilization is derived from the combinations of the universal law of development with the nation's individual condition of time and space'.[25] He continued:

All peoples necessarily develop, but each one develops in its own way, because development operates according to certain constant laws, within a strict subordination to the conditions of time and space. And since these conditions never occur in exactly the same manner, it follows that there are no two nations that develop alike. The individual mode of progress constitutes the civilization of each nation.[26]

For the avowedly positivist Chilean José Lastarria, an 'individual mode of progress' possessed its own history. Thus when

the United States won its political freedom, the people did not free themselves from the influence of English literature; they could use it and, in effect, did, for their new situation, because their sentiments and their ideas, their interests and their social necessities continued to be British, with the sole difference that their social organisation would be better served by a republican organisation; and it could be so because it was not a violent innovation, but progress, a natural development.

Hispanic America was different:

The entire interest of the political organisation, for example . . . was summarised in the magical word 'order', which for public opinion represented the tranquillity that facilitates the course of affairs, in addition to the peace which prevents unexpected attacks, by reconciling the peace of the home with that of the streets. For the statesmen and the petty politicians it signified the domination of an arbitrary and despotic power, that is, the political possession of an absolute power which in the peaceful times of the colony enriched the representatives of the Spanish king.[27]

Thus, 'while the habits and customs which the North Americans inherited from England led them through a natural evolution to political independence', wrote Lastarria, 'the habits and customs which the Hispanic Americans inherited from Spain led them into a new colonialism'. As a result, Latin Americans were tied 'to the point of departure, thus rehabilitating the colonial system'. In this way, Lastarria exemplified how the difference in history between North and Latin America corresponded to a distinction between what amounted to true and false or non-development.[28]

A persistent theme, again using Alberdi as an example, was that the difference between experiences in the past could not determine prescription for the future. Intentional development could not be read from a model of past development. To continue the life begun in independence 'is not to do what France and the United States are doing, but . . . to follow the natural development to acquire a civilization of our own, although imperfect, and not to copy foreign civilizations, although they may be more advanced'.[29] Yet how was the 'individual mode of progress' to be achieved? The first requirement was to establish an appropriate method of understanding historical change. An appropriate method, Lastarria argued, entailed the rejection of theories of history, such as those of Vico and Herder, which 'were founded upon a supernatural concept of history':

In order for there to be a science in history one must believe that human events are a natural phenomenon linked to each other and dependent upon human action and will; consequently, in order to discover the body of truths which because of their connection with a single object, humanity, form a body of doctrine or philosophy of history, it is necessary to investigate the relationship which those events have to each other and to man's activity, that is, to all his faculties.[30]

It was precisely this 'science in history' that Lastarria believed he had found in Comte.[31] So too, for instance, did the Mexican savant Gabino Barreda turn to Comte. Barreda's interpretation of Mexican history followed familiar lines. The revolution against Spain at the beginning of the century, although successful, had merely led to prolonged civil war and ultimately to the farce of Maximilian's abortive empire. In his 'Civic Oration' of 1867, given on the heels of Maximilian's execution, Barreda put Mexican history within a Comtean framework. For Barreda, Spanish colonialism had occurred just at the beginning of the historical moment when Catholicism had started to lose its ability to maintain itself as the organic force of order. Following this theme, the Mexican Revolution and civil wars were the emblematic struggles of positive science striving to be born. The triumph of Juarez, who had invited Barreda to make his 'Civic Oration', over Napoleon III was the 'triumph of the "positive spirit"'.[32] From a positivist standpoint, Mexico was the locus of the crucial battle between the forces of human progress and the retrogression of the defenders of the theological era. As such, the struggle against Napoleon III was a struggle taken up on behalf of all human progress.

Yet, in the wake of the victory over Napoleon III, 'disorder and anarchy reigned in every corner of the Republic'. The development of Mexico demanded that Catholic ideological hegemony be displaced by establishing the new positive order in the minds of the people, especially among the victorious forces of reform. Formal education was to be the vehicle for the positivist cause of development. Zea put the case well:

Positivism justified the Mexican revolution and it furnished an order according to the ideals of the revolution. It postulated liberty of conscience which was summed up by Barreda as follows:

'scientific emancipation, religious emancipation, political emancipation.' These were possible only through mental emancipation, that is, through the decay of old doctrines and their replacement by others. The old doctrines were in this case those of the Catholic clergy; the new doctrines that would replace them would be positivist doctrines. These would be the doctrines in which Mexicans would be educated. These positivist doctrines would eliminate the disorder that was provoked by a class that did not want to recognise the fact that its mission had ended. The men educated in this doctrine would assume power and implant the new order in all fields.[33]

Zea's reference to 'liberty' must be emphasised. In Mexico, as in Chile, Argentina, and ultimately Brazil, positivism was married to a specifically Latin American liberalism.

Indeed, positivism was repeatedly seen to be the necessary basis upon which an indigenous liberalism would flourish. The spiritual realm would be left to the priesthood in exchange for the acceptance of their removal from temporal affairs. Nor were the Catholic clergy to be replaced by the priesthood of the Comtean Religion of Humanity. Given the history of conquest, oligarchy, civil war and the theological dictatorship of the clergy, positivism in Latin America was only rarely to be equated with the dictatorial rule of either moral or scientific experts. Rather, positivist education was to be the force by which those who led society were to come to understand their own, as well as society's interests. The dictatorial altruism of trusteeship was pushed into the background. Hence, Barreda altered the Comtean motto of 'love, order and progress' to that of 'liberty, order and progress'. The substitution of liberty for love was not limited to Barreda but became, whether explicitly or implicitly, the common coin of late nineteenth-century Latin American positivist thought.[34]

The substitution of liberty for love was matched by the substitution of Herbert Spencer for Auguste Comte as the ideological fountainhead of Latin American positivism. In Spencer, Barreda's successors in Mexico, along with much of the rest of Latin American positivist thought, found a thinker whose 'social Lamarckian' precepts of the struggle for survival in social evolution better fitted their aspirations than Comte's Religion of Humanity.[35] As the author of one recent survey of Latin American history puts it:

In republics with large non-white populations, this kind of 'scientific' politics came very close to an officially sanctioned racism – Indians, mestizos and blacks tended to be regarded as irredeemably unskilled, indeed as obstacles to the nation's progress. Under such conditions, as the Mexican positivist Justo Sierra famously observed, it might be necessary to 'try a little tyranny' for the sake of development. Mexican *científicos* like Sierra, in fact, provided a modernizing rationale for the forty-year dictatorship of the liberal *caudillo* Porfirio Diaz.[36]

During the 'Porfiriato', as this period is known, 'Mexico would make more economic progress than at any time since independence' with the economy growing at the 'astonishing rate' of an average of 8 per cent per year. The achievement of this rate of growth had much to do with the 'modernization of the export-economy', which included expansion of the railway system opening up new areas of mineral and agricultural production for export. The latter, in turn, led to the 'consolidation of the hacienda' and to more rapid expropriation of Church and Indian lands. Industry expanded as well due, in part, to an 'open attitude to foreign investment'. One part of this 'open attitude' consisted of a well-armed paramilitary force largely for internal use. The deployment of this force in conjunction with 'repression of strikes and peasant revolts, and control of the press, saw to the political opposition'. All of this took place under the direct influence of the *científicos*, a 'clique of progressive technocrats' who, as the reference above to Sierra indicates, 'advocated the suspension of democratic rights while economic progress transformed a country otherwise vitiated by unenlightened values and racial deficiencies'.[37]

The influence of positivism, in Mexico at any rate, was interrupted and modified by two decades of civil war commencing in 1910. Significantly, this conflict was sparked by the 1907 recession which, 'compounded by a series of bad harvests', brought 'inflation, falling wage rates, and unemployment' creating 'unprecedented misery among the peasants and industrial workers'.[38] In the aftermath of the civil war, or as it officially came to be known, the Mexican Revolution, *científico* policies were officially discredited as part of an attempt to ideologically root the post-revolutionary state in a rhetorical fusion of Indian and mestizo traditions. Yet in the corporatism of the 'new' Mexico, much of the *científico* survived in the uneasy compromise between capitalist development and *indigenismo*. Between 1920 and 1940 state policy lurched between these two poles. In the late 1930s, for example, President Cardenas presided over the implementation of 'agrarian reforms' as a way of countering discontent arising from the depression. Millions of hectares of land was redistributed, cooperatives were formed, and an attempt was made to 're-create' the *ejidio*, a supposedly traditional form of Indian communal holding. The results were typical of many later variants of the agrarian doctrine of development. The land distributed was often of poor quality, credit and machinery were inadequate, peasants resented dependence on state officials, and the administration of the scheme was seen as being corrupt.[39] By the 1940s 'a political order stable enough to replace the Porfiriato had been found'. In consisting of 'a one-party state run by a strictly temporary autocrat, pledged to an ideology of revolutionary nationalism yet committed to a path of intensive capitalist development' it had more than a few echoes of the *científico* past.[40] Between 1940 and 1980, 'Mexico was transformed from an agricultural country into a predominantly urban and industrial society' in which 'high public expenditure on welfare, health and education' in tandem with a 'comprehensive apparatus of control' kept 'discontent from exploding into violence.' Significantly, the initial stimulus to this period of industrial growth has been ascribed to United States' technical and financial assistance for the exploitation of the country's mineral wealth to help in the war effort against Germany and Japan. Such were the roots of the post-Second World War 'modernisation theory' against which Frank was to rail.[41]

A COMMENT

Writing in 1949, the extraordinary Leopold Zea described the era of positivism in Latin America in terms which anticipated those of Gunder Frank by nearly three decades. It was a time, according to Zea, in which 'Hispanic-American man . . . became more and more aware of his *dependent* relationship with a world which he did not consider his own' and a period in which 'the *disarticulation* of Hispanic-American man becomes evident'.[42] Later, Zea pronounced his verdict on the positivist epoch in Latin American thought as a whole. Again, the judgement was strikingly similar to that of dependency theorists:

> The history of positivism in Mexico as well as in Latin America is the expression of an experience that terminates in a dead-end alley. Despite the spirit of this doctrine, and their supposed assimilation of it, Mexicans did not transform themselves into 'Yankees of the south.' Nor was Latin America converted into the United States of the south. If one form of subordination was destroyed, another was immediately created by the formation of Latin American societies hardly different than those created by three centuries of Iberian colonialism. A new colonialism was created by nations who were the leaders of progress. Mexican positivists became tools in this new order. Their social predominance was patterned after Iberian colonial exploitation.[43]

However, there are important differences between Zea's evocation of dependency and disarticulation and that of the 1960s dependency school; only part of *dependista* appraisal is reminiscent of Zea. Most importantly, Zea had directly recognised, as had Glade, the existence of the Latin American 'era of positivism'. In doing so, Zea directly identified the positivist era as a 'dead-end alley' of development – a false development – which had not allowed Latin

Americans to 'negate' their history in the sense of having fully 'assimilated' it. Here again Zea's discussion anticipates debates that were to take place decades later with little or no reference to his writing or, indeed, to the phenomenon of nineteenth-century Latin American positivism in general. Frank did initially engage with 'positivism' in his early article entitled 'The sociology of development and the underdevelopment of sociology' but this engagement was with the positivism of North American 'modernisation theory'.[44] As we shall see below, this preoccupation with a body of thought external to Latin America to the neglect of indigenous writing is characteristic of the dependency school.

One possible reason for this lacuna is the widely held view among historians of Latin America that positivism was, by the early twentieth century, a dead force which had been buried by the Mexican Revolution and analogous events in South America. Whatever the strength of this view – and we have contested it for Mexico – the absence of direct reference by dependency theorists to Latin American positivism has obscured the way in which their own analysis of the nineteenth-century economic history of Latin America constitutes a direct response to positivist thought. In doing so, it has also obscured the way in which the dependency riposte to positivism has hinged on the question of the 'internal' and 'external' determination of development. We will now turn our attention to underdevelopment theory and its critics to draw out these important aspects of this debate.

UNDERDEVELOPMENT THEORY AND ITS CRITICS

Almost as soon as his pen was dry in the early 1970s, Frank's idea of underdevelopment spawned elaborations of his theme and provoked barrages of criticism. Most of the criticism is well known: exchange was given priority over production; state–state and capital–capital relations were emphasised over capital–labour, class relations; the external was made dominant over the internal.[45] But it is apposite here to recapitulate some signal points of critique. Warren, for instance, pointed out that underdevelopment 'actually corresponds to no objective process' and footnoted the term as corresponding to 'the *fact* of different levels of living standards and of degrees of development of the productive forces – and to an *ideological* process'.[46] However, other than directly associating the ideology of underdevelopment with Third World nationalism, Warren did not spell out what 'an ideological process' might mean.

Banaji referred to Frank's *World Accumulation*, written as one of many responses to his critics, and drew out what the idea of development might imply. Frank, Banaji pointed out, held to an absolute idea of world capitalism. The idea remains where it started – as if development consists of nothing other than the repetition of some same formula. As such, Frank confused the conditions of accumulation created by the movement of capital with those belonging to the history which created capitalism.[47] Brenner, another critic, accused Frank of displacing 'continuity in change' from where it belonged, namely in the development of productive force, to an idea of the self-development of the capitalist class. If the development of productive force is the immanent and objective process of development, self-development is development by intention.

Here, Brenner begins to break open the *method* by which Frank arrived at underdevelopment as an idea. First, the theory merely inverted the progressivist thesis of Adam Smith. The world-wide expansion of trade and the division of labour had failed to deliver progress. If trade was the agency of progressive change for Smith, it had become for Frank the original means of underdevelopment. In other words, Smith's idea of progress had not been transcended by an idea of development.[48] Second, the way in which self-development is used to account for both the origin and the process of development is tautological. Imagine traders deciding to develop their material interest by turning from trade to production through deciding in what form they will dispose of the means of production and in what way producers will become sources of wage-labour power. To imagine that it is possible for would-be capitalists to have the capacity to so decide, Brenner wrote, is to assume that they function according to what has already been determined by the development of capitalism. It is to assume that they are already capitalists.[49] The argument is circular in the same way that the urge 'to empower people for development' assumes that the people are developed in so far as they have the capacity and power to exercise choice in the name of self-development. Brenner concluded by referring to the objective essence of economic development as opposed to the abstract form of the expansion of capital:

> At every point . . . Frank – and his co-thinkers such as Wallerstein – followed their adversaries in locating the sources of both development and underdevelopment in an abstract process of capitalist expansion. . . . As a result, they failed to focus centrally on the productivity of labour as the essence and key to economic development. They did not state the degree to which the latter was . . . centrally bound up with historically specific class structures of production and surplus extraction, themselves the product of determinations beyond the market.[50]

Therefore, Brenner concluded, Frank ended up with a conclusion which he did not intend. The 'notion of the "development of underdevelopment" opens the way to third-worldist ideology'.[51] By way of a different tack from that of Banaji, Brenner also concluded that underdevelopment theory had failed to distinguish the origin from the continuing process of capitalist development. The key to the problem is that underdevelopment was one way of expressing an idea that however much things may seem change, change is the means of preserving the old order of things.

An intimate connection between underdevelopment and the idea of 'continuity in change' through development has been acknowledged by, among others, Booth. In his study of Chile, Booth noted, Frank put emphasis on 'the continuity of capitalist structure and its generation of underdevelopment rather than on the many undoubtedly important historical changes and transformations that Chile has undergone within this structure'.[52] The problem, as Booth saw it, was that the emphasis upon continuity is a result of a curtailed 'conceptual apparatus', which in turn reflects '*the relative narrowness of the purpose for which it was created*'.[53] Frank's purpose was to attack the orthodoxies of both the ECLA school and the Latin American Communist parties. They had stressed that inner-directed development or 'development towards the inside' after 1929 had created a novel development predicament after 1945: consumer-goods import-substituting industrialisation created capital-goods bottlenecks which put the industrialising countries under the foreign exchange thrall of the US-dominated international economy. For Frank, however, this predicament was not new. It was, as Booth interprets it, 'a "contradiction" of capitalist development of the principle of "continuity in change" which meant in effect that, where underdevelopment is concerned, *plus ça change, plus c'est la même chose*'.[54] In the face of external domination of a world system, of capitalist expansion, development could not be true. If industrialisation appeared, if a proletariat was formed, then this was evidence of ragged or lumpendevelopment, the result of the 'lumpenbourgeois policies of underdevelopment which have been in operation throughout our [Latin America's] history'.[55] The general question, then, is that if the phenomenon of development, such as that of the post-1945 world, is deemed not to be new, to what extent could it not equally well be said that the method or conceptual apparatus, which gives rise to this appreciation of development, is also not novel?

THE 'INTERNAL' AND 'EXTERNAL' OF DEVELOPMENT

It was the question of the 'internal versus the external determination' of development that most provoked Frank to reply to his critics. After all, the major hypothesis of the first edition of *Capitalism and Underdevelopment in Latin America* was that: 'If it is satellite status which generates underdevelopment, then a weaker or lesser degree of metropolis–satellite relations may generate less deep structural underdevelopment and/or allow for more possibility of local development.'[56]

This is what disturbed Frank's earliest critics, including the most friendly such as Arrighi, who berated Frank for subordinating the analysis of internal class structure to the external domination of a colonial and neo-colonial structure. Arrighi quoted Mao Ze dong to sustain his criticism and Frank replied, at the beginning of his *Dependent Accumulation and Underdevelopment*, by also quoting Mao at length:

> As opposed to the metaphysical world outlook, the world outlook of materialist dialectics holds that in order to understand the development of a thing we should study it internally and in its relations with other things. . . . The fundamental cause of the development of a thing is not external but internal; it lies in the contradictoriness within the thing . . . its interrelations with other things are secondary causes.

Frank continued, quoting from Mao's 1937 speech entitled 'On contradiction', in order to rebut his critics:

> Does materialist dialectics exclude external causes? Not at all. It holds that external causes are the condition of change and internal causes are the basis of change, and that external causes become operative through internal causes. In a suitable temperature an egg changes into a chicken, but no temperature can change a stone into a chicken, because each has a different basis.[57]

This difference between the cause and basis of change was written by Mao as part of his criticism of 'metaphysical mechanical materialism and vulgar evolutionism', the kind of theory which had been so much part of nineteenth-century reflections on development. Mao's own idea of development, 'the materialist-dialectical world outlook', was presented as a doctrine of development whose immediate purpose was to explain why the Chinese Communist Party should serve the cause for the national development of China.

Mao delivered his 1937 speech at the onset of the Japanese military invasion of China. Against a body of opinion within the Chinese Communist Party, but at the behest of Stalin, the chairman sought to justify a united front with the Kuomintang (KMT), the party of warlords, landlords and the '*comprador* big bourgeoisie'. Hitherto, the 'internal foreigners' of *comprador* capital, which the external force of imperialism had imposed upon the mass of the nation, were regarded as external to both the internal of China and of the CCP.[58] An external force, as a condition of development, had no capacity for 'internal and necessary self-movement', the basis of 'development of things'.

In 1937, however, when conditions suddenly changed, it was possible for the external to become part of the internal of development. Because of internal change within the KMT, arising from mass dissatisfaction with the Kuomintang for not defending China, Mao argued, a united front could now become the internal basis of development. The basis of a new *identity*, which Mao took to be synonymous for the unity, coincidence, interdependence, and cooperation between the CCP and KMT, was thus developed as part of the internal process of development.[59] Thus, given that the political purpose of Frank's under-development theory was to decry the possibility of identity between the *comprador* bourgeoisie and any possible socialist force for authentic development in Latin America, there is much irony in Frank's appeal to Mao.

Frank's immediate political paradox, however, only hints at the deeper irony embedded in his appeal to Mao. To unravel the paradox, we need to examine more closely how Mao arrived at his meaning of the internal and external of development. However much it is recognised that Mao's political acuteness drove him to find a new way to 'describe the process of qualitative change in things', the new description rested upon an older positive conception of knowledge. Much has been made of Mao's unoriginality and plagiarism in 'On contradiction'. In addition to his acknowledged references to Lenin and Engels, his unacknowledged debt to Soviet Marxism, especially to the work of M.B. Mitin *et al.* whose *Outline of New Philosophy* and *Dialectical and Historical Materialism* were translated and published in China during 1936, have been noted.[60] The origins of Mao's thought on development are, however, deeper still.

It is in the various philosophical attempts to provide a bridge across the dualism of body and mind that the intellectual heart of distinctions between the internal and external of development are to be found. For Mao, the external of development was the world of matter which was independent of the thought of mind. Matter was naturally scattered in the separate, discrete and various existence of 'things', persons and events. It was the action of the mind, exercising practical or perceptual knowledge, which created the potential for a composite body of potentially unified things. The external was reflected in the concepts of mind upon the basis of rational knowledge. It was upon this theoretical basis that the composite body of knowledge found the potential to be realised in policy and programmes for action. Such was the basis of the internal of development. Through the 'bridge' of development, the external condition of matter was apprehended and changed. The extent to which change was deemed to be true or authentic, in that it conformed to the intent of development, was established by an iterative process in which rational knowledge was tested by practice. Sets of recursive sequences of perceptual and rational knowledge followed on another, through the mediation of practice, in a spiral of progressive movement. Today's perception was yesterday's rationality which had been tested in practice to become the basis for tomorrow's rationality, which was in turn the basis of further perception and so forth.[61]

The events of 1937 had made the process, one in which the external was apprehended by the internal of development, more pronounced. Mao's internal of development was based upon the CCP and its leadership while the external was embodied in the mass of people. 'In all practical work of our party', Mao wrote in the CCP 1943 directive 'On the mass line', 'all correct leadership is necessarily from the masses, to the masses'. This meant that the Party should:

> take the ideas of the masses (scattered and unsystematic ideas) and concentrate them (through study turn them into concentrated and systematic ideas), then go to the masses and propagate and explain these ideas until the masses embrace them as their own, hold fast to them and translate them into action and test the correctness of these ideas in such action.

Such a procedure, Mao continued, was to be repeated 'in an endless spiral with the ideas becoming more correct, more vital and richer each time'.[62] If, as Mao claimed, the procedure was 'the Marxist-*Leninist* theory of knowledge, or *methodology*', then it was also affiliated with socialist trusteeship. What was labelled 'true socialism' in the 1840s had been based upon the idea that the internal authority of development was the trust, held in the name of the mass, for the external body of the mass to be conditioned for the purpose of social improvement. True socialism was a socialist variant of trusteeship which was, as we shall show in the following chapter, the object of Marx's invective. Thus, the issue of the internal–external of development not only involved the direction of causation – whether or not it was the internal or external which primarily caused development – but lay at the heart of why and how intent could be brought to bear upon a process of development.

Mao's intellectual formation typified those members of an early twentieth-century generation of Chinese who rebelled against what they regarded as the stasis of official authority and tradition which, despite the 1911 revolution against the Manchu empire, was seen by them as preventing progressive development in China. Much has been made, by Li Yu-Ning and Martin Bernal, who noted the way in which Western radicalism and socialism 'played a critical role in twentieth-century Chinese history'. Saint-Simon, J.S. Mill, Henry George and the British Fabians strongly feature in accounts of how development doctrine came to China as a subversive source of authority against the old order. It was only in 1920, after Sun Yat-sen's national revolution subsumed the 4 May movement of 1919, the radical cause in which Mao himself was embroiled, that these radicals turned to Marx. Even then, however, despite Mao's reading of Marx, and Lenin on Hegel, it was the German idealist tradition of knowledge which provided the cornerstone, indeed the internal–external, of Mao's method and ideal of development.[63]

Yang Chang-Chi, whose daughter Mao was to marry and who had studied in Europe absorbing the British neo-Hegelian idealist doctrine of T.H. Green, was one profound source of influence upon Mao in the immediate period before the 1919 national revolution. We will meet Green's 'development' in detail in Chapter 5. Mao was unabashed about the extent to which he had absorbed another strand of the idealist tradition, that associated with neo-Kantianism. In particular, he had avidly studied Friedrich Paulsen's neo-Kantian and popular 1888 *System of Ethics*.[64]

Paulsen, while claiming that he had forsaken his youthful 'utopian socialism' for J.S. Mill's system of logic and thinking about policy,[65] had followed the general Kantian outline of establishing the extent to which conscience played a role in adjudicating the conflict between individual inclination and social duty. Duty, Paulsen wrote, was not 'rooted in the will of the individual, but seems something external to him, something opposing him with absolute authority'. It was 'custom', being equivalent to society, which formed the 'original content of duty'. If custom appeared to be part of individual instinct, this, according to Paulsen, was only because the content of duty was transmitted by purpose. The purpose which education served, as with Mill, was to do the work of the intent to develop. The content of conscience, through which obligation was made internal to the individual, was as varied as different societies, but the form of conscience was 'universally the same: a knowledge of higher will, by which the individual will feel itself internally bound'.[66] Paulsen went further, providing a basis for the distinction between the restricted and general or expanded domains of development. We shall meet this distinction again below.

Conscience, for Paulsen, developed through the development of 'mental life', providing the criterion for development itself. Instead of simply measuring the value of individual life, the worth of individual function for society, conscience measured 'the actual life by its special ideal'. The ideal, represented by Christ as the archetype of the authority for true development, was to create and renew the content of conscience. False development, whose archetype was the Faustian figure, and who we shall come back to in Chapter 7, destroyed the content of conscience. Faust destroyed society, Paulsen proclaimed, when he endeavoured to free himself from it. To destroy conscience's content by 'false' theory and education, Paulsen concluded, was 'the most serious injury which can be done an individual or community'.[67]

What did Mao make of all this? Interestingly, he reflected that Paulsen, because of his German state-socialist environment, had put too much emphasis upon duty and too little upon individual inclination, the source of self-conscious will which Mao's generation wanted to assert against the Manchu tradition.[68] However this may be, and despite the extent to which Mao substituted class for individual will after his turn to Marx, it is clear that class struggle, the arbiter of development for Mao, occupied the same role which conscience had played in Paulsen's neo-Kantian scheme.

Contradiction, as Mao defined it when he commenced his 'On contradiction' by quoting from Lenin's *Philosophical Notebooks*, was the 'unity of opposites'. A universal contradiction, for Mao, stemmed directly from Lenin's set of binary oppositions, as for example, between the positive and negative of mathematics or electricity of physics. The corresponding set of oppositions for 'social science' was class struggle.[69] Having made an immediate distinction between the universality and particularity of contradiction and, further, the 'principal' and 'principal aspect' of contradiction, Mao could conclude that: 'Contradiction and struggle are universal, absolute, but the method for resolving contradictions, that is, the forms of struggle, differ according to the differences in the nature of contradictions.'[70] In other words, Mao's 'forms of struggle' expressed what Paulsen had implied by the content of conscience and what Lenin had regarded as the 'clerical obscurantism' of 'philosophical idealism'.

Lenin defined development in the few pages he wrote 'On the question of dialectics' in 1915 and from which Mao had quoted. 'Development', Lenin wrote, was 'the "struggle" of opposites'. In making what amounted to the Hegelian distinction between natural development and the principle of development, a distinction to which we return in the following chapter, Lenin had spelled out the internal and external of development. What was significant about the internal of development, '*self*-movement, its *driving* force, its source, its motive' was absent from the conception of 'development as decrease and increase, as repetition'. The source of self-movement in the process of 'spontaneous development' was 'made *external* – God, subject, etc.'.[71] It was the second conception of development, what Hegel called the principle of development, although Lenin elsewhere thought principle to be 'a clumsy absurd word',[72] that furnished 'the key to the "self-movement" of everything else; it alone furnishes the key to the "leaps", to the "break in continuity", to the "transformation into the opposite", to the destruction of the old and the emergence of the new'. Consciousness, through knowledge, was the basis of the internal of development.

'Human knowledge', Lenin concluded from Hegel, 'is not (or does not follow) a straight line, but a curve, which endlessly approximates a series of circles, a spiral.' It was the transformation of a segment of the curve into 'an independent, complete, straight line' which characterised clerical obscurantism. If such could be taken to be the philosophical idealism of the idea of progress, then Lenin did not say so. Rather, Lenin suggested that it was class interest of 'the ruling classes' which turned one facet of development into its absolute.[73] Mao endlessly elaborated on Lenin's few pages but obscured the principle of development in the political immediacy of 1937.

The unity of opposites, as that between the CCP and KMT, was relative to the change brought about by the external condition of 1937. Struggle, Mao had insisted, destroyed one relative condition and replaced it by another, a new external condition which itself would be subject to the internal basis of development. Thus Mao gave the example of how opposing classes of landlords, 'the haves', and peasants, 'the have-nots', became 'interconnected' through agrarian revolution. Contradictory things, he repeated, change into one another. The haves, it was implied, would lose land in quantity but consciously gain the prospect of public property; the have-nots would 'become small-holders of land' and enter into proprietorship: 'Between private property and public property there is a bridge leading from one to the other, which in philosophy is called identity, or transformation into each other, or interpermeation.'[74] Mao's 'bridge' was what was known, as will be shown below, as the 'bridge of development'. Here, the bridge was between what Mao vaguely called 'certain conditions', the variation in social experience and the events of history, and the interior of development which was expressed by the force of class struggle. If, for Paulsen, the form of conscience was the existence of society in the consciousness of the individual, then Mao's construction of class struggle, despite his language of the dialectics of materialism, and one which inverted form and content, followed the same idealist route.

Mao's 1937 speech was underwritten by an attack on the Soviet philosopher, A.M. Deborin, whose 'idealism has exerted a very bad influence in the Chinese Communist Party' and who, the chairman expounded, was responsible for the 'doctrinaire' argument that there was no necessary 'movement of opposites' from the 'beginning to end' of 'the process of development of everything'. Deborin's school, wrongly identified by Mao with Nikolai Bukharin, was held responsible for the view that there were 'only differences but no contradictions between the kulaks and peasants in general'. In denying the universal principle of contradiction, Mao alleged that Deborin had returned to metaphysical doctrine because he put primacy on the external motive force of development and denied its 'internal' causes.[75] Before he was officially disgraced in 1930, Deborin's introduction to the 1927 German and English translations of Lenin's *Materialism and Empiriocriticism* had eulogised Lenin's contributions to philosophical and political practice 'amongst the Russian Marxists'. He further commented that it was unfortunate that 'matters' were 'different beyond the borders of the Soviet Union' and 'where Kantian scholasticism and positivistic idealism are in full bloom'.[76] Deborin may or may not have been conscious of Mao, but it was Mao, living in a glass house, who started to throw stones.

During the 1920s, dispute had raged between the Deborin School and the so-called Mechanists over the place of philosophy in science and official Soviet doctrine. Broadly, the Mechanists held to the positivist view that science had the potential to control, in David Bakhurst's words, 'the external forces of both nature and society' thereby ensuring that 'the communist individual' would 'be a truly self-determining subject, at one with the environment'.[77] Mechanists, including Aleksander Bogdanov and Bukharin, generally subscribed to the positivist view that laws of natural science provided the basis for the explanation of phenomena. Deborin and his followers, who finally but briefly found official favour against the Mechanists in 1929, attempted to renew the Hegelian Marxist doctrine of dialectical materialism by upholding Hegel's principle of development against sceptical questioning by the Mechanists. At issue was the question of whether society could be construed to be the external of development.

The question of the discontinuous change of quantity into quality, which was for Mao a law of development, was the focus of Mao's accusation against the Deborinists. In contradistinction to physical phenomena, in which change occurred at quantitative limits but resulted in different entities of the same essence, such as when water became steam or ice at its boiling or freezing points, Deborin insisted that there was a range of phenomena in which qualitative change generated new phenomena which had to be explained by way of new concepts.

As such, qualitative variation in the external world could not be simply reduced to the existing concepts of knowledge and internalised upon the basis of development.[78] One excerpt from Deborin, which Lenin had reproduced, ran as follows: 'Between the external and internal world there exists a certain distinction, and at the same time a definite similarity, so that we arrive at the cognition of the external world through impressions, but they are impressions produced by objects of the external world.'[79] The problem of development was 'the distinction' between the internal of 'inner experience' and the external of 'things', a distinction which was created by 'cultural development' itself. By 'cultural development' what was meant was the development of society, the phenomenon which interposed the internal and external of development.

According to Deborin, it was through development that the 'external world is constructed by us out of our perceptions, on the basis of those impressions evoked in us by the external world, by things in and for themselves'. As such, it was a fundamental 'provision' of cognition that 'the objects of the external world are in causal relation not only with us, but to one another, i.e., between the objects of the external world themselves there exists a definite interaction . . .'.[80]

This is the provision which Mao presumed to be Deborin's determining external force of development when he wrote, against Deborin, that 'contradiction within a thing is the basic cause of its development, while the relationship of a thing with other things – their interconnection and interaction – is a secondary cause'. There is little doubt that this was mere presumption by Mao. Deborin had made it clear that development involved the potential for 'a knowledge of those conditions which made it possible to foresee and predict not only the action to be exercised upon us by objects, but also their objective relations and actions which are independent of us, i.e. the objective properties of things . . .'.[81] To recognise that the external world is based upon 'definite *combinations of matter*' and that such matter changes discontinuously to create new phenomena is the injunction which Deborin took into the 1920s debate with the Mechanists. It is also what the 'Bolshevisers', such as Mitin who had learned their dialectical materialism from Deborin, used against their teacher when they dislodged Deborin's dialectical materialism in favour of their '*diamat*', the official doctrine of Stalinist state policy.

Accusing Deborin, the old Menshevik, and his associates of being both 'Trotskyite agents on the philosophical front' and 'Menshevising idealists', Stalin and his Bolshevisers plunged into planned industrialisation by stripping Deborinist doctrine down 'to the contentious idea that the material world is a self-developing unity of diverse relations' and by appealing to the '"autodynamism" of matter as the source of everything and as the justification of everything progressive'.[82] Mao's reliance upon the Bolshevisers in his attack on Deborin was therefore doubly pragmatic. He could attack the author of the doctrine which, through Lenin, he had imbibed while using the doctrine to invert the internal and external of development. Autodynamism, for Mao, resided not in the world of matter but in that of the mind, in belief, action and leadership. Socially organised experience, the motif of Bogdanov's version of Mechanism, provided a bridge of development for Mao, who had drawn upon the inheritance of the Soviet Russian dispute between the logic of Hegelian marxism and that of positivism.

Bukharin, a leading Mechanist, had proposed a 'Marxist Sociology' based upon the idea of a historical process which pivoted upon an equilibrium between production relations and technology and, more generally, social and natural order.[83] Bogdanov, Bukharin's mentor, had gone far further in rejecting Marx's use of Hegel's principle of development and turned towards replacing 'development' by the science of organisation. Development, as Bogdanov understood it, implied either the progress or regress of order or organisation whereas the dialectic could only properly be used to explain the progress from 'lower to higher forms' of organisation which followed from the struggle of opposites. If development was *the* struggle of opposites, its domain was restricted to accounting for phenomena which provided conditions for, but not the general basis of, development. Bogdanov contended that Marx's 'developmental' approach was 'unscientific', indefinite, inexact and too broad to provide the organising principle of knowledge. Rather, it was the organisational process itself which could provide a 'real' basis of dialectic, namely harnessing the external world of nature for the purpose of ordered human progress.[84]

Tektology, the science of organisation and a precursor to cybernetics, systems and control theory, was invented by Bogdanov, as Bakhurst has explained, to project 'the operation of natural laws' upon 'not just social but also *psychological* systems (belief sets, conceptions of the world) as evidence that the natural, the mental and the social are all aspects of a single self-organising structure governed by the same set of organising principles'. The material to which these organisational principles would ultimately apply was that of *experience*. Reality, by which Bogdanov meant the 'external world' of matter-in-motion, was thereby 'socially organised experience', and thus it was the organising principle or means by which the external could be constructed to depend upon the mind.[85]

In Chapter 1, we explored the way in which Comtean positivism, including the method adopted by J.S. Mill, rested upon the idea that while the science of the social could be derived from a general hierarchy of science, sociology itself could not be reduced to a more general, developed science, such as that of biology. A similar contention arose among the Soviet Russian mechanists in that there was a difference, as Bakhurst explains, between the strong view that all phenomena which could be reduced to entities were to be explained by the physical laws of science, and the weaker view that not all phenomena could be so reduced but that all could either be explained by physical laws or by the explanatory *model* of physical science.[86] By this reckoning, Bogdanov's reductionism was a strong and ultra-positivist version of Mechanism. Comte and Mill's method was analogous to the weaker, conditional view which regarded social experience as phenomena of development because they were not yet part of the natural order of things. Social experience could be ordered, to conform to what was naturally ordered about the world, through development of a social science which rested upon the explanatory model of the natural sciences.

Development, according to the Comtean standpoint, was the bridge between progress and order. In Mill's emendation of Comte, it was social experience which provided the material for development to do its work. Bogdanov's organising principle made development redundant because there was no external condition of development which was separate from the intent to develop and, therefore, no bridge of development between the internal and external of development was necessary. Mao's rejection of the Deborinists, who had renewed the dialectic of development to counter positivism, effectively stood as the means to justify why an older positive view of development could be recovered. Social experience was to be organised with the modified version of Bogdanov's strong premise of the self-organising principle. Mechanism, in treating 'the individual as a complex machine somehow capable of self-development and self-organisation',[87] was a far cry from Mao's neo-Kantian association with positivism. Rather, it was the Comtean source of trusteeship which gave direction to the basis of development because those who were 'developed' by social experience were supposed to understand that the facts of experience were conditioned by the external world of 'reality'. And then, the positive ideal was spliced, for Mao, into the principle of development as one founded not upon any external condition but upon the 'autodynamism' of social matter in motion. Society, with its own internal dynamic of development, would be developed by the body of socialists who had apprehended the external and who were thereby capable of managing the self-development and self-organisation of the social mass of individuals.

When Frank invoked Mao for the 1960s purpose of securing foundations for underdevelopment theory, he inadvertently also carried forward the provenance of Mao's views in the Soviet Russian debates of the 1920s. These were the disputes which formed the backdrop to Mao's emphatic comment that 'we are proponents of internal causation and oppose the theory of external causation'.[88] To attack the theory of 'an external motive force', as both the cause and basis of development, may have seemed to be a subject of keen debate in twentieth-century political economy but the roots of the theory lay, as we have tried to indicate, in ideas of conscience, duty and knowledge which were part of the cause and basis of development itself.

Curiously, Lenin had ended his 1915 notes on the dialectic of development by reflecting upon the roots of the 'subjectivism' and 'woodenness' which were to be found in the one-sided 'development' of 'idealism': 'clerical obscurantism (= philosophical idealism), of course, has *epistemological* roots, it is not groundless; it is a *sterile flower* undoubtedly, but a sterile flower that grows on the living tree of living, fertile, genuine, powerful, omnipotent, objective, absolute human knowledge'.[89] When he gave emphasis to the grounds of idealism, Lenin was pointing back to the many arguments which ranged over the conditions for 'development' in many areas of ideological and theoretical dispute in the nineteenth century. 'Clerical obscurantism' was far more literal, and less metaphorical than Lenin may have intended to imply. There was much clerical interest in, and dispute over, the idea of development in the nineteenth century.

THE NINETEENTH-CENTURY THEOLOGICAL DEBATE

The most pertinent example of such a dispute over 'development' was that of 'modern theology'. The first part of A.M. Fairbairn's *Place of Christ in Modern Theology*, which was published in 1893 and passed through twelve editions, started with 'The Law of Development in Theology and the Church'. Through referring, in his first chapter, to 'The Doctrine of Development', Fairbairn took John Henry Newman's 1845 idea of development as the source of the doctrine in English theology. Lord Acton, the historian, claimed that Newman's 1845 *Essay on the Development of Christian Doctrine* 'did more than any other book of the time to make the English "think historically, to watch the process as well as the result"'. And Mark Pattison, cleric, critic and early supporter of the Social Science Association, had written to Newman in 1878:

Is it not a remarkable thing that you should have first started the idea – and the word – development, as the key to the history of church doctrine, and since then it has gradually become the dominant idea of all history, biology, physics, and in short has metamorphosed our view of every science, and of all knowledge.[90]

Given these claims, Fairbairn attempted to prove that Newman's theory of development was a one-sided logical or abstract theory rather than a theory which was both scientific and logical.

Development, Fairbairn implored, must be 'concerned with real persons or organisms'; a theory of development must be 'scientific' or 'biological, i.e., it must study life as living, as lived, and as perpetuating life. It cannot be merely logical, i.e., proceed as if nature could be reduced, as it were, to the forms of the syllogism, or stated in its terms.'[91] By the 'logical', it was inferred that Newman was employing the 'dialectical', searching as a Hegelian for the meaning of development in the 'explication of the Idea, the Cause, or Force which unfolds or is unfolded into the system'.[92] This was development as an objective process whereby new forms arise out of the old and the more complex out of the simpler. An objective, immanent process was necessary but not sufficient to evaluate development:

Development may be defined as at once a subjective method and an objective process – as a method it seeks to conceive and explain a being or thing through its history; as a process it denotes the mode in which the being or thing becomes a mode of progressive yet natural change worked by two sets of factors, the inner and outer, or organism and environment.[93]

Fairbairn went further. Logical development defined conditions for proof but biological or scientific development was concerned with giving an explanation for historical change. It 'seeks to discover whether the great factors of change are inner or outer'. Inner factors are those of the organism; the outer are those of the environment which determines why and how one organism is different from another. It is evident, Fairbairn postulated, that 'development cannot be dealt with as if it has been governed entirely from within. The internal were indeed the creative forces, but the external were factors of form and of formal change.' And, 'while the formal factor is found in the environment, the material factor must be found in the organism, and the truth of one must be tested by its adequacy as a vehicle or mode of expression for the other'.[94] Fairbairn accused Newman of treating the Church, and in particular the Roman Catholic Church, as if it had developed entirely by virtue of the immanent process of an organism. Without taking regard of 'place and time', Newman had not given an account of why and how the Church was subject to the discipline of the environment and thus to the

Providential laws which governed all human beings.[95]

Fairbairn's immediate purpose in relation to Newman was different from that of Frank's critics nearly a century later. The question of the cause of underdevelopment in the Third World and its peculiar place in the general history of capitalist development is a question of historical materialism. Yet the difference between the internal and external, of what Mao had signalled to lie behind the basis and cause of 'development of underdevelopment', corresponds to the 'inner' and the 'outer' of Fairbairn's nineteenth-century characterisation of development. Fairbairn, as a Congregational Free Church apologist, took Newman's idea of development seriously as a doctrine to be criticised because it had been so influential within theology. What Newman had also done, however, and what Fairbairn had ignored, was to prefigure a general theory of underdevelopment in 1845.

NEWMAN'S IDEA OF DEVELOPMENT

J.H. Newman arrived at an idea of development when, as Fairbairn put it, 'development' was 'in the air, working consciously or unconsciously, in all minds'.[96] Although he called it 'corruption', Newman's idea of development contained within it a modern idea of underdevelopment. Corruption was the negative counterpart to development. Like development, it was an immanent historical process which happened simultaneously with development itself. It was not an original condition for development but a 'false' or 'untrue' development.

Newman's immediate purpose in writing his *Essay on the Development of Christian Doctrine* was to explain his conversion from the Anglican Church to Catholicism. He took as his spiritual task the defence of Christianity as an 'objective religion' whose principles were revealed through Providence. In doing so he sought to oppose Anglican theory, which held that the Anglican Church had succeeded to the Church of Rome, with credentials that were premised upon the original interpretation of the Apostles. For Newman, in 1845, the Roman Church remained the authentic continuing tradition of early Christianity because the doctrines which had been developed by the Church, such as the beliefs in the Virgin and Transubstantiation, and faith in the authority of the Papacy, were a series of historical additions to early doctrine which embodied Christological principles. The question was whether these additional beliefs could be interpreted to be 'new' doctrine which had authentically developed out of early Christian doctrine.[97] Doctrine, as a form of belief, could be said to have developed because it gave active life to the principles, the essential laws of religion which had been revealed but which did not develop in and of themselves.

Anglican doctrine developed as well, but in such a fashion as to undermine the authority of the one and single Christian Kingdom –the 'one body politic'. Protestants appealed to the Bible alone as a fixed standard of law which was contained in scripture. There was no appeal, as for the Roman Church, to tradition.[98]

As a foremost member of the Oxford Movement, Newman had sought to de-Protestantise the Anglican Church. Regarded as a state apparatus, a department of government which owed its origin and continuing existence to the exercise of state power, the Church of England was taken by Newman to be an artificial religion. The Anglican Church had not evolved gradually and spontaneously; it was devoid of a living ethic of belief.[99] Whereas Protestants had accused the Roman Church of being corrupt precisely because it had surrounded Christological principles with mystery and superstition, Newman turned the accusation on its head. It was the Anglican Church which was corrupt because its embodiment of principle had no vital force for the reproduction of doctrine.

For Newman, the capacity for belief, deriving from the immutable revelation of God through Christ and exercised by dogma, had decayed with the advent of liberal religion. In England the Church had become a dead mechanism, an 'engine created of statesmen in aid of the policeman' serving to domesticate belief and cultivating it in artificial forms which owed little or nothing to God's revelation.[100] Newman wrote:

> We see in the English Church, I will not merely say no descent from the first ages, and no relationship to the Church in other lands, but we see no body politic of any kind; we see nothing more or less than an establishment, a department of Government, or a function or operation of the State – without a substance – a mere collection of officials depending on and living in the supreme civil power.[101]

At once, we see what Newman meant by corruption and what criteria he was using to establish whether or not the phenomenon of the Church had developed. The development of the Church of England was corrupt because it had no continuity that stemmed from the design of God; it had no identity with the first Christian Church of God. The Church of England had not developed through natural life; it was there to keep the mob at bay and to instil belief in its members that their poverty was God-given as part of their condition in nineteenth-century England. As such, the Church had no source of sovereignty that was independent of the state – it had no capacity for self-development.

Newman's theory of development was the attempt to show why and how identity, in this case the immutability of revelation, could be reconciled to the historical variations in doctrine. To preserve identity through change was to develop and to dissolve was to corrupt.[102] Through the variation of history, Newman's case was that the active, but spontaneous, life of the Roman Church had preserved the identity of God's revelation in all periods of history.

Newman referred to development as a 'hypothesis to account for a difficulty'. The difficulty was the process of history. Different events happen and different doctrines and institutions emerge out of change. Yet principles of belief, if historical change is to have any meaning, are to be regarded as being immutable. How, therefore, are historical variations to be evaluated so that a principle of belief may maintain its identity through the course of historical change? The idea of development was the most influential answer given to this question in the mid-nineteenth century. 'Throughout the 1850s and well into the 1860s, for example', we are informed, evolutionary theory was commonly referred to as 'the Development Hypothesis'.[103] But the development hypothesis was not confined to a scientific theory of evolution and it was certainly not to be assimilated, as through Darwin, for instance, into the idea of 'progressive development'. There are two theories of development in Newman. The first, which is pre-eminent in the 1845 *Essay*, is a restricted meaning of development. The hypothesis was treated as if it belonged to the positivism of, for instance, Comte. Development, according to Comte's motif of 'Progress is the Development of Order', was what bridged the gap between the historical imperfections and variations of Progress and the desirably positive and naturally organic quality of state order. Development was what made Progress possible. Newman, however, expunged Progress from the arch of history. Given that 'Newman never believed in progress',[104] it became logically possible to lift from development the weight of its positive burden of a deliberate source of agency and action. Newman could maintain that 'To create is not to develop'[105] because development did not have to serve any constructivist purpose. Any such purpose, with primacy put upon static order, was devoid of life. If static order was to be established through a set of mechanisms then there could be no change and no development. To quote the commentator Grave, the meaning of development is that 'living things develop without losing their identity'.[106] Development must be life itself.

But Newman goes further than this, developing an expanded meaning of development. By the time he arrived at his 1870 *Grammar of Assent*, Newman had made the restricted meaning or domain of development, that which is confined to a world of organic objects, completely subordinate to what distinguishes the form of life which has subjective essence. The plant or the body or the political state, which is regarded as if it were a biological organ, is a mechanism without conscience. It is the form of life, imbued with conscience and the capacity for belief, which develops. On this ground, Newman excludes what he called 'mathematical', 'physical' and 'material' developments from his idea of development. They are excluded because they 'cannot be declensions from the original idea' of development.[107] And the criterion for development is the magisterial dictate of conscience, that which makes development possible but also adjudicates as to whether development has happened in history. Newman's development is the adjudication of whether 'living things' can maintain their identity in change. But it is not clear whether the source of adjudication can stand outside history itself.

The problem is knowing whether it is conscience which develops. If conscience develops in society and through history, then it is the belief in development as a tradition, the belief of conscience, which is the only plausible way that development is made possible. What, therefore, is immutable in the development hypothesis? It is not conscience itself, but what enables conscience which is the transcendental source of the capacity for belief. For Newman, it is the immutable belief in God, outside of history, which is what makes the adjudication of development possible. To see how the problem of development arises we need to go back to the first, restricted domain of development and then move on to the general or expanded domain.

Newman's original formulation of development took an analogy from the organic world of biology.[108] Yet, as was appreciated by Newman's nineteenth-century critics such as Leslie Stephen, but then misinterpreted by his early twentieth-century interpreters such as Barry, this was no simple anticipation of the biological doctrines of evolution.[109] It was not progressive development. Stephen referred to it as a genuine theory of development with the survival of the fittest left out.[110] Stephen, the father of Virginia Woolf, was a stern critic of Newman. 'The development of a system of belief', claimed Stephen of Newman, 'may be compared to the development of a species under natural selection'. Some beliefs survive through history; others do not have 'the vital force necessary to secure their permanence' and if they are to survive, they have to be kept alive and be reproduced artificially: 'As the gardener manages to preserve a hybrid plant in his hot-houses, the statesman preserves the artificial equilibrium of a body which left to itself, would split into its natural elements.' But Stephen immediately goes on to quote from Newman's Lectures:

> The life of a plant is not the same as the life of an animated being; and the life of the body is not the same as the life of an intellect; nor is the life of the intellect the same in kind as the life of grace; nor is the life of the Church the same as the life of the State.[111]

The inconsistency between the similarity of the plant and the state and then the avowal of their difference is the problem of development. It is, for Newman, development, as both a 'hypothesis' and a 'fact', to account for a difficulty. The difficulty is not that of biology but of history.

DEVELOPMENT, 'CORRUPTION' AND UNDERDEVELOPMENT

As a prelude to the 'Development of Christian Ideas', at the start of his *Essay*, Newman wrote a lengthy chapter entitled 'On the Development of Ideas'. He argued that 'Ideas and their developments are commonly not identical, the development being but the carrying out of the idea and its consequences.'[112] The development of an idea, Newman said, was

> the germination, growth and perfection of some living, that is, influential truth, or apparent truth, in the minds of men during a sufficient period. And it has this necessary characteristic that, since its province is the busy scene of human life, it cannot develop at all, except either by destroying, or modifying and incorporating within itself, existing modes of thinking and acting.[113]

In setting out destruction as an integral 'characteristic' of development, Newman was harking back to a cyclical conception of history. And like all others who thought seriously about development, he was unnerved by the implications of the logic that development was characterised by destruction. For if this was so, how could it be that an idea, such as Christianity, could be imagined which did not contain the capacity to destroy within it?

In attempting to resolve this problem, Newman pointed to the different meanings of development in order to counteract the 'confusion in our thinking' which arose out of the everyday use of the word 'development'. It was used, he argued,

> in three senses indiscriminately, from defect of our language; on the one hand for the process of

development, on the other for the result; and again generally, for a development true or not true (that is, faithful or unfaithful to the ideas from which it started), or exclusively for a development deserving the name. A false or unfaithful development is called a corruption.[114]

The problem of the distinction between development as a process and as a result has been noted above, in Chapter 1. But we must look more closely at what Newman refers to as the third sense of development, namely a result which does or does not truly meet the expectations held of an idea before the process of development proceeds. If such a result fails to meet these expectations, Newman refers to it as 'corruption'. For Newman,

the corruption of philosophical and political ideas is a process ending in dissolution of a body of thought and usage which was bound up, as it were, into one system; in the destruction of the norm or type, whatever it may be considered, which made it one; in its disorganisation; in its loss of the principle of life and growth; in its resolution into other distinct lives, that is, into other ideas which take the place of it.[115]

And: 'The corruption of an idea is that state of a development which undoes its previous advances.' Or, as he concludes, development is a corruption, 'which *obscures or prejudices its essential idea*, or which *disturbs the laws of development* which constitute its organisation, or which *reverses its course of development*'. But that development is '*not* a corruption . . . which is *both a chronic and an active state*, or which is *capable of holding together* the component parts of a system'.[116] Two implications follow from Newman's use of 'corruption'. The first is analogical in that twentieth-century 'underdevelopment' takes its categorical meaning from corruption through being used as a direct substitute for it.

We take an example from Celso Furtado's *Development and Underdevelopment*. Although Frank had criticised Furtado's dualism which was inherent in his earlier 1959 account of the economic growth of Brazil, he was sympathetic towards the former Brazilian planning minister's emphasis upon underdevelopment as one particular process of development.[117] Typically, Furtado defined a process of underdevelopment to be one in which 'economic regression', as evidenced by a reduction in 'production per capita', is 'due to disarticulation of the economic system'. Regression, Furtado pointed out, 'does not mean a movement symmetrical with that of progress or growth' because regress 'does not bring a reversion to the primitive forms of production, does not imply complete abandonment of the more advanced techniques' of production.[118]

'Atrophy' of the economy, Furtado argued, followed from 'development by external induction'. In explaining why regress should happen when modern capitalist enterprises were grafted onto 'archaic structures' of an economy, underdevelopment could be viewed 'as a state of factor unbalance reflecting a lack of adjustment between the availability of factors and the technology in their use, so that it is impossible to achieve full utilisation of both capital and labor simultaneously'.[119] An underdeveloped economic structure was one in which capital, however fully utilised, could not absorb 'the working force of labour' which was displaced from the archaic or 'backward' sector of the economy when capital entered the 'developed' sector where advanced technology predominated. Since, Furtado continued, the surplus population 'had to be absorbed by the backward sector at the level of productivity prevailing in that sector', and even though per capita income of the whole population may have risen, the failure to absorb new technology throughout the economy made the economy relatively underdeveloped.[120]

Disarticulation, asymmetry, unbalance and lack of adjustment were ways of expressing what made underdevelopment a corrupt process of development. 'A theory of development', Furtado insisted, 'must have as its basis an explanation of the process of accumulation of production.' A process of development was underdeveloped or corrupt according to this essential idea of capitalist development. Furtado's corollary was that intentional development, 'to further a policy of development', should 'take the form of positive guidance of the capital formation process'.[121] Given the motive force of the external of development, this is what made Frank so sceptical about the intent of development to do anything other than develop underdevelopment.

The second implication of Newman's term 'corruption' is that the changed meaning of corruption is part of the historical change in the meaning of development itself. Newman's use of corruption drew upon its seventeenth- and eighteenth-century meaning, a meaning that was associated with the new idea of progress but which harked back to the classical idea of a phase of decay and decomposition following the phases of expansion and maturation in the cycle of development.

Corruption is a word widely used to characterise the post-colonial development of many a country. It is used in a way that conveys the impression that state personnel execute policy when they are induced to do so by the favour of monetary bribes and other personal inducements. As such, the policy of development is negated by courses of action that are directed to particular ends that are not faithful to the 'stated' idea of development. By this token, politicians and state officials are urged to be pure and authentic in the name of a rational and positive pursuit of development. Corruption, it is claimed, distorts and refracts the authenticity of development and the supposition is that if state personnel were redeemed from corruption, the idea of development would be realised.

In the current literature on Third World development, there is controversy over whether corruption is the negation of development and the overriding characteristic of underdevelopment, or whether it is a condition for development which will only be eliminated after development has resulted. For instance, against the view that corruption is dysfunctional for development, it is argued that corruption is the primary accumulation of the process of capitalist development and not a result of underdevelopment. Theobald refers to 'the heart of the problem of corruption: that it is inseparable from the problem of underdevelopment and it is inconceivable that the former can be tackled apart from the latter'.[122] This current controversy ties corruption to under-development and, indeed, is one meaning of Newman's 'corruption'. What is interesting here is the change in the sense of corruption, a change which happened at the advent of modern development in Europe and which brought the modern meaning of underdevelopment simultaneously into the process of development. The change in 'corruption' mirrors that of the meaning of development itself.[123]

At the turn of the eighteenth century there were followers in the tradition of James Harrington, the author of the utopia of *Oceana*, who sought to reform the British state. They looked back to the ancient world of Greece and Rome and found that state instability could be explained by a determinate pattern of forces that relentlessly culminated in the destruction of the order of any one state at the very point at which that state appeared to be managed to perfection. Bolingbroke, a neo-Harringtonian, wrote that 'physical and moral systems are carried round in one permanent revolution, from generation to corruption, and from corruption to generation, from ignorance to knowledge, and from knowledge to ignorance, from barbarity to civility, and from civility to barbarity'.[124] In this view, corruption was a phase of a cycle of both classical development and early capitalist history. Here, if development was to be authentic, it had to pass through recurrent phases of 'corruption'. In the eighteenth century 'Luxury' became the equivalent of corruption when it was argued, from Petty to Cantillon and authors of the Scottish Enlightenment, that the relative advantages of the wealth of some nations would not be easily regenerated because, according to Cantillon, 'when a State has arrived at the highest point of wealth . . . it will inevitably fall into poverty by the ordinary course of things'.

Inasmuch as the idea of progress was an attempt to register the possibility that cyclical decline could be overridden through a linear rather recurrent series of stages of history, it was an attempt to overcome corruption. Progress required an extraordinary course of things. Foremost, it was concluded, was the injunction for the work of labour to be made productive, without some presupposed finite limit. The application of reason would counter the volatilities of trade and exchange, which were now taken to be global and not local. When poorer nations, as Sir William Temple remarked, would 'come "to set up" in competition with the rich "some must give over, or all must break"'.[125] The gist of the argument, resting on an analogy between trade and the flow of the sea as imagined by the ebb and swell of a tide which displaces water from one place to another, was that the advantage of a 'low price' of labour in a poor country would override the advantages of 'superior industry and skill' in the richer country, but as the price of labour was driven up by wealth, so production would be shifted from less poor to poorer countries.[126] This argument, as expressed by Hume for instance, is often taken to be an assertion of how free trade transmits progress from one area of the world to another.

When the argument was recast through the idea of development, in the image of the past progress of industry and skill moving from one country to another through the movement of capital, then there was no problem of whether, in Newman's words, development is faithful or not. Any poorer country in relation to another is undeveloped. In forgetting the cyclical conception of development and its phase of corruption, there is, for the richer country, no under-development. It is undevelopment that belongs to the idea of progress; underdevelopment to that of development.

There was another related connotation put upon the eighteenth-century word 'corruption', one which was linked both to its commonsense, contemporary usage and to the justification of self-regulated trade. Harrington, in the midst of the English Civil War, had resurrected the ancient Greek case, as so closely advocated in Aristotle's *Politics*, for what was called a mixed constitution. Through the tradition of 'civic virtue' or modern prudence, the aim was to pre-empt the inevitable phase of state disorder and decline by engineering a state which would rest upon a constitutional balance between sovereign power, the aristocracy and the mass of the population. Corresponding to this distribution of political power, property would be distributed to give economic independence to some, if not most, of the population so that they would be politically autonomous of the sovereign and a landed aristocracy.[127] Emphasis was put by Harrington upon the corrupting effects of the state maintaining a standing army instead of freeholding landed militia. His proposals for what he called the 'agrarian law' were designed to distribute land to maintain independent producers. Were the means of production to be widely distributed, then free exchange of the products of labour could be justified without the corruption of overbearing state power associated with the Crown and the aristocracy. Later, in the eighteenth century, 'old corruption' came to be used by radical reformers to argue that the English state rested upon patronage of the Crown and the Lords to corrupt the peoples' representatives of the Commons.[128]

Corruption, therefore, came to refer to cases where state order had become unstable and national power had declined because the balance of political and economic power had been shifted either against the few, of the aristocracy, or the many, of the people. Through its association with unbalance, in the constitution, or skewedness, in the distribution of economic and political power, corruption had long been cast as the symbol for the degeneration of state power. An ideal of political development was to regenerate the state through rediscovering the basis of balance in the mixed constitution.

Newman referred to 'developments of an idea' which 'may be called *political*'. He saw them in the 'growth of States or the changes of a Constitution':

> Barbarians descend into southern regions from cupidity, and their warrant is the sword: this is no intellectual process, nor is it the mode of development exhibited in civilized communities. Where civilization exists, reason, in some shape or other, is the incentive or the pretence of development.[129]

This distinction between 'barbarism' and 'civilization' was joined to that of 'order' of government and 'liberty' of the constitution. By way of one interpretation, it was Guizot, the historian and politician of the French Revolution and after, who was the source of these distinctions for Newman.[130] Guizot had written: 'France wanted a non-revolutionary revolution, capable of giving together order and liberty.'[131] Balance between order and liberty was to be imposed from above to ensure, commented Soltau of Guizot, 'the supremacy of the middle class, its natural monopoly of political wisdom, the absolute coincidence between its development and the rise of civilisation'.[132] When Guizot got it politically wrong, he did so, according to Soltau, because he could not see, unlike Louis Napoleon later, where the solid middle class lay – in the mass of peasant proprietors.

With the development of capitalism in the twentieth century, the connotations which hitherto had been conveyed by 'corruption' were now conveyed by 'underdevelopment'. In so far as underdevelopment is confined to a Third World ghetto, it is because corruption was so marked where the development of capitalism was alleged to be false rather than true. And, as in Europe of the past, the ideal of political development, no less than that of social and economic improvement, continue to be based upon the premise that it is corrupt government which has to be confronted to secure stable political order.

The extent to which the contemporary meaning of underdevelopment is prefigured by Newman's use of the connotations of corruption is remarkable. Newman made it perfectly clear that the only incorrupt development was what he called '*mathematical* developments'. A set of axioms, when developed for the purpose of proof, necessarily end in a set of conclusions according to how the axioms are defined. The conclusions, Newman remarked, 'cannot be declensions from the original idea'. In other words, the consequences which follow from the logical implication of an idea cannot be evaluated as decay or deformation, degeneration or dissolution of, or deviation from, the idea itself. Corruption is the first stage at which the idea becomes historically part of a system of thought and practice which is understood to have no 'vigour'. It has entered into decay and decomposition as a prelude to its extinction. As such, there is development of corruption: 'When a system really is corrupt, powerful agents, when applied to it, do but develop that corruption, and bring it more speedily to an end. They stimulate it preternaturally; it puts forward its strength, and dies in some memorable act.'[133] Carried across into the twentieth-century political economy of capitalism and socialism, this indeed is what is meant by the 'development of underdevelopment'.

The connection between corruption and underdevelopment is so close that we find a commentator on Newman referring to 'the underdeveloped or corrupt conscience' when he refers to the way that Newman used the dictate of conscience. If conscience consists of the dispositions of belief that are created by the authority of God, then it is only the true conscience which will allow adjudication between true and false development.[134]

THE BELIEF IN DEVELOPMENT

It was 'development' which offered, for Newman, the hope of filling the gap between the immutability of doctrine and the history of variation, contingency and diversity. To develop, in short, was to recognise what was so variable in history but to make what was so immediately contingent about experience conform to and become a part of the unchanging truth of revealed doctrine. This was the restricted domain of Newman's first idea of development, where tests are made of events and doctrines. History happens, in discontinuities, and it is evaluated by a set of criteria contained in the 'tests of development'. Development conveys immutable doctrine across the rifts of history. The 'hypothesis' of development is to examine the extent to which there is continuity in change.

Newman's seven tests of development are applied to the history of Christian doctrine but he illustrates them more generally. Thus, his first test, that the original idea must be preserved, gives examples of corruption and development: 'Judges are corrupt, when they are guided in their decisions, not by justice and truth, but by the love of lucre or respect of persons.' And:

> An empire or a religion may have many changes: but when we speak of its developing, we consider it to be fulfilling, not to be belying its destiny; so much so that we even take its actual fortunes as a comment on its early history, and call its policy a mission.

All developments are part of an original idea with 'what is inconsistent with it being no development'.[135] Likewise for the other six tests. In so far as doctrines expand and grow, like organisms, they do so under the particular circumstances of history through variation and difference. Newman draws on history widely to reinforce his criteria for testing whether doctrine has developed. Doctrine develops when it shows continuity with the past; when it has the potential to assimilate new ideas; when it preserves the original apostolic faith; when it flows logically as the result of that faith; and when additions to doctrine preserve the original creed.[136] All these tests evaluate development as a result of history. If the search for continuity of principles in the 'discord' of the discontinuities of doctrinal histories is uncontroversial, then the sixth and seventh tests reveal the extent to which Newman starts to go beyond the evolutionary nostrums of the development hypothesis.

For a doctrine to be a 'true development', it must have the potential to anticipate its future and that true development must show 'chronic vigour'. We are now being asked to think about development as more than an evaluation of something that has happened. How can we know whether the doctrine has the power to create its future other than the expectation that since the present is the result of a past which anticipated its future, development in the present will also anticipate its future? Mill's inverse induction or historical method could not allow this inference. Newman's method does allow it: Doctrine can be Dogma, that which anticipates a goal of knowledge before investigation begins to establish whether it is possible for knowledge to fulfil its purpose.[137] And the antithesis of dogma is Heresy, a doctrine which is chronically inconsistent with principle and which, when it dies out, acts to affirm doctrines which are consistent with principle and conform to what has been tested as a 'true development'. Doctrine, when corrupt, dies in one place but, when developing, gets vigorous life in another.

By the time that Newman had published his *Grammar of Ascent* in 1870, he had reworked his theory of development. The tests of development had became 'notes' on development in the second, 1878 edition of the *Essay on Development*. The 'development hypothesis' was torn from its positivist roots, as the starting point for empirical investigation, to become the hypothesis which stands in place of the gap of empirical evidence. The gap is of evidence which cannot be found and the hypothesis is to account for the absence of the evidence of experience. In a later edition of the *Essay*, Newman wrote: 'Physical facts are present; they are submitted to the senses, and the senses may be satisfactorily tested, corrected and verified. . . . But it is otherwise with history, the facts of which are not present.'[138]

Newman rejected the application of mechanistic theories of cause and effect, following an invariable sequence, as the only sound basis for scientific method.[139] If the *Essay* is said to have anticipated Darwin, then the *Grammar* has been recognised as anticipating some twentieth-century reactions against Newtonian mechanics as the basis of scientific method. Among others, Newman is seen to prefigure the philosophy of American Pragmatism, Wittgenstein and, probably through A.N. Whitehead, Keynes.[140] We will return to Whitehead and Keynes in our concluding chapter. Here it is sufficient to emphasise that it cannot be implied, Newman wrote in the *Grammar*, 'that we are appealing to experience, when really we are only accounting, and that by hypothesis, for the absence of experience'.[141] Development, as the 'hypothesis to account for a difficulty', was the hypothesis which bridged the gap of history. As such, 'development' stood as a substitute for the incomplete processes of history. The implicit in history is explicated through development.

With the *Grammar*, Newman arrived at the expanded meaning of development.[142] When asking the question of whether development has happened or not, the answer must be grounded in the faith that it is possible to believe that development can happen. In other words, development cannot be evaluated without making assumptions about those who make development and those who evaluate it. Assumptions involve questions of belief and faith. Truth was found by acting believingly and, as Robert Pattison has interpreted Newman, it is the complement of belief and action which establishes the basis of truth. True development depends, therefore, upon the 'correctness' of belief and the 'rightness' of action.[143]

If development is to happen it must be given assent. Assent cannot be given simply through the formal proofs of reason. It has to be based upon evidence which, at best, can only give a probable guide to action. Evidence gives grounds for what Newman referred to as *antecedent* probability. Probability refers to the relationship between one set of propositions, which contain incomplete evidence, and a set of propositions which give the grounds for why action should be undertaken. Newman wrote: 'Life is for action. If we insist on proofs for everything, we shall never come to action: to act you must assume, and that assumption is faith.'[144] The relationship between evidence and the action is not founded upon the scientific procedures of reason but upon what Newman called the '*illative sense*'.

Scientific procedures of reason, which Newman associated with one logic of 'mathematics', worked upon self-enclosed sets of propositions which were associated with logical and self-referential relations between the cause and effect of phenomena. Logically, the procedures governing the relationship between propositions gave grounds for certainty. Certainty was to be regarded as a quality of the propositions and depended upon the formal demonstration of proof. These procedures could not be applied to reasons for human action. When rational reasons were so applied without the presupposition of belief, Newman saw the application of science to sources of human action as the baneful influence of utilitarianism, which he referred to scathingly as 'Liberalism'. Newman, Pattison wrote, 'did not care if science proved he was a monkey's uncle so long as he was a monkey's uncle who believed correctly'.[145] Probability established the extent to which the belief was correct.

Probability, for Newman, was antecedent rather than consequent because it could not involve calculation which set out to establish what consequences would follow from undertaking courses of action. Rather, probability involved bringing together disparate but cumulative sources of incomplete evidence which then converged, non-linearly in terms of cause and effect, to create propositions. States of mind stood between the 'facts' of evidence and the sources of action. A state of mind gave grounds for 'certitude' and not for certainty; probability could only be transformed into certainty through the faith in certitude. While certainty was inferred from being grounded in events of action, Newman's concept of certitude, as Selby has noticed, was grounded in the beliefs, the acts of assent, of persons. 'Certitude, Newman wrote, is a mental state; certainty is a quality of propositions.'[146] Reason, for Newman, would only work if it was subsumed by faith. Newman's illative sense was derived from Aristotle's 'phronesis'. Assent is given, according to Newman, through the state of mind which has the capacity to judge and to act. The illative does not necessarily work through the apparatus of propositions. It moves through opinion, influence, conduct and convention. 'The heart is commonly reached', Newman wrote, 'not through reason, but through the imagination, by means of direct impressions, by the testimony of facts and events, by history, by description.'[147] To quote Gilley on Newman, 'the peasant can read the clouds, and the multitude can judge all sorts of practical issues with complete assurance. The mind in these matters has passed from fact to conclusion without notional, propositional or linguistic intermediaries.'[148] Newman referred to the difference between the 'practical' and the theoretical as sources of judgement, as that which made 'simple' differ from 'complex' assent. Simple assent was given without formal reflection; complex assent could embrace science and its procedures. But, however complex the assent, the mind's capacity to judge whether phenomena were true or false rested upon faith. Without faith, there could be no true source of judgement and reason could not work properly.

The faith with which Newman was concerned was the Christian belief in God. God revealed truth through Christ; the early Christians gave simple assent to the revelation because of a prior belief in God. Revealed truth was embodied in doctrine which developed, and continues to have the power of development, through the dogma of the Roman Church. Early dogma did not have a language or a propositional structure. Language appears through the history of the Church and according to what Newman called 'economy', the method whereby language partially represented revealed truth.

Selby has shown how Newman counterpoised 'reserve' and economy in the process by which truth was revealed. Reserve, according to Newman, was the withholding of truth from believers to safeguard the sacredness of knowledge. Words, Newman maintained, were too clumsy to express truth. However, truth was revealed by language, but in bits and pieces and as approximations to objective truth. Economy, from its original meaning of 'husbandry', consisted of the means, from stories to mathematics, by which phenomena were represented as something more real than themselves. As premise or the *a priori*, and what Newman had in mind was the regulative principle of knowledge, the economical brought the external world of God, that of immutable truth, to the internal determination of the self. Condescending to the 'infirmity and peculiarity of our minds', economy gave insight to truth and permitted the explication of dogma.[149]

Dogma was what made implicit doctrine explicit through the formulation of language and all that gave grounds for complex assent. Doctrinal development, when true, made explicit what was implicit in early Christian belief. How is it known that doctrine is explicated by development rather than merely added, in details, and, therefore, is capable of corruption? Since reserve and economy work symmetrically, claimed Selby for Newman, the answer is that in being made explicit, doctrine is taken back to the implicit, 'from what we know to what we wish to know'.[150] True development consists of doctrine that is expected to be developed. Belief is the standard for evaluating the truth of development.

Doctrine, to repeat, embodies principle as the essence of belief and principle is immutable; principle does not develop. However, for Newman through his twofold meaning of conscience, it is clear that the principle is the motive force, as Lash puts it, of belief.[151] Conscience, for Newman, reflects both the belief in the existence of the transcendental object of God, as Yearly points out, and the human desire to believe in the sacred object.[152] Without conscience, in Newman's words, both as 'moral sense' or 'judgement of reason' and 'sense of duty' or 'magisterial dictate', development cannot be judged and cannot be deemed to be true. Without the prior desire to believe, conscience is incomplete. If conscience regulates dogma, then the dogma of the Church gets real assent from its members because it is believed, by virtue of the prior desire to believe, that doctrine has the potential to develop. When development in general explicates the implicit it does so, for Newman, upon the basis of belief.

Newman's idea of development was formulated to confront not merely the idea of Progress, so closely intertwined with the belief in reason, but also to reassert the primacy of revelation against what he called the natural truth of liberalism and liberal religion. However much the genesis of positivism was traced out as a counterpoise against liberal doctrine, Newman must be understood as a reaction against both. Natural truth, for Newman, had no means of appreciating the extent to which revelation was true. In taking on Progress, Newman had also dismissed Providence, or that version which had appeared in the work of William Paley. Newman rejected this attempt to resuscitate the old belief in Providence against that which had displaced it, namely the idea of Progress. Development, with its portent of a future, and not Providence, with a reaction from the past, is what Newman posited against Progress.

Paley, for example in his 1802 *Natural Theology*, had argued that the providential design of God was not to be understood through the rationalism of 'evidences' and, as such, upon the basis of natural truth.[153] For Newman, however, doctrine developed through revealed and not natural truth, and whereas natural truth rested on 'knowledge' to give ground for altruism, revealed truth could only be judged on the basis of conscience.

Liberal religion substituted knowledge and altruism for conscience.[154] Upon the humanist assumption that human nature was universally good, liberalism gave vent to the obligation of private judgement; took revelation to be a manifestation of natural truth rather than the mystery of dogma; put a primary value upon useful goods and thereby value in exchange in so far as the 'given' was transmuted by the 'made'.[155] Newman contested these major principles of liberal religion in his major argument that liberalism would culminate in atheism.

Natural truth had the attributes of the human. It had reason, literalness and circularity in thinking upon the basis of logical assent, sensitivity to historical variation and 'corruption', care for the transitory and the external world. Natural truth was the antinomy of revealed truth, with its associations of the divine, revelation, the mystical and belief in the unity of an Idea upon the basis of supernatural assent, awareness of the immutable and the transcendent and care for the spiritual and the eternal.[156] But natural and revealed truth were in contradiction and not opposition to each other; natural truth could be incorporated into revealed truth. All the attributes of natural truth could become the means to reach revealed truth if it was supposed that what was revealed by evidence and formal proof was the result of faith. Evidence, as it is obtained and presented through procedures of natural truth, is not the groundwork of faith but the result or reward of the faith that evidence is true; faith, as it is put, interprets evidence.[157] And, in so far as natural truth became part of revealed truth through faith that truth could be revealed, doctrine could be said to have developed. When it is asked, 'What evidence can you produce to attest that development has happened?' Newman's answer is that any evidence of development is ultimately the result of the belief that development is possible.

Newman's formulations have been subject to criticism. For instance, it is argued that assent may be absolute or conditional; that Newman has no secure explanation of conscience; that it is difficult to appreciate how the original revelation is elaborated through development into new doctrine without the prior belief that the new is more than simple addition to the old. There is also the criticism to the effect that Newman could not satisfactorily deal with corporate authority, namely that he confused the individual and the institutional, as for the Church, as sources of ultimate authority. On philosophical grounds, there is dispute over the extent to which Newman was a 'modern' and whether he did break from the empiricism of Hume and Locke.[158] Some of this criticism found its way into the general elaboration of Newman's idea of development and into political theory in particular.

But it is as metaphor that Newman's expanded meaning of development became so powerful. Development is the working out of the unconscious in the conscious; it is making the implicit explicit. It is, in his renowned aphorism, to keep things constant through change in so far as faith preserves identity through change. Above all, it is through Newman that development has generally come to be represented as 'continuity in change'.

NEWMAN'S INFLUENCE

Stephen, at the start of his critique of Newman, wrote:

> Dy. Newman and J.S. Mill were nearly contemporaries; they were probably the two greatest masters of philosophical English in recent times. . . . And yet they move in spheres of thought so different that a critic, judging purely from internal evidence, might be inclined to assign them to different periods.[159]

How could the same period of history produce different 'spheres' of thought?

Stephen's answer was simple: Newman's avowal of Roman Catholicism was unjustified because he had made a mistaken attempt in using Mill's method of historical or inverse induction to prove that the Roman Church could fit every, including the latest, period of history. This result of development, Stephen insisted, could only be justified if development was taken to be part of the phenomenon it was intended to explain. Hence, the Roman Church, in Newman's mind according to Stephen, had grown naturally according to the natural law of development as a principle. As such, Newman had disregarded the fundamental premise of historical induction in so far as the method presupposed that historical evidence, to confirm that change had occurred in the direction that would result in development, was the groundwork for the belief that development was possible.

Through his commitment to the progress of positivism, Stephen, for instance, refused to accept Newman's claim for development: the Roman Church belonged to an earlier period or stage because, while Catholicism was consistent as a creed, it could not be proven, by induction, to be true. The proof which Stephen adduced was that as people became more 'civilised', as they gained more knowledge, science and technology, they were more inclined to dismiss Roman Catholicism as an ethical source of conduct.

Likewise, Fairbairn took Newman to task for using 'development' to justify the Roman Church rather than to explain its genesis. The development hypothesis was 'without reason or function'. Newman, Fairbairn wrote, claimed that Catholicism 'helped to organise modern civilisation' in the medieval period but for the modern period, 'The countries which most suffer from revolution are the countries where its rule is or has been most absolute; the countries where it has least authority most represent order and progress.'[160]

In limiting 'development to a process of formal without substantial change', Fairbairn accused Newman, 'the being and truth' of the Church and its doctrine must be assumed before the development process 'can be called into action, and even then it can act only under their superintendence'.[161] In other words, the criticism was that the authority of the Church, as expressed by the 'truth' of conscience, was without development.

Stephen is not taken particularly seriously, not even by his biographer who takes pains to show that Newman's critic ignored his debt to Comte. In his attempt to marry Mill with Darwin through a theory of evolutionary ethics, Annan concluded that Stephen had evaded any theory of obligation or conscience.[162] Despite this, Stephen is of interest to us because he illuminates the focus which development puts upon secular agency. Although this focus grew stronger at the turn of the twentieth century, it did entirely blur the religious association between evolution and development. Barry, in his 1904 *Newman* for instance, suggested that Newman used natural selection as his image for the 'supernatural selection' of institutions which are developed to meet new human conditions.[163] In Stephen's Victorian mind, on the other hand, the mechanism of natural selection was treated as the means by which it was hoped to justify why it is possible for human life to be regulated by progressive change, as if human life was akin to that of a plant. Annan commented:

> Stephen was making the social sciences do the work of religion. Evolution replaces God. Evolution is the Creator. Man is his child; Evolution is an immanent God or Process at work within the world. Just as a belief in God comforted Newman and reconciled him to the spectacle of evil in the world, so Stephen was comforted by the belief that morality was created and sanctioned by Evolution.[164]

And for Evolution, Stephen meant Development. With development went the trusteeship of the intelligentsia as the source of agency for development. 'If the [human] race is to progress', Stephen concluded his critique of Newman,

> men must not be content to bow to the first authority at hand, even if it shows signs of strong and prolonged vitality. . . . Our principle must be to place ourselves in that direction which is shown to have the greatest promise by the general set of opinions of qualified thinkers.

Qualified thinkers were those 'who themselves acknowledged no law but reason, and have not been propagated by ignorance, blind submission to arbitrary rules, and reluctance to believe unpleasant truths'.[165] Little could be further from Newman, who had put the ultimate burden of development upon what came to called self-development.

'Of all the men of the Victorian age', J.N. Figgis wrote of Newman in 1914, 'none was more universal or had a wider influence, or was more fruitful of ideas and principles which still endure.'[166] Figgis has recently been described as a 'political radical and an Anglo-Catholic theological conservative', as the major inspiration for the present attempt to resurrect the theory of 'English political pluralism'.[167] Figgis continued:

> It may be doubted whether any contemporary has made more difference than John Henry Newman. . . . His ideas have so deeply permeated the thought and feeling of the age that his books are read less than they ought to be. His principles have become so general that the author is sometimes forgotten.[168]

Newman's own influence upon Figgis seems to have been forgotten.[169] This is surprising, given that the principle of development played the key part in both Figgis' criticism of contractual theories of indivisible state sovereignty and in his programme for the associative power of plural governing societies.

In his appreciation of Newman, Figgis quoted from Newman's 1873 *The Idea of a University*: 'Nothing great or living can be done except where men are self-governed and independent; this is quite consistent with a full maintenance of ecclesiastical supremacy.'[170] The link between the potential for internal self-government and the external, given source of supreme authority was the problem of development. It was, as we have seen, the source of the tension between progress and order, or what has been called growth and structure. In the *Grammar of Assent*, Newman had put the weight of progress upon individual development:

> [Progress] is committed to the personal efforts of each individual of the species; each of us has the prerogative of completing his inchoate and rudimental nature, and of developing his own perfection out of the living elements with which his mind began to be. It is his gift to be the creator of his own sufficiency; and to be emphatically self-made. This is the law of his being, which he cannot escape.[171]

Figgis, as if to berate Newman, had to declare that the isolated individual does not exist in the real world. Individual 'personality can develop only in society, and in some way or other he always embodies some social institution'. The distinctiveness of the individual 'can function only in a society'. In insisting that societies 'have their effect and limit and develop your life', Figgis was extending the ideas of corporate personality which he had imbibed from the theories of Otto von Gierke and F.W. Maitland. Associations, such as the Church or the club, the school or the union, have a 'real life; they act towards one another with a unity of will and mind as though they were single persons'.[172] But in all that he meant by way of 'develop', he had been permeated by Newman's ideas.

'Self-development' is the title of Chapter 4 of Figgis' *Fellowship of the Mystery.* As elsewhere, Figgis here attacks Comte's positivism on the grounds that it attempts 'to retain in human relations, ideals which are by implication Christian, i.e., the worth of every man, the ideal of love'.[173] Yet although he had scorned these Christian ideals, Nietzsche was worthy of sympathetic attention because he was somebody who had attacked the doctrine of 'self-centred development'.[174] Self-centred development, in becoming so fashionable before 1914, subverted for Figgis the meaning of development because it denied the merit of self-renunciation. Pain, stress and suffering, through self-renunciation, is necessary 'for all beings who are developing'. The end of life, supposed Figgis, was 'the development of the personality for a society beyond this world'; self-renunciation was to be experienced and reflected upon as the means to the end of development – 'the condition *sine qua non* of all development'.[175] This indeed has been how modern, secular development has been experienced and reflected upon, through the two meanings of the word.

When development is taken to be immanent process, as in the development of capitalism, then the suffering of the proletariat which bears the burden of social dislocation and industrialisation is to be redeemed by revolution in the name of development. The question is whether suffering could have been avoided without the development of capitalism. When development is taken to be intentional practice of state policy, then pain is expressed as a possible, if not probable, choice of development. Pain may be taken to be the enforced reduction in present consumption to maximise future consumption; the degree to which pain is avoided is established through a utilitarian calculus.

To calculate the extent to which net benefit could be gained, with pain netted out, was precisely what Figgis, inspired by Newman, abhorred about utilitarianism. But what interests us is the link between the pain of immanent development and the 'development' of intentional practice. Our argument has been that the 'development' of state policy was taken, from the early nineteenth century, to be the means by which 'suffering', made necessary not by choice but by the development of capitalism itself, had been recognised as an object and principle of policy. This was the condition of development. As a condition of development, policy towards suffering was constructed upon the basis of state sovereignty, in the name of national development.

When Figgis attacked the basis of undivided state sovereignty, he did so on the grounds that the contract theory of the state was an early form of utilitarianism. As such, the 'organic character of the state' was denied. Hobbes, for instance, Figgis claimed, treated the state as if it had no 'quality of life' and 'no principle of internal development.'[176] So did the tradition of popular sovereignty, associated with Rousseau, in failing to recognise that the state

> has distinct laws of development, which may not be transgressed with tinkering with it, as a machine. The logical issue of the popular theory is to treat the state as a lifeless creation of the popular will with no power of development and with no source of strength in sentiment or tradition.[177]

The state that does have development laws and the power of development is the state which is constituted as if it were a 'society' in the sense of being an 'association'.

As an association, the state, like others which have 'quality of life', expresses the purpose of self-development. The incorrupt Church, from Newman, was one such association with its distinct power for development. Another would be an incorrupt state, albeit at the apex of a hierarchy of all other associations. Imbued with the power of development, including the capacity for self-renunciation, the state would be justifiably empowered to regulate the relations between all other associations. It would recognise the independent power for self-development of lower-order associations. If it is asked why should other associations give consent to the power of the state to regulate their capacity for self-development, then the answer is drawn from Newman. Conscience, for Newman, was the source of authority both for regulating development and for establishing whether development has happened.

When making use of Newman's concept of conscience, Figgis and his disciples stretched it beyond the state of belief that Newman had imagined. From Figgis it can be inferred that the state would have to renounce sovereign power if it were to possess the capacity for development because state power, on the basis of undivided state sovereignty, cannot be self-limiting. If self-renunciation is the condition for development, it follows that power must be renounced in favour of other subjective sources of self-development. Figgis had influenced the equally influential Harold Laski, another proponent of political pluralism. 'The works of both Maitland and Figgis', Michael Newman has written, 'influenced Laski enormously, as he frequently acknowledged.'[178] Laski waxed lyrical on Newman's 1875 'Norfolk letter', a response to Gladstone's forced resignation as Prime Minister over the Irish question and involving the issue of whether Catholics could be undividedly subject to British state authority. Calling the letter 'a masterpiece', Laski claimed: 'It remains with some remarks of Sir Henry Maine and a few brilliant dicta of F.W. Maitland as perhaps the profoundest discussion of the nature of obedience and sovereignty to be found in the English language.'[179]

In his 'Letter', Newman inferred that in so far as members of the Roman Church were empowered with the development of conscience, they were not obliged to obey the undivided authority of the British state. For Laski, whose concern was then to support First World War resistance against conscription, Newman's 'conscience' was one source of sovereignty which was counterpoised against the demands for allegiance and obedience stemming from state sovereignty.

Further into the twentieth century, pluralist concern with the conscience of development was extended. P.Q. Hirst, for instance, mentioned British state-imposed limitations upon trade unions' rules of association and, following Laski and G.D.H. Cole, referred to how associations were 'thus denied the capacity to decide for themselves how to develop'. Associations, to repeat, were 'means of self-development'.[180] In so far as the model for the non-state association is the Church, Hirst followed Figgis in using the example of the 1900 Free Church of Scotland legal case. The House of Lords decided that the overwhelming desire of members of the Church to merge with the United Presbyterian Church was *ultra vires*.[181] R.B. (Lord) Haldane, in defending the merger, had used the doctrinal 'tests' of development, particularly that of continuity of life. 'Development', Haldane had written, 'is in all cases the realisation of what was not there at the beginning of the process.'[182] Figgis had complained that the judgement denied the Free Church 'the power of defining and developing its own doctrine'. Behind the complaint stood an associative model of the state, resting on self-limitation of the association. A limitless exercise of state sovereignty was the negation of true development. The British state, it was inferred, was corrupt because it did not exercise conscience.

The problem was that when Laski interpreted Newman's conscience to be the general 'arbiter of conduct', he created the opening for the arbiter of conduct to become, as it appears in Hirst, the maker of 'the rules of conduct' by which associations are to be regulated.[183] Conscience, for Newman, had the power to develop, but the power, which regulated development, was the immutable and prior belief in God. Laski quoted Newman on conscience as being 'a messenger from Him'. And, in any case, Newman had advised Gladstone in 1844: 'Mr. Gladstone had said the State *ought* to have conscience – but it has not a conscience. Can *he* give it a conscience? Is he to impose his own conscience on the State?' No, answered Newman, the state 'has a thousand consciences, as being in its legislative and executive capacities, the aggregate of a hundred minds – that is, it has no conscience'.[184] It is difficult to see how conduct may be arbitrated, in the name of development, by a state without conscience.

At first sight, these 'facts', as Newman called them, might seem congenial for pluralist belief. There is no single sovereign and omnipotent body of 'the state'. Thus, Laski, for instance, had rejected the idea that the positive law of the state was to be obeyed on the grounds of '*command*'. Although he did 'not give a precise circumscription of the concept of conscience', according to one commentator, Laski's early theory of what could justify consent to state power was based upon a moral theory of judgement. State law was to be judged by individual citizens of the state, given that individual judgement was to be based upon individual '*conscience*'.[185] Laski had been so taken by Newman that the title of his major work, the *Grammar of Politics*, was inspired by the *Grammar of Assent* and by its theme that personal will is not binding on individual consciences.[186]

Yet it had been stressed by Figgis that the state, as one of many associations, has personal will in an organic entity. If the state is to arbitrate conduct through rules, who frames the rules if not the 'hundred minds', according to the 'thousands of consciences'? For Figgis, conscience arrived through his belief in God and developed the person, through society and the state. Development was both means and a transcendental end. But conduct, when freed from conscience and its immutable foundation in the belief in God, has no necessary power of development. Development, when regarded as the purpose of association, becomes mere name.

There was a historical model for how conduct could be arbitrated. Figgis drew the analogy of the Dominion or self-governing status of states within the British Empire. He pointed to South African immigration law, which banned Asian labour migration. The Westminster Parliament refused to intervene, and quite rightly so, according to Figgis, 'even though they were doing a manifest injustice to their fellow subjects'.[187] Here, the British state, as part of an imperial relation to Dominion governments, was exercising self-limitation. Likewise, Laski used the imperial analogy, drawn from the pre-1931 model of British Dominions, as the basis for his proposals for an economic federation of associated producer interests.[188] The Dominions, Figgis had observed, assumed active but delegated authority from the passively nominal sovereignty of Westminster. In such a way, Figgis concluded, 'the local body', meaning South Africa in his case, 'had a real independent life'. In other words, it had the power to develop.

The major hypothesis of underdevelopment theory, to repeat, is that imperial authority, either of pre-1931 British imperial model or of the US dominated post-1945 international order, did not enable true development. Indeed, it did the contrary: imperialism created and developed underdevelopment. Nominally independent countries were denied the capacity to decide for themselves, as if they were corporate associations, how to develop. To assert the contrary of a proposition of development, in the name of underdevelopment, is not to criticise the basis upon which development is deemed to be true or false. The method by which development was to be determined, through the relationship between internal and external causes, was that used by Frank, for instance, following Mao. It was a method which had figured in the tradition associated with Newman.

Newman was criticised for positing the origin, and end, of development in the prior belief of God, but then allowing conscience to develop or to corrupt as part as the process of development. This was the criticism which was made of Frank – that the origin of the process of capitalist development was taken to be part of the process itself. In his 1913 *Studies in Modernism*, Fawkes, like Figgis, criticised Newman for oscillating between evolution and explication as the means by which development might be understood. But Fawkes, who paid attention to both Newman and development in theology, went further. The internal, or the 'interior', in Fawkes' words, 'signifies' the unity of Christianity; it is 'direction and life'. Newman subsumed the internal under 'external creed', 'external revelation' and 'external court of appeal'.[189] All the external sources of change, including the Catholic idea of the Church, Fawkes argued, are abstract and formal ideas. By giving priority to the abstractions which gave authority to development, Newman took life out of development.

The trajectory of the idea of 'development' in theology bears a striking resemblance to the path that this idea has followed in the realm of development studies. During the 1960s, we find development becoming a 'buzzword' of theology and 'widely seen as Newman's gift'[190] to the Catholic Church, making Rome, as a result of Vatican Council II in 1965, 'accept the demands of historical development'.[191] Yet, by 1973, the theologian Misner could state that doubt was being expressed about 'the whole paradigm of development'[192] and then, in 1990, Lash could claim that '"development" had had its day'. It was 'not the solution to a problem, but simply a name for that problem'. It was 'a name invented to come between immutability and change, with the latter betokening corruption' or what we have argued came to be called, in a later, albeit immediately different context, underdevelopment.[193] Coeval with the growth of theological doubt about development, we note the rise of 'liberation theology' as the positive corollary of that doubt. Liberation theology was one of the starting points for the present-day substitution of 'empowerment' for development in self-described theories of 'alternative development'. We shall return to such theories in our conclusion. Here it is enough to note the convergence of the logically parallel lines of the development debate in these two superficially different spheres.

CONCLUSION

In this chapter we have argued that, rather than is usually claimed, the idea of underdevelopment, like that of development, was very much a part of nineteenth-century thought. As with the origins of development, those of underdevelopment are of more than mere antiquarian interest. It is our view that the significance of the twentieth-century development debate can only be fully understood when informed by an understanding of the nineteenth-century debate within theology.

Central to this understanding is a comprehension of the ways in which Frank's critique of development and the subsequent criticism of that critique both parallels and diverges from the nineteenth-century debate. Both Frank and Newman railed against positivist thought, in Frank's case against modernisation theory, in Newman's by detaching development from progress and thus from positivist thought. But the results of the two critiques were very different. For Frank, the attempted departure from positivism led to an obsession with the 'external' of development and a rejection of the 'internal' determination of development which had so preoccupied the nineteenth-century Latin American positivists. This rejection of the internal subsequently became a central focus of much of the criticism of Frank's writing. We must now turn our attention to another nineteenth-century figure much preoccupied with the problem of development who shared his rejection of positivism with Newman. It is to Marx that we now turn.

3

THE DEVELOPMENT OF PRODUCTIVE FORCE

Marx and List

RECAPITULATION

In Chapter 1, we tried to show how the modern *idea of development* emerged in the first half of the nineteenth century out of the disillusionment with the promise of potentially infinite improvement. From the Saint-Simonians to Comte and Mill, the general positivist postulate was that industrial expansion had happened in periods of social disorder precisely because there had been no constructive coincidence between self-interested investment of capital and the need of dislocated populations for employment. As a result, the growth of potential social wealth was accompanied by the actual poverty and unemployment of its producers and potential producers.

The idea of development took as its immediate aim the amelioration of the social crisis that had accompanied the rapid movement of population towards urban centres of industrial production. Given their concept of crisis and disorder, the positivists argued that progress could only be sustained through the intentional constructive activity of development. Industrial production and organisation was accepted by the positivists to be a historically given part of the movement towards an organic, positive or natural stage of society in Europe. The burden of development was to compensate for the negative propensities of capitalism through the reconstruction of social order. To develop, then, was to ameliorate the social misery which arose out of the immanent process of capitalist growth.

However, the positivists' faith in the potential contained within industrial society for the reconciliation of progress and order was only to be actualised, or so they believed, through trusteeship. It was through the trusteeship of social wealth that the positivists believed development could transform a critical period of history into the ideal organic or natural condition of human improvement.

Trust that industrial society would confer social benefit, rather than promote self-development of the few, had to be made active through the agency of trusteeship. Thus, the trusteeship of the few who possessed the knowledge to understand why development could be constructive, and were accepted as trustees because they were understood to be already developed, became integral to the intention to develop those who remained undeveloped. Active trusteeship was necessary for what made the positivist *idea of development* become a *doctrine of development*. Without trustees as the active agents of development, and a system of constructive social order behind it, development remained only a latent possibility.

Development doctrine brings the intention to develop to bear upon the processes of history. As doctrine, development came to be associated with the plans and intentions of state and public policy in such a way as to give the intention to develop priority over the actual processes of change. We have argued that positivism, in both India and Latin America, was brought to bear upon policy. These two instances show both the logical and historical tensions between the *idea* and *doctrine* of development. The positivist idea of development was realised as doctrine in India by virtue of trusteeship. In Latin America, by way of contrast, the idea of development came to rest upon the distinction between the external conditions for, and the internal basis of, development. This distinction, as we have argued in our discussion of Newman, was governed by the nineteenth-century foundation of theories of underdevelopment and, as such, remained distinct from the emergence of an active doctrine of development. Having recapitulated the emergence of a distinctive nineteenth-century positivist lineage of development and clarified what makes the idea of development distinct from doctrine, we now turn to the developmental thought of one of the chief antagonists of positivist thought.

MARX AND DEVELOPMENT

Marx's early, scathing critiques of idealism and positivism were prompted, in part, by an intuitive rejection of doctrines of development that rested upon trusteeship and that, in so far as they were contained within capitalism, were seen by him to continue to inhibit the realisation of human potential for full development. Nevertheless, Marx's writings, from the 1840s through to his drafts for the later volumes of *Capital*, are suffused with reference to development. While it may be difficult to find consistency in his replete use of the term 'development', it is possible to distinguish between *two domains of the idea of development* in Marx's writings. The first is that of a restricted domain of the idea of development. A domain of development is restricted when the intent to develop is subordinated to an immanent process of development and the process in question, as in Marx, is that of the development of capitalism.

It is to this restricted domain that Gerry Cohen refers in his discussion of what he calls Marx's 'development thesis', the thesis which Cohen uses to argue that Marx sought to capture what was significant about the process of capitalist development for progress. Cohen, to whom we shall return later in this chapter, goes so far as to claim that the thesis of historical materialism can only be properly defended on the ground that 'productive forces tend to develop throughout history'.[1] It is our argument that this development thesis belongs to the restricted domain of the idea of development precisely because it subordinates the intent of development to that of immanent process.

The second domain of the idea of development that may be distinguished in Marx's writings is that of an expanded domain of development in which the intent to develop prevails over an immanent process. But, within the general or expanded domain, there is no doctrine of development of the kind we have identified with nineteenth-century positivism and which, we will argue, stemmed from German philosophy and that of Hegelian idealism in particular.

One of Marx's first uses of 'development' as a fundamental idea of thought and action occurs in an 1837 letter in which Marx explained to his father why he wanted to forsake the practice of law for the study of legal theory. Here, Marx, impelled by the mid-nineteenth century metaphor of the organic, made a distinction similar to that which we have shown Newman made in 1845 – between an abstract geometrical algorithm that was incapable of development and the 'living world of ideas', whose definitive being in time carried the capacity to develop:

> A triangle gives the mathematician scope for construction and proof, it remains a mere abstract conception in space and does not develop into anything further. . . . On the other hand, in the concrete expression of a living world of ideas, as exemplified by law, the state, nature, and philosophy as a whole, the object must be studied in its development.

Arguing that this 'object' could not be arbitrarily divided because the 'rational character of the object itself must develop as something imbued with contradictions in itself and find unity in itself', Marx wrote that he wished to examine 'the development of ideas in positive Roman law, as if positive law in its conceptual development (I do not mean in its purely finite provisions) could ever be something different from the formation of the concept of law'.[2]

In 1843, bridling against press censorship, Marx restated the matter in a different context noting that,

> Since legal development is not possible without development of the laws, and since development of the laws is impossible without criticism of them . . . it follows that a loyal participation of the press in the development of the state is impossible if it is not permitted to arouse dissatisfaction with the existing legal conditions.[3]

This principle – that conceptual development, being a part of the development of the whole object of the law and the state, could only proceed through the movement of critique – was to inform Marx's analysis and critique of political economy. When his object of analysis became capital and in generating his critique of capitalism, particularly after his immersion in Hegel, Marx came to the idea that the true or full development of humanity hinged on the intent of the human subject to develop, not within the restricted development of labour power for capital, but through the expanded freedom of activity in general. This, for Marx, was the full development of freedom.

HEGEL'S PRINCIPLE OF DEVELOPMENT

As far as Marx was concerned, Hegel's 'principle of development' rested upon the difference between the development of *mind* and of *nature*. This difference rested both upon a categorical distinction between time and space and upon the general implication of what was understood by him to be the internal and external determinants of development. Despite Marx's inversion of Hegel's relation between mind and nature to explain the unnatural basis of capital, he remained indebted to this Hegelian principle. Thus it is necessary to explore this principle here. Later, in the course of Chapter 5, we shall return to Hegel when we look at how British Hegelian Idealism attempted to resolve the problem of intent in development.

For Hegel, all organic entities, including those of the natural world, were capable of development in so far as they contained an 'inner determination'. Since determination referred to what an organic entity or being ought to be 'in-itself', a being developed because it contained a potential which had yet to be realised. Natural entities, however, developed without change because their inner principle of development remained unchanged. Thus, when a seed became a plant, it developed only because the potential of the different parts of the plant were in the essence of the seed. An individual organism, as Hegel put it when he gave a final summary of his development principle in 1830, made 'itself what it already was potentially': 'the development of natural organisms takes place in an immediate, unopposed and unhindered fashion, for nothing can intrude between the concept and its realisation, between the inherently determined nature of the germ and the actual existence which corresponds to it'.[4]

Natural development was 'a harmless and peaceful process of growth' since natural being could not develop through and in time. Change was confined to the continuous repetition of phases, of expansion, maturation and decay, within the cyclical process of *physis*. 'In nature', Hegel wrote,

> the life which arises from death is itself only another instance of particular life; and if the species is taken as the substantial element behind this change, the destruction of particular things will appear as a relapse on the part of the species into particularity. Consequently, the survival of the species consists purely in a uniform repetition of one and the same mode of existence.

Without any change in the essential principle of being, there was no true development for Hegel because nature could not 'comprehend itself'. Natural being, which 'retain[ed] its identity and remain[ed] self-contained in its expression', always remained as being in-itself and could not develop into being for-itself.[5] True development, the process by which being in-itself became that of for-itself, could only be realised by mind or what Hegel called *spirit*. What distinguished 'the world of the spirit' from that of nature was the process whereby 'inner determination' was 'translated into reality as mediated by consciousness and will'.[6] Scott Meikle has recently put the matter thus:

> The basis of the difference between the two processes of development lies in the different character of the nature undergoing formal change. With natural species the process is one of the reproduction over time of the species; the preservation of an identical generic nature by its transmission through successive numerically distinct individuals, each of which equally embodies that single nature. The historical process, however, does not *preserve* a nature through successive generations; it *develops* a nature through successive forms.[7]

It is in the process of, as Meikle would have it, historical development that the internal/external or interior/exterior relations of development were, for Hegel, fundamental. Hegel acknowledged that one of Kant's great contributions to philosophy was to distinguish between 'relative or *external*, and *internal* purposiveness; in the latter he has opened up the Notion of life, the Idea'.[8] Thus, Hegel construed the Idea to be the Absolute idea of freedom. This was the essence of development, the end of development towards which logic in both thought and reality moved in history.

History, Hegel repeated time and again,[9] 'takes its source' from development; history is the development of the consciousness of the spirit which naturally inhered in the human being. The essence of consciousness was freedom and it was in being conscious of development that the 'internal determination' of purposive freedom was realised through the external or immanent process of development. However, since he had praised Kant for providing the means to eliminate the received dualism between an idea of a given 'reality' external to thought and the internal 'knowing' of the mind, Hegel was highly conscious of the danger of reproducing the real–thought dualism through the binary opposition of the internal–external determinations of development.[10]

Freedom, in being the absolute Idea for Hegel, was 'nothing more than a knowledge and affirmation of universal and substantial objects as law and justice, and the production of a reality which corresponds to them – i.e. the state'. The state was both the end and bearer of development: 'Every state is an *end* in itself, – *external self-preservation*; its *internal development* and *evolution* follow a necessary *progression* whereby the rational, i.e. *justice* and the *consolidation of freedom*, gradually emerges'.[11]

We note in passing that Hegel's logical state, the external embodiment of the Idea of freedom, the end of development and, therefore, the source of history, was realised for him in the absolutist Prussian state. It was just this absolutism which exercised the censorship that gave rise to Marx's appreciation that both the external and internal of development were not only limited in the form of the state but fundamentally misplaced. Marx's 'inversion' of Hegel arose from his claim that Hegel's logic was tantamount to an inversion of the external and the internal in the sense that the logic of the state, as the being for-itself of free citizens, was nothing more than a being in-itself, an objective force of *internal self-preservation* acting externally to throttle the progressive development of freedom. As such, the state could not be for Marx the bearer of development. Rather, the burden of development had to fall upon a proletariat who produced material reality. Moreover, the end of development, being for-itself in free activity, could only arise out of the dissolution of this same class. This was so because the proletariat's only forms of freedom – freedom from property and freedom to sell its labour power – were the very conditions which both created its class being in-itself and restricted its development to the external end of production. In grappling with what confined development to the end of production, and a limitation which was associated with Hegel,[12] Marx was compelled to take Hegel's 'development' back to its roots.

Echoing Adam Ferguson, Hegel maintained that historical change of development was necessarily slow, gradual and 'a hard and obstinate struggle' of humanity 'with itself'.[13] Equally, he rejected the basic implication of belief in progress, that the change of development could be 'a peaceful process of growth' towards 'a better and more perfect condition' since the goal or end of progress could not easily be defined other than in quantitative terms:

we are offered no criterion whereby change can be measured, nor any means of assessing how far the present state of affairs is in keeping with right and the universal substance. We have no principle which can help us exclude irrelevant factors, and no goal or definite end is in sight; and the only definite property which remains is that of change in general.[14]

Without any *principle* of development, progress was 'invariably' understood to be expressed 'in quantitative terms' which were 'devoid of intellectual content'. As such, historical change, in general, might as well be assimilated to the cyclicality of the natural world and the concept of *natural* development. In the natural world, Hegel commented, there was 'nothing new under the sun and in this respect its manifold play of forms' produced 'an effect of boredom'.

Hegel had pointed earlier, in the preface to his *Phenomenology*, to 'an expanded science' of the growth of material knowledge, which he later referred to as the world of nature:

But a closer inspection shows that this expansion has not come about through one and the same principle having spontaneously assumed different shapes, but rather through the shapeless repetition of one and the same formula, only externally applied to diverse materials, thereby obtaining merely a boring show of diversity. The Idea, which of course is true enough on its own account, remains in effect always in its primitive condition, if its development involves nothing more than this sort of repetition of the same formula.[15]

The now commonplace distinction between economic growth and development owes much to Hegel's idea of development, but the original import of this distinction has been lost. True development for Hegel as much as for Marx meant that development, however much it encompassed the quantitative change of growth, had to 'be presented and recognised as variations in quality'.[16] Given that development had purpose, the problem of development was to understand what the variations in quality meant when change was more than simply quantitative. Interwoven within the problem of development, the bridge between variation and the principle of development, was the contradictory question of order in relation to development.

If, as within the Comtean motif, progress consisted in the development of order or 'social stability', then progress, for Hegel, was the development of consciousness. 'Since progress consists in a development of consciousness,' Hegel wrote, 'it is not just a quantitative process but a sequence of changing relationships towards the underlying essence'.[17] While he also understood 'stability' to be a value of progress, arguing that 'stability' was 'a value which must certainly be accorded the highest respect, and all activity ought to contribute to its preservation', Hegel steadfastly maintained that the 'underlying essence' of development conveyed the idea of freedom. Freedom was the goal and end of development, the purpose of which unfolded in the course of time as expressed by the 'activity of the spirit'. Stability was inconsistent with this unfolding.

From the standpoint of the unfolding of potential, Hegel argued, stability did not 'appear as the highest value; on the contrary the highest value' was 'that of change itself'. Doctrine which reduced the internal determination of development to the external development of natural reality and then asserted the purpose of development to be that of constructing perfect order, confined change to the tedium of recurrence of a stable condition. Thus, Hegel could conclude that 'the idea of progress' was 'unsatisfactory simply because it is usually formulated in such a way as to suggest that man is perfectible'.[18] Progress of the positivist kind could not create 'real change' but only a one-sided development, either because the knowing of mind was reduced to real matter or because the real was apprehended as if it conformed to an inner determination of will.

By contrast, the genuine purposeful change of development was both gradual and a 'hard and obstinate struggle' precisely because both the real and knowing were divided in themselves or 'doubled' as the human potential for freedom unfolded and culminated in the form of the state, in which the intent to develop was not limited by external reality and thereby was made fully conscious.[19] The doubling or two-sided unfolding of development followed from Hegel's *logic* to denote the logical movement of the 'pure concept' or 'abstract Idea' where 'Nature and Spirit constitute the reality of the idea, the former as an *external* existence, the latter as *self-knowing*'.[20] Development, for Hegel, was 'progress of the mind' in which the external existence of nature became an internal process of development while mind developed through its subdivision in three stages.

The 'mind subjective', of the first stage of development, possessed the potential for the ideal of freedom (being 'self-contained and free') but had yet to open up the substantial world of nature through knowledge. In the second stage, the 'mind objective', mind opened up the force of nature by externalising itself. However, the development of mind remained limited because 'freedom present[ed] itself under the shape of necessity'. It was in the third stage of development, the 'mind absolute', that the subjective mind 'of ideality and concept' and the objective mind which brought the material world into consciousness, were united.[21] Mind now realised the ideal of its own progress in the absolute knowledge, which only now produced stable and perfect order.

Hegel attempted to make the three stages of development of the mind correspond to logical concepts of history. The development of the mind, from the concept of the family to civil society and then to the state was 'the return of the essence into itself from its externalisation'.[22] What did Hegel mean by 'externalisation' and the return from it? Mind, unlike nature, was always mediated by consciousness and will which were 'immersed at first in their immediate natural life'. The concept of family represented the first, immediate stage of development in which potential for the idea of freedom was infinite but where its 'absolute substance' was implicit because the form of ethical life was the piety of familial love and natural emotion.[23]

To develop meant the self-knowing mind would make the idea of ethical life explicit. In being free in-itself from trust, obedience and subordination to the particular patriarch – the only 'one' who was free – the potential for 'some' to be free unfolded through the development of consciousness and will. Potential unfolded doubly because while it was 'the primary object and aim' of consciousness and will 'to follow their natural determination', that of 'the will of the spirit' was 'to fulfil its own concept', which was freedom, the true end of development.[24] The struggle of self-knowing was that of modifying and changing the concept, the internal determination of development. Nevertheless, being continued to exist externally as part of immediate natural life, as the purpose of recurrent change was to maintain the identity of the particular self. This was why the spirit was divided against itself and why the process of development necessarily contained imperfection.

Until its culmination in the state, real change could only happen through abstraction or alienation. The self had to be alienated from immediate natural life and find existence externally in a non-natural world of free particulars. From childhood, the first stage of development, followed youth, the second stage, during which the struggle against dependence upon the family took a partial and imperfect form. Hegel referred to this second stage as a 'dark and impenetrable intermediate zone' of development where neither nature nor spirit were open or transparent.[25] In the *Philosophy of Right*, this hazy zone of development was represented by civil society.

In civil society, individual wills were pitted against one another as if 'the other' was presupposed to be part of the immediate life of nature. Each individual stood as a natural object in relation to the other, a being in-itself which could not develop for itself. The spirit (and here Hegel was concerned with the nation-being as an individual subject) knew 'how to bring the unreflected – i.e. the merely factual – to the point of reflecting upon itself': 'It thereby becomes conscious to some degree of the limitation of such determinate things as belief, trust and custom, so that consciousness now has reasons for renouncing the latter and the laws which they impose.'[26] What matters here is the phrase 'to some degree' by which belief, trust and custom become limited by the progress of conscience.

Presupposition of one concept, which contains 'trust and custom' of 'family', according to the real positing of another, was a necessary delusion of self-knowing. It was the means by which the 'old' concept was retained within the new during the course of development. If consciousness or will was driven to 'fulfil its own concept', then 'at the same time it obscure[d] its own vision of the concept, and [wa]s proud and full of satisfaction in this state of self-alienation'.[27] For Hegel, each concept was encapsulated within another to unfold the potential of the idea of freedom in an iterative progressive spiral. As it was internalised, the external essence of one concept became the inner determination of another, creating a new concept with its own external limit of essence.

During the third stage of development, quaintly called 'manhood', the subject was able to posit an inner end for itself, to enable the self to find the 'recognition' of self-esteem through its work while coming to know, and thus bringing the external essence of the same being for-itself of the other. Consciousness here was made general out of the alienation of the particular 'self', with its own desires. However, the universal had only a form; it had yet to find an external essence.

Universal freedom, and its external embodiment in the state, could be envisaged in the final stage of development. Clouded by the subservience of self-alienation, the concept of the state struggled amidst the reality of civil society to find expression in the idea of the free individuality of all, the idea that 'man as such is free' could be truly expressed. Working on the presupposition that individuality could only find expression in the obligation of service to the state, Hegel made what might be called an anticipatory side-swipe at positivist doctrine by commenting that 'the personality of the individual and service towards the universal stand in opposition'.[28] Universal freedom, as a concept, was to be realised only when the opposition between the interior drive for freedom and the external world, which mind had now fully developed out of nature, came to an end. As 'the divine spirit has come into the world and taken up its abode in the individual',[29] the 'objective' spirit of the external world and the 'subjective' spirit of consciousness were reconciled. The internal and external became a part of each other and development had no further purpose.

Natural life was also doubled. Through the development of concept, natural life found a new, different external form of existence such that nature was doubled in a way that spatial variation entered into that of time. Spatial qualities, such as those of territory or place, were now determined abstractly as part of the abstract qualities of development, and immersed, so to say, in the internality of development. Again, however, the particular qualities of place were not obliterated by the abstraction of development but made more real, more complex, by the 'internal reflection' of consciousness. What reflection did, in opposition to nature, was to make nature more complex so that it became a means for freedom – the purpose of development.

HEGEL'S FLAW OF DEVELOPMENT

Hegel, to repeat, founded development upon a 'relationship' between nature and consciousness: 'all development involves a reflection of the spirit within itself in opposition to nature, or an inward particularisation of the spirit as against its immediate existence, i.e. the natural world'.[30] To be able to reflect upon nature, the original basis of 'liberation' from the force of nature and the course of natural development, human beings had to become relatively distant from nature. 'But', Hegel reflected, 'where nature is too powerful, his liberation becomes more difficult'. Where was nature so powerful that its solid force could not be opened up by progress of the mind? Hegel's answer was that this realm lay in the zones of extreme climatic conditions where:

> The power of the elements is too great for man to escape from his struggle with them, or become strong enough to assert his spiritual freedom against the power of nature. The frost which grips the inhabitants of Lappland and the fiery heat of Africa are forces of too powerful a nature for man to resist, or for the spirit to achieve free movement and to reach that degree of richness which is the precondition and source of a fully developed mastery of reality.

Hegel continued:

> In regions such as these, dire necessity can never be escaped or overcome; man is continually forced to direct his attention to nature. Man uses nature for his own ends; but where nature is too powerful, it does not allow itself to be used as a means. The torrid and frigid zones, as such, are not the theatre on which world history is enacted.[31]

A decade later Friedrich List was to reach a similar but more specific conclusion regarding the development of productive force in manufacturing industry. When Hegel claimed that Africa had no history, he meant that no inner determination of development could be found in Africa and, as List was to do far more self-consciously, he opened up an inextricable link between dependency and the internal/external of development. Without an interior drive for development, the south would be dependent upon internal determination of the north.

When Hegel counterpoised the 'south', as he called it, to the 'north' of the temperate zone of nature, and then founded the logical genesis of development upon the capacity of the mind to fulfil development's purpose through self-knowledge of the real, he banished the objective force of nature from the course of development which it was his intention to explain. In so far as Hegel understood true development to be the potential of subjective force, that of consciousness and will of the spirit, he had presupposed that nature had done the work necessary for development to begin. In other words, the precondition of the natural, that nature was sufficiently dense or rich for material need to be met as a condition for the mind's development, was a presupposition of the kind which Hegel himself had characterised as the necessary delusion of self-alienation.

Hegel's principle of development rested upon the phenomena of nature. The development of nature had no principle. For it, only empirical explanation was possible since the phenomena in question were deemed to be quantitative. The original precondition for development could only be supposed. Since the precondition was present in the final concept, the end of development, universal freedom, could likewise only be presupposed – again, as a delusion of vision of the concept itself. As long as there was the delusion of abstraction, there was necessarily movement and fluidity. There could not be, in principle, a fixed point, purpose or end of development. This was the flaw in what came to be the Hegelian system of method and thought.

Time and again, the flaw is revealed by the analogy and typology which Hegel employed to characterise his genesis of development. For example, the model of progressive development from childhood to old age, which Hegel used to characterise stages of consciousness, is transposed, in *World History*, to that of national development. We are so used to hearing of 'the young nations' of a Third World that is not yet really independent that we forget that for Hegel, at any rate, the source of the metaphor was early nineteenth-century national consciousness in Europe and, in particular, Germany. Calling the final stage of development 'the

Germanic age', Hegel commented: 'If it were possible to compare the spirit's development to that of the individual in this case too, this age would have to be called the old age of the spirit.' Perhaps mindful of his own 'old age', Hegel stopped in his tracks:

> But it is a peculiarity of old age that it lives only in memories, in the past rather than present, so that the comparison is no longer applicable. In his negative aspects, the individual human being belongs to the elemental world and must therefore pass away. The spirit, however, returns to its concepts.[32]

Spirit was externally embodied in the nation but the internal determination to develop derived from the vigour of youth, the stage Hegel called 'the Greek World'. How, then, could the interior of development be maintained in the course of progressive development towards the logic of old age? Hegel's development encompassed destruction, decay and decomposition. But, unlike for Newman and in the twentieth century for Schumpeter, these 'negative aspects' were not a part of the principle of development.

In view of Hegel's critique of the idea of progress, it may seem surprising to find Hegel's principle of development devoid of a sense of destruction. In the *Phenomenology*, Hegel referred to 'the seriousness, the suffering, the patience, and the labour of the negative'. Such was the necessary condition of self-development for the human being implicated in the movement of being in-itself to for-itself. Hegel wrote: '*In itself*, that life is indeed one of untroubled equality and unity with itself, for which otherness and alienation, and the overcoming of alienation, are not serious matters.' But Hegel associated the negative with abstract universality through which particulars of the natural world are made general by being painfully destroyed to become 'indifferent diversity'. This was 'the movement of positing itself' by which the 'living substance' came into the 'being which is in truth *Subject*' and this subject is 'pure *simple negativity*'.[33]

Hegel's 'labour of the negative' is ambiguous, particularly in the case when labour is posited as the subject of negativity. Cohen, for example, takes labour to be the 'hard' work of self-interrogation and 'negative' because it is 'destructive', presumably of natural diversity.[34] For Hegel, there was purpose in the abstraction of work from the diverse, particular forms of activity. Chris Arthur quoted Löwith's interpretation of Hegel's meaning of work as 'not a particular economic activity, to be contrasted, say to leisure or play, but the basic way in which man produces his life, thereby giving form to the world'. Work, for Hegel, 'is neither physical nor intellectual in a particular sense, but spiritual in the absolute ontological sense'.[35] The upshot of Hegel's meaning of work, as Arthur further indicates, is that it is tantamount to the labour of love, the work of 'the intellectuals of his day, academics, specialists and artists of various kinds', work as idealised free activity.[36] A labour of love, that which abstract labour could not be, might be what captures the suffering of labour in that the hard work of self-interrogation is not motivated by the absolute need for work but the purpose of the work and, thereby, the ideal existence of mind. However, it was the suffering Hegel was most conscious of stressing, that of 'love', in the name of God, which could not be transmuted simply into the immediate world for the positive purpose of humanity. In other words, the concept of universal freedom, giving recognition to different but complex diversity, involved the penance of the negative, but through knowledge, the negative would be redeemed in the positive purpose of development.

When Hegel criticised Condillac and his apostles, he pointed to their 'principle' of progress:

Their ruling principle is that the sensible is taken (and with justice) as the *prius* or the initial basis, but that the latter phases that follow this starting point present themselves as emerging in a solely *affirmative* manner, and the negative aspect of material activity, by which this material is transmuted into mind and destroyed *as* a sensible, is misconceived and overlooked.

Hegel emphasised here that development, as progress of the mind, proceeded through destroying the external world of nature, namely 'the sensible'. He went on to make clear that as material activity destroyed, it re-created the conceptual product, the content of mental activity:

Consequently, the activity of mind, far from being restricted to a mere acceptance of a given material must, on the contrary, be called a creative activity even though the products of mind, in so far as mind is only subjective, do not as yet receive the form of immediate actuality but retain a more or less ideal existence.[37]

Mind, to repeat, was of creative activity. Its purpose, as thinking activity, could not be destructive and, therefore, the decay of the old could not be inferred from the meaning of the principle of development – the progress of the mind.

The Greek world had been destroyed; so had the 'Roman World' which, corresponding to the phase of 'manhood', followed it in the manner of Hegel's relay race of world nations. In this relay one state, in a cycle of development, passed on the spirit of development to another in a spiral of progression towards universal freedom. Hegel contrasted 'several great periods of development which have come to an end without any apparent continuation' with the 'unbroken processes of development, structures and systems of culture' and then concluded:

> The formal principle of development in general can neither assign to one product superiority over another, nor help us to comprehend the purpose which underlies the destruction of earlier periods of development; instead it must regard such happenings – or more precisely, the retrogressions they embody – as external contingencies, and it can only evaluate the gains [of past culture which have been destroyed] by indeterminate criteria (which, since development is the ultimate factor, are relative rather than absolute ends).[38]

Contingencies are the events of history. All that could be said, in principle, was that 'something' was permanent if the potential of an event not to happen did not 'encroach upon' or impede the event which had 'a positive existence for us'. But the principle of development could not explain why the potential of an event was or was not realised.

We are faced with a fundamental asymmetry in Hegel's principle of development. His scheme of doubling or internal division in the movement from one concept to another was designed to explain why the particulars of a concept were not 'lost' as a new concept developed by the encapsulating the old in itself. 'All this is the a priori structure of history', Hegel proclaimed, 'to which empirical reality must correspond.'[39] This structure, to repeat, was to provide a basis for progressive development. Unlike the linear image that the idea of progress evoked, the course of development was curvilinear or spiral-like, always impeded or arrested within its own logical structure. But the purpose of the principle was to explain the potential for improvement, the potential to overcome imperfection and to point 'forward to something which will eventually attain reality'.[40] This, indeed, was ascent, and development provided the source of the recognisable events of progress. It could not do the same for descent, the events which negate development by destroying the very structure which Hegel presupposed to be a natural condition for the purpose of development.

If there is a doctrine of development in Hegel, then it is a part of the immanent course of development and not of the idea of development. We shall return, in Chapter 5, to see how Hegel was understandably baffled by the problem of poverty in so far as he postulated that service to the state was intrinsic to his second stage of development in which trusteeship implied the alienation of individuality. As Kain has interpreted Hegel: 'This alienation produces the development of the modern state to universality, but it implies the estrangement of state power.' This estrangement meant 'that the individual [wa]s confronted by an independent and hostile power which ha[d] an objective life of its own'.[41] As such, a doctrine of development could be assimilated into Hegel's principle of development but only with the consequence of making the state fall short of its own Hegelian ideal. Hegel himself later came to defend 'the rational monarch' who expressed the thought of freedom for the individual who must defer and subordinate inner determination to the monarch's authority.

Marx had accused the young Hegelians of failing to criticise the 'innermost root' of Hegel's 'principle' when they accused the philosopher of having 'accommodated' himself to the Prussian state:

> Therefore, if a philosopher has actually accommodated himself, his students have to clarify this out of his *inner essential consciousness*, that had *for him, himself*, the form of an *exoteric consciousness*. In this way, that which appeared as progress in conscience is likewise a progress in knowledge.

And, Marx continued: 'The particular conscience of the philosopher is not placed under suspicion, but rather his essential form of consciousness is reconstructed, raised into determinate shape, and thereby, is at the same time gone beyond.'[42]

In attempting to reconstruct Hegel's principle, Marx was scathing about the way in which a doctrine of development was thus grafted on to the core of Hegel's thought.

TRUSTEESHIP

During their early period of collaboration, Marx and Engels did not refer to socialist trusteeship by name. However, much of their writing of the mid-1840s was concerned with showing how trusteeship was implanted in Germany through a subversion of Hegel's principle of development in which a one-sided regard for spirit resulted in a disregard for the substance, body or 'mass' of nature. Marx and Engels commenced *The Holy Family* by arguing that,

> *Real humanism* has no more dangerous enemy in Germany than *spiritualism* or *speculative idealism*, which substitutes '*self-consciousness*' or the '*spirit*' for the *real individual man* and with the evangelist teaches: 'It is the spirit that quickeneth; the flesh profiteth nothing'.[43]

This 1845 polemic, subtitled 'against Bruno Bauer and company', asserted Hegel's unity of mind and body in the true 'reality' of the person by scorning the Young Hegelian creed of 'critical criticism' which, in its left-radical disposition, sought to recompose the basis of progress in Hegel upon the primacy of consciousness of mind. Bauer and company had interred sensuous matter in the body of the manual person. Likewise, they had embodied the consciousness and will of spirit in the person of the intellect whose mind was conscious because it had developed and whose self-conscious burden was to direct, because it determined, the body of the mass.

How did this separation of mind and body, with its replication of the internal and external of development in the mental and manual, come about? According to Marx, the Young Hegelian critique of Hegel, and therefore 'critical criticism', was based upon an inversion of the relationship between matter and self-consciousness. In Hegel, the absolute mind, or spirit, united the elements of Spinoza's 'substance' and Fichte's 'self-consciousness' by the 'antagonistic' way each 'was *falsified* by the other' in the course of development. Both mind and nature, as we have attempted to explain, were doubled by the principle of development.

Bauer and company, Marx pointed out, had *separated* substance from self-consciousness by making one become the mirror-like reverse image of the other, thus transposing the internal and external of development. Bauer, for instance, took self-consciousness to be self-expansion of the mind as if such expansion was the internal process of development. Since Hegel had taken trouble to explain why 'expansion' had pertained to natural development, Bauer had gutted what was distinctive about the principle of development. David Strauss, a Young Hegelian antagonist of Bauer, did the same to Hegel but in the reverse direction by treating substance as if it were the active interior of development. Both Bauer and Strauss, wrote Marx and Engels, 'carried each of these elements' – substance and self-consciousness – 'to its *one-sided* and hence consistent development':

> Both of them therefore go *beyond* Hegel in their criticism, but both also remain *within* his speculation and each represents *one* side of his system. *Feuerbach*, who completed and criticised *Hegel from Hegel's point of view* by resolving the metaphysical *Absolute* Spirit into '*real man on the basis of nature*', was the first to complete the *criticism of religion* by sketching in a grand and masterly manner the *basic features* of the *criticism of Hegel's speculation* and hence of all metaphysics.[44]

Marx's intention was to set out what the development of 'real man' might mean by engaging in a critique of idealism from the standpoint of Hegel's own principle of development. As early as 1845, Marx had arrived at the distinction between an immanent, unconscious process of development and the intention to develop. This intention to develop, however, was not that of the doctrine of development associated with positivist 'science' or Hegelian metaphysics but rather something quite different.

Marx alleged that the Young Hegelians had committed the tautology of defining mass as non-spirit in opposition to spirit by separating the mass of material matter from absolute spirit. Mass was 'spiritlessness', 'indolence', 'superficiality' or 'self-complacency'. Spirit, inversely, was deemed synonymous with, identical to, and axiomatic of activity, energy, essentiality and perpetual restlessness or the reverse of the inertia and passivity of material matter.[45] It was circular logic to define mass by spirit when spirit was the ideal of what mass was not. Bauer and company had formally followed Hegel in his commitment to the idea of development being progress of the mind. However, since progress was unequivocally progress of the mind whose movement was embedded in the identity between self-consciousness and the perpetual motion of mental energy and activity, the Young Hegelian idea of progress was equated with that of the spirit embodied in people with developed minds in opposition to an adverse 'mass' of people who personified the indolent and self-complacent qualities of material matter.

Marx ironically commented of Progress:

In spite of the pretensions of '*Progress*', continual *retrogressions* and *circular movements* occur. Far from suspecting that the category '*Progress*' is completely flat and abstract, Absolute Criticism is so profound as to recognise '*Progress*' as being absolute, so as to explain retrogression by assuming a '*personal adversary*' of Progress, *the Mass*.[46]

A neo-Hegelian reformulation retrogressed the idea of progress to its original liberal conception and drove 'communist and socialist writers' to associate it with '*progress against the mass of mankind*, driving it into an ever more *dehumanised* situation'. Marx praised the insights of 'utopian' socialists such as Fourier and Owen for recognising that the 'abstract *phrase*' of progress revealed 'a fundamental flaw in the civilized world'. Although he did not here refer to Fourier's sense of the 'repugnant' and 'repellent' work which accompanied progress, Marx did contrast the positive intention of development with that 'development' which had happened in the name of progress.

'This communist criticism', Marx commented first, 'had practically at once as its counterpart the movement of the *great mass*, in opposition to which history had been developing so far'. He added, second, that one 'must know the studiousness, the craving for knowledge, the moral energy and the increasing urge for development of the French and English workers to be able to form an idea of the *human* nobility of this movement'.[47] In other words, Marx imbued what was for the Young Hegelians the indolent mass of labour with the restless energy by which the Hegelian inheritance had characterised spirit.

Mass had mind because it had become part of the inner determination of development. Moreover, mass had the potential to be conscious because the essential, absolute unity of human being was that of mind and body, whose capacities for knowing material need and making material need meet the human purpose of freedom could be developed by the mass itself. It was in this sense that Marx put Hegel back upon his feet. As we shall see, Marx's fundamental objection to Hegel was not the contrariety between Hegel's idealism and an alternative materialism, but rather that Hegel had alienated the source of active agency of the intent to develop from the purpose of true development. For Marx, the Young Hegelian antithesis between spirit and matter was a caricature because it misread the motive force of development in Hegel's *Phenomenology*. Hegel had written there, as we have seen, of 'the doubling which sets up opposition' between mind and matter as 'thought does unite itself with the being of substance', and as the abstract subject as 'living substance' or 'pure simple negativity' has 'power to move' and thereby become being for-itself because 'unmoved' natural phenomena have been negated.[48] If Bauer and company had evaded the message of the *Phenomenology*, it was because, Marx pointed out, they had avoided what Hegel called 'abstract universality', the presupposition that made the development of being in-itself to for-itself possible.

What the Hegelians had 'learnt from Hegel's *Phenomenology*' was '*the* art of converting *real objective* chains that exist *outside* me into *merely ideal*, merely *subjective* chains, existing merely *in me* and thus of converting all *external* sensuously perceptible struggles into pure struggles of thought'.[49] Hegel, Marx suggested, was responsible for this Young Hegelian evasion because he had been 'half-hearted' in his claim for the determination of development by and through the 'absolute spirit'. Spirit makes 'history only *in appearance*', wrote Marx of Hegel, but 'becomes *conscious* of itself as the creative World Absolute Spirit' and thus resides in the '*actual philosophical individual*' only after the events in history.[50] The philosopher speculated through opinion and imagined a course of history but could not determine development.

In reducing Hegel's movement of mind to that of pure thought, embodied in the subject of the person, Bauer and company thought that development could be determined by thought in opposition to 'the spiritless mass' by the progress of mind. Hegel's squeamishness could now be overcome:

> No longer, like the Hegelian Spirit, does he make history *post festum* and in imagination. [Bauer] *consciously* plays the part of the *World Spirit* in opposition to the mass of the rest of mankind; he enters into a contemporary *dramatic* relation with that mass; he invents and executes history with a purpose and after mature reflection.[51]

A '*handful* of chosen men' could now assume the mantle of the 'active spirit' to become the inner determination of development. Trusteeship was thus given its justification. There was, however, a further twist. In the early 1840s, a doctrine of 'true socialism' appeared as an import from France and Britain when Saint-Simonian and other doctrine was bedded down in Hegel's thought. Marx was equally scathing about this attempt to make constructivist critiques of 'free competition' conform to an ahistorical, and therefore non-developmental, conception of human essence.

It was in the latter pages of the *German Ideology*, written in 1845–6, that Marx and Engels undertook a critique of German or 'true socialism'. Karl Grun and Georg Kuhlmann – a police informer for the Austrian state who agitated among German artisans – were the main targets of Marx's polemic. Significantly, true socialism appealed, as did Proudhonism in France, to self-employed artisans and small farmers as a programmatic reaction to industrial unemployment. Broadly, the economic ideal of true socialism was the search for a unity of production and consumption upon the basis of the philosophical unity between life and the enjoyment of 'happiness'. For Marx, this philosophical idea illustrated again the way in which Hegel's 'development' had been subverted by an evasion of the necessary 'abstract universality' which presupposed the potential of true development.

A cornerstone of true socialism, as Marx recounted from Grun, was Saint-Simon's deathbed pronouncement that 'all men must be assured the freest development of their natural capacities'.[52] This injunction was spliced by the true socialists into Hegel's philosophy in such a way that the attempt to find conscious purpose in development eradicated the perpetual distinction which Hegel had made between natural development and the principle of development. Yet, Hegel's conception of the internal/external remained the framework for true socialist development. Grun, in the following passage quoted by Marx, started with particular phenomena:

> Every one of these phenomena, every *individual life*, exists and develops only through its *antithesis*, its *struggle* with the external world, and is based upon its *interaction* with the *totality of life*, with which it is in turn by its nature linked in a whole, *the organic unity of the universe*.[53]

Grun's 'antithesis' flattened out what Hegel had meant by antithesis, namely that the external world was abstracted by the mind, which both posited it as a concept, thereby providing the condition for development, while presupposing the external as if it was separate from the internal of development. There was no idea in true socialism of self-alienation. Hegel's murky zone of development in which the ideal 'individual' subject was both of, and apart from, its 'totality', was banished from true socialism.

For Grun, 'Interaction with the totality of life' was the setting of development. 'By reason of my nature, he wrote, 'it is only in and through community with other men that I can develop, achieve self-conscious enjoyment of my life and attain happiness.'[54]. When Grun also pleaded that 'the social idea did not fall from heaven, it is organic, i.e. it arose by a process of gradual development',[55] he fell into the obvious tautology of development. The 'social idea' or 'community with other men' was both the result of a process of development and the precondition for the intent to develop. No recourse could be made by the Young Hegelians to the Saint-Simonian 'universal love of humanity' in founding the ideal of community upon the 'love for other'. Comte had asserted love to be the means of development, the very altruism which Hegel in the *Phenomenology* as elsewhere had proclaimed to be implicit, and thus not developed, until spirit was embodied in the beneficent community of the state.[56]

Marx, in the *Holy Family*, recounted the Young Hegelian intent to expunge love from the course of development. Of Bruno Bauer's brother, Edgar, Marx commented ironically that, 'The passion of love is incapable of having an interest in *internal* development because it cannot be construed *a priori*, because its development is a real one which takes place in the world of senses and between real individuals.'[57] True socialism, Marx insisted, was tantamount to unreal humanism. Socialists such as Grun presupposed a natural human essence for liberation from an external world, arguing that there was a natural, unalienated disposition for activity of 'true life' to be made free from 'dependence' upon the external.[58] On behalf of human life, Grun demanded 'from society that it should afford me the *possibility* of winning from it my satisfaction, my happiness, that it should provide a battlefield for my bellicose spirit'. And, to drive home the extent to which he had lost Hegel's principle of development, Grun continued:

> Just as the individual plant demands soil, warmth and sun, air and rain for its growth, so that it bears leaves, blossoms and fruit, man too *desires* to find in society the *conditions* for the all-round development and satisfaction of all his needs, inclinations and capacities.

It was in society that 'the opposition of individual life and life in general' became 'the condition of conscious human development'. Society, like nature, was of the external world; unlike nature, society was the conscious external world against which the individual had to struggle for natural capacities for happiness and enjoyment. Grun's conclusion was that society '*must* offer' the possibility of happiness but how the individual 'will use that chance, what he will make of himself, of his life, depends upon him, upon his individuality. I can alone determine my happiness.'[59]

Marx commented that Saint-Simon's 'free development of the capacities' was 'correctly expressed' but the logic by which it was to be achieved was 'absurd'. Individuals in society wanted to preserve their individuality 'while they demand of society a transformation which can only proceed from a transformation *of themselves*'.[60] In Marx's view different stages of development were here collapsed into an eternal essence of development. Historical development was made to consist of the metaphysical abstractions of 'individuality' and 'universality of society', which effectively eclipsed the potential for development because the interior of free development did not progress while the external of development was conditioned by the natural essence of human desires. As such, and according to Hegel's logic, there was no principle of development at work in what was claimed to be the Hegelian construction of development.

MARX'S INTERNAL AND EXTERNAL OF DEVELOPMENT

Marx's 1845 critique of the Hegelian inheritance of development, including his polemic against true socialism, was founded upon a view of development which has been given the generic name of historical materialism. This critique incorporated the outlines of what Cohen has set out as Marx's development thesis. By 1857–8, when he compiled the notebooks for the *Grundrisse*, Marx had attempted 'self-consciously to change, and to apply the method' of Hegel's logic.[61] As Nicolaus has astutely observed, while Marx changed the fundamental premise of Hegel's developmental principle, he did not abandon the method Hegel had used to make the principle move through a course of development.

Hegel's premise was that the principle moved from nature, the original external world of natural being without the consciousness of mind. Marx, in his reconstruction of Hegel, made the principle pivot upon an essential unity of the natural essence of the human body and mind as embedded in the head of that human body. Hegel's illusion was that he conceived of 'the real as the product of thought concentrating itself, probing its own depths, and unfolding itself out of itself'[62] as if the mind could both find purpose in development and realise the realities of historical development. If this Hegelian mysticism had found its way into true socialism, Marx concluded, it was because of the way in which human instinct and energy of self-consciousness was imparted to nature and then reflected back upon mind:

[I]n order to pay nature back for finding *its* self-consciousness in man, man seeks his, in turn, in nature – a procedure which enables him, of course, to find nothing in nature except what he imputed to it . . .

He has now arrived safely at the point from which he originally started, and this way of turning around on one's heel is now called in Germany – *development*.[63]

Yet, if Marx was no 'vulgar idealist', he was no 'crude materialist' either. David Ricardo, the British economist, best exemplified what Marx meant by crude materialism. In his treatment of fixed capital and in particular of the proper basis upon which the distinction between means of consumption or production should be made, Marx argued that 'social relations of production among people' came to be treated by Ricardo as '*natural properties* of things'. Second, no matter how much Ricardo voiced an appreciation that 'the development of the productive forces of social labour is capital's historic mission and justification', Ricardo came to be preoccupied by the ground rent of land and supposed natural barriers against the interior determination of development.[64] Last, but not least, Ricardo reduced the productive force of labour to a stream of physical effort in production. Marx's well-worn aphorism, 'that what distinguishes the worst architect from the best bee is that the architect builds the cell in his mind before he constructs it in wax', might well have been directed at Ricardo who failed to understand that the productive force of mind of the person in production was embedded in the bodily capacity to labour.[65]

In the course of his polemic against true socialism in the *Holy Family*, Marx made two objections to the way in which Grun supposed that consumption and production, or need and capacity, could be unified. First, Grun had argued, on the basis of a supposedly natural desire to consume, that production took place in the act of consuming. What Grun forgot, Marx reminded, was that 'demand must be *effective*' and that to be effective, the consumer 'must offer an equivalent for the product desired, if his demand is to cause fresh production'. Grun had omitted 'the connecting link, the cash payment'. Second, although Grun had understood that in order to produce, '*material*' must be consumed in production, he 'forgot', Marx also reminded,

that the bread which is produced today by steam mills, was produced earlier by wind-mills and water-mills and earlier still by hand-mills; he forgets that these different methods of production are quite independent of the actual eating of bread and that we are faced, therefore, with an historical development of the productive process.[66]

These two points about exchange and the means of production reappeared in Volume 1 of Marx's *Capital* in 1867 as the *two* elements of the 'production process'. The capitalist production process, Marx repeated, consisted of both valorization through exchange and the labour process. By the time he had completed the first volume of *Capital*, Marx had established his system of exchange based on self-expanding value or valorization, had worked out the fundamental distinction between labour and labour power, and had explicated the labour process as the conjunction of labour power and the means of production. The production process could not be reduced to the labour process of capitalism. Labour power was brought to the means of production on the basis of valorization while the purpose of production, to repeat, was to serve self-expanding value as an end in itself.

This distinction between valorization and the labour process was the culmination of Marx's use of the Hegelian principle of development. Labour power, the capacity to produce and create, developed through its use in work. Hegel's 'rational kernel' appears time and again in *Capital*. There was, for Marx, a general process of labour, independent 'of any specific social formation'. Labour, the 'process between man and nature', was the movement in which 'materials of nature' were confronted 'as a force of nature'. Human being, by acting upon 'external nature' and changing it, simultaneously changes its own nature. Human being 'develops the potentialities slumbering within nature', Marx wrote, and makes 'the play of its forces' subject to human 'sovereign power'.[67] While acknowledging the Hegelian principle, but altering its premise, Marx went on to write that:

> We are not dealing here with those first instinctive forms of labour which remain on the animal level. An immense interval of time separates the state of things in which a man brings his labour-power to market for sale as a commodity from the situation when human labour had not yet cast off its first instinctive form. We presuppose labour in a form in which it is an exclusively human characteristic.

Work involved purposive will, whose purpose was presupposed and therefore developed by mind in the body of labour as it acted upon 'external nature':

> At the end of every labour process, a result emerges which had already been conceived by the worker at the beginning, hence already existed ideally. Man not only effects a change of form in the materials of nature; he also realises his own purpose in those materials.[68]

This purpose, it is to be noted, was realised with the transformation of material production. Equally, however, this action upon the external nature of development itself acted to change the consciousness of the purpose of work which, for Marx, was the internal of development. Development was, therefore, to change the purpose of work, but such change could only be realised through the activity of production itself.

What was involved here was not the simple idea of an historical period during which necessary work was compelled to be followed by another characterised by the ideal of free time or the 'enjoyment of life' as the true socialists had maintained. In his polemic against Grun, Marx had recounted what the true socialist had written about how 'a perverse world' had 'thrust' the 'concept of *value* and *price*' between '*activity*' and '*enjoyment*' and thus torn apart the natural unity between production and consumption. Then, Marx commented sarcastically that by doing so the concept of value had torn 'man *asunder*' as well. 'Not content with this, it thereby tears society, i.e., itself, asunder, too. This tragedy happened in 1845.'[69]

The point of Marx's sarcasm, which he later put down in the more sombre tones in Chapter 7 of *Capital*, was that activity and enjoyment were a part of each other in the labour process to the extent that work was purposeful and therefore demanded attention. It was the kind of work which mattered to the worker:

> The less he is attracted by the nature of the work and the way in which it has to be accomplished, and the less, therefore, he enjoys it as the free play of his own physical and mental powers, the closer his attention is forced to be.[70]

The potential for free development or 'the free play' of 'physical and mental powers' existed in work as a general attribute of the labour process. Under a capitalist labour process, however, attention to work as purposeful activity had to be directly 'forced' by supervision over and above the activity of work, while work itself was enforced by the need for valorization. Moreover, the extent to which the potential for enlarging different kinds of work and thereby making the general 'nature of work' more attractive, was limited by the valorisation of activity.[71]

PRODUCTIVE FORCE

Much of the controversy surrounding Marx's development thesis has centred on the questions of whether or not it is a technologically determined thesis reducible to development of the instruments of production and the explanation of the degree to which productive force is developed. However, this thesis, as enunciated by Cohen, is about the labour process as a whole and includes, other than work itself, 'the object of work' and 'the instruments of that work'. Here, productive force is equated with the labour process as a whole and thus about much more than the development of productive objects and instruments. Even so, however, the thesis remains within a more restricted domain of development than that which was inferred by Marx when he recovered it from the neo-Hegelian morass of the 1840s.

Entangled in just that morass, Grun had complained that 'consumers have been uneducated, uncultured', and did not 'consume in a *human way*'. He proclaimed: 'Preach the social freedom of the consumers and you will have true equality of production.' True socialists such as Grun, Marx complained in turn, were either unbothered 'about the real living conditions and the activity of men' or did not take real conditions of life and activity to be the starting point of theory. And, it was 'precisely those economists who took consumption as their starting-point' who 'happened to be reactionary and ignored the revolutionary element in competition and large-scale industry'.[72] Development of large-scale industry was a process of development through which real conditions had to be understood.

It must be emphasised that Marx's focus on development was an early reaction to the constructivism of intentional development and, thus, development doctrine. Marx's contention that it was the capitalist process of development, what he then called the development of 'competition', which made it necessary to presume that production was the starting point for an understanding of conditions of life, was penned a year after *Economic and Political Manuscripts* of 1844. It was the same presumption which continued in the pages of *Capital* written in the 'age of progress' after 1850, and the years of the supposedly economistic 'later Marx' who has been so often juxtaposed against the early writings of the mid-1840s.

What Marx called 'competition' in the mid-1840s was later named the capitalist mode of production. Along 'with its methods, means and conditions', Marx explained, the capitalist mode of production 'arises and develops spontaneously on the basis of the formal subsumption of labour under capital. This formal subsumption is then replaced by a real subsumption.' The spontaneous development of capitalism, together with the supercession of formal by real subsumption of labour, was presupposed by 'a long process of development', by 'the natural and spontaneous product of a long and tormented historical development'. Development was immanent but in the sense of Hegel's natural development. There was no purpose in what Marx supposed the unconscious, spontaneous history of development to be. In Chapter 7 of *Capital* Marx referred to the most basic instrument of production, land.

The earth itself is an instrument of labour, but its use in this way, in agriculture, presupposes a whole series of other instruments and a comparatively high stage of development of labour-power. As soon as the labour process has undergone the slightest development, it requires specially prepared instruments.[73]

Yet, if the process of development was natural, the development of labour power, when engaged in a labour process of the 'slightest development' was the result of intention – the attention and activity, both mental and physical – which was attributed to work within the labour process. The process of development remained natural but the instruments of work were developed purposefully through the internal determination of that natural process.

The genesis of capitalist development followed the same principle but with one vital difference, namely that purpose of development was abstracted from producers in whom labour power was embodied, and was presupposed to be part of the 'natural' process of development. The purpose of development was not lost but made abstract by virtue of becoming a part of the spontaneous movement that was a part of the new concept, self-expanding value, whose ideal form, that of freedom, possessed the attributes of the movement of a natural force. Capital, which had been part of the external of development in that it established conditions for objects and instruments of work to be developed, now came to embody the internal determination of development as if it were a natural force devoid of principle and thus the perpetuation of natural development. Marx's intention was to explain this grand delusion of capitalist development, a delusion he also found in Hegel and which was his rationale for reconstructing Hegel's principle of development.

When, in the *Grundrisse*, he discussed Wakefield's theory of colonisation, Marx made the point, which he reiterated time and again, that the advent of the capital–wage labour relation in agriculture was 'initially created only by modern landed property, i.e. by landed property as a value created by capital itself'. This presupposed '*a total restructuring of the mode of production*' (and agriculture) itself:

it therefore presupposes conditions which rest on a certain development of industry, of trade, of science, in short of the forces of production. Just as, in general, production resting on capital and wage labour differs from other modes of production not merely formally, but equally presupposes a total revolution and development of material production.

He continued:

Although capital can develop itself completely as commercial capital (only not as much quantitatively), without this transformation of landed property, it cannot do so as industrial capital.

And concluded:

It must be kept in mind that the new forces of production do not develop out of *nothing*, nor drop from the sky, nor from the womb of the self-positing Idea; but from within and in antithesis to the existing development of production and the inherited, traditional relations of property.[74]

It is the explanation of why and how forces of production develop that is addressed by the question of the development thesis.

Marx offered a theory of history, according to Cohen in his original and powerful defence of historical materialism, that contained two theses. The first, that 'the nature of the production relations of a society is explained by the level of its productive forces' is the *primacy* thesis. The second, or *development*, thesis stated that there was a tendency for productive forces to develop throughout history. When he later modified the development thesis, Cohen argued for 'an autonomous tendency' for productive forces to develop and explained what he meant by this '*full* development thesis':

Since productive forces are not unmoved movers, the autonomy here assigned to their tendency to develop is not an absolute one. The tendency's autonomy is just its independence of social structure, its rootedness in fundamental material facts of human nature and the human situation.[75]

Here, to explain why primacy should be given to the development of productive forces is simultaneously to explain why productive forces develop. Social and economic structure or 'form' explains how, and at what rate, the forces are enabled to develop but the material content, 'the matter', of development is conditioned by social form. Cohen explicated what he understood Marx to mean by 'forms of development':

Humanity in social organisation thrusts itself against its environment, altering it and its own human nature, for it develops its own powers and needs in the course of the encounter. The development of the productive forces is expressed in the transformation of nature, and socio-economic structures are the forms in which this development proceeds, its 'forms of development'.[76]

In the *German Ideology*, Marx did set out the human premise 'of the materialist conception of history', referring, by way of the difference in the level of the development of productive forces and division between town and country among nations, to the 'various stages in the division of labour' which were 'just so many different forms of property'.[77] But later, in the *Grundrisse*, as if to re-emphasise Hegel, Marx cast 'social forms' in terms of Hegel's three concepts and historical stages of development.

Productive capacity developed only slightly in the first 'social forms', those of 'personal dependence'. Marx continued: 'Personal independence founded on *objective* dependence is the second great form, in which a system of general social metabolism, of universal relations, of all-round needs and universal capacities is formed for the first time.' Capitalism was to be followed by communism: 'Free individuality, based on the universal development of individuals and on their subordination of their communal, social productivity as their social wealth, is the third stage.'[78] If 'subordination' of the communal and social may not have been clarified, then it is significant that Marx started with 'free individuality' and its basis in the development of all individuals.

Cohen concurs with Marx in his insistence that the 'second stage creates the conditions for the third'. The problem arises in the way that Cohen interprets Marx to have made fundamental dichotomies between the *material*, with its synonyms of 'human', 'simple', 'nature', 'matter', and the *social*, synonymous with 'economic', 'historical', 'form'. This interpretation has been much contested, not least from the standpoint of Marx's appropriation of Hegel's principle of development, the very principle which Marx used to inveigh against true socialism. Cohen seeks to emphasise a basic dichotomy between the material and the social for several reasons.

First, Cohen suggests that the purpose of the 'explanation of social history' is to serve 'material development'.[79] More importantly, he wants to assert that while Marx was right to claim that human essence is creative, Marx was mistaken in inferring that the work of production is governed by human essence and is, therefore, part of productive force. Rather, Cohen argues, the 'philosophical anthropology' in Marx must be divorced from historical materialism. Marx's 'theory of human nature' cannot be a basis for historical materialism because, Cohen insists, 'people produce, historically, not because it belongs to their *nature* to do so, but for the almost opposite reason that it is a requirement of survival and improvement in their inclement *situation*'. For Cohen, Marx was too 'materialist' and too little concerned with identity: 'It is the need to be able to say not what I can do but who I am?' Last, but not least, Cohen maintains that Marx conflated a vision of *full* development of the individual with *free* development of society. There is no possibility, for Cohen, of *anyone* being able to develop in any direction alongside the condition that there can be a free society in which '*everyone* is free to develop in any direction'. In other words, form would eternally contain creative human expression because the natural possibility of being human, to do anything one can imagine, is limited by the force of nature itself. Human capacity to do anything is limited and the logic is that the restricted form of the social is a substitute for what human capacity cannot be – for the unhuman possibility of everyone doing everything that can be imagined to be done. As Cohen puts it, revealing his purpose, no social form is imposed upon communism which 'is not a society' but which 'does have a form', the form being '*now just the boundary created by matter itself*'.[80]

This boundary created by matter is the external of development. When Cohen, for instance, sets out the reasons why there is progression in the tendency for productive forces to develop, he refers to the 'historical situation of humanity' being one of 'scarcity'. Circumstances of scarcity necessitate work of 'repugnant labour'.[81] However, repugnant labour, for Marx, was not work itself but work that was relatively unattractive in that the 'free play of mental and physical powers' was circumscribed. For Marx, but not for Cohen – because he excludes 'relations of work' from the panoply of productive forces – the external of development limited the internal determination of development. It is important to be clear about what scarcity means and why Cohen deliberately makes it historical.

'Even if we leave aside the question of the level of development attained by social production', wrote Marx in *Capital*, 'the productivity of labour remains fettered by natural conditions.' Natural conditions included 'the nature of man himself' and 'external' conditions or 'natural wealth' of both the 'means of subsistence, i.e. a fruitful soil, waters teeming with fish' and 'instruments of labour, such as waterfalls, navigable rivers, wood, metal, coal, etc.'. Marx inverts Hegel's implication of a niggardly force of nature for human development, but he does so in connection with the development of capitalism. Capitalism 'presupposes the domination of man over nature', and Marx continued:

> Where nature is too prodigal with her gifts, she 'keeps him in hand, like a child in leading-strings'. Man's own development is not in that case a nature-imposed necessity. The mother country of capital is not the tropical region, with its luxuriant vegetation but the temperate zone.

Thus, after claiming that 'the most fertile soil' is not necessarily 'the most fitted for the growth of the capitalist mode of production', Marx goes on to point to water control and irrigation schemes in Egypt, Holland, India and Persia: 'It is the necessity of bringing a natural force under the control of society, of economizing on its energy, of appropriating or subduing it on a large scale by the work of the human hand, that plays the most decisive role in the history of industry.'[82] In this latter case, it is implied that there is a directed purpose in bringing natural force under control to develop productive force. For capitalism, however, there is no such directed purpose and yet, with the development of modern industrial capital, the development of productive force accelerates to an extent that the natural conditions of scarcity can scarcely be imagined in the old historical form. Scarcity becomes a part of the internal determination of development to make labour as a relation of capital repugnant because the productive force of human creativity is fettered by the force of capital itself.

Capital appears here as a part of Hegel's natural development, self-expanding in diverse and myriad forms of accumulation but reproducing its inner essence without change. Always being in-itself, capital cannot find a purpose for development. Yet equally for Marx, Hegel's principle is at work both in the genesis of capitalist development as a phenomenon of history and in the logic by which the productive force of labour appears to be that of capital: 'Thus both the historically developed productive forces of labour in society, and its naturally conditioned productive forces, appear as productive forces of capital into which that labour is incorporated.'[83] How, in short, does capital, the external of development, become part of the inner determination of development? Marx finds the answer in Hegel. As Uchida puts it: 'Marx explicates what Hegel has expressed only implicitly.'[84] Put differently, Marx develops Hegel's principle of development to explain the conundrum of capitalist development.

It has been shown by Uchida and others that Marx centres Hegel's presupposition of the concept upon the person. Marx supposes that the alienation of the ideal subject is involved in the person of capital. Marx embodied Hegel's ideal subject of 'living substance' in the person who possesses money capital, 'the *money* of the spirit', and who posits the 'other' to be the body of labour power, presupposed to be natural and objective in form. Then the 'value subject' unites mind and body in the production process through exchange, the relation between generally independent persons. Capital, Marx wrote, is a simple 'direct unity of product and money or, better, of production and circulation. Thus it is something *immediate*, and its development consists of positing and suspending itself as this unity.'[85] However much capital self-expands and acquires empirical complexity, the unifying principle remains simple because the mind of labour cannot find its universal expression of freedom in the production process. Mind and body, Marx claimed of Hegel, were presupposed to be separate and in Hegel's second stage of development the body of labour in physical labour is indeed divided from the mental powers of the mind. It is through abstract universality that the particular attributes of labour are made abstract with regard to the purpose of exchange.[86]

In so far as the relation between persons came to be expressed by exchange, the ambiguity of 'the social' and social form was revealed by Hegel's hazy and intermediate zone of development. This zone of civil society was none other than the bourgeois society of the nineteenth century, for Marx, because it was here that 'society' had been presupposed in a bourgeois form and it was in this form that new productive force developed: 'While in the completed bourgeois system every economic relation presupposes every other in its bourgeois economic form, and everything posited is thus also a presupposition, this is the case of every organic system.' Typically, Marx used the organic metaphor to encapsulate what was a closed system for Hegel. Marx continued:

> This organic system, itself, as a totality has its presuppositions, and its development to its totality consists precisely in subordinating all elements of society to itself, or in creating out of it the organs which it still lacks. This is historically how it becomes a totality. The process of becoming a totality forms a moment in its process, of its development.[87]

This development, however, was restricted to that of capitalism itself. The expanded domain of development is that of Aristotle's 'active reason', innate in human being and not the 'living substance' of the mind. Hegel, as it has often been pointed out, subverted Aristotle when he substituted the latter for the former to emphasise that the negative, the natural decay and death of 'merely immediate and individual vitality' is the *emergence of the spirit*, the 'true subject' which lives as substance.[88]

Although active reason, for Marx, was at the heart of the progress of productive force, productive force is often treated by Marx as if it belonged to the stages of progress that governed, as we saw in Chapter 1, the economic progression from agriculture to commerce and manufacturing. Whereas the eighteenth-century idea of progress rested stages of progression upon subsistence, Marx substituted production for subsistence when he claimed that 'progress consists in a development of production'. That Marx acknowledged the superior productive force of capitalism over past 'stages' of production cannot be denied. However, the acknowledgement that capitalism was historically progressive cannot justify the claim that the advance of productive force, in and of itself, can form the basis, in the form of the potential of capitalist production, for entry to, as Marx himself put it, the 'true realm of freedom' as opposed to the 'realm of necessity'. It is through active reason that productive force develops and active reason is a part of the 'free play of physical and mental power' in work of the labour process.[89]

Scarcity was only the first premise of Cohen's explanation of Marx's development thesis. Reason forms the basis of two further premises. Cohen argues that 'people have the intellectual and other capacities needed to discover new resources and to devise productivity-enhancing skills and tools' and 'they are rational enough to be able to seize the occasions their capacities create to make inroads against the scarcity under which they labour'.[90] Self-expanding value, the spontaneous process of valorisation, which Cohen treats separately as an economic form, makes the search for enhanced productivity compulsive, and therefore promotes the development of productive force. However, this search is limited to what is 'socially necessary' in production; in other words, it is determined by exchange value, 'the *social existence* of things'.[91]

Social existence is not simply the social and economic form which Cohen postulates as distinct from the material basis of life. Scarcity is created, by the 'social existence of things', in relation to the social labour of productive force in the course of historical development. Relative poverty, for example, expresses the social scarcity of things. As far as Marx is concerned, scarcity arises out of limits placed on human creativity in relation to the potential of 'intellectual and other capacities', which the 'social existence of things' diminishes through the productive purpose or form of capital. Knowledge and all its social attributes are, for Cohen, a part of productive force. By virtue of an internal teleology of development, it is rational for knowledge to be expanded for productive purpose. Yet this purpose, contained in the economic form of the rationale of profitability, is negated by an external teleology of development. From the vantage point of Marx's external teleology, it is sensuous activity, of the mind in the body of work which perpetually expands beyond the limits of the economic form and which has, in the name of 'development', to be contained within what is possible for labour capacity to produce.

At the end of one of the rare instances in which he deliberately wrote of his method, Marx set out in the *Grundrisse* a series of '*Notabene*'. The sixth of these noted the dialectical relationship between the concept of productive force and that of the relations of production, 'a dialectic whose boundaries are to be determined, and which does not suspend the real difference' between them.[92] While Cohen's analytical approach certainly does not suspend the difference between force and relations of production, he has been criticised by Shamsavari for not grasping the significance of what Marx meant by 'dialectic'. Shamsavari maintains that if 'productive force' is a concept of the content of matter, and 'relations of production' that of form, then the history of capitalist development is one in which form and content pass into each other. Following Hegel's principle of development, therefore, the content of productive force contains the form of property which has been developed in the historical course of capitalism. It is only in early capitalism that productive forces stand in an external relation to the new social form. Thus it is for early capitalism that Cohen's defence of Marx is appropriate. However, in later, more 'mature' periods of capitalism, productive forces come to be interior to the economic form of capitalism. Thus underdevelopment is not a historical condition in which the development of productive force has been negated by the external of development but one in which new matter has not yet been assimilated. It is the principle of development, the valorisation of the exchange process as much as development of the labour process, which causes capitalism to develop in space as well as time.[93]

This critique admitted, the force of Cohen's argument is not that small-scale property is universally incapable of generating productive force because it is small scale but because it is of property, which is governed by a social existence of things in which there is no compulsion to develop. Where there is compulsion, as under capitalism, then as Marx stated time and again, the producer is rendered free of non-capitalist conditions of production, to be propertyless, possessing only 'living labour capacity'. As such, whatever the standard of life of relative poverty, the 'free labourer' is made poor, 'is a pauper: virtual pauper',[94] in the sense of a poverty which confines the free play of mind and physical powers to that of the purpose of valorisation.

Cohen accompanies Marx in seeing the 'development of productive power' as an 'advance in the "mode of self-activity of individuals"' and, equally, affirms, with Marx, that the 'productive power of *man*' is development which happens 'at the expense of the creative capacity of *men* who are the agents and victims of that development': 'They are forced to perform repugnant labour which is a denial, not an expression, of their natures': it is not 'the free play of their physical and mental powers'.[95]

Marx's philosophical anthropology, based upon essential human creativity, could be made consistent, Cohen believes, with historical materialism. History, Cohen pronounced, is a substitute for nature. This is the same logic that made social form, for Cohen, a substitute for the limited human capacities to create what could be freely imagined on the basis of universal self-development. But when the scarcity of nature is made internal in the social form, to be a part of what makes productive force develop as a principle in itself, then the intent to develop, we may infer, is to be found in a social doctrine of development.

Indeed, Cohen quoted approvingly from another influential proponent of the analytical approach, Jon Elster. Elster has argued that, contrary to economic doctrine, there is no social need to create incentives for production because, following Marx, there is a 'natural creative urge of the individual "in whom his own realisation exists as an inner necessity"'. Elster continues:

> Special incentives are needed only under conditions of scarcity and poverty, in which the needs of the individual are twisted and his capacities developed only in a one-sided way. In the early stages of capitalism there was a great deal of scarcity and poverty. . . . Under those conditions, capitalism was the best and most progressive arrangement, even though it subordinated progress to profits.[96]

If we leave aside the way in which the social here is reduced to the 'individual', we find the import unmistakable. One-sided capacity and twisted need are precisely what the true socialists had noticed in the 1840s and what had prompted their attempts to make development doctrine work under the aegis of a socialist trusteeship. Marx would have no truck with it then. It is difficult to make sense of why he should be forced to do so now by critics such as Elster.

SURPLUS POPULATION AND DEVELOPMENT DOCTRINE

Marx followed Hegel's procedure in which a new concept encapsulated an old one, one concept in opposition to, but becoming part of, the other. In the process of development, the old was not lost but made more conscious by the new. Thus, for Marx, when industrial capital developed out of merchant capital, the rationale of merchant capital enclosed that of the production process, with capital becoming more conscious of the search for money-profit. Marx referred to the development of credit, of modern landed property and, most importantly, the development of wage-labour in like fashion. Marx used 'development' to denote consciousness in the change of the form of the wage when he referred, as in *Capital*, to the distinction between time-wages and piece-wages, writing that 'one mode of payment may well favour the development of the capitalist process of production more than another', in that young apprentices, when paid in piece-wages, overwork themselves for the benefit of their masters 'in the very period of their own development'.[97]

Another clear example of how Marx uses Hegel's principle can be found in the *Grundrisse* where, in notes on Wakefield's theory of colonisation, he dealt with capital and land: 'Capital arises out of circulation and posits labour as wage-labour; takes form in this way; and developed as a whole, it posits landed property as its condition as well as its opposite.' Marx then refers to the transformation of 'rural labourers' into 'wage-labourers' through clearing of the estates and to the 'double transition to wage-labour' to arrive at 'its double purpose':

> (1) industrial agriculture and thereby the development of the forces of production on the land; (2) wage-labour, thereby general domination of capital over the countryside; it thereby regards the existence of landed property itself as a merely transitional development, which is required as an action of capital on the old relations of landed property, and a *product of their decomposition*; but which as such – once this purpose achieved – is merely a limitation on profit, not a necessary requirement for production. It thus endeavours to dissolve landed property as private property and to transfer it to the state. This is the negative side.[98]

The negation of landed property, for Marx, is '*wage-labour*'. Wage-labour also 'demands the breaking-up of large landed property'; wage-labour 'wants to posit itself as independent' to achieve positive purpose, 'to escape wage-labour and to become an independent producer – for immediate consumption'.

Marx interpreted the real purpose of Wakefield's doctrine to be that of increasing the capitalised rent of land. Independent production, however, would thereby be made impossible. To increase the capitalisation of land, in the name of what was later regarded as positive development would be 'to make capital act as capital, and thus to make the new colony *productive*; to develop wealth in it, instead of using it, as in America, for the momentary deliverance of the wage-labourers'. In other words, the purpose was to 'transform the workers into wage-workers' instead of merely transferring a surplus population of wage-workers from Britain to Australia.[99]

In the case studies of Australia and Canada which follow in Chapter 4, we argue that the local origin of development doctrine is to be found in just such an attempt to invest landed property in the state and make wage-labour 'independent' for the purpose of its immediate consumption. Our argument is that the elaboration of this development doctrine was not independent of capital. As the case of Kenya, discussed in Chapter 6, indicates, independent production was valorised through a course of capitalist development. In none of these cases was the application of development doctrine about the negation of capital, a negation, which Marx insisted, must be the negation of wage-labour itself. Rather, development doctrine posited a form of production, whether that of household small-scale production on the land or that of artisans in manufacture, as an alternative to large-scale production. The positing of this alternative arose historically in each of these cases out of the condition of unemployment of wage-workers, the creation of a relative surplus population of potential wage-workers who could not find wage-work in production.

In both the *Grundrisse* and *Capital*, Marx castigated Malthus for having supposed '*overpopulation* as being *of the same kind* in all the different historic phases of economic development'.[100] He denigrated Malthus' law of human population as being tantamount to the demographics of 'baboon' population, and whose course was dictated by a process of natural development.[101] Marx argued that 'what may be overpopulation in one stage of social production may not be so in another'. Thus, while emigration to the colonies of the Roman Empire may have been a response to overpopulation, these emigrants were not 'paupers'. Nineteenth-century emigration from Britain was a different phenomenon: 'Only in the mode of production based on capital does pauperism appear as the result of labour itself, of the development of the productive force of labour.' Productive force of labour, according to Marx, was socially necessary labour, which, as a result of being objectified in the process of the exchange of money capital for labour power, became mere 'living labour capacity'. Labour capacity was made 'socially necessary' only if surplus labour had 'value for capital', if it could 'be realised by capital'. The predicament of the *'free labourer'* was that while the worker was free to possess the capacity to work without the need to possess means of production, the need to possess socially necessary means of subsistence was dictated by what made labour socially necessary. As such, the *'free labourer'* who was 'freed' from non-capitalist means of production, and all the social constraints which were exercised to confine the whole person to the place of work where labour was exercised for necessary subsistence, faced capital to exercise the potential but not actual productive force to produce both what is necessary for subsistence and surplus to it. Valorisation of work simultaneously enhanced the productive force of labour and made workers superfluous. This is why Marx claimed that 'invention of surplus labourers, i.e. of propertyless people who work, belongs to the period of capital'.[102]

Under capitalist conditions of production, surplus labour is not distinct, either in time or space, from the application of labour to work. Therefore, surplus labour is labour capacity which cannot find the means of necessary labour to realise its subsistence in exchange against money-capital:

> the relation of necessary and surplus labour, as it is posited by capital, turns to its opposite, so that a part of necessary labour – i.e. of the labour producing labour capacity – is superfluous, and this labour capacity itself is used as a *surplus* of the necessary working population, i.e. of the portion of the working population whose necessary labour is not superfluous but necessary for capital.[103]

Thus, a relative surplus population is that which is unemployed 'relatively' to that part which is engaged in what has been deemed to be socially necessary work only by virtue of 'the social existence of things' in exchange. A relative surplus population does not work and thus constitutes a surplus of labour capacity.

Marx noted that one implication of the various schemes for the state nationalisation of land put forward as a remedy to the growth of a relative surplus population was that in converting ground rent into 'universal state rent' or a 'state tax', 'bourgeois society reproduce[d] the medieval system in a new way, but as the latter's total negation'. The principle of development, we must remember, reproduced the old form of one concept within the matter of another. What applied to modern landed property, posited by capital, applied to labour capacity, also posited by capital. The labour capacity of a surplus population under such nationalisation schemes was to be funded 'out of the revenue of all classes'. The state tax on land was to support the necessary conditions for the unemployed to invest necessary labour in their own subsistence.

Such was the scheme, Marx claimed, in which 'society in its fractional parts' was to undertake 'for Mr. Capitalist the business of keeping his virtual instrument of labour – its wear and tear – intact as reserve for later use'.[104] Under what we now know as development schemes, work was to fall 'out of the conditions of the relation of apparent exchange and independence' and into forms of production in which the state mediates between exchange, making possible the subsistence of a surplus population. In this guise, development doctrine put together the two aspects of capitalist development, namely the negation of old social forms and surplus labour capacity, to do the work of intentional development. As such, the work of development was a far cry from 'the free play of mental and physical powers' which Marx had imagined to be the virtue of the idea of development.

Development doctrine purported to put this relative surplus population to work within the integument of the nation. It was, for List, in the name of the national, that the potential productive force of labour came to be defined. Here, a national policy of development came to assume the inner determination of development in which capital, positing itself as the internal force of development, was assimilated to be the external of development. Labour, by this reckoning, was no more than 'living labour capacity' or Hegel's 'living substance'. Yet, as Marx had argued when he inveighed against the projects of true socialism in the 1840s and as he had amplified in the *Grundrisse*, a national policy need be no more than that which was posited by capital as the means for development. Moreover, in the case of List's national system, it was a thoroughly historically redundant means to true development.

MARX AND LIST

Although never short on invective, Marx reserved some of his most vicious criticism for Friedrich List's *The National System of Political Economy* published in 1841.[105] Marx, in his 'draft of an article' on List's book, accused the latter of hypocrisy, plagiarism and the falsification of sources.[106] We will later return to the charge of hypocrisy. In support of his charges of plagiarism and falsification, Marx placed selections from Ferrier, Pecchio and Louis Say, to name only three, side by side with passages from List's book to effectively make his case against List. However, missing from this part of Marx's attack on List was another source which List had 'leaned upon' heavily. The source was the industrial programmes of the American Alexander Hamilton.

Hamilton's *Report on the Subject of Manufactures*,[107] presented to the United States House of Representatives in 1791, has come to be accepted as one of the key inspirations for List's work. List's *National System* and Hamilton's *Report* have both remarkable similarities as well as some important differences. One key difference which has been discussed, in part, by Keith Tribe, is to found in the differing use of the concept 'productive powers' by Hamilton and List.[108] Moreover, despite his vitriolic criticism of List, Marx's own idea of the development of productive force has been repeatedly conflated with the Listian doctrine of the national development of productive power. It is for these reasons that we centre the discussion which follows on this concept.

HAMILTON AND PRODUCTIVE POWERS

The concept of productive power which Hamilton used in the *Report on Manufactures* should be located within his larger political programme. It was in September 1779 that Hamilton, of *Federalist* fame, became the first Secretary of the Treasury of the United States, the then 'young' nation which had emerged from a revolution in which the American inability to manufacture the sinews of war had become painfully obvious.[109] Hamilton sought to solve the manufacturing problem within a larger programme of political centralisation that accompanied the national need to provide for the economic reorganisation of the 'new' republic. One of Hamilton's first objectives was to bring order to the chaos of the public debt. Both foreign and domestic debt, totalling $50 million and consisting of rapidly depreciating paper issued by the states, the central government and various agencies, imposed a crushing burden upon the economy of the United States and was responsible for a shortage of credit. Congress, after a good deal of political horse-trading, agreed to Hamilton's proposal that outstanding public debt, and especially external debt, should be put on a firmer footing in which a premium of creditor's interest would be exchanged for rock-solid guarantees of payment.[110]

Hamilton then moved to establish the Bank of the United States upon the model of the Bank of England but with a direct government proprietary interest. The Bank was to serve 'as a bank of deposit, to act as the fiscal agent of the government, and to loan the government money' as well as to provide a circulating medium to replace the existing chaotic forms of circulating currency. Again, after much debate, Congress authorised the central bank in February 1791 and before Hamilton's *Report on Manufactures* was placed on the table for Congressional approval at the end of the same year.[111]

Many of the nineteenth-century antecedents of mid-twentieth-century under-development theory, whom we met in Chapters 1 and 2, were prefigured by Hamilton's *Report on Manufactures*. Hamilton wrote in the *Report* that, because of the 'injurious impediments' which other nations had erected, it was impossible for the United States to 'exchange with Europe on equal terms'. He also complained that 'importations of manufactured supplies' seemed 'invariably *to drain* the merely Agricultural people of their wealth'. Such was the premise for Hamilton's plea that government assistance be afforded to promote manufacturing in the new republic.[112] The objection that government support would be 'unproductive' of wealth, the view voiced by representatives of agricultural states, was demolished by Hamilton who summoned Adam Smith's *Wealth of Nations* and its refutation of physiocratic theory.[113]

In admitting that agriculture had 'intrinsically a strong claim to pre-eminence over every other kind of industry', Hamilton denied that it had an 'exclusive predilection' for the United States economy. Agriculture, he argued, would 'be advanced' by the 'due encouragement of manufactures'. Against the 'unproductivity' charge, Hamilton answered that the annual produce of the land and labour of a country could be increased in only two ways 'by some improvement in the *productive powers* of labour, which actually exists within it, or by some increase in the quantity of such labour'. Hamilton argued that output was increased through higher average labour productivity since 'the labour of Artificers being capable of greater subdivision and simplicity of operation, than that of Cultivators', it was 'susceptible, in a proportionately greater degree, of improvement in *productive powers*, whether to be derived from an accession of Skill, or from the application of ingenious machinery'.[114]

Having shown that agriculture was not the sole form of productive labour, and that its claim to superior productivity was unproven, Hamilton concluded that mutuality between manufacturing and agriculture was the principal cause of the increased wealth of 'Society'. However, there were 'some principal circumstances' in which it could be inferred that manufacturing 'establishments' not only augmented the 'Produce and Revenues of the Society', but made national output greater than if 'such establishments' were absent from the economy. Among the circumstances augmenting national output were an increase in the division of labour; an extension in the use of machinery; increased opportunities for employment; the promotion of emigration from foreign countries; the creation of a greater scope for the employment of a diversity of talents; an extension of the 'field of enterprise'; and the creation of new, or enhanced existing demand for the 'surplus product of the soil'.[115] Given that all of the above were deemed desirable by Hamilton, then the case for the expansion of manufacture was established. It only remained for Hamilton to make the case for government intervention to support the promotion of manufacturing enterprises. In doing so, Hamilton now stood Adam Smith on his head.

'If a system of perfect liberty to industry and commerce were the prevailing system of nations', Hamilton hypothesised, then 'arguments which dissuade a country in the predicament of the United States, from the zealous pursuit of manufactures would doubtless have great force'. Since this hypothesis was so implausible, it followed that the United States was 'to a certain extent in the situation of a country precluded from foreign Commerce'. Manufacturing incapacity partly explained why the United States was disabled from 'exchange with Europe on equal terms'. In the absence of its own manufacturing capacity, the United States would be always dependent on Europe for its manufactures while Europe would have merely 'a partial or occasional demand' for American agricultural production. Agricultural growth in the United States would thereby be retarded immeasurably.[116]

After having arrived at the heart of the matter, Hamilton put forward his programme through addressing what he considered to be the foremost obstacles to the establishment of manufactures in the immediately 'post-colonial' United States. Foreign competition and the want of labour, capital, machinery and skill were the constraints on industrial expansion. To remedy the shortage of labour, Hamilton believed that successful state action to promote manufacture would induce both immigration and, for existing population in the United States, 'the employment of persons who would otherwise be idle (and in many cases a burthen of the community), either from the byass of temper, habit, infirmity of body, or some other cause, indisposing, or disqualifying them from the toils of the Country'. Hamilton commented that 'in general, women and Children are rendered more useful and the latter more early useful by manufacturing establishments than they would otherwise be'. He noted approvingly that '4/7 nearly' of those employed in 'the Cotton Manufactories of Great Britain' were 'women and children; of whom the greatest proportion are children and many of them of very tender age'.[117] Moreover, Hamilton continued, 'the husbandman himself' would 'experience a new source of profit and support from the encreased [*sic*] industry of his wife and daughters; invited and stimulated by the demands of neighbourhood manufactories'.[118]

To ameliorate the force of foreign competition, Hamilton urged a series of measures including the lowering and elimination of duties paid on imported manufacturing inputs, or the instituting of 'drawbacks' on such inputs. Although he recommended tariffs on competing imports of final goods, Hamilton deemed that 'Bounties', to be paid on manufactured goods, were a more effective form of protection. Government action to promote already proposed manufacturing investment would alleviate the shortage of capital, machinery, and skill and encourage 'new inventions and discoveries, at home, and the introduction into the United States of such as may have been made in other countries, particularly those that relate to machinery'.[119] Hamilton was to take a direct hand in what is now called industrial research and development.

While still Secretary of the Treasury, and when drafting the *Report* Hamilton proposed and furthered the Society for Useful Manufacturers. The purpose of the Society, in which Hamilton held equity, was the establishment of an 'industrial empire' near the Great Falls of the Passaic river in the northern state of New Jersey. Hamilton obtained much of the Society's capital through what would now be called, post-colonial, Third World 'corruption'. Both he and a number of the members of the New Jersey state legislature were members of the Society. Hamilton, it is reported, 'stipulated that capital stock of the S.U.M. should consist largely of government bonds and shares of the Bank of the United States', the same bank which he, as Treasury secretary, had created. Although Paterson in New Jersey was later to become an important industrial centre, Hamilton's scheme for industrial development was a grand failure.[120]

Despite failure, Hamilton's late eighteenth-century programme and endeavour to finance an industrial scheme of development prefigures the Saint-Simonian banker-trustee, the development bank and Joseph Schumpeter's twentieth-century *ephor* of development whom we meet in Chapter 7. What is significant here is that the inspiration for Hamilton's Society spawned the creation of like-interested, even if much more modest bodies. One such body was the Pennsylvania Society for the Promotion of Manufacturers and Mechanical Arts, an important venue for the exposition of the thought of Friedrich List.

LIST, PRODUCTIVE FORCE AND THE NATIONAL DOCTRINE OF DEVELOPMENT

Gavin Kitching has written that List's *National System of Political Economy* provided the 'first full-scale defence of industrial protectionism, a defence which has been reproduced with slight variants by every advocate of protection since that time'.[121] Yet, however much List, like Hamilton, was an advocate of industrial protectionism, he was also the fountainhead of 'national development', a doctrine which would also come to be seemingly endlessly reproduced.

Born during 1789 in Württemberg, it was here in the land of Faust that List agitated for constitutional reform, political freedoms and a German customs union. Only after his emigration to the United States was List to formulate his *National System*. Forced to emigrate in 1825 due to his political agitation, List travelled as a part of the entourage of the French hero of the American Revolution, Lafayette, and thereby gained access to the higher circles of American political society. In 1827, while editor of the German language, Pennsylvania-based *Redlinger Adler*, List wrote his *Outlines of American Political Economy* which prompted a dinner to be held in his honour by the Pennsylvania Society. List's protectionist essay, in part due to the flowering influence of Hamilton's *Report*, which he read along with the American economists Raymond and Carey, was much appreciated by the Society who had reprinted the *Report*.[122]

It was, therefore, the influence of Hamilton and the American economists, which seems to have made List abandon the slow gradualism of transition between the progressive stages, 'from the condition of the mere hunter to the rearing of cattle – from that to agriculture, and from the latter to manufactures and commerce' which, according to Adam Ferguson and the apostles of progress, had taken untold centuries in Europe. List's observation that this progressive transition in the United States had happened fast, 'before one's eyes', was to be acutely significant for the formation of a Listian constructivist doctrine of development.[123]

List returned to Europe as an American national becoming, in 1834, the United States Consul to Leipzig.[124] In 1837 he wrote the *Natural System of Political Economy* and in 1841 published his *National System of Political Economy.* It was in this *National System* that List formulated the justification for his protectionist views through the creation of constructivist doctrine of national development. List arrived at this doctrine of national development primarily through a reworking of Adam Smith's concept of 'productive powers', the concept which was central to Hamilton's *Report*. In particular, List repeated Hamilton's argument that it was manufacturing which conveyed the greatest potential for realising the improvements of productive powers. Through reworking the concept of productive powers, List extended the concept beyond Hamilton's mainly restricted technical sense of 'powers' by furnishing the power of productive *force* with moral or 'spiritual' meaning. While there has been debate over the immediate source of List's spiritual morality, one plausible candidate might be Adam Muller, the turn-of-the-eighteenth-century German 'economic Romanticist', disciple of Edmund Burke and exponent of a return to medieval corporatism.[125] Whatever his source, List's moral dimension of productive force is crucial for making sense of the various ways in which he uses the concept in his *National System*, for an understanding of the basis of the system and, not least, for why productive force has entered into the expanded domain of development.

Although Adam Smith, List argued, had understood that the welfare of nations depended upon their productive power and had set out an objective or 'materialist' basis of the division of labour to explain how productive power could be enhanced in production, he had neglected the 'subjective' dimension of the division of labour: 'Industrial production depends much less on the apportioning of the various operations of a manufacturer among several individuals, than on the moral association of those individuals for a communal end.'[126] Therefore, following Smith's example of making pins, List distinguished between the objective division of labour, the distribution of productive tasks of labour activity between different individuals, from the prior association of individuals whose purpose was to engage in the production of a single activity such as the making of pins. The reason why 'ten persons united in that manufacture' could produce an 'infinitely larger number of pins than if every one carried on the entire pin manufacture separately' was that 'the division of commercial operations without combination of the productive powers towards one common object could but little further this production'. In a preceding paragraph, referring to Smith and his followers as the 'popular school', List had noted that:

The essential character of the natural law from which the popular school explains such important phenomena in social economy, is evidently not merely *a division of labour*, but *a division of different commercial operations between several individuals* and at the same time *a confederation or union of various energies, intelligences and powers on behalf of a common production*. The cause of the productiveness of the operations is not merely that *division*, but essentially this *union*.

And List continued:

Adam Smith well perceives this himself when he states, 'The necessities of life of the lowest members of society are a product of *joint* labour and the co-operation of an number of individuals.' What a pity he did not follow out this idea (which he so clearly expresses) of *united labour*.[127]

Productive power did not reside in the autonomous individual, fulfilling a single task of production. Nor were productive powers united upon the universal basis of free exchange whereby, according to the popular or 'cosmopolitical' school of Smith, productive tasks of both individuals and enterprises were made commensurate. Rather, the division of labour, according to List, started in the associated labour of the factory and then ended in the 'national associations' of the nation, the highest ordering of 'united labour' where productive powers were enhanced by virtue of production by associated labour.[128]

It was through the particular nation, List also argued, that both 'social' and 'moral conditions', entailing 'legal security' for 'persons and their properties', the free exercise of minds requiring political freedoms and 'high moral culture', developed to make possible 'economical development'.[129] Any idea of the universal association of nations by intent, what List called the 'civilization of the human race', was 'only conceivable and possible by means of the civilization and development of the individual nations'.[130] Universal association of nations had to be based on 'equivalence' rather than 'subjection' and 'dependence' and could only follow after the development of all nations. 'Philosophy', according to List, was what might determine the associated labour of nations since the ideal of a reciprocal process of exchange had to rest on the 'communion of nations'. For the 'particular people' of a nation, however, the uniting of labour to enhance productive powers, through the development of the ideal of social and moral conditions had to be governed by 'government' or 'Policy'. History was what made possible the reconciliation of philosophy and policy.[131]

Nowhere was the national need for associating labour more urgent for policy, List urged, than the need to unite labour of manufacturing and agriculture. Here, List based his argument on a premise that Hamilton, mindful of his political colleagues from agricultural states, dared not make. List argued that an

> agricultural population lives dispersed over the whole surface of the country; and also, in respect to their mental and material intercourse, agriculturalists are widely separated from one another. One agriculturalist does almost precisely what the other does; the one produces, as a rule, what the other produces. The surplus produce and requirements of all are almost alike; everybody is himself the best consumer of his own products; here, therefore, little inducement exists for mental intercourse or material exchange.

List's purpose, it should be noticed, was to stress the image of a set of negative conditions of agriculture, the conditions which were the opposite of what made manufacturing beneficent for productive power. He continued:

> The agriculturalist has to deal less with his fellow-man than with inanimate nature. Accustomed to reap only after a long lapse of time where he has sown, and to leave the success of his exertions to the will of a higher power, contentment with little, patience, resignation, but also negligence and mental haziness, become to him second nature. As his occupation keeps him apart from intercourse with his fellow-men, so also does the conduct of his ordinary business require but little mental exertion and bodily skill on his part. He learns it by imitation in the narrow circle of the family in which he was born, and the idea that it might be conducted differently and better seldom occurs to him. From the cradle to the grave he moves in the same limited circle of men and circumstances. Examples of special prosperity in consequence of extraordinary mental and bodily exertions are seldom brought before his eyes. The possession of means or a state of poverty are transmitted by inheritance in the occupation of mere agriculture from generation to generation, and almost all that power which originates in emulation lies dead.[132]

Conversely, the social and moral conditions of involvement in manufacturing were described by the 'spirit of striving for a steady increase in mental and bodily acquirements, of emulation, and of liberty'.[133] For List, agrarian societies were economically backward *and* tyrannical, while societies which had been developed by urban manufacturing were *both* economically progressive *and* free because labour was more associated in manufacture than agriculture.

It is easy to find in List an example of the modern dilemma of development, an example which we meet time and again. Manufacturing had developed, by history, to provide the ideal source for development, the source which is the premise of development policy. That which it is intended to develop is present before development begins. Thus, to extend the Listian tautology of development, we can examine the way in which List generally used the concept of 'power' aside from the more specific use made of productive powers. State power, in an international arena, was evidenced by the capacity of the state to defend itself against other states. Defence capacity, in turn, depended upon both the extent to which armaments, and the capacity to produce military arms, had been developed. Productive powers, the capacity to produce arms, permitted a successful defence by the state of its 'integrity', the very condition which was required before the enhancement of national productive powers could take place. State power, to repeat, was both the product and the source of productive powers.

We can now understand what List meant by the immediate purpose of policy, the aspiration to reach '*balance*' or '*harmony of the productive powers*' containing the unity and mutuality of agriculture and manufacturing. A purely agricultural nation, according to List, could never achieve balance because in trading with a manufacturing nation, it would be like '*an individual who in his material production lacks one arm*'. Trade might satisfy material needs but trade alone would never enhance individual *and* national productive power to realise the harmony of united labour within a nation. Although 'the support of a foreign arm' was useful, the productive body of the nation would be governed by the 'caprice' of a foreign head.[134]

Adam Smith's cosmopolitical economy, according to List, was besotted by a 'dead materialism' because the Smithian 'theory of exchangeable values', so essential for associating labour within the nation, dissociated labour from the nation when the theory was essentially applied to trade between nations. International trade, between more and less developed productive powers of nations, traded away national productive force for the mere 'exchange values' which were embodied in imported manufactures. List went so far in his plea for protected international trade as to argue that even were manufactures offered free they should be refused.[135] Equally, to protect agriculture, to maintain productive power, was mistaken since the only secure source of protection for agriculture was the development of manufacturing, 'the basis of external and internal trade, of navigation, of an improved agriculture, consequently of civilisation and political power'.[136] This was the purpose of policy.

While Smith, List maintained, had written *as if* the nations of the world did not exist, the Saint-Simonians with 'remarkable talent at their head, instead of reforming the old doctrines' had cast Smithian doctrine 'entirely aside' and 'framed for themselves a Utopian system'. Saint-Simonian doctrine contained 'truth' in so far as the '*principle of confederation and the harmony of productive powers*' was acknowledged. However, the 'weak side' of the Saint-Simonians was their 'annihilation of individual freedom and independence'. The Saint-Simonians, List wrote, forced 'the individual' to be 'entirely absorbed in the community, in direct contradiction to the Theory of Exchangeable Values, according to which the individual ought to be everything and the State nothing'.[137] In contradistinction to both, List's belief was that his own doctrine established the true relation between the individual and the nation, by which both simultaneously realised their developmental potential through the development of each other.

Given his belief, List's true mission of political economy was to make policy 'furnish the economical education of the nation' and thereby do the work of development to 'prepare' the nation 'to take its place', through philosophy, in a future peaceful universal association of peoples. For List, like the Saint-Simonians, universal association was a speculative prospect of the future, whereas the followers of the cosmopolitical school, List repeated time and again, treated an uncertain prospect as if it was the certain truth and reality of the present. Smith's political economy, List asserted, was devoid of the political because it considered that political economy 'ought to yield to universal economy' of free trade. The real reason for the absence of the political and policy in 'English' doctrine was that, for England, the cosmopolitan and national principle were one and the same thing. Since the two principles were so different for mid-century Germany, the ideal German state had to intend the development of productive powers through tariffs and other means to promote and protect domestic manufacture.[138]

In the absence of German policy, List argued, British industrial advance meant that the caprice of an English foreign head would act to protect its own global domination and, as a still predominantly agrarian nation, the German body of productive power would regress. Without German state action to promote industrial development, there would come a time, List foretold, when the mass of the German people would have nothing to furnish an English world other than: 'children's toys, wooden clocks, philological writings, and sometimes an auxiliary corps, who might sacrifice themselves to pine away in the deserts of Asia or Africa, for the sake of extending the manufacturing and commercial supremacy, the literature and language of England'.[139] Emigration of German population to work for the British empire represented a loss of productive power which would otherwise be employed in Germany where 'natural resources' were unemployed. Development, for List, was about 'transforming the unemployed power of the nation into a material capital, into instruments of value, and productive of income'.[140] To enhance the productive in Germany was the true means of creating the universal potential of associated labour. Without German industrial policy, List claimed, development towards the future possibility of the universal association of nations would be forestalled.

In contrasting the 'civilised' case for constructivist protectionism with the free trade creed of minimal state intervention, List argued that nowhere was the intervention of government less than in a 'savage state'.[141] That 'savage states' of the world's 'torrid zone' continued to exist in the face of more capital, which List defined to be 'the instrumental forces' consisting of 'a mass of power of mind and body,'[142] was proof that nature did not intend:

> industry, civilisation, riches, and power, to be the exclusive portion of any single people, so long as a large portion of the surface of the earth suitable for tillage shall be inhabited by savage animals, and the greatest part of the human race shall be plunged in barbarism, ignorance, and misery.[143]

Yet, if List's precepts for industrial policy were designed to promote the productive power of countries, such as Germany, of the temperate zone, policy could and should not be applied to and by 'savage states' of the torrid zone. What Hegel and Marx had suggested, for contrary reasons, was what List reinforced as the exceptional case for deliberate non-industrialisation. He wrote:

> A country of the torrid zone would make a very fatal mistake, should it try to become a manufacturing country. Having received no invitation to that vocation from nature, it will progress more rapidly in riches and civilization if it continues to exchange its agricultural productions for the manufactured products of the temperate zone.

This exemption of the exceptional case, made on natural grounds, was turned by List to the advantage of Germany. If countries of the torrid zone were in and of themselves unfitted by 'nature' to associate and augment their productive powers through manufacture, then they, nevertheless, might assist in the industrialising process in the temperate zone. England, for List, had shown the way by augmenting the savings of its landlords and farmers with funds derived from overseas colonisation when saving was used to establish manufacture. Germany, too, must acquire colonies to break the English domination of the world. Imperial policy would, it was true, make the 'tropical countries sink thus into dependence upon those of the temperate zone'. However, colonial dependency would be mitigated by the competition between imperial nations and competition would 'ensure a full supply of manufactures at low prices' and thereby 'prevent any one nation from taking advantages by its superiority over the weaker nations of the torrid zone'.[144]

Finally, the one exceptional but permissible case of agrarian nation–industrial nation trade was justified on the ground of colonial trusteeship. For List, exceptional trading would permit the 'mission of political institutions to civilize barbarian nationalities, to enlarge those which are small, to strengthen those which are weak, and above all, to secure their existence and duration'. Colonial trusteeship could be accomplished by trading because: 'The economical education of a country of inferior intelligence and culture, or one thinly populated, relatively to the extent and fertility of its territory is effected more certainly by free trade, with more advanced, richer, and more industrious nations.' If the free trade in commodities was insufficient to carry out the task of trusteeship, then the surplus population of the temperate zone could itself be exported to carry out development through colonisation.[145]

MARX'S CRITIQUE OF LIST

Entwined with, and added to, Marx's biting criticism of List's plagiarism and dishonesty was the chief charge of hypocrisy. Put simply, Marx argued that List's *National System* amounted to a case of special pleading on the part of the German bourgeoisie, who, in losing out to English industrial competition, speciously cloaked the phoney trappings of a noble idealism around their self-interested desire for material wealth. More specifically, Marx's condemnation of List turned round the moral claim for productive powers, which is rendered, in the English translation of his draft critique, as productive forces.

List's concept of productive forces, Marx argued, was far from being the advance on the writings of Smith which List claimed. Rather, Marx suggested, List's concept represented theoretical retrogression in so far as List regressed Smith's treatment of exchange value. There was a particular passage of List, commenting on what he reckoned to be an important, though limited, contribution of Smithian economics, that provided the opportunity for Marx to break open List's system. List had written, Marx recounted:

> that by means of the theory of exchange values 'one can establish the concepts of value and capital, profit, wages, land rent, resolve them into their component parts, and speculate about what could influence their rise and fall, etc., without in so doing taking into account the political conditions of the nations.[146]

What List, Marx argued, had really 'established' from Smith, without referring to his own theory of the confederation of productive forces within the nation, was the 'reality' of 'labour' as 'unfree, unhuman, unsocial' activity, as a 'commodity'. In short, Marx revealed, List had established what 'exchange value' really meant but he had not developed Smith's concept because he was not conscious of explicating what was implicit in exchange value. As such, what List had really established, again with no necessary reference to what Marx described as List's 'idealistic eye-wash' of the historical necessity of the nation, was the basis of 'objectified labour' or private property. This was the logical key to the understanding of the development of capitalism.[147]

It was no accident, Marx contended, that List found it necessary to make the 'theory of exchange values' essential for associating labour *within* the nation while he invoked the essence of the nation to suspend operations of the theory for exchange *between* nations. To have done otherwise, Marx alleged, would have been to challenge the capitalist system itself. When List summoned up the national, he did so as a German bourgeois claiming the absolute right to exploit his own workers while denying the same right to his foreign competition. Marx summed up List:

> What does the German philistine want? He wants to be a bourgeois, an exploiter, inside the country, but he wants also not to be exploited outside the country. He puffs himself up into being the 'nation' in relation to foreign countries and says: I do not submit to the law of competition; that is contrary to my national dignity; as the nation I am superior to huckstering.

However, since the German bourgeoisie had reduced their own workers, through the workings of the theory of exchange values, to the 'free slavery' of 'self-huckstering' labour, List the bourgeois had consequently reduced himself to a huckster as well.[148]

Industry could be understood, Marx wrote, 'from a completely different point of view than that of sordid huckstering interest'. For instance, the Saint-Simonians, in their 'dithyrambs', had glorified the 'productive power of industry' even if they had committed the error of confusing the force of industry itself with the forces which industry called into being. What redeemed the Saint-Simonians despite their error, and unlike the 'German philistine', was that they had 'attacked exchange value, private property' and the 'organisation of present-day society'. While the Saint-Simonians, Marx continued, were 'punished for their original error', in that their 'glorification' of the 'productive forces of industry' became 'a glorification of the bourgeoisie', they made it possible, to repeat, for industry to be regarded 'from a completely different point of view'.[149]

Marx suggested that industry could be regarded as the 'great workshop in which man first takes possession of his own forces and the forces of nature, objectifies himself and creates for himself the conditions of human existence'. However, to regard industry in this way was to '*abstract*' industry 'from the *circumstances* in which it operates today and in which its exists *as industry*'. Marx then entered into the expanded domain of development. 'One's standpoint' had to be '*not* from within the industrial epoch, but *above it*.' As such, industry was not to be regarded 'by what it is for *man* today, but by what present-day man is for *human history*, what he is historically'. In the mid-nineteenth century, it was not the present '*existence*' of industry as such which ought to command attention 'but rather the power which industry has without knowing or willing it and which *destroys* it and creates the basis for *human* existence'.[150]

Any assessment of industry which recognised that industrial power had been developed would 'then at the same time' recognise that the hour had come 'for it to be done away with, or for the abolition of the material and social conditions in which mankind has had to develop its abilities as a slave'. To maintain, as Marx argued of List, that each nation had to go through the development of industry 'internally' was 'absurd'. 'What nations have done as nations', Marx wrote after Hegel, 'they have done for human society.' The 'whole value' of what nations had done consisted

> only in the fact that each single nation has accomplished for the benefit of other nations one of the main historical aspects . . . in the framework of which mankind has accomplished its development, and therefore after industry in England, politics in France and philosophy in Germany have been developed for the world . . . their world historic significance, as also that of these nations, has thereby come to an end.[151]

Marx would have no truck with List's constructive doctrine of development as it was set out in his *National System*. Behind the doctrine of productive force of this 'idealising German' lurked the 'dirty reality' of the 'present system' within which 'crooked spines', 'twisted limbs', 'intellectual vacuity', and 'one-sided development' were all 'productive forces'. To complain about these ills of industry or, indeed, about the estrangement of industry from one's native soil, or of a growth of population which did not correspond to the exploitation of the soil, or still more about the conversion of peasants into poor, and perhaps, idle proletarians was rank hypocrisy from one who accepted the 'theory of exchange values'.[152]

While Marx attacked the basis of List's system, Frederick Engels, in an 1845 speech in Elberfield, addressed 'practically' the results that might be expected from List's protectionism.[153] If, Engels argued, a German state were to adopt free trade, thereby ignoring List's advice, nearly the whole of its industry would be ruined and mass unemployment would result. However, Engels maintained, if 'protective tariffs' were adopted, as List advised, the same result would still eventually ensue. Engels' argument is lengthy and there is no room or need to rehearse it here, but its burden was to say that tariffs could neither indefinitely suspend the workings of capitalist competition nor forestall its results.[154] If these 'conclusions' were 'correct', Engels maintained, the 'necessary result' would be 'social revolution and practical communism'. And if this latter proposition was true, then, Engels continued, 'we will have to concern ourselves above all with the measures by which we can *avoid* a violent and bloody overthrow of' the social conditions which culminated in violent revolution. For Engels, there was 'only *one* means' of recourse, 'namely the peaceful introduction or at least preparation of communism'. 'If', Engels warned,

> we do not want a *bloody* solution of the social problem, if we do not want to permit the daily growing contradiction between the education and the condition of our proletarians to come to a head, which, according to all our experience of human nature, will mean that this contradiction will be solved by brute force, desperation and thirst for revenge, then, gentlemen, we must apply ourselves seriously and without prejudice to the social problem; then we must make it our business to contribute our share towards humanising the condition of the modern helots.[155]

CONCLUSION

We have seen, in the course of this chapter, how Marx simultaneously struggled against both the trusteeship of 'true socialism' and bourgeois trusteeship of List's *National System*, in order to open up space for his expanded domain of development. Yet, despite his struggle against these two versions of trusteeship, neither would disappear. Both 'socialist' and bourgeois trusteeship would return again and again in the years down to the present to complete the intent of development doctrine. Two variants of development doctrine, the constructive and the palliative, could be set out in both versions of trusteeship.

The constructive intent of socialist and bourgeois trusteeship would emerge respectively in the permutations of the doctrine of 'socialism in one country', which needs a book of its own, and in endless reiterations of List's system of which we provide an example in Chapter 5. Palliative intent would be expressed, and practised, by its bourgeois trustees in 'progressive' approaches to the social welfare of the poor and in attempts to manage subsistence for relative surplus populations. Case studies which follow in Chapters 4 and 6 focus upon attempts to ameliorate the official problem of unemployment by way of an agrarian doctrine of development. The same palliative intent would also be expressed, and according to similar programmes for development schemes, by those who took up this variant of socialist trusteeship. As we will show in all of the chapters that follow, all four of these intentions to develop would be combined and recombined in bewildering kaleidoscopic patterns during the rough-and-tumble formulation of development doctrine as state policy and practice. Since constructive and palliative intent have crossed the boundaries of the two versions of trusteeship, the varying arrays of development doctrine only aggravate the immense confusion of understanding what development means.

Ironically, one element of Marx's own thought would be itself, in part, responsible for the repeated re-emergence of trusteeship. If both 'true socialist' and Listian bourgeois trusteeship had a commonalty through their immediate origins in Saint-Simonian positivism, then what has come to be known as marxism is also partially rooted in Saint-Simonian 'scientific socialism'. The attempt by Marx, which we noted in this chapter and to which we return in Chapter 7, to distinguish between what he believed to be the genuine contribution of the Saint-Simonians from their errors in doctrine and failures of practice, opened the door of marxism, time and again, to the re-evocation of the Saint-Simonian connection. One result has been persistent Marxist attempts to reconstruct Marx in the cause of the trusteeship of development. Indeed, what has passed for a 'Marxist theory of development', as a position within recent development studies debates, is little else than a variant of development trusteeship. That such temptation has been so powerful, even in the midst of Marx and Engels' struggle against it, is witnessed by Engels' ambiguous dictum to the 'gentlemen' of Elberfield, namely that 'we must make it our business to contribute our share towards humanising the condition of the modern helots'. The 'we' who would take up this cause, in place of that of Marx's expanded domain of development, face this temptation time and again.

Part II

4

IMMANENT AND INTENTIONAL DEVELOPMENT

The origins of development doctrine in Australia and Canada

The difference between development as that which happens in history and what is deliberately willed to happen has often been mentioned in the literature on the history of the idea of development. H.W. Arndt, for instance, noted that the distinction between immanent and intentional development lies at the heart of economic development:

> Whereas for Marx and Schumpeter, economic development was a historical process that happened without being consciously wished by anyone, economic development for Milner and others concerned with colonial policy was an activity, especially though not exclusively, of government. In Marx's sense, it is a society or an economic system that 'develops'; in Milner's sense, it is natural resources that are 'developed'. Economic development in Marx's sense derives from the intransitive verb, in Milner's sense from the transitive verb.[1]

And then, interestingly enough, Arndt found that economic development entered the English language neither in Britain nor the United States, where 'economic development *happened*', but in Australia where it had to be made to happen:

> In Australia's hostile environment, where settlers from the earliest convict days had to contend with drought, flood, pests, distance, and more drought, economic development did not happen. It was always seen to need government initiative, action to 'develop' the continent's resources by bringing people and capital from overseas, by constructing railways, and by making settlement possible through irrigation and other 'developmental' public works. So well established did this notion become in Australia by the 1920s it was referred to as the 'doctrine of development before settlement'.[2]

Arndt's argument was that development had to be made to happen to stimulate the flow of labour power into a colony.

In Arndt's case, that of Australia, the point of 'the doctrine of development before settlement' was that immigration, primarily from Britain, would not occur unless the state set to work on development. Our argument is the opposite. The idea of intentional development, through the state practice of development, had arisen in Europe to deal with the problem of a surplus population in Europe. Likewise, in the colonial cases of Australia and Canada, which we examine in this chapter, development, as an intention to develop 'resources', arose during a period when both colonies faced the experience of mass unemployment.[3] In Victoria colony, the government faced the prospect of the mass *emigration* of population after 1856. During the late 1840s, Quebec had experienced the mass emigration of a French-Canadian population to the United States. In both cases, new governments of self-governing colonies were determined to act, in the name of development, to prevent what List had called the 'loss of productive force'. It was not development to attract a gain in population. It was the intention of development to make the negative, a loss of population from 'national' territory, positive for an immanent process of capitalist development.

Our argument is that policy for development cannot be inferred from what might have happened during a subsequent course of development. We do not dispute that there was a 'doctrine of development before settlement' and we show, in the Canadian case, why and how the doctrine was developed before the mid-nineteenth century. What we do dispute, however, is that this doctrine provides an overarching principle for a practice of development which marks off a colonial world from that of Europe. Development, as an immanent process, had created a surplus population which was given official and quasi-official recognition in mid-century, precisely the same historical period in which 'development' had been invented in Europe. The doctrine of development which informed policy of this period was, as we will show in detail for the Canadian case, an immediate and direct response to the problems of past settlement.

It might seem that a 'doctrine of development after settlement' is perverse in the same way that is argued, by way of underdevelopment theory, that the process of capitalist development is perverse and not 'true' in the imperial world. Again, we dispute that development policy can be deemed to be perverse in relation to a process of development. If policy was perverse, in that it did not stem emigration and the loss of productive force, it was a corruption of the ideal of development. The mid-century national ideal of development, for the two colonial cases, was to assert the new self-governing status of colonies by confining population to what had become, in effect, national territory. This ideal of development, 'the desire to develop' on the part of the state, was what marked off the experience of development of the new states in Australia and Canada. As such, however, much of it sat uneasily between what was a dictate of national development and what was a predicate of capitalist development, namely the object of making productive force available for capital; the doctrine of development was consistent with both.

THE CASE OF VICTORIA, AUSTRALIA

Arndt pointed to Charles Mayes, a civil engineer who, significantly imbued with mechanics rather than biology, was the first to coin the phrase 'economic development'. Mayes did so in his prize-winning 1860 *Essay on the manufacturers more immediately required for the economical development of the resources of [Victoria] colony.* Economical development, here, was directly associated with the manufacturing industry rather than agriculture or infrastructure and it is no surprise to find the use of 'economical development' in List's 1841 *National System*, which was translated into English in 1856 and was widely read in Australia.[4] It may be entirely reasonable to suppose that a doctrine of development would be found in the colonies of Australasia, or Canada for that matter, where colonial governments were about to embark on a fledgling nationalist stance with regard to the 'development' of their own resources and were less inclined to regard themselves as agencies serving a British imperial interest in trade and raw material extraction.

However, if we look more closely at Mayes' essay, we find that the purpose behind its subject matter was precisely the opposite of 'development before settlement'. The Victorian gold fields, discovered in 1851 after the colony was founded in 1836 and became self-governing in 1856, attracted the immigration of population on a scale such that the population of Victoria increased from 97,500 in 1851 to 500,000 in 1858. The collapse of the gold boom after 1856 led to a marked reduction in immigration and, indeed, emigration in 1861. Thus, the annual average of net migration, of 40,000 persons between 1852 and 1857, fell by nearly three-quarters to 12,000 between 1858 and 1860, while in 1861 there was net emigration of 9,000 persons. The decline in gold production had been dramatic. Production fell from its peak of 3 million ounces in 1856 to no more than 2.2 million in 1860; employment in gold mining, of both independent diggers and wage-workers, fell from its 140,000 peak to 100,000 in 1860. Estimates show that GDP per capita in 1858 was no higher than it had been before the gold bonanza.[5]

In May 1860, a Melbourne newspaper commented on 'an entire stoppage of immigration, a stationary population, a falling supply of gold, a long-continued commercial depression, and a diminution of credit at home'.[6] Mayes was concerned with the unemployment of labour and capital after settlement and deliberately set out to advise the government of Victoria as to how labour power might be prevented from leaving Victoria in the face of depression and destitution. As in the Canadian case of a decade earlier there was a conjuncture between the political advent of responsible government and economic depression. Mid-century development doctrine was a necessary official response to an immediate and recognised predicament of impending social disaster.

Mayes' prognosis for development is an example of a case in which the supposed origin of a term or concept is part of the concept itself. What became known as 'a doctrine of development' in Britain, fifty years later at the beginning of the twentieth century, was a deliberate attempt to respond to the threat of unemployment which accompanied de-industrialisation. A British doctrine of development was enunciated to make the British state developmental and so make it modernise manufacturing industry in Britain. Moreover, Mayes' thesis was about an explanation for unemployment in Victoria with reference to trade unions which interfered with a 'natural' labour market and made workers refuse to work for lower money wages than those that would cause the demand for labour to rise and absorb the unemployed into manufacturing employment.

During the gold rush, the average wage rate for recently unionised artisans increased from 8 shillings per day in 1851 to 40 shillings in 1854. With the collapse in gold production after 1856, money wages for artisans remained constant at this rate but fell substantially for semi-skilled and unskilled workers. The average wage rate for unskilled workers fell a half, from 10–13 shillings in early 1854 to 5–6 shillings in 1861, while the rate for 'building mechanics' and farm labourers fell by the same proportion, from 20–35 shillings to 10–20 shillings.[7] Mayes had wanted a reduction of one-quarter to 30 shillings in the average rate of artisans, protected by trade unions, so that manufacturing, through the processing of raw material supplies available in Victoria, could compete with imports. Serle, in his standard history of the 1850s, described the condition of manufacturing:

> Manufacturing, which developed to a limited extent from 1856, was still confined almost entirely to those products which were naturally protected or were readily served by local raw materials, for Victoria's high wage-rates as well as the inefficiency of infant industries made it almost impossible to compete with nearly all imports.[8]

A century earlier, Mayes had written:

> It will be seen, from the list of 'raw materials' produced in Victoria, how much our success, as a manufacturing people, depends upon the economical production of agricultural produce; but even supposing we produced abundance of raw materials at a cheap rate, we must also possess cheap labour and good machinery for the successful conversion of such raw produce into manufactured articles used in the every day purposes of life.[9]

When this doctrine of development was transported back to Britain, and then onto colonies in Africa, it was never stated with as much clarity. Mayes set out all the elements for the doctrine, bar one, the predisposition to make the labour market function 'flexibly'.

Much of the existing labour power in Victoria was exerted in independent gold pegging. Mayes acknowledged that 'workmen generally prefer their own trade to that of any other' but held that the returns from digging were subject to 'great uncertainty'. Moreover, he maintained that the workers had been 'injured' by working in the gold fields arguing that 'as long as any inducement is held out, either real or imaginary, for men to work at the gold fields, with the most distant prospect of acquiring vast sums of money by some happy stroke of good fortune, they would not settle down to more legitimate and, as a rule, more profitable employment offered by trade, agriculture and manufacturers'.[10]

The majority of gold diggers, whose average earnings were 12 shillings per week, were destitute and as badly off as the majority of artisans and labourers, unemployed in Melbourne. Destitution for women was worse than for men, according to Mayes, who felt that destitute women 'might profitably be employed in factories, instead of seeking relief from the parish or a *much worse alternative*, which, hitherto, seems to have been their *only alternative*'.[11] Agricultural enterprises were failing because unskilled labour was 'dear at any price' and farm workers worked with far less effort than they expended in gold digging. Those who worked in agriculture, Mayes maintained, only did so because of the unlimited free ration of food paid by the farm employers.[12]

Economic development, by this reckoning, was about regularising the use of labour, shifting it out of independent work, and creating the minimal means of subsistence for labour by work in manufacturing industry or agriculture whose enterprise would be freed from dependence upon the day wage-worker.

Mayes' prognosis for development followed from an early, but later repeatedly argued, labour aristocracy thesis. Workers as a whole were 'obstinate' and blind 'to their true interests'. Money wages were fixed by trade unions at rates between two and four times those prevailing in England for similar jobs and industries but the cost of living was only one-fifth higher in Victoria than in England. A minority of the labour force had permanent jobs at the relatively high core wages rates while the majority of workers, whether skilled or unskilled, were either self-employed or destitute in peripheral work. Mayes observed that nothing was more 'humiliating' to workers than that they would ultimately succumb 'to such reduction in their wages as will enable them to compete with imported goods'. It is a difficult task, the civil engineer maintained, to make workers believe that they are as well off with lower money wages and that 'they ought cheerfully to submit to the inevitable laws of supply and demand'. Yet, if wages were reduced by a proportion below the then average protective tariff of 25 per cent on most manufactures, Mayes argued, all workers would be 'better off' than they 'could be while upholding a contest alike destructive to their own interests and that of the country at large'. Moreover, to ensure that labour effort increased at the same rate that money wages were reduced, Mayes called for the introduction of piece rates, to which unions were opposed in principle but not in practice: 'I have every reason to believe that the manufacturing interests in Victoria will succeed in proportion as the "piece-work" system is successfully adopted.'[13]

The mid-century argument over the significance of the piece-wage was nothing other than what is now known as 'labour flexibility'. Labour market deregulation was to be pivoted on development policy. 'It must be made perfectly clear', Marx wrote of the piece-wage 'that the way *in which* wages are paid out does not affect the situation in the least, although one mode of payment may well favour the development of the capitalist mode of production more than another.'[14] Given that there was a surplus population, the residuum of labour power in primary production, pressing upon the price of labour power for manufacturing industry, Mayes understood that the piece-wage would be one means by which money-wage earnings would be reduced to the average price of labour power in Victoria while the intensity of labour effort in production was increased. As far as Mayes was concerned, there was no argument about trade protection; it was the 'artificial' protection of labour through trade unions which was to be undermined by bringing piece-work, from mining and agriculture, to manufacturing industry.

Mayes published a builders' price list for Victoria fixing rates of pay for specific timed tasks for his own construction industry.[15] When Mayes followed Ure's *Dictionary of Arts, Manufacturers and Mines*, listing the basic materials for industrial production, from alcohol to wine, the constant barrier against 'economical' development was the fixed, time-wage rate. Mayes assumed that protection should be adopted for the 'principal productions of Victoria, an infant colony, whose rapid growth and development is a source of wonder to the civilised world'.[16] If gold had set off development which had then created destitution, it was the intention that manufacturing should become the basis of economic development to maintain subsistence and prevent the loss of the productive power of labour for capital in Victoria.

'The problem for Australia', Schedvin has written, 'has been to maintain the brilliant success of the mid-nineteenth century. Staples of the productive efficiency of gold and wool were not readily replaced, and diversification around the original export base has been a protracted and only partly successful process.'[17] This course of development, as an immanent process of development, cannot be attributed to a simple prognosis for positive development, in the manner of Arndt, but must incorporate the policy of development whose intention was to replace what had been destroyed and lost in the course of capitalist development itself.

For Victoria, Mayes may well have been satisfied by the immediate course of development after 1860. Output in construction, Mayes' own industry, rose rapidly with house building increasing by 50 per cent in the early 1860s. Manufacturing output, with direct linkages to construction, also increased, albeit at a slower rate.[18] Employment in manufacturing was 26,000, or one-quarter of that in mining, in 1861. It more than doubled between 1861 and 1881 and quadrupled from 1861 to 1901.[19] Thus, whereas the value of output of manufacturing and construction contributed only 6 per cent to Victoria's GDP in 1852, its share had risen to more than 10 per cent in 1860. The share of mining fell from one-half to one-quarter of GDP over the same period. This trend continued through the 1870s and, however fitfully, beyond.[20] Railway construction, during the 1860s and 1870s, contributed to the absorption of the relative surplus population which had been engendered by the decline of gold mining. Last, but not least, average money wages in industry declined during the 1860s and then were stabilised during the 1870s and 1880s. Mayes might well have approved of the fact that the average industrial money wage of the 1890s was less than what it had been thirty years earlier.[21]

The movement of labour power into construction and manufacturing was accelerated by the doctrine Mayes had enunciated, a doctrine which commanded popular support. As Sinclair has repeatedly written, the decline of the gold bonanza 'left a residue of immigrant labourers who strongly favoured governmental interference with market forces to permit them to remain employed in the colony'.[22] Protection during the 1860s was as much the result of pressure from labour unions as from industrial capital itself.[23] In a fashion similar to the mid-century Canadian case, the revenue from the tariff on industrial imports was partly, to quote Sinclair, 'a device used by the Victoria government to obtain the revenue to match the New South Wales railway building programme', a programme which had been funded, in this neighbouring state, by the sale of nationalised land.[24] The Victoria government had no such option, a significant fact which also explains why an agrarian doctrine of development was such a minimal part of the official intention of development there.

An agrarian doctrine of development, namely what was a maximal aspect of 1849 development policy for the Quebec case which follows, was an attempt to develop agriculture and ancillary industry for the immediate purpose of settling a surplus population on the land. Its direct purpose was neither to maximise agricultural production nor the surplus of agricultural output which could be realised either in private profit and/or state revenue. Rather, the purpose was to maximise employment, by settling unemployed labour on the land where it could directly secure subsistence, as an alternative to direct state assistance to the unemployed through welfare programmes.

This is not to say that an agrarian doctrine of development could not be consistent with the development of capitalism through smallholding schemes of agriculture. We will later show how this was to happen for mid-twentieth-century Kenya. Equally, an official agrarian doctrine could be compatible with the expressed desire of producers, whether unemployed or not, to escape from the throes of wage-labour of early industrial capitalism, a vision of independent production on the land. Indeed, the popular desire for agrarian development was strongly expressed in Victoria, and probably more so than in mid-century Quebec where doctrine was driven by the clergy and urban professionals.

It was in September 1860, at the beginning of the second Parliament of self-governing Victoria, that the Nicholson Land Act was passed by a government which was forced to accede to popular clamour for an agrarian doctrine of development. In a criticism of the staples approach to development, Fogarty has written of mid-century Victoria: 'Labour, far from being the scarce factor of production, was now relatively abundant, but a government sensitive to popular opinion was not prepared to wait passively for the factor imbalance to adjust itself through low wages, unemployment, and emigration.' In fact, this is precisely what did happen, through manufacturing and construction, according to Mayes' doctrine of development, and with continuing emigration to New South Wales and South Australia, but without the presumption of government passivity. Fogarty recounts the aspiration for an agrarian doctrine of development:

The opening up of the land to closer settlement was regarded as essential to the fulfilment of the aspirations of both the government and the mass of the people: 'it was the way in which radicals formulated their essential political-economic demands'. The opening up of the northern plains in the early 1870s appeared to meet the people's demand for the unlocking of land.[25]

Serle has been more careful in defining 'the people's demand' for land. Under the influence of British Chartism, the Victoria Land League had been founded in 1857. Elsewhere, we have shown how one influential version of Chartism unsuccessfully sought to return British workers to the land through mid-century smallholding schemes of agriculture.[26] T.H. Irving has linked the appearance of 'mass unemployment and rudimentary living conditions', in Melbourne and other Australian cities, with the demand for land reform as the answer to urban grievance: 'Significantly, the appearance of Melbourne's Land Convention in 1857 coincided with rising unemployment, and in Sydney the unemployed demonstrated almost daily for six months. In Brisbane, too, demands from urban workingmen for land reform were barely audible until Queensland's economic crisis of 1866.'[27] The oft-repeated slogan, that 'a man would rather work for himself for £1 a week than go to a master for £6 a week', was heard most loudly in mid-century.[28]

Between 1857 and 1860 in Victoria, there was strident 'urban clamour for land-selection on free and easy terms'. Land reform became both an urban and goldfield cause: 'Independence – self-employment – was the great aim of the migrants of the fifties', Serle suggests of the diggers, 'and as the independent digging life began to pall for many, the obvious second chance of taking up land came to be discussed more and more round the campfires.' A Victoria colonist's 1856 idea, that the only 'requisite' for human equality under capitalism was 'to permit labour to have free access to land', became a nostrum which, as Serle has concluded, 'was to prove one of the most tragic Australian delusions'.[29] This conclusion was amply confirmed by the course of the agrarian doctrine of development in mid-century Victoria.

Mid-century land reform movements were based upon a struggle against large- to middling-scale pastoralists, known as 'squatters', whose economic power stemmed from their capacity to develop wool production during the 1830s. Called the 'pure milk of rural capitalism', the squatters' social origins varied widely, from 'the wealthiest of the land' to 'shop-boys' and 'old shepherds'. Squatting referred to the unauthorised occupation of Crown land without legal title, or to occupation, which was legalised from the late 1830s under a license or lease which was bought at minimal cost by the squatter.[30] Land reform was an attempt to break the concentration of land in the hands of squatters by opening, as McMichael has put it, 'colonial wastelands to selection by small farmers', thus known as 'selectors'. McMichael continued, stating that for Australia as a whole:

> Under the progressive mantle of universal rights to landed property, the reforms created land markets in regions that squatters considered to be in their productive possession. . . . Whereas urban politicians sponsored such reforms in the interests of a *social* ideal of an industrious yeomanry, ignorant of the practical *economic* measures necessary for its realization, squatters easily found loopholes to engross the best of the land for themselves.[31]

McMichael has summarised the economic problems involved in settling small arable farmers on pastoral land in the face of unremitting squatter resistance. Serle has described how the same problems in Victoria caused the Nicholson Land Bill and 1860 Act, originally designed to sell relatively small, 320-acre land blocks at favourable terms, to favour 'the large capitalist at the expense of poor man'. The 1860 Act was left 'mutilated and unrecognisable' and made 'few new settlers'.[32]

Squatters got bank credit and finance, took advantage of relatively cheap transport rates, engrossed selectors' land either through dummy companies or through cheaply buying land forfeited by bona fide selectors, and used these advantages to transform the technical basis of large-scale sheep farming. Fencing and dam construction alone did much to induce heightened construction activity in Victoria during the late 1860s. Subsequently, from the late 1860s, after railway construction and the capitalisation of squatters' production, smallholding production was more firmly established on the basis of mixed farming, the offer of seasonal wage-work to selectors by squatters and, especially, substantial state assistance for smallholding agriculture.[33] Indeed, it is reported that Victoria legislation in 1869, to correct for the bias of the 1860 Act, attracted immigrants from South Australia to the wheat production in the northern plains.[34] Nevertheless, Sinclair insists that 'the most that can be said for the selection legislation is that it may have increased wheat production up to the limits permitted by the restricted scale of the market'. He confirms that, in so far as there was 'a general association between the establishment of democratically elected parliaments and action resulting in rapid output increase', agrarian doctrine was not an essential part of development doctrine in Victoria.[35]

The significance of 1860 development doctrine for Victoria is clear. Serle has concluded that:

> 'Probably a proletariat was never created so quickly, or out of human material so unmalleable', as in Victoria in the late fifties and early sixties. Hopes for independence, for self-employment on the land or in trade, for permanent, self-respecting middle-class status, were dying away to ashes, as the diggings contracted, land was only slowly released, and opportunities narrowed.[36]

In his criticism of the staples approach, Pomfret has shown that the outcome of a similar gold rush and bust, from 1858 to 1865 in British Columbia, was very different. There, the mass of diggers emigrated to the United States. 'In Australia most of the immigrants remained to find other jobs, while in British Columbia few remained and the colonial government was left to face fiscal crises.'[37]

Economic development in British Columbia, Phillips wrote, 'was intimately connected related to the development of transportation routes'. The first major road project at the onset of the gold rush, we are informed, was 'launched for the dual purpose of solving the problem of communication and easing the burden of unemployment'.[38] However, accelerated programmes for road construction 'were completed only when the supply of placer gold was becoming exhausted and many miners were leaving'. Colonial government indebtedness was thus due to the failure to raise increased revenues which were assumed to follow economic expansion and thereby redeem the loan costs incurred in the prior construction of the road system. Immigrants who did not emigrate turned to agriculture 'to provide a population base which indicates', as Phillips commented, 'a misunderstanding of the role of agriculture in the existing economy': 'What in fact appears to have been envisioned was a self-sufficient agricultural community; they did not realize that this could not produce a dynamic growth or an investment frontier upon which to build an expanding economic base or sustain a growing population.'[39]

One of the leading advocates of agrarian doctrine in British Columbia was the 1858 immigrant, Nova Scotian William Smith, who adopted the name of Amor De Cosmos to express 'what I love most, viz: love of order, beauty, the world, the universe'. Significantly, it was De Cosmos, the opponent of colonial rule, who led the local campaign for democracy and self-government in the colony while upholding his view that the only purpose of Canadian confederation for British Columbia was to 'protect the farming interest'.[40] The agrarian doctrine of development in British Columbia was associated with the claim for self-government, a claim that had been realised in both the other Canadian case of Quebec and in Australian Victoria before the advent of Canadian confederation and Australian federation. British Columbia, however, without self-government before Canadian confederation in 1867 and getting self-government as part of its conditions for entry into the confederation in 1871, was bereft of official development doctrine to lock population within this colonial territory.

COLONIAL SELF-GOVERNMENT, NATIONALISM AND DEVELOPMENT DOCTRINE

Colonial nationalism was a concept coined by Richard Jebb, the British Edwardian imperialist and social reformer who, at the turn of the twentieth century, sought 'Britannic cooperation' with Canadian and Australasian nationalisms. Motives behind Australian nationalism, involving policies demanding 'the exclusion of Asiatics, the promotion of "secondary industries" by a protective tariff, and the creation of an Australian navy instead of the payment of naval subsidies to Britain', were founded upon 'a policy of self-respect'. Self-respect, self-government, culminating in the Commonwealth of Australia Act of 1900, and 'nationalist aspirations' were compatible, Jebb urged, with 'an imperial alliance'.[41]

An opposing view, more recently reiterated by Benedict Anderson, to whom we return in Chapter 6, was that there was an 'inner incompatibility' between nineteenth-century 'empire and nation'. Anglicised Australians, Anderson argued, could not serve, unlike Britons, to promote the extension of Britannic, or English, nationalism beyond their colonial territory and were forced thereby to found a self-governing and national basis for the administration of colonial territory.[42] There is merit in this view but only for the nineteenth century, in the cases of Australasia and Canada, and it does not capture the fuller significance of colonial self-government.

In the course of his seminal essay on the conceptual problem of colonial nationalism in Australia and Canada, Douglas Cole returned to the question of compatibility between nation and empire:

> An Australian nationalism, based upon a consciousness of ethnic differentiation, would have been incompatible with an imperial ideology based upon the unity of blood, language, ancestry and tradition. An Australian patriotism, based on loyalty to state and territory need not be. They could be 'independent Australian Britons' because pride of race 'did not conflict with love of country'.[43]

Cole's fundamental conceptual distinction was between nation and state, which he identified with 'patriotism', and seemingly needlessly so were it not for the recent neo-populist interpretation which effectively identifies patriotism with 'consensus' when it affirms the claim that modern state power rests upon 'the consensus of the citizen in its rule'. Such is the view of Alastair Davidson, who finds that self-government in Australia was, and continues to be, a singular case of 'democracy without substance' because it was the form of government whereby the state defined 'the people' and constructed consensus on the basis of 'their' citizenship.

Davidson's argument is that 'a modern political space', as in England in 1688, France in 1789 and the United States in 1766, 'was opened out by a revolutionary system in which popular reason was dominant over the different reasons of the administration, law and judiciary'. A 'complete' imposition of 'the national popular will' upon the state, Davidson continued, would include 'a non-imposed self-definition as the people' to direct the state. Australia's self-governing democracy, however, was incomplete. 'The people, in this active mode, had not emerged in the Australian colonies', wrote Davidson, and thereby 'were accorded a passive role by both the administration and the judiciary, as recipients of their decisions'.[44] We will soon see that there was little singular about administrative rule in Australia and that if there was historical singularity in colonial self-government it was because, by the nineteenth century, the modern state form had been established as a condition for 'a national people' rather than the other way around.

Cole, for instance, emphasised that '"nation" almost invariably means a self-governing state, recognised as such, and there is little or no ethnic consciousness to it'.[45] Self-governing colonies were the basis for a state-nation rather than nation-state according to the precept, which Eric Hobsbawm has reiterated by referring to both the United States and Australia, that 'nations are more often the consequence of setting up a state than they are its foundations'.[46] Self-government, for the state of Victoria or the province of Canada, the entity including Quebec before confederation, was the official reason for fostering administrative autonomy for the colonial territory while extending the reach of a state apparatus to regulate population and property. Colonial self-government, Cole concluded, was 'a simple desire for expanded self-government'.[47]

When searching for the immediate Australian, as opposed to imperial roots of nineteenth-century self-government,[48] reference is often made to the contending ideologies of W.C. Wentworth and J.D. Lang. Wentworth, who represented the interest of squatters, proposed self-government to regenerate, as Blackton explained, 'a new model England in the South Pacific, ruled by a pastoral aristocracy, a loyal Anglo-Australian oligarchy. Its foundation was property, its structure was crowned by a responsible elite, and its prophet was Burke.'[49] Self-government was the means, as we have seen, for an oligarchy to license its own extent of landed property. Davidson makes much play of the struggle for self-government by squatters, against the 'despotism' of colonial governors, and understood why they should be followed by merchants and professionals as 'liberals' but then gives little social or political space to those who inherited the republican tradition of 'national popular will' in Australia.[50] Lang, a Scots Presbyterian, offered a different prospect of self-government by fusing elements of 'Chartism, British nonconformist radicalism, and continental republicanism' variously inspired by 'Paine, Jefferson, Tocqueville, and the American revolution'. Although charged to be communism and socialism, Lang's programme for self-government was limited, in his own words, to 'universal male suffrage, perfect political equality and popular election'.[51] A democratic rhetoric, itself founded upon the right to property for the securing of subsistence, proved to be the basis whereby the economic domination of pastoralists and merchants might be ameliorated.

Indeed, Lang's republicanism was assimilated in, and directed towards, a wider movement of what was called the drift to 'social melioration'. Social 'sympathy' and 'society' building, as Nadel has shown in detail, was to counter the 'colonial degradation' of individualism and materialism in Australian colonial territory. Relieved of material poverty in Britain and Ireland, so it was assumed, ex-convicts and free immigrants gave vent to money-making as an end in itself. Before self-government and from the early 1830s, Lang and his associates established institutions, the foremost being the idea of Mechanics Institutes borrowed from Britain, to equally propagate knowledge of science *and* social understanding for the ideal of small property-owning artisans in industry. Australia was to be the home of the 'educated and morally elevated artisan'.[52] Such an endeavour might have run counter to Mayes' proposal to undermine the material basis of artisanship, through making labour markets flexible to provide for an undifferentiated supply of labour power for industry. But the doctrines of 'mental' and 'economical' development shared a common conjuncture between self-government, and the capacity for the administered regulation of labour power.

Social amelioration, after the sudden emergence of mass unemployment during the late 1850s, came to mean melioration of poverty. Through the gamut of measures, from the selection agriculture of smallholdings to state banks making loans to the poor, which were proposed in New South Wales and Victoria by many in Lang's tradition, the presupposition was that self-government encompassed responsibility for meliorating poverty.[53] Expanded self-government was expansion in the means for administering populations, within the territories of new states, and independently of each other.

However much the focus of politics was to become 'parochial' after self-government, and whatever the extent to which infrastructure was developed by governments upon the focal points of industry within each colonial state, the 'national' of colonial nationalism rested upon the positivist basis of social meliorism. This was what Nadel called 'the palladia of nationality'.[54] Victoria, Blackton pointed out, was the template for extending and constitutionally recomposing the palladia for the twentieth-century federal state of Australia:

> [Victoria] became the director of national sentiment, committed neither to the Anglo-Australianism of the pastoralists nor to Lang's doctrine of total independence. Victorian democracy, radical in theory, moderate in practice, focused on the compromise concept of an autonomous nation of Australian Britons. Hence Victoria's latter role as a center of enthusiasm for federation, economic nationalism, Australian imperialism in the Southwest Pacific, and as the birthplace of the Australian Natives' Association.[55]

In having described self-governing Victoria as a 'digger democracy', with Chartists and radicals sitting in the Legislative Assembly to counteract the pastoral oligarchy, Blackton understated the extent to which this 'compromise' allowed Mayes' doctrine of development to run its course through the development of capitalism. For at least two decades after 1856 it was a majority of 'urban mercantile, trading and professional men' who commanded a majority in the Assembly while the Legislative Council was dominated by capitalists and squatters. What the national compromise of self-government did represent was that trusteeship of a 'paternalist state' was associated with the official agency of development. Intentional development was to be both the palliative of social melioration and the simple positive purpose of developing 'roads and bridges' of infrastructure.[56]

In 1868, Charles Dilke, the literary critic, had remarked that Victoria 'probably represents an accurate view, "in little", of the state of society which will exist in England, after many steps towards social democracy have been taken, but before the nation as a whole has become completely democratic'.[57] Dilke might well have been commenting on a series of 1860s articles which had appeared in the *Westminster Review* bemoaning the 'paternal despotism' of self-governing Victoria, which, it was alleged, had become 'a scene of scramble and pillage' for public money. Candidates for political office had to contend according to 'who will make the best bargain with the Ministry of the day for the sale of their vote, and get the utmost possible in the shape of government monies for the district'. Elected politicians got public money for public works and a slice of public money for private benefit. Inveighing against 'bribery in the worst form' and corruption, the politician was cast as 'a kind of broker of claims, indemnities and grievances'.[58] Davidson, who recounted this critique of colonial self-government to be justified because it represented democratic form 'without the substance' of popular will and participation, concluded: 'Only half of the sum devoted to "Public Works", "a gigantic system of outdoor relief" over which parliamentarians scrambled, ever reached its destination, the roads and bridges, charitable institutions, lunatic asylums and public gardens.'[59] However, such were the national programmes of social democracy, noticed by Dilke as the foundation for allocating public money according to social need but which were deemed, equally, to be part of a corrupt political process. Public money was made possible by belief in doctrines of development which accompanied both self-government and the expansion of state administration.

It is the extent to which state administration was made more complex and more professional from the simpler original policing function which forms the impressive historical body of Davidson's account of state formation in Australia. His account, starting in the 1830s, parallels what Kay and Mott have written about Britain:

> The 1830s and 1840s saw the first establishment of the great bureaucratic structures of state administration to deal with poverty, disease, trade, insanity, etc. And once this had been achieved it was possible to eliminate the welfare and comfort functions from the police. This allowed the police to stand simply and formally for that department of state endowed with coercive powers. Conceptually this is the recognition that law is positive and derives from existing political conditions.[60]

Positive law is the conceptual key to what self-government meant generally. As we will see, the same concept of positive law appeared in the Province of Canada during the same period and, like Australia, formed the basis for development doctrine.

Furthermore, as we show through the case of Kenya in Chapter 6, the same associations appeared a century later but were now wholly defined by 'development'. Following political independence, development effort was popularly defined to be local schemes of self-help in which the emphasis was put upon what is now called political entrepreneurship. Positively assumed to be the medium of cooperative and community action for development, political entrepreneurship, the other side of the coin of political brokership by 'bribery', as in Victoria, was both hailed and condemned on the same ground of development. Political entrepreneurship could be vaunted as one form of popular participation and reason as the only reasonable means by which the wherewithal for development could be got. In confronting the administrative reason of bureaucracy as the medium of development, *community* development was claimed to be authentic development. But, abhorred as corruption, the same phenomenon would be condemned for arresting the true purpose of national development. Again, the foundation of the local impetus for development, together with the allocation of public money for schemes of development through political brokership, was development doctrine. Behind the emphasis on self-help was state administration and what was presupposed to be, as Davidson called it, 'the substantial nature of the State'[61] at the onset of colonial self-government and political independence of colonial territories.

Little was more oppositely apposite than George Woodcock's historical description of British Columbia administration at the beginning of confederation in 1866. Vancouver Island had an elected assembly with limited powers and no approach towards responsible government. 'In British Columbia there was not even a semblance of democracy; no appointed council of local worthies and no elected assembly.' Governor James Douglas 'ruled in Cromwellian style through his ten gold commissioners with their constables'[62] and whilst he might have built roads, he had neither intent nor apparatus to develop according to doctrine which had entered the system of self-government elsewhere.

Responsible government was the keynote of colonial claims for self-government. Apparently coined in 1833, by W.L. MacKenzie, the colonial reformer in Upper Canada, the concept of responsible government was employed, as by De Cosmos thirty years later in British Columbia, to contest what was called the 'family compact' in Central Canada, or 'family-company compact' in British Columbia to signify ties of affiliation between the governorship and the Hudson Bay Company. Family compact, as Graeme Patterson has shown, came to symbolise the negative connotation of a personal body of rule, embodying the court of appointees and favourites who surrounded the governor in the executive and legislative councils, of what the critics of colonial despotism regarded as arbitrary government.[63]

Once, however, rule was disembodied in representative government, the positive connotation of responsible government ceased to have any meaning other than as a surrogate for self-government. What Patterson emphasised for colonial self-government, namely that the executive was 'responsible' only to parliament, was the source of critique of government within Britain.[64] As we saw in Chapter 2, the associationist model of government, in the Newmanian tradition of Figgis and Laski, was based upon the idea that Dominion government, the culmination of colonial self-government, represented responsible government because it asserted the autonomy of parliamentary assemblies in relation to Westminster.

THE CASE OF QUEBEC, CANADA

In the eastern half of then British North America, unlike British Columbia, there was no gold rush, but there were the beginnings of what was to become mass emigration from Quebec to the United States. Emigration only became an issue of government, spawning development doctrine, when francophones formed an equal part of the first government of the self-governing colony of Canada. Development doctrine accompanied what has been called the development of the *Colonial Leviathan* of Canada.[65] A decade before self-government came to Victoria, the same principle of intentional development was at work in Quebec. While promoting railways and infrastructure, the government was supposed to develop, as one account has it, the associative means to protect 'the victims of the dominant economic system' of capitalism.[66] Given the longevity of the French presence in Canada, this was a case where there was ethnic association between self-government and nation because the purpose of development doctrine was to keep a francophone population within Canadian territory. Notwithstanding the failure to prevent continuing emigration to the United States, a francophone proletariat was created, but more slowly and in a more complex way than in Victoria.

And, akin to the invention of 'Australia', the invention of Canada, as Suzanne Zeller has stressed, owed much to mid-nineteenth-century ideas of science, the very medium through which development was invented, by way of the positivist approach to knowledge. It was in central Canada, Zeller has argued, that 'insecurity and retreat' was the premise for the positive view of development, especially when spelt out by the national idea that self-government was a constructive basis for Canadians to have 'a sense of direction, stability, and certainty for the future'.[67]

The case of Quebec, in the same period as that of Victoria, is treated extensively here because the immediate solution which was proposed to deal with the problem of the emigration of French Canadians from Quebec was what we have referred to as the agrarian doctrine of development. The problem of development was the same – the emergence of a surplus population which was officially recognised to be a national problem because it prompted the fear of social disorder and the loss of a population with the potential for productive force. But, whereas the general direct solution, as proposed by Mayes for Victoria, was industrialisation, in the case of Quebec, the solution was seen as state promotion of agricultural colonisation and settlement to soak up the surplus population and thereby forestall emigration to the United States. At the outset of the modern development problem, therefore, there may have been one problem but there was already variation in the proposed solution.

Emigration of French Canadians to the United States was regarded, by the first 1849 Report on the problem, as 'much more considerable' than generally believed, and threatened 'to become a real calamity'[68] for the Lower Canada Province of Quebec.[69] A loss of French-Canadian population was regarded by the governing class of Quebec as calamitous because it threatened what was called *survivance*, the slogan for the national survival of a French population in North America. The year of the Report, 1849, was significant because, amidst economic crisis and social distress, it signalled that there was both the desire and constitutional capability for the state to attempt to lock the French-Canadian population within Canada. There was, thus, a national idea of development. The question was whether there was economic capacity to solve the development problem.

State action to promote the national survival of the French population in Quebec was constitutionally possible in 1848 because the Canadian province was made self-governing through a political coalition which included the moderate wing of the nationalist movement, the *Patriotes*, who had sparked off rebellion in Lower Canada a decade earlier in 1837-8. Self-government came to Canada under a theory which was to be applied to the Australasian colonies during the 1850s and, however differently, to British colonial African colonies a century later.[70] Reflecting on this theory in 1853, Earl Grey, who as the British Colonial Secretary between 1846 and 1852 presided over the granting of Canadian self-government, admitted that self-government was,

> probably the best plan hitherto adopted of enabling a Colony in an advanced stage of its social progress to exercise the privilege of self-government; it may therefore be regarded as the form which representative institutions, when they acquire their full development, are likely to take in the British Colonies.[71]

National development for Canada was not the evolutionary metaphor of progressive development, which was used so characteristically by Grey, an apostle of free trade in an empire of colonies which had the potential for political independence. It was the culmination of political development, in Grey's restricted sense of development, which was the premise that made a doctrine of national development possible.

Different explanations have been offered as to why Canada got self-government in practice in 1848. One explanation argues that post-1846 British free trade policy made direct colonial rule a redundant form of securing imperial economic

Map 1 Lower Canada *c.* 1850

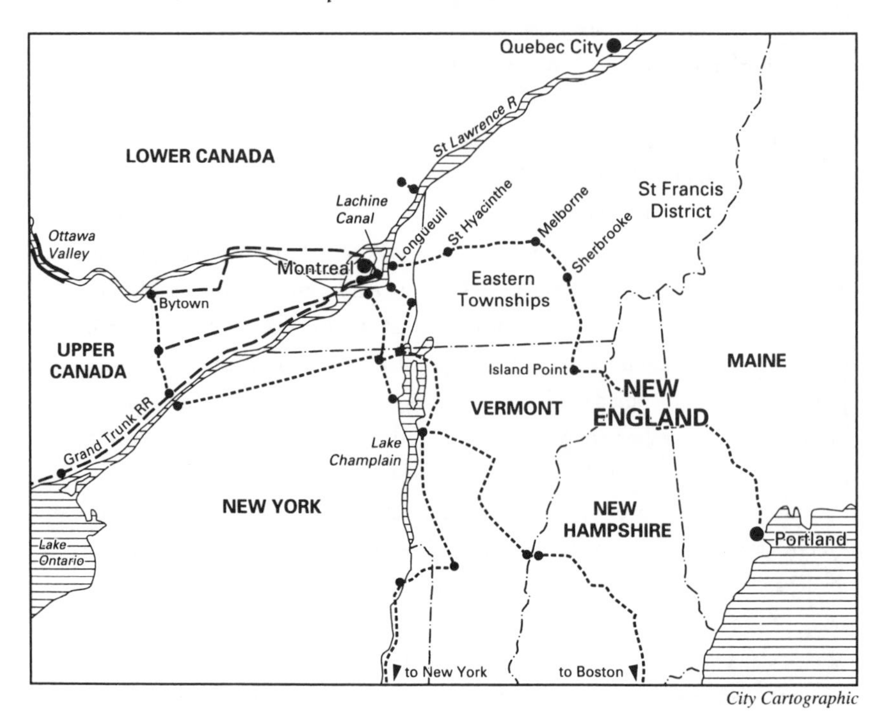

City Cartographic

control over Canada. Another maintains that the threat of United States aggression towards British North America imposed a defence expenditure burden which could be shifted to a Canadian government. A third holds that given nationalist demands, self-government was the only way to maintain 'collaboration of colonial elites in the perpetuation of Imperial rule'.[72] Yet, whatever explanation is used to explain Canadian accession to self-government, the two struts of national development – agrarian doctrine through systematic colonisation and the development of infrastructure with attendant trade protection – rested upon the political power which was exercised by the anglo-francophone Canadian establishment.

The politically established, moderate wing of the francophone movement had made a historic compromise both with anglo reformers, who agitated from the old Upper Canada for Canadian self-government, and with an anglophone bourgeoisie in Quebec, largely based on an amalgam of commercial and emerging large-scale industrial capital in Montreal. But, however much the francophones compromised to secure the reins of state power, the political establishment, known as the *Bleus*, stood as one with the oppositional republican wing of the movement, the *Rouges*, on the issue of culture and language. The official 'national idea' was to prevent assimilation of the French population through anglicisation, to aggressively maintain the French language and Roman Catholicism, and to secure the social equality of francophones with, if not in domination over, the anglophones of Quebec.

Pierre-Joseph-Olivier Chauveau chaired the 1849 select committee on emigration which had been sponsored by the Legislative Assembly and, given his literary bent, presumably wrote its Report. As an advocate, author of the first and 'classic *Canadien* agrarian novel',[73] member of the assembly, and then later, in 1867, as the first Premier of Quebec Province, Chauveau typified the establishment wing of the nationalist movement. In denouncing absolutism and colonialism and, especially, domination by mercantilism, Chauveau was 'a liberal imbued with the idea of progress for whom the French revolution constituted "one of the progressive developments of christian societies" and for whom the constitutional monarchy "united order with freedom"'.[74] Chauveau, it has been written, 'may have wanted to be a "nation builder", but for Charles Guèrin, the hero of his novel by the same name, building a nation was equated with building a parish north of Montreal'.[75] In this 1846–7 novel Chauveau espoused a model parish 'where industry played a crucial role' in a rural setting. In other words, as commentators have pointed out, agrarian bias in this case was no simple bias against manufacturing industry or categorical refusal of modern urban life.[76] Bearing in mind the liberal belief in progress which typified Chauveau, it is not amiss to suggest that his desire to sponsor and support rural colonisation was, to translate, 'the rationalisation of the check against the entry of a bourgeoisie into a continental world of affairs dominated by foreign capital'.[77] Or, to quote Little:

> In sum, the French-Canadian nationalists were not reacting against the rise of industrial capitalism so much as they were attempting to accommodate themselves to the challenges it presented. . . . They did not question the essence of the capitalist system, but they did wish to resist its offshoots – English-Protestant economic domination and massive emigration to New England.[78]

Liberalism was in play for a bourgeoisie in Canada. But, by the mid-nineteenth century, liberalism was no longer predicated upon the universalist postulate of a free world market which, when governed by the universal equivalence of exchange for all market agents, would guarantee the spontaneous outcomes of improvement for all, including the francophones of Canada.

For a French-Canadian bourgeoisie, as for Friedrich List, the universalist postulate of liberal theory had become a rationalisation for the domination of British capital in world trade and, as such, it had been reduced to the relativism of a national domain. To compensate for the loss of trade, production and productive force which followed from the particular national gain of British capital, the state had to complete the extension of freehold land tenure across the Province of Canada, to erect tariffs, actively support the construction of infrastructure and reform education. Compensation for unequal world exchange became the hallmark of development by intention and design. Only thus could the universal postulate of exchange guarantee the universal improvement of progress. In the Quebec case, where anglo capital was regarded as if it were foreign capital, the national arena of development, perforce, had to be reduced to the boundaries of a particular francophone entity if its bourgeoisie was 'to enter the world of affairs'. If it was to trade, let alone to command production, then a French locus of accumulation had to be protected according to a highly restricted conception of the national domain.

Because the concept of a national domain was so restricted, the idea of a modern national economy could not provide, at the outset, the capacity to succour an exclusively francophone focus upon development. It follows, therefore, that the *Bleus* had to believe that it was through colonisation of land that a francophone population would subsist in Quebec. A body of interpretation, including Little, assumes that it was possible to treat the domain of development *as if* it could be restricted to a francophone national economy in Quebec. There was, Little contends, 'a basic conflict between the forces of French-Canadian nationalism and those of large-scale English-language capitalism'.[79] This interpretation, as we shall see, is the 'two nation' thesis, which has strong affinities with both the retardation and staple theses of Canadian economic history. There is, however, another interpretation which finds a blend between the accumulation practices of anglo capital and the French national aspiration for development. Young expresses this 'blending' thesis clearly:

> The businessmen and French Canadian political leaders had no difficulty in establishing common goals. Economic growth was quite compatible with faith and nationalism. Railway expansion promoted colonisation: the industrialisation of Montreal encouraged the repatriation of French Canadians from the United States. English-speaking entrepreneurs provided jobs for the surplus rural population while not interfering with French Canadian mores or traditional power structures.[80]

Before we unravel these divergent development theses, it must be noted that they both mistake the intention of development for the immanent process of development. Little, for instance, contests the idea that the systematic colonisation of land was designed, as Young and others have suggested, to create an industrial labour force for capital in Canada.[81] The intention to develop through systematic colonisation, that is the deliberate movement of population for organised settlement on frontier land, was, for Little, 'a grim determination to survive by gradually expanding the traditional frontiers of their society'. We may agree that the intention was neither 'empty rhetoric' nor 'aggressive optimism'.[82] We can understand how the emigration of a French surplus population, from town and country, threatened the livelihood of French traders, lawyers and priests and how the founding of new peasant colonies, exercising demand for goods and services, could redeem the entrepreneurial and professional propensities of a nascent bourgeoisie.[83] But, the logical outcome of colonisation, as a part of a process of development, did not resolve the problem of development.

Little concludes, from his study of Eastern Township schemes, that colonisation was a 'dynamic but flawed response' to the problem of surplus population and emigration. For Little, colonisation was flawed not because the intention to develop on the basis of a 'peasant-style agricultural economy' was wrong but because the wrong model for such an economy was chosen. It was mistaken policy for the state to choose to settle poor households on poor quality land, with minimal expenditure on infrastructure, and little access to sources of non-agricultural revenue. With a different policy, Little concluded, it need not have followed that, in the second half of the nineteenth century, the result of colonisation would be a mere transfer of surplus population – from Britain to Canada or from one part of Quebec to another. A different policy, but with the same 'dynamic' intention, could have attenuated, rather than accentuated, emigration and the growth of a potential wage-labour force which was created for industrial capital, both in the United States and Quebec.

On the other hand, Young's 'blending' thesis, in taking development to be an objective process of the development of capitalism, fails to capture the subjective intent behind the original practice of modern development. There was an intention on the part of the French-Canadian establishment to develop Quebec and the colonisation movement did not arise out of an ideology, whether clerical or agrarian, which was separate from this intent to develop. Ramirez, for instance, has reflected on the problem of 'the extent to which the strictly clerico-agrarian version of the colonization ideology coexisted with a secular and developmental one'.[84] Colonisation ideology was a part of the secular intent to develop because the intention of development was more than ideology. The mid-century intention to develop was there because it had played a central part in giving an exclusive definition to a francophone population, a growing proportion of which lacked the means of subsistence long before the manifest economic crisis of the late 1840s. What made the intention immediate was welfare, the recognition of a now politically responsible government that it had to be responsible for the means of subsistence of a French surplus population. If colonisation was a policy of development, it was so on the basis of welfare, that of making a surplus population bear the direct burden of its own subsistence.

In 1848, before the Report on emigration, Chauveau moved a motion before the Quebec City branch of the *Associations de Townships*. The 'lack of work and famine' in Quebec City, Chauveau is reported as saying, 'made it essential to send the surplus labouring population to the Townships as colonists'.[85] Chauveau wrote this report with a general theory of systematic colonisation in mind. He might have been referring to emigration from England in an early paragraph of the Report when he wrote that:

> It is true that when a country is sufficiently peopled, when the whole extent of its territory is in a state of cultivation, when all its agricultural, industrial manufacturing, and commercial resources have become exhausted in nourishing an overflowing population, without any other means of existence left than begging, it is true that under such circumstances, an emigration which has the effect of transforming families, which were a burthen to the older community into founders of a new one, is a benefit to the country thus afflicted and to mankind in general.[86]

The Report continued, in like vein to Mayes in Victoria, now referring to Quebec:

> But in a new country, a portion of whose territory is in a state of cultivation, whose population is nowhere numerous enough to present the painful spectacle of pauperism; in a country which, instead of having an excess of population to cast off, calls on the contrary, to its assistance the strength and capital of foreign emigration; the double fact that this foreign emigration leaves in it very few settlers, and that the inhabitants themselves emigrate in great numbers to foreign countries, must arouse the attention of the Legislator, and lead him to inquire, whether all is right in the order of things which it is his duty to direct and modify; whether it is nature herself which does not offer to the inhabitant of the country, sufficient advantages to induce him to remain; or whether it is not rather society, which has neglected to turn to advantage the field which nature has offered.

Chauveau moved on to distinguish between spontaneous emigration and systematic colonisation:

> It would undoubtedly be absurd to attempt to prevent or even discourage those inhabitants of the country who can find more accessible, better cultivated, and cheaper lands elsewhere, from seeking

out of their country for that which their country denies them. . . . But Your Committee sincerely believe . . . it would be possible for the Legislator to adopt efficient means of settlement, which should be alike advantageous to the settler and to the Government, whose duty it is, at all events, to ensure for every part of this Province the best possible chance in the race of rivalry, by which the different countries of America are incited to advance in population, in riches, and in material progress.[87]

Chauveau had taken the core of the influential theory of systematic colonisation, a theory developed to plan the movement of population from Britain to colonies, and applied it to the movement of population from Quebec to the United States. Whereas the theory in its original guise had been a means to foster a centralised system of British imperial control, it was now used to foster development for an exclusive French national endeavour in North America.

SYSTEMATIC COLONISATION

Edward Gibbon Wakefield was responsible for the theory of systematic colonisation which Chauveau appropriated. This theory is regarded as the source of the model for the post-mercantile doctrine of the new imperialism of the mid-Victorian period.[88] Wakefield's theory has been taken to have been part and parcel of constructive development of 'positive programs of Empire' in contrast to laissez-faire doctrine whose sway, however much derided as 'myth', made it so difficult for constructive development to form the basis of official British colony policy in the nineteenth century.

'It is the great merit of E.G. Wakefield', Marx wrote in the renowned last chapter of the first volume of *Capital*, to have said 'not something new *about* the colonies, but *in* the colonies, the truth about capitalist relations in the mother country.'[89] What Marx alluded to was that Wakefield had established the fact that without wage-labour, neither money nor means of production could become capital. Wakefield referred to the colonies,

Where land is very cheap and all men are free, where every one who so pleases can easily obtain a piece of land for himself, not only is the labour very dear, as respects the labourer's share of the produce, but the difficulty is to obtain combined labour at any price.[90]

It was contrariwise in Britain, where the only property of labourers was 'their labour', where there was an excess of sellers over buyers of the property of labour on the market so that, whatever the merits of English agriculture with its combined labour employed by farmers who concentrated exclusively on agriculture, 'the English agricultural labourer' was 'a miserable wretch, a pauper'. Rural misery, accompanying growth in surplus population, had led to agrarian revolt in the 1820s and 1830s and presaged, so Wakefield feared, general social revolution.[91] Wakefield's solution was deceptively simple: transfer the surplus population from where 'the field of production' was limited, both in relation to capital and labour, to where 'the field' was relatively expansive. Colonisation would build up labour and diminish the colonial field of production.

A system of protection, whether for manufacturing industry or agriculture, is usually seen as the hallmark of development doctrine. The 'National Policy' of Canada, for instance, as enunciated by the federal prime minister Sir John Macdonald in 1879 has been interpreted as being a policy of constructive development based on the protection of manufacturing industry. However, a system of colonisation, as we shall later see, is not only equivalent to a system of protection for constructive development but is logically prior to it since it starts from the concept of population rather than from a theory of trade.

The theory of colonisation may be set out as follows. Voluntary emigration from Britain, insisted Wakefield and his many supporters such as John Stuart Mill, was a 'natural' economic process through which the active individual would become less poor by moving from a region where the price of labour was lower to where it was higher. Colonisation, on the other hand, was a positive 'duty' of the state, an obligation which had been neglected by officials in both Britain and the colonies.[92] To plan the movement of population was to plan the use of land. Systematic colonisation would be self-financing if the state nationalised waste land in the colony. Land would be sold by the state at a 'sufficient' price to the colonist. The colonist would have to earn and save wages to pay this sufficient price of land.

With land as property, the colonist would employ new migrants, the poorest of the poor from the old country, whose passage would be state-financed from revenue gotten from the sale of state land. The sufficient land price, Wakefield proposed, was to act 'as a check, almost a bar, to the appropriation by persons not able, or not willing to, use their property'. Each person, according to a 'social compact', would appropriate 'the right quantity of waste land', the quantity that the person would be able to use to meet subsistence.[93] Here was the ideal scheme for disposing of the relative surplus population, or in Marx's words, 'the production of redundant wage-labourers, redundant that is, in proportion to the accumulation of capital'. Marx's comment on Wakefield was: 'Today's wage-labourer is tomorrow's independent peasant or artisan, working for himself. He vanishes from the labour market – but not into the workhouse.'[94] Wage-labour, Marx concluded of Wakefield, was the certificate the redundant worker had to earn to buy the means of production which he would use to meet his own means of subsistence.

Systematic colonisation, as a model, was criticised on a number of empirical counts. How was the sufficient price of land to be determined?, it was asked. Wakefield could only say it was the 'price as will keep the wages of labour and the profits of capital at a maximum'.[95] The model could not apply, it was said, to relatively dry, timbered or hostile land, the colonisation of which would require capital beyond the scale of even an association of colonists.[96] Thus, the model was adduced to be successful for Wakefield's colony of South Australia but not for the timbered, hostile Quebec frontier lands. Notwithstanding these criticisms, it was the logical outcome of Wakefield's theoretical conception which was at fault. As relatively easily cultivable land was taken up by a regular supply of labour, an essential condition for the model, the price of land would be bid up. On average, each new wage-worker would have to work longer to buy the land certificate and, without a perpetual state subsidy to capitalise land, the outcome would be that the worker would spend a working lifetime in wage-labour. Thus, it would be impossible to realise Wakefield's ideal of, 'a class who, as labourers should become capitalists and landowners, would fill their place in the market of labour; becoming themselves, in time, capitalists and landowners and having their place filled, in turn, by immigrants of the same class'.[97] Wakefield understood his own problem. He proclaimed in 1836, before a House of Commons Select Committee, that if the state land price be raised above the sufficient price, 'you would condemn the whole class of labourers to a long term of service'.[98] This is why the logic of systematic colonisation led to the creation of a wage-labour force in Canada and other territories of the Empire.

Wakefield's role in Canada was as much political as economic. Although he had been an unofficial adviser to Durham and was alleged to have written the appendix to the *Durham Report* which dealt with land policy, Wakefield dissented from Durham's 'two nations' comment that he had 'expected to find a contest between a government and a people' but had found instead 'two nations warring in the bosom of a single state' finding 'a struggle, not of principles, but of races'.[99] Durham was reflecting on the 1837–8 revolt, the very reason why he had been commissioned to inquire into the political and economic conditions of Lower and Upper Canada. As a colonial reformer, and supreme wire-puller, Wakefield had argued for colonial self-government, played a key role in furthering the union of the Canadas in 1841, and assisted in founding the first municipal institutions for the Province of Canada. He organised the Anglo-French political coalition, that of the Baldwin–LaFontaine establishment, which, whether in government or opposition, carried the Province to responsible government in 1848. Lafontaine triumphed over the French radical opposition because, according to the historian Monet, a majority of French Canadians were persuaded 'of the fundamental premise that their nationality' was 'best preserved and insured by the organic vitality of the British Constitution' in Canada. Chauveau, the typical anti-colonial anglophile, was part of this coalition and of the French-Canadian coterie which was in constant contact with Wakefield.[100]

Described as 'half genius', 'half charlatan',[101] Wakefield was not averse to playing at land speculation in Canada while he accompanied Durham in making what became a long-lasting critique of colonial land policy in Canada. During 1835, Wakefield took over the shell of the North American Colonial Association of Ireland and turned it into a fully fledged land company which was to colonise land according to his model of systematic colonisation. Buying the Beauharnois seigneury in 1838, after it had been sacked during the rebellion, Wakefield sought, between 1839 and 1842, to sell divisions of its land to immigrants who would work for wages on public works. Public works were to be financed by loans from Wakefield's company, acting as a banker as well as land agent for the Provincial government, the legislature of which, Wakefield urged, should dispose of all state land sales revenue. Despite opposition from the Colonial Office, Wakefield got most of what he wanted, including the diversion of the new canals of the St Lawrence system through his Beauharnois estate.[102]

'Why have so many English and Irish labourers who had emigrated to Canada', Wakefield asked in *England and America*, 'removed from Canada to the United States?'[103] The answer was that population had moved to territory where the average land price was higher, employment more regular and the average wage rate higher because, all other things being equal, land policy in the United States conformed more to the Wakefield model than it did in Canada. Pointing to the 'remarkable contrast' between the different sides of the Canada–United States border,[104] Wakefield bemoaned the lack of a systematic land policy on the Canadian side.

Waste land had been nationalised but then disposed of in such a way that speculators had appropriated land to an extent that land settlement was dispersed rather than concentrated. There was conflict between the principle that waste land should be held in trust by the state to meet the imperial and colonial government need for finance, and thus sold expensively, and the principle that it should be granted cheaply to settlers to speed up settlement.[105] The upshot was that land was sold cheaply but to speculators rather than settlers. The historian Macdonald has amplified Wakefield:

> In theory, the Imperial Government visualised Canada as a land of peasant proprietors, but its indiscriminate policy of free grants, interspersed with various reserves, scattered the settlers over too wide an area which precluded mutual assistance, imposed crushing economic burdens upon all, and threw large tracts into the hands of speculators.[106]

To make closer settlement possible, along the lines of United States practice, Wakefield inspired the 1841 Land Act. By providing free grants of up to 50 acres of land, and providing larger grants for sale, the Act was designed to fight speculation. Macdonald has shown how the 1841 Act was too little, too late. Indeed, he remarked that the 'salutary lesson' to emerge in the 1841–6 period was 'the ease with which a redundancy of population could be created in a new country, bringing in its train many of the social evils that were usually associated with it in older established societies'.[107] Whether or not it was the case that the official recognition of poverty was confined to the seasonal dimension of hardship in winter, urban unemployment in Montreal was open enough for a workhouse to be established there in 1818. Teeple argued that a relative surplus population had emerged by the 1820s:

> It is therefore somewhat ironic that, by the time Wakefield's policy of systematic colonisation was instituted, the problem of labour for industry was largely solved. By 1841, there was no shortage of workers who were landless and in search of jobs. The question had become one of how to keep the immigrants and settlers in the Canadas, to prevent their emigration to the United States.[108]

This was the question that generated development doctrine in Quebec in 1849, but it did so, bearing Wakefield in mind, to confront the emigration of the francophone population from an old established colony of North America.

THE SIGNIFICANCE OF 1849

The year 1849 was a momentous one, at the end of what has been called 'the most critical decade of the colonial period', while near the onset of the 1850s which 'really saw – for better and worse – the beginnings of modern Canada'.[109] It was a violent year. There were French-Canadian rural riots against taxation for compulsory schooling.[110] In Montreal thousands of members of the old anglo Tory Party went on rampages in protest against the Rebellion Losses Bill which extended state compensation to French Canadians whose property had been destroyed during the 1837–8 rebellions.[111] During the sacking of part of the Assembly, most of the Chauveau committee records were destroyed.[112] The year of 1849 was also the crucial moment in an immediate economic and social crisis which arose out of the final stages of the British abandonment of formal trade protection in 1846 and the mass immigration of Irish as a result of the Great Famine. 'From the middle of 1847 until the end of 1849', report Easterbrook and Aitken, 'the commercial life of Montreal and Quebec was practically stagnant'. Over 1847–8, 'the extent of urban poverty and suffering was . . . overwhelming'. In 1847, following fires which virtually destroyed Quebec City, 16,000 Irish immigrants and untold numbers of Canadians in Quebec and Montreal died from cholera.[113] Since total government revenue had fallen by one-quarter between 1845 and 1848, the Canadian government, we are told, 'was critically short of cash in 1848, and the situation continued bleak throughout 1849'.[114] Disaster was in the air.

The year 1849 was also a significant turning point in the pattern of immigration to Quebec. Between 1827 and 1847, nearly 45 per cent of British immigrants arriving at the ports of Quebec City and Montreal crossed the border into the United States; the proportion for the twenty years after 1849 was only 15 per cent. In these later years British migrants moved directly to the United States because, after the collapse of the timber trade in the late 1840s, they could no longer be induced onto British timber ships.[115] The collapse of the timber trade was also the precipitate cause of the deep depression which struck Quebec between 1847 and 1850. Furthermore, a new 1849 statute tightened immigration control by requiring the captain of each arriving ship to report destitute and/or disabled passengers. It was hoped, thereby, to encourage 'a more healthy and useful class of emigrants' from Europe.[116]

Last, but most importantly, diverse historians generally agree that 1849 represented the end of a long period of mercantile colonialism, old commercial capitalism, and forms of labour organisation which have been associated with proto-industrialisation. Ouellet has claimed that the economic structure of the mercantilist system 'remained in place until 1849, as did the doctrines that justified it', while Acheson has referred to 1849 as the year of 'the final crisis of mercantilism'.[117] Innis, of staples thesis fame, used Alexander Galt, the Canadian Minister of Finance who said in 1849 that the 'only hope lay in the fact that people had at last the management of their own affairs', to illustrate his thesis that 'the autonomous capitalist state replaced the old colonial state'.[118] Bryan Palmer has written of the 'vast transformations' of economic and political life between 1846 and 1848. The 1840s had been a decade of 'contradiction and ambiguity' of relations between capital and labour, characterised by paternal order and 'a delicate balance' between defiance and submission on the part of labour. However, 'the insurrection of labour', over 1853–4, was the result of the extent to which the distinction between 'rough' and 'respectable' workers had rapidly diminished.[119] At this turning point of mid-century, a new order was being created. 'The state', Palmer reflected, 'once little more than a style of rule, was created to address substantive questions of national development and political consolidation.'[120]

But if there is much agreement on the significance of 1849, there is a major fault line in the interpretation of what the substantive question of development was afterward to become. This division is around the way in which the place of Canada was, and is, seen in relation to the United States, from both historical and historiographical perspectives.

From an historical perspective, the most aggressive response, on the part of both anglo- and francophone commercial capital in Quebec, to the end of British imperial protection in 1846, was organised pressure for annexation with the United States. The Annexation Movement of 1849, inspired by the desire for revenge among anglo Tories who made the charge that the British state had abandoned Canada, drew in the support of the radical French-Canadian intelligentsia centred on the *Institut Canadien*. However much this alliance of anglo merchants and French radicals might seem to be an absurdity, it rested on an impeccable principle of liberal logic. What Cobden and the Free Trade Movement had done against the aristocracy in Britain, so the argument ran, the United States would now do for business in Canada. Seigneurial tenure, church tithes and clergy land reserves would all be finally abolished. 'By the early 1850s', Young has written, 'the process of establishing freehold land tenure across Lower Canada had been set in motion.' Positive law, including state regulation of contract, corporation and bankruptcy, had been extended to 'organize, publicize, and standardize private business relations'. It was through the alliance of francophone radicals and anglo merchants that the 'positive state' was firmly established by mid-century.[121]

The Wakefield ideal would be realised, and canal and railway systems would be constructed at lower overhead costs, particularly in relation to the cost of self-government in Canada.[122] 'We can see no way to get out of the scrape', the Annexation Manifesto declared, 'but by going to prosperity, since prosperity will not come to us'.[123] For the francophone radicals, the United States offered the republican tradition of a constitution for liberty, and thus for them the culmination of ideal government. 'The United States', wrote one radical, 'is the only country in the world which is really able to be called a pure democracy.'[124] Above all, annexation to the United States would dissolve the economic boundary which made the southward emigration of population a national problem.

Ranged against the Annexation Movement was the panoply of Canadian political force which had come to define the modern origin of the Canadian state as being coeval to national development for a population which officially refused annexation to the United States. The established reform coalition of Baldwin–LaFontaine fought off annexation with the support of the Catholic clergy according to 'the fundamental premise', as we have noted, that French-Canadian nationality was best guaranteed by the British constitutional connection. Lord Elgin, the Governor-General, fended off the annexationist movement and British liberal pressure to relinquish the Canadian connection by telling Earl Grey, the Colonial Secretary that he could place:

> the connection on a surer foundation than ever, if I could only tell the people of the Province that as regards the conditions of material prosperity they would be raised to a level with their neighbours. But if this be not atchieved [*sic*] – if free navigation, and reciprocal trade with the Union be not secured for us – the worst I fear will come, and at no distant day.[125]

Canadian capital was freed from the navigation laws, which, in 1849, confined seaborne traffic from the St Lawrence to British-owned ships. The Reciprocity Treaty with the United States, enabling the mutual abolition of import duties across a range of 'natural products', was finally signed in 1854[126] with the most intense Canadian pressure for the treaty being applied during and after 1849. For the 1848 Baldwin–LaFontaine government, as Tucker records, the 'manufacturing districts of the United States would often be the best market for Canadian agricultural products'.[127] Reciprocity lasted until 1866, when it was abrogated by the United States. The Canadian response was confederation in 1867. Elgin, therefore, achieved the immediate objective of keeping Canadian self-government tied to a British political connection while accelerating the force of trade with the United States to appease the economic thrust of the annexationist movement. In doing so, however, official policy was to lay claim to development, a policy which went far beyond tariff and trade. Development was, in its Quebec integument, the means to replace the export of population down south with the means to export things to the United States.

Unlike his predecessors, Elgin was forced to admit to the necessity of development. Dalhousie, for instance, however much he was later to be a proponent of development for India,[128] had been an apostle of progress and not development for Canada. Opposing grants of Crown land to the Canada Land Company, Dalhousie wrote in 1826 to the Colonial Secretary:

> The settlement of a young country like Canada could not be forced; it must be progressive and slow, one step must follow another in regular succession, and by the cumulative power of a population, prospering in its own wealth, and independent of the monopoly or the means of any great company.[129]

Not in favour of development in Canada, three decades later Dalhousie was to become an apostle of development for India. In 1854, *before* the Indian Mutiny, he urged for Indians the 'development of vast resources of their country' through the central planning of railways, roads, communications and education. Likewise, in the name of securing the rights of occupants to property, land reforms were so swiftly and broadly implemented that even relatively small landowners who rented out land were aggrieved. By 1849, Dalhousie's later developmental injunction for India was being applied to Canada. Francis Hincks, the government Inspector-General [Treasurer] of 1849, who later succeeded Baldwin as the leader of the anglo part of the government coalition, had long been in favour, from the vantage point of his rural political base in Upper Canada, of development. In 1839, Hincks wrote:

> I am a zealous advocate for all practical measures of public improvement, which are calculated to develop the resources of the Province, giving a preference, however, to the improvement of the public Roads, on a systematic and uniform plan, as the measure of most importance to the agricultural population.[130]

In 1849, his zeal for development was extended from roads to canals and railways. Like Wakefield, Hincks had been a moving force behind the 1841 union[131] and his development plan of the 1840s was little more than the Wakefield model of systematic colonisation.

Hincks has been used as an example to contest the proposition that self-government and 1849 represent a revolutionary or radical break in policy. It is an argument which has had a long and general pedigree, associated with the idea of 'the imperialism of free trade'. It holds that there was no necessary correspondence between a policy of free trade, founded upon British commercial domination with London as the world trade centre, and colonial political freedom.[132] The specific Canadian argument[133] is that Hincks in 1848 believed, in spite of free trade, that a British surplus population could be 'removed to a country where under a system of free commercial intercourse the products of the soil' would 'be exchanged for British manufactures'.[134] To make this policy possible, British capital had to be procured to finance colonisation and the construction of public works. Sales of Crown land would fund the repayment of loans. This policy, referred to as 'a coherent development strategy' for 'a process of development',[135] was that of Wakefield, and was adopted by Lord Sydenham, the first governor of the Canadian Union in 1841. Why, therefore, should 1849 be seen as so significant for development doctrine in Canada?

The answer is that, however much Hincks may have been convinced of development before 1849, it was only in the midst of disaster in 1849, and after self-government, that the state was compelled to maintain subsistence for a Canadian population. Development *practice* arose out of the conjuncture of 1849. A development strategy was mounted to fend off the political threat of annexation to the United States by providing the economic wherewithal for Montreal to be maintained as a North American commercial centre. Development was also the means of maintaining a surplus population in Canada by making that population produce its own subsistence in agriculture. There was no necessary inconsistency between the mounting of a development strategy for commercial capital in Canada and an agrarian doctrine of development. Both the strategy and the doctrine put the emphasis upon 'public works', in particular roads and railways, the development of which made industrialisation possible. Public works may have been justified, along the lines of the Wakefield model, to promote the employment of immigrants from Britain, but the compulsion for development was equally dictated by the idea of agricultural colonisation as the means by which the emigration of population from Canada would be contained.

The first major railways were built, not from East to West across continental Canada but from North to South, across the St Lawrence and the United States border. For Elgin and Hincks, the transcontinental Canadian railway was the first desire of their favoured development design. It took until 1885 for the trans-Canada railway to be completed, well after Confederation in 1867, when the idea of the railway became synonymous with that of Federal Canada itself. But, in 1849, a previous colonial era of canal construction had limited the financial capacity of the post-colonial state to raise either official finance or private loan capital from Britain. The St Lawrence canal system had neither capitalised land nor increased trade to the extent that had been imagined when it was planned. Therefore, when a railway system was deemed necessary to complete what the canal system was set out to do – to lower transport costs so that the trade in primary products in the trans-Atlantic trade would pivot upon Montreal rather than New York – the economic solution was railways to connect with United States systems. Equally, railways running across Quebec, and south across the border through the Eastern Townships, served the French-Canadian doctrine of development – to stimulate the export of agricultural and other products to United States markets in place of the export of population for manufacturing across the border.

In 1850, 9000 miles of railway had been laid down in the United States, whereas only 66 miles had been constructed in Canada.[136] Canada's first and only railway until 1846, the Champlain–St Lawrence, which was completed in 1836 to serve the Eastern Townships and the export of lumber to the United States, was partly sponsored by French-Canadian merchants. The extension of the railway to connect with the New England system, and thereby to Boston and New York ports, was only completed in 1851. Likewise, the impetus for the competing railway connecting to Portland in Maine, the St Lawrence–Atlantic, which was completed in 1853 and later became the basis of the infamous Grand Trunk, came as much from business and land companies in the Eastern townships as it did from commercial capitalists in Montreal and Portland, both competing with business in New York. Most private finance for the railway came from 'small investors', including French Canadians, who bought shares in the railway in 1846.[137] Rail links across Canada and to the North of the St Lawrence were inspired by the French-Canadian colonisation movement.

Young has looked at how 'French Canadians, attracted by the potential for profits and economic growth, joined enthusiastically in the scramble to link their villages' to railways and how 'clerics served as directors of colonisation railways that might aid their communities'.[138] Thus, although the Montreal Colonization Railway, running westward from near Montreal to the Ontario border and which later became a central part of the Trans-Canadian Pacific Railway, was the enterprise of Hugh Allen, a leading anglo capitalist, it commanded strong French-Canadian support. It was George Etienne Cartier, 'a Montreal bourgeois' friend of Chauveau and main legal mover for successive reform governments while being the Grand Trunk's solicitor, who is described as the 'father of the Canadian railway system'.[139] The French-Canadian bourgeoisie acted within the interstices of competition between rival railway companies. To claim, as does Young, that 'rhetoric of the colonization movement' was no match for 'the realities of transcontinental commerce'[140] is no less true than our claim that it was the reality of the idea of development in 1849 which made the state development of infrastructure possible for commerce. As Tulchinsky confirms, Canadian railways 'were from their very beginnings specifically and unequivocally designed to feed traffic into (and later lock into) the northern United States railway system'.[141] And, after all, the blending of 'colonization and economic development'[142] is nothing other than Young's argument about development in this period writ large.

It was the Guarantee Act, passed in 1849, which made it possible to complete the construction of the cross-border and colonisation railways. The 1849 Act guaranteed state subsidies to pay interest on up to half of the value of authorised bonds held by private investors provided that one-half of a line had been constructed and that it was more than 75 miles in length. In addition to Canadian government grants, local municipal authorities, under the Municipal Corporation Acts of 1849 and 1852, were empowered to make grants for railway construction. During the 1850s, provincial and local government contributed nearly one-third of the cost of finance for the Grand Trunk.[143]

The Elgin–Hincks schemes for a transcontinental railway had been stymied by the reluctance of the British Treasury to guarantee City of London loans for railway development on Canadian, as opposed to British, terms. Thus in 1849, as in 1852, the following year and in each year, 1857–9, imperial authorities refused Canadian terms; over 1850–1 and 1861–4, the Canadian government refused British terms for loans. 'Colonists', we are told, refused direct taxation to pay for imperial designed projects which would benefit imperial defence and commercial interests: 'Many preferred to commit their limited collective resources to railways to United States markets.'[144] Hincks complained in 1849 about the bias of the City of London against Canada: 'British capitalists do not choose to place the same confidence in [Canadians'] honour that they do in that of the people of the United States'. He argued that Canada's 'present difficulties' stemmed from the fact that since 1845, the 10 per cent surplus of state revenue over expenditure, with most revenue raised by tariffs, had been applied to canal construction rather than to debt redemption.[145]

If the refusal to redeem debt was met by the refusal of banks to extend further credit, then, argued Hincks, this was tantamount to a British abdication of trusteeship for Canada, impelling annexation to a protectionist United States which would box out British manufacturers from Canada. Self-government for Canada, Hincks continued, meant nothing other than the control over revenue by the government of Canada. The 1849 Guarantee Act 'is not one of mere parchment, but the ways and means' which 'have been provided beforehand to enable the Government to fulfil its obligations'.[146] One such obligation was the railroad, 'no mere luxury but a necessity', as Morell pointed out.[147] Baring Brothers, the bank which was the prime source of loan finance from London, had turned Hincks' development strategy on its head: 'We may think it improvident on the part of the legislature to have engaged in such extensive guarantees before the credit of the Colony was firmly established, but we are aware of the state reasons for the promotion of public works & especially railroads.'[148] To assuage Baring Brothers, the liberal reform government forsook state ownership of railways, the principle which the colonial government had adopted for canals,[149] but it did not accept the bank's questioning of its development imperative. In the event, guarantees of finance to private companies were to fuel the 'corruption' that made railways synonymous with 'an orgy of development and speculation' in Canada.[150] 'To contemporary minds', Easterbrook and Aitken reflected, 'it was not easy to distinguish between short- and long-run difficulties.'[151] Although the immediate impression of economic and political crisis faded fast after 1849, the belief in development fastened on in 1849 stuck hard and made the development of capitalism in Canada possible.

Corresponding to the 1849 historical fault line are two foci of historiographical interpretation. The most influential interpretation, following the contours of the staples thesis, because it inclines to the national of development, puts the focus on the east–west dimension of Canadian development. The second, focusing on a north–south dimension, emphasises the continentalism of America. Whereas the first sees Canadian development as being determined along a parallel but different line to the United States, the second sees it as being a replication, a delayed sequence following after the development of the United States. The first emphasises the external determinants of development, stresses dependency, and sees Canada as a unique case of development; the second takes Canada to be a case, but not necessarily a unique case, of capitalist development in which the emphasis must be on internal forms of change which create the essential conditions of capitalism.

THE NATIONAL DIMENSION: THE STAPLES THEORY OF DEVELOPMENT

There is no one staples theory of development, although what is referred to as the staples thesis has been identified as that of the Canadian economic historian Harold Innis. In Canada, the 'frontier' theory of Turner and Myint's 'vent for surplus' theory of trade and development have both been closely associated with staples theory. Innis was neither the original proponent of the staples thesis as an account of colonial development, nor the first to apply the theory to policy. The Keynesian economist W.A. Macintosh[152] was, for instance, influential in integrating Keynesian ideas with staples theory in the 1930s.[153] If the staples thesis has become irrevocably tied to Innis it is because he was so fertile in his attempt to develop a general theory of development, a theory which was propounded through the experience of capitalist development in Canada. Starting with Adam Smith's evaluation of progress in British colonies, Innis ended up with a prognosis for planned development in Canada and, thereby, gave general stress to the *values* of development. Innis quoted Smith: 'Compare the slow progress of those European countries of which the wealth depends very much upon their commerce and manufacturers, with the rapid advances of our North American colonies, of which the wealth is founded altogether in agriculture.'[154] But for Smith, as for Innis, the rapid improvement in agriculture in the colonies as in Europe, was an effect and not the result of commerce and manufacturing. Smith referred to how 'water carriage' had opened up North American markets and Innis turned Smith's point into a general hypothesis: 'The improvement in transport facilitated the expansion of external and internal trade.' Improved technique was the result of advances in manufacturing and the improved techniques of transport facilitated new techniques for the extraction and production of primary products.

Innis studied the staple products of fishing, fur and base minerals to prove his hypothesis; his followers studied wheat, oil and hydro-electric power. From these studies, the staple thesis emerged: 'The technique of production of these various commodities involved sharply differentiated economies.'[155] This was the staple thesis which was exported across the world, particularly to Africa, where Hopkins' influential *Economic History of West Africa*, for instance, is founded upon a distinction between plantation and peasant 'economies', a distinction which is determined by the physical qualities of the commodity and the technical form of its production.[156]

Hopkins' gross and objectivist staple theory is a one-sided interpretation of Innis' thesis, an interpretation which stems from the powerful influence of Thorstein Veblen upon Innis. Veblen had sketched out a general theory of development which was largely inspired by a version of Darwinian theory after Comte and Spencer. Innis, in objecting to the then prevalent view that Veblen had derived his idea of development from Hegel and Marx, was struck by the emphasis which the philosopher-turned-economist had put on 'the growth and decay of institutions and associations'. Innis took his theory of development, the process of growth and decay, from Veblen.[157] Canadian development was the means by which there was adaptation to the decline in one form of commodity production through innovation and expansion in another form. Thus, fishing was replaced by the fur trade, the decline in fur by the rise in timber, the fall of timber exports was compensated for by the rise of wheat export production during the nineteenth century and so on. Successful adaptation, through adapting the organisation of production to the specific commodity in a sequential series of staple products, made development possible. A failure to adapt, as in the case of wheat in Quebec, as opposed to Ontario of the same period and so highlighted by Ouellet as the cause of the 'agricultural crisis and the development problem of emigration', meant underdevelopment.

However, once the organisation of production is given focus, the Innis version of the staples thesis cannot be a gross objectivist theory of development. Innis, as Patterson has shown, increasingly came to regard the mechanistic model of change with disdain. Borrowed from Newtonian physics, the mechanistic metaphor of social science was used to postulate atomic-like entities which maintained their existence, as independent entities, during the course of change. Like Keynes, but probably separately from him, Innis came to employ what Aristotle had called 'formal causality'. Given the geographical space of the St Lawrence river, and following his distinction between different forms of the St Lawrence communication system and the differing content of staples which followed each other in time, Innis' staples thesis could be interpreted mechanistically. For instance, the change in the content of each staple, in the series of staple products, could be adduced to changed forms of transport, as from canal to railway, or vice versa, as when wheat was substituted for timber, and timber for fur, when the ship-canal and road supplemented and then replaced the river-*bateau*. However, the thesis was not only that a new form of transport gave new content to the staple, or the other way around in that content gave form to communication, but that different forms of communication interacted with each other to change the forms of media by which communication was organised. It was the interaction between forms of phenomena, which gave new content to form and, in particular, that of the state which organised communication for the purpose of creating the possibility of development.[158]

Furthermore, as Marshall McLuhan has explained his later work, Innis' 'attention shifted from the trade-routes of the external world to the trade-routes of the mind'.[159] Communication was interpreted to be the media of both material staple and that of the understanding of mind. Yet, before he arrived at 'mentalism' and as will become clearer later, Innis put primacy upon the subjective intent to develop communication media, including that of mid-nineteenth-century transport in central Canada, in governing both what had made development possible and what he meant by values of development. The subjective meaning of phenomena, according to Innis, could not permit belief in the linear postulates of progress, especially when fixed in space and time.[160] Bias of space, Innis might have suggested, was what had made nineteenth-century thought confine manufacturing to temperate zones in the same way that bias of time, which he did emphasise, made progress a linear function of time and, therefore, followed the sequence of stages necessarily slowly. 'Space and time', Innis later quoted from one possible account of the new twentieth-century physics, 'and also their space-time product, fall into their places as mere mental frameworks in our own constitution.'[161] States of mind, inhering in organisation, were what ultimately determined capacities for decision and capacities to decide were what the internal of development was all about.

There are a number of ways in which Innis' theory of development can be shown to be development by intent and enveloped by what, later, was called dependency and underdevelopment theory. First, the staples thesis was rooted in dependency in so far as it was both the external demand for commodities and the external source of techniques of production which initially determined the organisation and pattern, through time, of natural resources production in a case such as Canada. Second, the distinction between trade and industry, as different spheres of economic and social activity, was made emphatic by the thesis. Third, Innis was fundamentally interested in the institution as the means through which adaptation was organised. Last, the end of development rested upon its values, namely the balance between innovation and tradition in the course of historical change.

Dependency

Smith had referred to the progressive order of manufacturing industry and agriculture as that which 'being contrary to the natural course of things, is necessarily both slow and uncertain'. Innis repeated Smith but interlocked the order within a Canadian national frame of reference.[162] Manufacturing and agriculture were interlocked in time, as in the 'course of things', but separated in space, across the Atlantic. Manufacturing and agriculture were located in different places:

> Throughout the economic history of Canada, the dominance of water transportation in the Maritime Provinces and the St. Lawrence has accentuated dependence on Europe for manufactured products

and for markets of staple raw materials. . . . Concentration on staple commodities was accentuated by the migration of technique from the United States.[163]

Thus, there was an idea of dependency within the Innis of 1937. It was a small step for Innis to be interpreted within the full gamut of underdevelopment theory after Gunder Frank was brought to Canada in the 1960s and to treat Canada as if it were the world's richest underdeveloped economy, always dependent, first upon France and Britain, and then upon the United States.

The generalised staples thesis has given ground for both positive and negative concomitants of development which follow from a common assertion of dependency. 'The nation that today is Canada', wrote Aitken, 'has never been master of its own destiny; as a satellite staple-producing economy, it reflected, and still reflects, in its rate of development, the imperatives of more advanced areas.' But the response of the Canadian state, at all levels of government, to external pressure has been 'positive or creative'.[164] Aitken's explanation of a positive response was that state policy in Canada had always been development policy.

Durham, in the *Report*, had suggested that the primary function of the state in Europe had been defence and that military spending had always been greater than the immediate means of the state. Aitken repeated Durham to say that in North America it was development again for which spending was greater than the state's 'slim fiscal resources could afford' which held primacy.[165] Canada, with Aitken following Innis and Creighton, is then distinguished from the United States according to Creighton's claim that government in Canada had normally followed the injunction that it 'must clear and prepare the way for the beneficent operations of the capitalist'.[166] The bias towards public works, for instance, is explained as an adjustment and an adaptation to the serial rise and decline of staples. 'Each phase of expansion in Canada', Aitken wrote, 'has been a tactical move designed to forestall, counteract, or restrain the northward expansion of American economic and political influence'.[167]

In other words, development was oriented through the subjective will of the state. The state had both the desire and the ability to decide upon the strategy and tactics of policy. It was only a matter of degree, Aitken concluded, to establish how much the capacity to decide was limited by the external economic environment in which the demand for staple products was exercised, and how much the state was able and willing to generally act through raising, for instance, finance for public works. All 'relatively less-developed countries' faced the same development problem.[168] According to this interpretation, whereby the process of development is one of adaptation by the agency of the state for the purpose of expansion, Canada was a case of positive development because its adaptation had been relatively successful.

It is in nineteenth-century Quebec that we find the Canadian case of negative development. Ouellet's account of the 'agricultural crisis' in Quebec, the crisis which forced population off the land and into emigration, is based upon a case of the failure to adapt. As forest land was rapidly depleted during the 1820s, the French Canadian *habitant*, or peasant producing household, failed to adapt to the export production of the new staple, wheat, which would compensate for the decline in timber that could be cut by household labour to secure its main source of income. Given that the anglophone farmers of Ontario had successfully adapted to wheat production for export, the burden of Ouellet's argument came to rest upon the specific cultural determinants of investment in technique. Francophone peasant producers responded to a rise in the person/land ratio in the old *seigneuries* by switching from wheat to potatoes for direct consumption; they failed to introduce crop rotation, improved equipment and livestock and failed to perfect weeding, ploughing and harrowing practices. New techniques were known but were not used.[169]

Ouellet's account has been subject to much controversy and we will return to it again. Here, it is sufficient to note that the object of Ouellet's account of the crisis was to criticise the nationalist idea that the underdevelopment of Quebec was due to the British conquest of Quebec and the economic domination of francophones by anglo commercial capital. Rather, Ouellet's focus upon culture came to rest on the power of the Catholic Church which had 'spread from religion to education, health care, social security, and eventually to all aspects of life in rural and urban communities'. Ouellet continued: 'This development, along with the extraordinary weakness and timidity of the state in dealing with the urgent problems produced by industrialization and urbanization, suggests that both the francophone community as a whole and its ruling class were extremely reluctant to face the realities of the modern world.'[170] Experiencing 'anxiety felt in the face of the demands of progress', the ruling class in Quebec failed to adapt to the objective development of capitalism. Quebec became underdeveloped in the nineteenth century because 'the petty and middling bourgeoisie' lacked the desire to respond positively to the pressures of an external, international environment, pressures which had been exaggerated by the clergy and petty bourgeoisie.[171] As such, Quebec became a case of non-development.

Ouellet absolved anglo capital in Quebec from the responsibility of causing underdevelopment in Quebec. However, other adherents of the staples thesis, who also stress the negative concomitant of development which follows from the presumption of Canadian dependency, make 'foreign capital' the object through which underdevelopment arose in Canada. They also make the state, which refuses to act to prevent the foreign ownership of capital, the subject which perpetuates underdevelopment. Thus, Watkins, who is both taken to be the authoritative source of a general staples theory and has been responsible for pushing the thesis in a neo-Marxist direction, latches onto entrepreneurship as 'the key factor' which explains the degree to which the opportunity for investment, created by the expansion of staple exports, is realised. 'Economic development', Watkins maintains, 'will be the process of diversification around an export base.' Foreign capital which 'is difficult to distinguish from foreign entrepreneurship' concentrates on export trade and prevents diversification into manufacturing industry. For Watkins, therefore: 'An adequate supply of domestic entrepreneurship, both private and governmental, is crucial. Its existence depends on the institutions and values of society, about which the economist generalizes at his peril. But the character of the staple is clearly relevant.'[172] Beginning from an earlier distinction of the institution, or form, of economic organisation that was defined by the character of the staple and technique, that of the plantation and family farm, Watkins later supposed that this fundamental distinction hinged on whether the staple was to be regarded as an activity of trade or of industry.

Where the state was to act towards staple production as if it were trade for export, as was the Quebec case with trade controlled by foreign capital, then all to which Ouellet referred, concerning the non-developmental state of Quebec, followed logically. Household production, in so far as it is production for the market and assumed by Watkins to be 'independent commodity production', 'reinforces merchant capital' and 'tends to retard the development of mature industrial capitalism'. However, where 'the staples-region' was governed by the character of industry, as for timber and minerals, 'a class of capitalists emerged' to create 'a state structure and a "national policy" in its own image'.[173] In this case, Watkins implied, development would be authentic.

In 1952 and after exploring the genesis of the commercial 'dependence' of American publishing upon advertising, Innis proclaimed: 'We are indeed fighting for our lives.' The development of 'American commercialism' threatened 'Canadian national life'. By the end of his life, Innis had become a fervent critic of cultural imperialism:

We can only survive by taking persistent action at strategic points against American imperialism in all its attractive guises. By attempting constructive efforts to explore the cultural possibilities of various media of communication and to develop them along lines free from commercialism, Canadians may make a contribution to the cultural life of the United States by releasing it from dependence on the sale of tobacco and other commodities . . .[174]

Whilst warning of the 'dangers of exploitation through nationalism', including 'our own', and in concluding that 'to be destructive under these circumstances is to be constructive',[175] Innis associated intentional development with an effort to confront the power and money-profit making propensity of commerce. National policy was both truly national and policy, so it was implied, when it conformed to the values of industrial development. Values of development, as Innis made explicit, should be founded on 'cultural' or 'social heritage' and, in particular, that of Europe.

Later, in this chapter, we will turn to Innis' concept of heritage and, in our concluding chapter, we will show the same associations being made a generation later, by Dudley Seers and others, who endeavoured to recapture 'development' for Europe according to the premise that a dependent European culture was at imminent threat from American cultural imperialism and what Innis had called 'the jackals of communication systems'.

Trade and industry

The staples thesis implies a distinction between trade and industry, a distinction which Innis emphasised, and which is at the heart of Veblen's theory of development. In his critique of 'the new Canadian political economy', Panitch perceptively noticed that the fundamental distinction between trade and banking, in the sphere of circulation, and industry, in the sphere of production, owed much to Joseph Schumpeter's distinction between 'capitalists and entrepreneurs'. However, Schumpeter, whose general theory of development is the substance of what follows later in Chapter 7, made banking the *internal* condition, and the banker the supervisor, of the process of capitalist development. Canadian proponents of underdevelopment theory are fundamentally mistaken when they assert that it was Schumpeter who treated production as the basis of capitalist development and, therefore, regarded production as being prior to, or dominant over, circulation.[176] Rather, it is the debt to Veblen which should be made explicit.

Veblen, in his *Theory of Business Enterprise*, pointed to the inherent conflict between 'business' and 'industry' which were counterpoised by him in a series of dichotomies. The financial interest of pecuniary capital in business was directed solely towards money-profit; the functioning of both business enterprise and the state was dictated by the predatory habit of 'exploit' and expressed through the concept of 'ceremony'. Ceremony, for Veblen, embodied the idea that the drive for money-profit, on the basis of private property, made consumption a matter of emulation rather than what was 'serviceable' for the work of industry. By contrast, capital in industry was, for Veblen, a force directed towards production of the 'machine process'; workmanship, 'parental bent' and altruism characterised the habit of industry. The conflict between business and industry explained, for Veblen, the origin and course of economic development. Change in technique was understood to be the social result of the progressive force of capital in industry since it led to change in institutions that were socially necessary if the 'machine process' was to be improved. Business, on the other hand, acted as a regressive force against change because of the social waste which was entailed by 'ceremony'. If business pervaded industry, the result was perverse development. Ideally, therefore, industry was to be liberated from business so that real development might take place.[177]

Veblen's distinction between pecuniary and industrial capital has often been confused with Marx's account of the difference between merchant and industrial capital. Yet, unlike Marx, Veblen repeatedly criticised the failure of classical political economy 'to discriminate between capital as investment and capital as industrial appliances'.[178] For Veblen, like the staple theorists who followed later, there was no abstract relation between capital as means of production and capital as the social means by which labour power was organised coercively in production. Thus, for Watkins, the independent commodity producer is independent of wage-labour but is 'a capitalist because he uses capital',[179] as if capital were simply a set of appliances. Moreover, the distinction between business and industry was reduced by Veblen to a concrete distinction between different institutions. Business was embodied in the bank or merchant trading company; industry in the industrial firm. For Marx, merchant and industrial capital were different, abstract relations between money and production, which might be embodied in the same institution. Merchant and industrial capital were logical categories which could not simply be reduced to an empirical difference between trading and industrial companies.

It is Veblen's distinction which, when conflated with that of Marx, has been carried over into the Quebec 'retardation debate' over whether or not the domination of merchant capital caused and perpetuates underdevelopment.

Veblen's idea that industry was a progressive force was based on the proposition that new technology, innovation and the rational search for improvement in the machine process, reduced the value of capital assets. Business was regressive because the financial interest of pecuniary capital was to inflate asset values and thereby retard potential investment in production. Whereas industry created a social product, business appropriated the surplus from the product and wasted it in the private 'ceremony' of consumption. Where and when pecuniary capital controlled production, then, the result was economic crisis, depression and underdevelopment.

Even though leading proponents of the Canadian retardation thesis take their bearings from a reading of Marx, they make similar arguments to those of Veblen. Naylor, for instance, has argued, that industrial capital investment is long term, high risk and faces uncertain outcomes; while merchant capital is 'directed towards short-term, relatively safe, investment outlets'. Innis, following Veblen, had characterised 'the commercial' as opposed to the industrial in exactly the same way.[180] The 'maximisation of the mercantile surplus', Naylor maintains, 'will minimise the industrial surplus'. And, in the same way that Innis separated manufacturing from industry across the Atlantic and the St Lawrence, so Naylor argues that the domination of merchant over industrial capital served 'to maximise the surplus appropriated by the metropole, and consequently minimised the amount of local capital formation'.[181] From the earliest period of sixteenth-century French occupation to United States-based investment in twentieth-century industrial plant, merchant capital, according to Naylor, has held sway over the state and the economy. Underdevelopment is clearly marked out by Naylor: 'Far from being paradoxical, this continuity in the face of change reveals the fundamental attribute of Canadian capitalism. In its consistency of development lies the key to its need for subordination and its ability to adapt.'[182] Just as the phrase 'continuity in the face of change' cannot but summon up Gunder Frank, so the reference to the 'ability to adapt' cannot help but evoke Innis. Moreover, the retardation thesis has been disputed on grounds which are not far removed from Innis' own ambiguity concerning the relationship between business and industry.

Much has been made of Innis' conclusion to his *Fur Trade in Canada* which he ended by observing that Canadian 'economic development' had 'been one of gradual adjustment of machine industry to the fur trade'. C.R. Fay, an economic historian who studied the British Empire through the staples thesis has argued, for example, that the purpose of Innis' conclusion was not so much to theorise on the origins of development but to offer a commentary on twentieth-century

Canada.[183] Whatever the case, Innis' reflection on Canada in 1930 was focused on explaining what made a positive and progressive account of development in Canada possible. Sudden and rapid growth following wheat production and other staples was, for Innis 'the result of the machine industry' and unlike the United States, Canada came 'under the sweep of the Industrial Revolution at one stroke'.[184] Yet, if industrialism came in one swoop, how could development have occurred through a 'gradual adjustment' of industry?

Innis proposed that what bridged this logical gap, and what accounted for the fact that Canada was a unique case of development, was that the trade in staples did not make Canada 'an economically weak country'. The business of trade had not been a regressive force because of the 'diversity' and 'elasticity of institutions'. Innis referred to the way that the 'elasticity of institutions facilitated the development of the compromise which evolved in responsible government'.[185] But in doing so he meant more than the political basis of the anglo–francophone compromise which established the platform for the active development policy of 1849. Elastic institutions, for Innis, substituted for the rigidity of the market price system.

Veblen had not been Innis' only inspiration. From his mentor in economics, J.M. Clark, Innis had learned that the consumption of capital goods, as much as the demand for consumption goods which was exercised by ceremony, implied social waste. The lumpiness, durability, irreversibility and specificity of capital goods made investment in capital equipment relatively fixed for the individual enterprise. The character of fixed investment was such that it required that the potential capacity to produce be greater than actual production for all enterprises in an economy. The result was unused capacity and social waste. 'Excess capacity', wrote Ian Parker of Innis, 'simultaneously involves a potential or actual drain on the strength of an organisation and an incentive and the resources to promote innovation'. A market price system, comprising a set of decentralised enterprises, was rigid because the input costs of fixed capital for any one enterprise were arbitrarily related to the expected profit of that enterprise. As such, the market price system, in itself, could not capture the social incentive for innovation. Thus, the idea of social overhead capital arose as an attempt to make fixed capital more flexible. Capital would be fixed for the economy but be more flexibly used by each enterprise in the economy. It was supposed that an industrial, as opposed to market price, system would share out costs according to the more flexible use which each enterprise made of the overhead capital.[186]

The institution which fixed capital socially was the state. 'In Germany, Italy and Japan and in the British Dominions', wrote Innis, 'the state became capital equipment.'[187] J.M. Clark, as Parker has noticed, extended the idea of social overhead from capital to labour. If capital had to be reproduced, so did labour power. From his familiarity with Marx, Clark argued that 'there is a minimum of maintenance of labour which must be borne by someone, whether labour works or not'.[188] The 'someone' was the state. While there is no evidence that Innis followed Clark's extension of social overhead to labour, Innis did see the importance of the government of the state being elastic to the extent that it made fixed capital flexible.

Government in Canada was necessarily interventionist, and therefore elastic, because it had to bear the burden of debt on the cost of money capital that was raised to finance expenditure on transport and communications, namely the overhead capital for all enterprises engaged in the trade and production of staples. Thus, a single enterprise, as part of a process of competition, could be deemed to be elastic in so far as it may be forced to introduce the latest technique, even in a region devoid of industrialism, by borrowing technology. The problem was how to fund the innovation. Through his analysis of overhead capital, Innis implicitly formalised Hincks' public works conundrum of the mid-nineteenth century. Neill has summarised the analysis:

> The long-term capital commitment necessary to implement modern industrial techniques was ill-fitted to the price cycles of primary product exports, particularly wheat exports. Output prices fell, but debt and other costs did not. The adjustment to equilibrium was slow and uneven. . . . When governments were called upon to relieve those who were victims of the uneven adjustment in prices, they did so by further interfering with the adjustment in prices.[189]

This analysis is now well used, especially in Latin American and African case studies of twentieth-century development. Indeed, it formed the cornerstone of the ECLA (Economic Commission for Latin America) school's 'structural economics'.

Innis, in his attempt to formalise the problem for nineteenth-century Canada, had argued that the effective 'combination of government ownership and private enterprise' guaranteed Canadian development.[190] Again, Veblen lurked behind the emphasis which Innis put upon the institutional organisation of merchants, production and the state. Innis followed Veblen in his critique of the market price system and saw the ideal in an industrial price system which would balance between planning and the system of market prices. Veblen had criticised the utilitarian basis of price theory on the grounds that the calculus which established the balance between the pleasure and pain in different forms of activity was based on a biased hedonism. In establishing the equilibrium between the gain in utility from consumption and the increase in disutility of work effort that came from a marginal change in any form of activity, and then in stipulating that economic equilibrium should be reached by equating net marginal utilities across all activities, the purpose of economic activity was, for Veblen, belied.

While the utilitarian procedure was agreeably scientific, to Veblen, in that it expressed action through a sequence of cause and effect, no final cause of action was stipulated for the system. In associating work effort with disutility, and therefore drudgery, the utilitarian scheme was nothing other than the implied rationalisation for the ceremony of pecuniary gain, called happiness, as an end in itself. There was, in short, no value or no significant content of what the presumed purpose of economic activity might be other than the teleological implication that the end of action was nothing other than the means, the net gain in utility, by which equilibrium should be reached. Therefore, Veblen set about an attempt to impart human motive to activity and did so on the basis of the institution.[191]

Institutional change

For Veblen, the institution was defined, as Roll effectively puts it, 'as the principle of action about the stability and finality of which men entertain practically no doubt'.[192] For the industrial enterprise, therefore, the principle of action was workmanship and so on; for the trading company it was money-making. These principles, about which there was no doubt – instincts, habits or conventions – were subjective principles of action. Understood as such, by Innis and other critics of Veblen, their subjectivity undermined Veblen's endeavour to create 'institutional economics' on an objective, scientific and non-teleological basis. Innis attempted to go back to where his mentor had gone wrong.

For Innis institutions were to be classified according to values rather than instinctive behaviour. These values were revealed by the interaction between institutions. Furthermore, it was in the interaction between institutions that Veblen's cyclonic theory of development acquired meaning. Innis did not end up with an objective meaning of development but rather one in which it was meaning, as belief, or belief in creativity, that made development possible as much as the mechanical transmission of technique on the basis of progressive change.

As Innis understood it, Veblen, like Schumpeter, parted company with the old idea of Progress when, in Innis' words, Veblen 'attempted to outline the economics of dynamic change and to work out a theory not only of dynamics but of cyclonics'.[193] There were still laws at work but they were of the 'growth and decay of institutions' and their purpose was 'the study of processes of growth and decay'. Decay was as much a process of development as growth, and development, whether as growth or decay, was swift, sudden and unpredictable. Innis asked: 'What are the results of a situation in which the profit index ceases to operate or operates too efficiently, and the engineer is allowed to run loose, so that the swiftness of development and its unpredictable character are beyond the scope of normal economic theory?'[194] The 'situation' was Canada wherein 'the profit index' would make the production of any one staple decline as rapidly as it had been established.

Innis did not call, as Veblen had, for the liberation of 'industry' from pecuniary capital. Rather, as Neill has pointed out, Innis wanted 'planned growth' to ensure that the introduction of a new staple would be accompanied by a planned rate of decline in the old. In this formula conservation was emphasised as much as innovation.[195] The agency which was to balance this growth and decline would be 'nationalism with intelligence'. 'An intelligent dictator' (e.g. civil service), with the politics left out, was 'preferred'.[196] Canada, he wrote, must be able 'to defend herself against exploitation, against the drawing off of her large resources and against violent fluctuations which are characteristic of exploitation without afterthought'.[197] The exploitation to which Innis referred was that of a satellite facing 'a highly industrialised country', first Britain and then the United States.

Not only had Innis, after Veblen, brought decay and swiftness into his concept of development. In the late 1940s, when he formally engaged with Marshall McLuhan, Creighton and Easterbrook at the University of Toronto to discuss values and communication, Innis brought back the classical conception of cyclical time to bear on development. The rigidity of linear time, which characterised the idea of progress, was counterpoised against, and balanced by the flexibility of cyclical time. Ancient Greece epitomised, for Innis, balance between written and oral traditions or, as Crowley has interpreted Innis, complementarity between 'the hierarchy of organised life in the polis' and 'direct discursive forms of reason'. Innis associated cyclical time with the aural and oral as flexible means of communication; linear time with the written and the visual. Balance between the oral and the written was Innis' means of proposing, as Patterson has indicated, the order of stasis in which everything is made subject to change.

Associating the frontier of development with the oral tradition, Innis made much play of correlation between the oral and democratic and revolutionary 'tendencies' against the conservative, autocratic tendencies of writing and print. Development, he argued, was abrupt and sudden because it happened through cyclical time. Aural and oral means of communication enabled innovation to escape from the linear confines of progressive change. Simply because it had taken a given period of time for a progressive change in a particular technique to happen, it did not follow that it would take the same amount of time for that change to be reproduced. Knowledge of an innovation was transmitted orally as much as by written record. Moreover, there was often little need to follow a precisely elaborated rigid sequence when an innovation was reproduced.[198]

Innis, in his earlier work, implicitly used the concept of cyclical time to explain how it was that machine industry had come to Canada in one swoop. In doing so he denied that the economy was necessarily doomed to regression and underdevelopment because its development as a 'satellite' or 'marginal economy' was that of a staple producer under the sway of commercial capital. The theory of cyclonics was used to argue that the sequence of stages of progress, from hunting to agriculture, from commerce to manufacturing, could be reversed in an area which came into contact with countries which had reached a more advanced stage of progress. Neill has summarised the argument precisely: 'The satellite, therefore, receives the last form of things first.' New technology entered the marginal area on the basis of institutional change so that,

> economic advance in marginal areas takes place at an ever increasing pace until, at an advanced stage of modern industrialism, an entire satellite economy with all its technical and organisational equipment is built in one brief moment of social excitation that can only be described as an economic storm of cyclonic intensity.[199]

However, this 'brief moment' hides much. What appears to be the change of an instant is in fact the result of the cumulative capacities of institutions to change through accommodation, adaptation and adjustment over time. These were the metaphors of social evolution that Innis had got from Veblen and that, in turn, Veblen had derived from the idea of 'progressive development'. However much Veblen might have objected to Herbert Spencer's justification of *laissez-faire*, this was the idea which linked the Lamarckian social evolutionary theory of Spencer to Darwin's theorem of natural selection which Veblen adopted by substituting the institution for the species. Adaptation, for Veblen, was selective.[200] In addition, the idea of cyclonic change was not to be confused with the chaotic disorder of spontaneous change. The satellite economy was to be commanded by the agency of the state. As has been noted above, this agency which was to regulate change was to be vested in the ideal of an 'intelligent dictatorship'. Thus the secular clerisy of positivism which Veblen had gotten from the Saint-Simonians and Comte, and which was perpetuated by the authoritarian American socialism of

Edward Bellamy, was embedded in the Innis–Veblen scheme.[201]

The overwhelming empirical body of Innis' historical studies were about the mechanics of institutional change. *The Fur Trade in Canada*, for instance, largely consists of detail about what Neill refers to as the 'struggle between competitive and monopolistic forms of enterprise'. Monopoly is treated as the centralised form of economic organisation; competition as a decentralised form. Neill's précis of Innis ran as follows:

> During the French period heavy social overhead in the form of military facilities for defence against the British led to the extreme centralization of economic activities. Following the conquest this overhead was unnecessary, and a more decentralized, competitive organization of traders moved north from the New England colonies.[202]

But competition between traders led to the exhaustion of supply. The frontier of supply could only be pushed forward by new investment in overhead capital in the form of new supply lines. The increased costs of overhead led, in the 1820s, to a new period of centralised organisation and monopoly, by which time timber was replacing fur as the major staple export.

Development has become associated with only one phase of this sequence, namely that of the centralised form of organisation. This follows from the premise that it was only the state which could undertake the organisation and funding of the social overhead. Thus, in the Quebec case, the origin of 'development' is located in the centralised state practices of the French colonial administration. Talon is remembered as the *intendant* who created, in the 1660s, a model of development practice through the detailed planning of the subsidised settlement of soldiers who had been sent out to subdue the Iroquois.[203] The Talon model of development is referred to as 'royal socialism' and an authentic '*Canadien* policy' which was submerged by British colonial experience and only re-created, without memory, by the Hincks public works conundrum of the late 1830s.

This meaning of development, so often the butt of free market ideology which associates 'development' with state monopoly and central organisation, has only one dimension, and is not what Innis meant by development. Development can also refer to the way in which the phase of one sequence and sequences of organisational change are folded, concertina-like into one other. Here, development is intended to be compensation for the exhaustion of a resource in a given area, the resource being conserved, through development, at a new frontier of supply. In the Quebec case, the exhausted area of fur-bearing animals became an area of supply for timber so that first phase of the new staple, timber, requiring the overhead of the St Lawrence system, coincided with the third phase of the old, fur-staple sequence. It is through the cumulative coincidence of sequences, a coincidence which is not planned, that the intended development of resources creates overhead and overhead becomes an autonomous source for industrialism.

Decentralised organisation was, for Innis, the source of creativity and flexibility. In the case of the second phase of the fur trade, competitive traders innovated in the means of transport; the canoe being replaced by the flat-bottomed *bateau*. Innovation, however, accelerated the exhaustion of supply and, in Ouellet's later account of the fur trade, led to a reduction in wage-employment of rivermen.[204] Competition had to be counterbalanced, therefore, by the central organisation of monopoly and the state. In his work on wheat, Innis associated decentralisation with democracy and central organisation with the rigidity of bureaucracy. Bureaucracy controlled the rate at which a resource was to be depleted and another expanded.

The whole problem of Innis can now be stated simply. How could one form of organisation, a centralised bureaucracy for instance, be counterbalanced by another decentralised form, competition, and yet also be the pivot upon which planned change was balanced against innovation? The question could not be answered through the positivism and scientific method which Innis inherited from Veblen. Innis' account of development, with its affinity to theories of under-development, was of the kind that had been offered by Newman in the nineteenth century.[205] Innis' 'almost automatic response to propositions and conclusions', Berger has noted, 'was to ask why men do the things when they do them, and why they believe the things they believe when they believe them'.[206] Development was a matter of belief and could only be expressed through, but not explained by, the values of development.

Values and 'culture' of development

Values, for Innis, included not merely the objective value of the commodity but the value of opinion which was put upon the form of enterprise and the state itself. Opinion stemmed from 'culture' and 'civilisation'. Civilisation, as Neill interprets Innis, was 'the biased organisation of values'.[207] Values in themselves could not be the object of scientific explanation since they were indeterminate. Nevertheless, an inexplicable principle of action could be the ultimate, independent source of explaining change and development. Innis took this inexplicable principle to be individual creativity. Civilisation was then reduced to the form of organisation which incorporated and succoured belief in the value of individual creativity. The capacity for the individual to be creative through exercising choice had been the utilitarian criterion for development. But 'in welfare economics', as Neill pointed out, 'the choice' was 'vacuous and static, whereas Innis' creativity implies all the crude productivity, the destructive power, and all the dynamism of the restless human spirit'.[208] Creativity was not only an absolute, universal value of development, the principle of civilisation, but also a relative basis for establishing how one economy could be developed and marked off from another by way of culture.

Innovation, for Innis, happened in the space which was contained by the intersection of two cultures and at the point of time in which one culture balanced another. It was an early but lapsed Fabian, Graham Wallas, from whom Innis drew much of his thinking about culture and oral tradition.[209] Wallas, whose *Human Nature in Politics* had led none other than Sidney Webb to comment upon Wallas' 'querulous discontent with Democracy itself', had made 'social heritage' the reason for his pessimistic regard for the possibilities of reform through the piecemeal change of institutions. It was the social heritage, the learned habits and conventions passed on through time differently from one society to another, according to Wallas, which had to be engineered through education if reform was to be possible for achieving what he called the 'great society'.[210] Innis assimilated Wallas' social heritage to the old utilitarian end of development, the capacity to exercise choice, and called it culture:

> Culture . . . is designed to train the individual to decide how much information he needs and how little he needs, to give him a sense of balance and proportion. . . . Culture is concerned with the capacity of the individual to appraise problems in terms of space and time and with enabling him to take the proper steps at the right time.[211]

The capacity to appraise and decide was the source of innovation. In the same way that the concepts of capital equipment and fixed capital had been equated with the state and rigid 'habit', capital depreciation became cultural 'depreciation'. '*Depreciation* of the social heritage', Innis wrote in the conclusion to *The Fur Trade*, would be mitigated by 'an appreciable dependence on [and "heavy material borrowing from"] the peoples of the homeland'.[212] Later, Innis generalised. When one culture made an impact upon another, then innovation was possible because, whereas each was characterised by the organisation of monopoly and bureaucracy, the area of interaction between the two created the opportunities for creativity. Thus, innovation happened at the fringes of monopolistic organisation, at the frontier where a new alien culture intruded upon the old, where there was freedom from the domination of monopoly.[213] Although Innis had arrived at these general propositions in his *Empire and Communications* and applied them to all empires, it is clear that he had Canada in mind. Cultures, in this case, referred to more than those of native peoples and the francophone population. Innis was inclined towards a culture which made Canada both inclusive, and distinct from the United States.But what made Canada inclusive was nothing other than the state, the epitome of what was construed to be 'the biased organisation of values'. How could there be balance between bureaucracy and democracy, or the associated dichotomy between monopoly and competition, when that balance was pivoted on bureaucracy? It was only possible if the state was regarded as imbued with the capacity for creativity, as if it had the subjectivity of the individual person. Characteristically, Innis wrote the following of the mid-twentieth century: 'North American exploitation of natural resources has reached the stage in which exhaustion through competition between Canada and the United States has necessitated concentration on conservation and, in turn, co-operation. . . . The task of conservation is not one of technology but of culture'.[214] We are not told what the status is that is given to 'Canada' and the 'United States' in competition nor what 'competition' implies here. It may be competition between Canadian and American-owned enterprises or competition between the two nation states or both.

We do know that Innis had two concepts of monopoly. The first belonged to the enterprise which kept supply of the resource scarce to earn rent; the second to the control of the means of communication between enterprises.[215] Of the second, and with regard to the physical means of communication, we have shown how Innis stressed that it was state control which made Canada distinctive. Given the dichotomy between competition and monopoly, we can infer that the same two concepts applied to the creative force of competition. In the course of a comment on the 'implication of an idea' and 'belief in an idea', Innis referred to Schumpeter's distinction between those factors 'determining' a monopoly position and those 'leading' to the breakdown of monopoly. Unlike latter-day Canadian 'nationalists', Innis understood Schumpeter's implication that a breakdown of monopoly followed from the need to produce new ideas 'in capitalism'. To underline that belief that a 'single idea' of monopoly could be avoided, Innis noted that the 'rate of production of new ideas' held 'against socialism' as well.[216]

Innis' belief in the creative force of competition is confirmed by the thrust of his argument that the ideal organisation of values was one in which creativity, necessary for democracy, was incorporated within the same organisation that expressed the force of democracy, as it did the force of bureaucracy. The ideal state, therefore, was both bureaucratic and creative, and could thus act as the pivot in the balance between monopoly and competition.

The Canadian state, Innis sought to show, was developmental because it was elastic, and it was elastic because it was both bureaucratic and creative. If conservation rested upon 'culture', then culture itself must be conserved. Transport and infrastructure, it is stressed time and again, were created in Canada during the nineteenth century by the state as part of a competitive process with the United States. It was the belief that competition with the United States was possible which made the state act to ensure the survival of merchant capital in Montreal.

Innis' own explanation of why and how Canada counteracted competitive pressure from the United States rested on the meteorological metaphor of cyclonics.[217] The north-east of the United States, with its focal point of New York, was a high-pressure zone attracting trade and west–east bound traffic through a system of canals and railways. Technical advance had lowered the actual costs of transporting bulk commodities from the North American mid-west, and the new system, especially after the completion of the Erie canal in 1825, had led to the diversion of trade away from the St Lawrence system with its own focal point of Montreal. However, the technical advance in transport followed the settlement of population which exercised the demand for consumption and intermediate goods that were initially imported and/or produced on the eastern seaboard, and which were moved in an east–west direction to balance west–east traffic. Balanced freight traffic was what made the initial private investment in infrastructure profitable.

In Canada, however, the fitful settlement of population made market demand much smaller and, given the more inhospitable climate and terrain of Canada, potential private investment in infrastructure had no prospect of profit. Therefore, as had happened in the case of the Canadian Maritimes fishing industry, the American low-pressure zone threatened to suck in trade and production activity from Canada. The American threat had to be matched by a west–east, competing system of canals and railways. With the creation of a competing system, merchant capital in Montreal was conserved. By so conserving merchant capital, the cyclonic effects were reversed and the basis was laid for the development of industrial capital in Canada.

The culture of Canada conserved merchant capital because the state 'took the proper steps at the right time'. Necessary overhead capital was created socially through the state. Overhead capital was created to an extent that there was an overhang of infrastructure in relation to actual trade and traffic. Excess capacity would then be eliminated as the potential for increased trade was created by this overhang in overhead capital. This was the moment of 'cyclonic intensity'. Private enterprise occupied the gaps which excess capacity had created. The construction and the operation of railways induced investment, which through sets of linkages, created industrial capacity and increased production in agriculture. Innis used the accelerator, another concept drawn from J.M. Clark, to explain why investment was induced by the prospect of the relatively certain profits that could be expected to accrue from the demand for capital goods.[218] In his account of cyclonic intensity, Innis had the experience of the 1880s in mind, namely the phase of industrialisation which followed from the National Policy of 1878. A similar account can be given of the stress and policy of 1849.

Canal construction in the 1830s created an excess capacity of overhead capital, especially in Upper Canada, whose government was forced into union with Lower Canada to pool the burden of debt that had been incurred by mobilising credit to finance that construction. The canal system spawned the first large-scale industries in Quebec City and Montreal. Further expansion was deemed to be necessary after the union of 1841, but further excess capacity, predicated upon the perpetuation of British protection for staple exports, was not eliminated until self-government of 1848 and the first phase of railway construction thereafter. In principle, there is no difference between the 'development' of 1878 and that of 1849. An intention to develop was expressed by policy; a historical experience of development followed. The question is why it is supposed that the experience of development, whether real or not, follows from the intention to develop. It might be supposed, as did Innis, that there is a recursive relationship between the intent and the experience. Policy follows from a previous experience and has to be re-created for development to happen.

The policy of 1849 followed a period of industrialisation in Quebec. It was a reaction to the development of capitalism in the United States. 'Development' was predicated upon culture – the specific culture of the French-Canadian population – and not upon technology. Yet, the abstract culture of rational procedure, 'taking the proper steps at the right time', was deemed to be a deformed one by, for instance, Ouellet. Quebec, if not Canada, was a case of abnormal development because policy did not follow the socially rational procedure that was expected of a developed capitalist society and economy. Policy followed the doctrine of agrarianism, itself a particular vestige of French-Canadian culture and society, and was tantamount, as we shall see, to an improper step at a wrong time. Resources in Quebec could have been developed in the nineteenth century if a different policy had been applied. This argument, of Ouellet and others, only has merit if development is understood as the one-sided development of natural resources within which population is regarded as if it is itself an abstract resource, de-naturalised by virtue of the development of capitalism.

The fundamental flaw of the generalised staples thesis, the thesis which was inspired by Innis' approach to development, was that it assumed that the conditions for capitalist development had been established to make the development of natural resources in Canada possible. In other words, by the nineteenth century, historical conditions for the development of capitalism had taken root in Canada in that classes of wage-labour and capital were extant. The abstract culture of policy, namely that which would have properly developed Quebec, was one of a developed capitalism.

When Innis referred to civilisation as the 'biased organisation of values', he identified civilisation with western or European civilisation: 'Fundamentally the civilisation of North America is the civilisation of Europe.'[219] Innis made this sweeping statement in order to create the premise for his conclusion that the development of capitalism in Canada was about re-creating that civilisation through trade and production in North America.

'Population', Innis wrote, 'was involved directly in the production of the staple and indirectly in the production of facilities promoting production.'[220] But population had neither been so wholly nor so abstractly involved. A part of the population, because it was not involved in trade and production, had to emigrate. Moreover, the francophone population had been enjoined to extract itself from the production and trade of staples in favour of the ideal of independent household production. This was Ouellet's complaint about the policy of development, which was tantamount to a reaction against the ideal of capitalist development that Innis had portrayed, and that made Quebec a deformed case of capitalist development. However, the ideal of capitalist development in Canada, whose history Innis recounted as the domination of 'the discrepancy between the centre and the margin of western civilisation',[221] depended upon the closure of that discrepancy by the rapid spread of new technology into Canada. New technology came to Canada from Europe and the United States according to the socially rational presumptions of a developed capitalism. The paradox, therefore, was that Canada was no unique case of capitalist development.

THE CONTINENTAL DIMENSION: CANADA AS A NORMAL CASE OF CAPITALIST DEVELOPMENT

Two curious facts about Quebec's economic development – that an agrarian policy of development appeared when the groundwork for industrialisation was being laid down and that the dominant presence of externally oriented commercial capital in Montreal retarded the development of industrial capital – have been used to assert that development in Quebec, and by extension, Canada, was unique and/or perverse. However, both curious facts can be equally well interpreted as consistent with a normal process of capitalist development.

The problem arises from the way in which development, as intention and design, comes to correspond with the process of development as it happens historically. The paradox of Canadian development is nothing other than the equivocation of the generalised staple thesis and the way it has been attached to dependency and underdevelopment theory. Aspirations for national development, through the development of transport, infrastructure and land colonisation, expressed a desire to develop Canada. But the process of capitalist development, including industrialisation, was an unintended consequence of the desire to develop. The unintended consequence was that development happened across a north–south dimension and thereby struck at the heart of the national ideals which deemed development to be authentic only if it were independent of the United States. It was, thereby, not the process of development that was perverse but the national ideal of development, the true purpose of which in Quebec was that of maintenance of the welfare of a surplus population, the existence of which was itself a consequence of capitalist development.

Perversity in development can be explained neither by the retardation thesis nor by the agrarian policy that was to separate agriculture from trade and industry, nor by the French-Canadian mentality of an anti-commercial and anti-industrial ethic. To repeat, policy can be perverse only in relation to an ideal of development, namely the national desire to develop, and not in relation to the process of capitalist development itself. L.R. Macdonald, for instance, has severely criticised the retardation thesis by arguing that merchants moved between trade and industrial production in the nineteenth century and did so because in long-distance and international trade returns on money were long and uncertain, whereas in industrial production, especially brewing, returns were relatively safe.[222] Pushing back to the eighteenth century, it was none other than Ouellet who argued that it was a 'quaint notion' to suppose that 'trade is a conspiracy against growth'.[223] In a critique of Greer's *Peasant, Lord and Merchant*, in which merchants are understood to be the active agents of underdevelopment, Ouellet used the case of the fur trade to maintain that there was a 'symbiotic pattern whereby subsistence agriculture and commercial production were inexorably linked, each providing the other with elements that it was not strong enough to obtain on its own'.[224] The fur trade merchants exercised, as did their analogues in forestry later on, a seasonal demand for wage-labour, the income from which made the economic survival of household production in agriculture possible. Likewise, for the nineteenth century, Analogues and McCallum have argued for a 'modified staples thesis' with emphasis upon the degree of linkages between industry and agriculture.

The concept of linkages, associated with the idea of balanced and unbalanced economic growth as an explanation of the possibility of rapid industrialisation, owes much to Albert Hirschman's examination of the course of development in twentieth-century Latin America. Hirschman's *Strategy of Economic Development* can be understood as a modification of Innis' staples thesis, in much the same way that Douglas North's account of United States development is strongly affiliated to that of Innis for Canada.[225] If the staples thesis has been projected forward by Hopkins to explain the course of twentieth-century development in Africa, then the theory of unbalanced growth, together with Arthur Lewis' economic dualism, has been projected back to explain the experience of nineteenth-century Quebec. From the standpoint of post-1945 development theory, there are three facets of mid-nineteenth-century industrialisation in Quebec that stand out in explaining why it was not a perverse case of capitalist development.

First, early industrialisation in Quebec is seen to be modelled on the experience of the north-east United States. Whereas development in Ontario replicated the balanced growth pattern of the American west, with a marked degree of backward and forward linkages between agriculture and industry, Quebec's experience has been understood as one of unbalanced growth. In the case of unbalanced growth, urbanisation, which is predicated upon the existence of a focal point in a trade and transport network, comes before industrialisation. In New England, the textile industry developed upon the trade in cotton inputs from the south and textile output for a wide American and international market. Labour, including migrant labour from Quebec, was drawn in from a commercially unproductive agriculture. Thus, unlike the case of balanced growth, where farmers used the consumer and capital goods of industries developing upon a local thriving base of wheat production, unbalanced growth was marked by a disjuncture between the labour supply emerging out of agriculture and the growth of local markets for industrial products. Quebec's economy in the nineteenth century is therefore deemed to be unbalanced because it was composed of a 'small enclave manufacturing sector' facing 'an unlimited supply of labour'.[226] But – and this is the case against retardation – unbalanced development is not necessarily underdevelopment.

Large-scale industrialisation in Quebec after 1849 was driven by a set of backward linkages. This course of development, the second facet of industrialisation, happened after the desire to develop had been established. In expressing the desire to develop Canada, merchants and the political establishment, as we have seen for Quebec, drove forward the expansion of canal and railway networks. The intended, but negative, consequence of the desire to develop transport infrastructure has been interpreted, as we have mentioned, as corruption because the canal and railway developers privately appropriated the social benefits that followed both from the means by which canals and railways were financed and constructed as well as from the unearned increments in land values that accompanied the expansion of the canal and railway network. Even when Platt and Adelman criticise the retardation thesis as 'the failure to see how the frustration of development *increased* Canadian self-control', they still follow the familiar ground of argument:

> Financed in Great Britain, the Grand Trunk drove a wedge between the goodwill of investors and the ambitions of colonial developers. The reasons for the discord say much about the contrary interests of bankers and local developers. They also reflect the difficulties Canadians encountered in turning the opportunities for development into sustained growth.[227]

Apart from corruption and the categorical enmity between developer and banker, the rail network, across the Quebec–United States border in particular, accelerated the emigration of population as much as it facilitated the movement of agricultural and forest products to the US market. 'In the larger scheme of things', MacInnis argued, 'railroad connections with the United States in the 1850s probably did more to foster French Canadian emigration than agricultural depression in years before then.'[228] Through negating the supposed national potential of development, it is easy to see why the desire to develop should lead to underdevelopment. Equally, however, the unintended, but positive, consequence of the desire to develop was to make development happen.

Development happened through the large-scale industrial plants established to service the canals and railways, while the canal and railway network induced industrial expansion. Linkages extended back through timber, steel fabrication and mechanical engineering works and eventually into a myriad of industrial enterprises that produced a range of products for export markets as much as single-purpose products for the rail network. Tulchinsky, in tracing the genesis of industrialisation in mid-century Montreal, has shown how the construction of the Lachine Canal in 1846, with its hydro-power and quay facilities, spawned large-scale marine and engineering works, iron processing plants, and a host of factories producing multi-use, intermediate products such as paint and twine. By 1854, after the second stage of the Lachine Canal had been completed, the development of large-scale textile and leather production had reached 'its full potential'.[229]

Likewise, railways – and the Grand Trunk, in particular – created the basis for vertically integrated, large-scale industrial plants whose size and extent of technological innovation were the equal of any in Europe and America. The Grand Trunk workshop in Montreal, which had its own sawmill and steel fabrication plant as early as 1856, was regarded as an industrial model for labour-saving innovation, product standardisation and quality control. 'The distinction between "commercial railways" and "industrial factories"', as Craven and Traves are led to comment, 'is plainly absurd, and should be abolished'. Railway companies 'owned and operated some of the largest and most sophisticated manufacturing plants in the Canadian economy from the early 1850s on'.[230] Railways turned Montreal into an industrial centre, as had happened to towns in the eastern townships, in that industrial inputs, from iron to cotton and hides, were transformed into products by manufacture for a trans-Canadian and North American market. Above all, this was no simple enclave manufacturing centre of production and exchange.

The third facet of industrialisation is that both the source of capital and labour for industrial production were cosmopolitan. Tulchinsky has paid particular attention to the origin of industrial capitalists. In mid-century, and half a century before American capital invaded the industrial arena of Quebec to set up branch plants of American-owned corporate companies, it was Americans, as much as resident or immigrant anglo Canadians, who were the driving force behind the first wave of large-scale industrial investment.[231] While poor French Canadians migrated south hoping to find a demand for their labour power, relatively rich Americans moved north to invest money-capital in industrial enterprise which they established, owned and controlled as individual capitalists.

French Canadians were not absent, either as capitalists or workers, from commerce and manufacturing in Quebec. Despite an apparently endless debate over the extent of francophone capitalist enterprise and Ouellet's contempt for his lack of statistical analysis,[232] Tulchinsky has provided some invaluable case studies of francophone participation in the Montreal nexus of merchant-industrial capital. 'French Canadian enterprises', Tulchinsky concludes, 'were characterised by a verve and resilience, by an unwavering pursuit of profits.' Young has carried this work further in his various studies.[233] What is significant, according to Ouellet, is that the proportion of francophones in Quebec who lived in Montreal declined from 7 per cent in 1765 to less than 4 per cent in 1851. If the early nineteenth-century labour force in Montreal became overwhelmingly non-francophone, through the 'ruralisation' of the French-Canadian population, then in the later part of the century it was the case that the urban–rural movement of francophones was reversed. However much the proportion of Quebec francophones in Montreal declined relative to that of non-francophones, the absolute level and rate of migration to the city increased after mid-century. In any case, Ouellet has emphasised the singularity of Montreal in relation to other towns:

> As industrialization took root outside of Montreal, the working class in the smaller centres became, with certain exceptions, more and more francophone, and was either subject to the control of French Canadian entrepreneurs who ran small and medium-sized enterprises, or worked for a small number of large companies owned by British or American capital.[234]

If the 'national idea' was that French capital should meet French labour, it had done so in small-town space. But however much the national aspiration to extend the idea to Montreal may have been stymied by the cosmopolitan character of both capital and labour, the process of industrialisation itself in Quebec cannot be seen as perverse.

The process by which linkages were extended backwards into general industrial production may have been unplanned and relatively slow, but the scale of industrialisation was such that industrial capitalism had emerged in Canada by the early 1860s, well before Confederation of 1867 and the 1879 National Policy.[235] 'In general', McCallum has written, 'the tumultuous economic and political conditions of the second half of the 1840s were perhaps not conducive to large-scale industrial development.'[236] However, it was precisely those conditions which gave ground for the 1849 development policy accompanying self-government. This was the policy that accentuated what Pentland, in criticising Creighton's idea of a then non-developmental Canadian state, registered as the passionate belief in 'an independent and protected economy'. This belief, which was already strongly held by 1840, had been stimulated by a 'government that had committed itself long ago to development'. By 1850, the belief had become a 'faith in the gospel of development'.[237]

Tulchinsky, in criticising Aitken's idea of development as defensive reaction against the United States, has claimed that 'the general role of the state was to stimulate economic development, not necessarily because of the need to fight a competitor but because it was desirable and right'.[238] If Pentland and Tulchinsky have protested too much against the prevalent orthodoxy, they have done so appropriately. But they have mistaken the form for the content of what subsisted in the gospel of development.

There was a profound belief in development in mid-century Canada. But the belief did not extend to the purposive and direct planning of industrial production. Despite the French integument, there is no general evidence that this gospel of development was either Saint-Simonian or Comtean in inspiration and conviction, though Etienne Parent, the leading proponent of French-Canadian industrial power, was familiar with the works of Saint-Simon, Fourier and Louis Blanc.[239] Rather, the convictions of the French Canadian ideologues of development were based on the liberal foundations of political economy, and in particular the progressive postulates of the physiocrats and Adam Smith. But liberalism was associated with 'a national idea' for Quebec and Canada, in the manner of List's 'national economy' rather than following what List had called Smith's cosmopolitical principle, and which, as we saw in Chapter 3, List had suggested was a rationalisation of the universal benefit of free trade in being made particular for the benefit of British productive force.

From an anglo-Canadian view of intentional development in nineteenth-century Canada, the foundation of state policy rested upon the complementary sum of industrial protection and subsidies for the development of transport. Neill has shown how the Upper Canadian economics of Isaac Buchanan and of John Rae, who was to inspire J.S. Mill's theory of protection, brought the 'court whig' rather than 'county' version of Adam Smith to Canada from Britain. Whereas the county version was axiomatically opposed to state patronage of industry, the Whig version, based on the four stages of economic progress, took a historical and empirical view of the transition to manufacturing industry. Protection was justified on the grounds that the potential reduction in average costs of production, through state development of infrastructure, should be larger than the average increase in prices which followed from tariffs on imports.[240] Tariffs, as we have mentioned and as economic historians continue to dispute, were applied to secure large-scale economies of production for infant industries and to secure revenue to pay transport subsidies.[241] As such, there was a sequence of development whereby the failure of currently operating canals and railways to pay off loan capital through current revenue made the state impose a tariff to pay the transport subsidy.

The French-Canadian view of intentional development was different since it consisted of the complementary sum of systematic colonisation and the transport subsidy. The systematic colonisation of French Canadians – rather than of immigrants from Britain and Ireland, the major source of labour power for early industrialisation in Quebec – was carried out to prevent their emigration from Quebec. There was no aversion to protection, but the impulse behind the gospel of development was an agrarian doctrine of development. 'For Francophones,' as Neill put it, 'railroads, forestry and agricultural expansion within Quebec became the inseparable trinity in promotional literature' of development.[242]

Transport, intended to comprise *local* roads and railways, was regarded as a condition for systematic colonisation. Despite the apparent difference in intent, the difference in history and culture that was so emphasised by Innis and Ouellet, there was no fundamental difference of purpose as far as the logic of development was concerned. Indeed, it was none other than Isaac Buchanan, the 'vigorous exponent of protection in the business community' of Upper Canada who tried to persuade farmers in 1858 that '*The employment of our own people*' could only be guaranteed in factories:

To find employment of the people, is just the very thing, which is supremely difficult, as to be often pronounced impossible. It is the problem remaining for the true political economist to solve; its solution will be an event not less brilliant and far more important to mankind, than the discovery of the solar system.[243]

Factories, no less than farms, proved not to be a solution, but unemployment remained the general problem of development.

In Quebec, local colonisation railways secured the core for the trans-continental railway system. Regarding the difference between tariffs and colonisation, it was also none other than Marx who pointed out, in his scathing critique of Wakefield, that systematic colonisation was equivalent to protection: 'Just as the system of protection originally had the objective of manufacturing capitalists artificially in the mother country' so systematic colonisation 'aims at manufacturing wage-labourers in the colonies'.[244] Development, as an artificial process unlike the 'natural' process of progress, was what made possible the manufacture of both capitalists and wage-labour in Canada.

The particular intent to develop through systematic colonisation was a different means to the same end of development, but the end did not have to be intended to be a consistent result of the means of development. Systematic colonisation was inspired by a desire to keep productive force in Canada, the same inspiration behind List's policy of protection. However, what has to be stressed about this case, and what it tells us about development, is that the policy to keep population in the territory of Quebec was a policy of agrarian bias in so far as productive force was to be confined to agriculture. As such, the 1849 doctrine became an agrarian doctrine of development.

THE AGRARIAN DOCTRINE OF DEVELOPMENT

The 1849 Chauveau Report on emigration articulated the constructive purpose of agrarian development. Agrarian development was associated with the French-Canadian ideology of agrarianism but the doctrine was not reduced to the ideology itself and could be made to be compatible with an industrial policy. Following from this first point, agrarian doctrine was not promulgated by farmers as a popular doctrine, but was enunciated and practised by Chauveau and others in the political establishment who took on the burden of trusteeship for what they understood to be the authentic course of development for the French-Canadian population. The practical application of the agrarian doctrine was directed at the systematic colonisation by a French-Canadian population of the new lands of the Eastern Townships and the Ottawa Valley in an endeavour to overcome what has been called the agricultural crisis of the first half of the nineteenth century. It was this crisis of agriculture in the old lands of the seigneuries that was deemed to be the major explanation for emigration to the United States.

If the desire for land colonisation was the basis for a popular agrarian movement in Victoria, Australia, than there is little evidence that this was so for Quebec. It has been suggested by Ramirez that the 'colonization of the hinterland' and migration to the United States 'constituted radically diverging responses to the problem of rural impoverishment'. Here, the response to impoverishment is presented as being the result of a choice made by impoverished farm families. However, Ramirez admits that 'the factors leading Quebecers to choose one or the other of these options', of colonisation or proletarianisation, is a matter of 'historical investigation' which has yet to produce conclusive results.[245] Were it to be shown that, as in the Victorian case, there was a popular movement for colonisation, the significance of agrarian doctrine would not be diminished. It was official doctrine which created the basis of systematic colonisation and thus made it possible for choice to be exercised. The question of whether or not choice was exercised is one which only follows from a prior conception of development and the existence of a doctrine which was framed within the imperative of trusteeship.

Agrarian ideology

Agrarianism has been identified by French-Canadian historians, from Brunet to Ouellet, as the dominant theme of ideology in nineteenth-century Quebec. Despite differences in interpretation, the broad argument is that the national idea, as it was defined between 1840 and 1860, was associated with the aspiration to maintain Quebec as an agrarian society. What appears to be exceptional about Quebec was the fact that it should have been a young generation of urban-based professionals, especially lawyers, journalists and state officials who espoused agrarianism. The national idea, when espoused by a 'new bourgeoisie', was normally associated, so it is believed, with aspirations for a modern industrial and urban society. Here, in Quebec, the aspiration was to suppose that French-Canadian population could be sealed in a static social order which was to be immune from urban and industrial society.

Different reasons have been advanced to explain the phenomenon of mid-century agrarianism. Brunet, for instance, sees agrarianism as a reaction to the domination of anglo-Canadian capital which is seen as having blocked the economic aspirations of the new bourgeoisie that had become politically established in the aftermath of the 1837–8 rebellions: 'Unable to start factories or stable banking and trading institutions, the leaders of French-Canadian society used every endeavour to promote rural settlement.'[246] Ouellet, on the other hand, has questioned the existence of any social aspiration for modern economic advancement and has directly associated agrarianism with political reaction and Catholic doctrine and practice. The implication of Ouellet's culture-bound and subjectivist explanation is that this new bourgeoisie had the historic potential to exercise a Saint-Simonian or Comtean positivist basis for industrial trusteeship but was disinclined to do so: 'For their part the liberal professions played the role both of an *Ancien Régime* elite and of a liaison agent between their society and the British.'[247] This characterisation of an 'auxiliary bourgeoisie' has a long pedigree in post-colonial dependency theory.

Other explanations have been sociological in that it has been emphasised that the new bourgeoisie had the dense rural roots of the newly urbanised; or economic, in that this new class of professionals expressed the interests of an old class of rural merchants whose commercial profit would be eclipsed by the economic force of rural depopulation; or political, in that a political career could only be fashioned through the opportunist cultivation of a rural political base.[248] The last, and most recent, explanation is anthropological, in that the colonists themselves created a 'social project' by expressing 'the duty to fulfil the destiny of the French-Canadian race'. Through 'kinship, cooperation and tradition', colonists aspired to re-create the community of the old *seigneury*.[249]

Whatever the different implications of these interpretations, it is clear that agrarianism was more than either a romantic reaction to an industrial present by those who had little direct experience of the soil, or a physiocratic expression of would-be improving capitalist farmers who wanted to mobilise state expenditure on agriculture for their private benefit. Chauveau, himself, is an example of how both romantic and physiocratic themes could be expressed and then be coupled with a policy for a state doctrine of development. As the author of the first Quebec romantic agrarian novel, Chauveau might well have put the physiocratic case in the words of the enormously popular 1862 sociological novel, *Jean Rivard*, by the son-in-law of Etienne Parent, Antoine Gérin-Lajoie, who wrote: 'Agriculture is the prime source of durable wealth; it is the mother of national prosperity, the only really independent occupation. There is nothing much as solid as agricultural wealth.' Gérin-Lajoie was anti-urban rather than anti-industrial in so far as his character Rivard called for 'the establishment of factories in the middle of our countryside' where all 'that is false, exaggerated and immoral' about 'the urban world' would be left behind.[250] This creed of Gérin-Lajoie, whose method was indebted to LePlay's practice of participatory observation for the sociology of development,[251] stood in descent from what Monière has written of Papineau, the 'father of the nation': 'Economic development was to be anchored to the sole resource that was accessible to French Canadians – agriculture.'[252] From the 1849 Report itself, we find Chauveau writing the following:

> The right of property is certainly sacred and inviolate: but the soil only belongs to man on condition that he works and cultivates it; and possession carries with it the obligation to make use of what one possesses in such a manner as not to injure others. Property should have its duties and obligations as well as its rights. It gives to the proprietor the rights of a citizen under the constitution by which we are governed; it imposes on him the obligation to contribute to the support of the state.[253]

Here, Chauveau appealed to the 'great proprietors', the old *seigneurs* and newer large landowners, mainly anglophone, who had appropriated so much Crown land in Quebec, in an attempt to persuade them of the necessity of state expenditure on agricultural improvement.

However, Chauveau and his associates made the appeal for raising state revenue to be spent on agriculture for a specific purpose. Their appeal may have been cast in the general language of John Locke in its physiocratic inflection but its purpose was to mount development through agricultural settlement and colonisation to constructively prevent the movement of the population off the land and to restrain the exodus of the French population to the United States. Mid-century agrarianism was an inextricable part of the recognition of this exodus. The agrarians 'did not understand', Brunet argued, that 'Quebec was an under-developed territory':

> There was no real alternative for the French Canadians; either they accepted an inferior standard of living or else they went to the foreign capitalists for their 'white bread'. They emigrated because post-conquest French Canadian society lacked the minimal economic leadership needed to ensure remunerative work for all the people in it.[254]

The question is what 'minimal economic leadership' means. Angers, for instance, in his critique of Brunet's evaluation of ideology, has suggested that it was not agrarianism which was the source of failure to find a 'solution of the problem' of emigration but that it was rather 'the absence of the national idea' itself, the absence of 'a national industrial policy', which would assert French-Canadian national control over anglophone industry in Quebec.[255]

Virtually all other interpretations of agrarianism make the same link between the emigration of population and the awareness of what emigration might mean for the internal of development. What is not so readily appreciated is that this link between the phenomenon and the consciousness of emigration created the basis of an agrarian doctrine of development, a doctrine which also rested upon the necessary conditions for trusteeship. If anything, the new bourgeoisie was too burdened by trusteeship.

Trusteeship

There can be little question that it was from the standpoint of agrarianism that trusteeship in mid-century Quebec was positioned. National figures of authority such as Parent, the campaigning zealot for French-Canadian participation in industry, were not immune to agrarian doctrine. Parent's motif was 'that the industrialists' were 'the lords of America'.[256] His widely heard appeals to francophones of the late 1840s, implored that they should forsake careers in the professions for those in industry. From his own readings of Smith, Ricardo, McCulloch, Say and Proudhon, Parent waxed lyrical on the national virtues of manufacturing industry. Like Joseph Chamberlain in England a generation later, he mounted campaigns against landlordism and for national education. He argued against the annexation movement and in favour of francophone enterprise establishing itself on an economic basis that would be independent of anglophone capital but politically affiliated to British imperial rule so as to protect the right of a distinct nationality.[257]

Yet, Parent held no anti-agrarian bias. In campaigning for the state-sponsored improvement of agriculture, Parent asked in 1846:

> Again, what has been done to turn to our advantage the exploitation of our vast stretches of uncultivated land. What organisations do we have for making access to such land easier for our surplus rural population in their old farm-houses, and for helping them to expand and settle there, as is done for the English-Canadian settlers.[258]

In deploring the exodus of the 'surplus rural population' to the United States, Parent followed what was to become a well-trodden path towards agrarian colonisation. Neill has summarised Parent's 1852 lecture 'Considerations on the kind of working classes':

> [Parent] advised the workers to refrain from breaking the laws of economics as established by God and set out by Adam Smith. They should adhere to their Catholicism, because in the last analysis, they would be unable to raise wages by striking. If their situation became too desperate, they could return to the land. The answer to the labourer's problems was government support for new agricultural settlements, and the frugality of the workers themselves.[259]

It was likewise for Georges-Etienne Cartier, the leading French-Canadian lawyer and proponent of railways. Ouellet, in order to affirm his argument that these figures of authority suffered from an inferiority complex, were ambiguous in their liberalism and devoid of 'underlying bourgeois virtues', concludes his *Economic and Social History of Quebec* with a long quotation from an 1855 Cartier lecture:

> French Canadians, let us not forget that if we want to ensure our national existence we must cling to the land. Each one of us must do all in his power to preserve his territorial heritage. He who has none must employ the fruit of his labour to acquire a part of our soil, however tiny it may be, for we must bequeath to our children not only the blood and the language of our ancestors but also the ownership of the soil.[260]

Ouellet interpreted the agrarian creed to be evidence of the persistence of what he calls the *ancien régime* into the world of mid-century capitalism. However, it is equally possible to see agrarianism as a mid-nineteenth-century response to the contemporary development of capitalism. There is nothing necessarily atavistic about the doctrine that development should rest upon an agrarian prescription to make a surplus population responsible for its own subsistence.

In the same way that there was no general francophone aversion to capitalist enterprise, there was also no aversion to development doctrine. Dever, in a now widely quoted thesis, has shown that opposition of French Canadian members of the Legislative Assembly to supply motions governing expenditure was based on constitutional and not development grounds. Regarding the allocation of funds for canals, roads, railways, other public works and agricultural support, Dever's examination of the voting record for the early 1830s showed no ethnic bloc voting and a high degree of unanimity in agreement to motions for state expenditure.[261] By 1849, Hincks' development proposals, such as the 1849 Guarantee Act, commanded unanimous francophone support in the Legislative Assembly.[262]

Furthermore, there was little francophone resistance to protection. In the late 1840s, the *Rouges* positively advocated protection. 'Free traders', Forster has recounted, 'were accused of being concerned with meaningless abstractions rather than practicalities' and 'of fostering anarchy rather than harmony and social order.'[263] While the *Bleus* may have been more agnostic 'they were willing to impose a tariff for revenue purposes and for incidental protection, to use Hincks' inspired phrase'.[264] Yet Hincks, 'who as late as 1845 had been an avowed free trader within the imperial context, was by 1852 an equally firm believer in the tariff as retaliation against recalcitrant American politicians and as protection against foreign industrialists'.

In 1852, Hincks introduced protection for infant industries to signal 'the emergence of a nationalist economic policy'. Galt's 1859 tariff, which provoked the wrath of British manufacturers, followed Hincks' somersault in that the purpose of the tariff has been adjudged to have been one of effective protection with its revenue element being considered incidental. *Bleus* gave support to the Galt tariff.[265] Parent, despite the free trade injunctions of his political economy mentors, was also a proponent of infant industry protection.[266] Chauveau's 1849 Report followed Parent's proposals:

> Your committee cannot . . . be blind to the fact that Lower Canada, by its geographical position, its wants, its natural advantages, is destined, as well as the Northern States of the American Union, to become a great manufacturing country; and all that can tend to encourage the establishment of local manufactures, provided at the same time too narrow limits be not prescribed to our commercial relations, will have the effect, not only of retaining in this country the labour and capital which are leaving it, but also that of attracting those of foreign countries.

Chauveau pointed out that protection, 'for procuring employment to the superabundant population', had been debated in the Legislative Assembly. Over 1847–8, revenue-based duties, at a general rate of 7.5 per cent, replaced all previous tariffs, with the 1847–8 rates being amended in 1849.[267] Given the significance of Hincks' somersault of 1852, it is clear that Chauveau had assumed that the principle of effective protection had been accepted. Therefore, the promotion of manufacturing industry to secure employment was incidental to the main strut of the Report's programme – the agrarian doctrine of development.

Second, as the 1849 Report makes it clear, the francophone political establishment urged state expenditure for agricultural improvement through roads, railways, agricultural societies to diffuse the latest agricultural techniques and, above all, subsidies to settlers for agricultural colonisation. All these demands were governed by the principle that a surplus population, if it was to be responsible for its own subsistence, had to be given the conditions which would make direct production of means of subsistence possible. If, as Ouellet and others have suggested, a new bourgeoisie substituted itself for the seigneurs and clergy of the *ancien régime*, it did so on the grounds of welfare. Ouellet has written:

> For the clergy, of course, promoting industrialization did not constitute an acceptable solution to rural problems. From their perspective, creating new cities or stimulating the growth of existing ones would simply develop new and larger foci of misery, where morality, the family and religion would be in danger.

What had been the perspective of the clergy before the 1830s was assumed as the mantle of trusteeship for the new mid-century secular authority. Ouellet continued, in referring to the clergy:

> Seen in this light, the social question had to be resolved in the countryside, where regions still existed that had not yet been opened to agricultural colonization and which could therefore receive the population surpluses of the old, well-established parishes along the St. Lawrence.[268]

In Montreal, for instance, the Church was the main provider of social assistance and welfare. For example, the Seminary's privileged financial position, allowing it to become both a major urban rentier and investment trust in industrial enterprise by 1870, owed much to the assumption that the Church would bear the burden of welfare.[269] Mid-century interpenetration between the *Bleus*, such as Cartier and Chauveau, and the Church led to the 'treatment of poverty' being extended to the realm of state authority and across Quebec as a whole.

Clearly, the social question was as much urban as rural. A significant proportion of emigrants to the United States came from urban centres. The first of eight 'classes of emigrants' analysed in the Chauveau Report referred to the 'very numerous' class of 'workmen of the Cities of Quebec and Montreal'. One-third of all emigrants were estimated to have come from the two cities where, during the later 1840s, money wages had decreased relative to those in the north-east of the United States.[270] Since it was supposed that the cost of living was equally higher across the border, such that real wages were equalised, the agrarian doctrine of development was deemed to have an appeal to emigrants who had faced the experience of urban unemployment and poverty.

Welfare was not, as it was to become in the twentieth century, a direct and centrally administered state scheme for the provision of subsistence. Rather, welfare was construed to be the means whereby subsistence was to be met indirectly through state support for the acquisition of land and then, upon the basis of administered land, households would directly produce their own subsistence. Thus, Chauveau accepted Wakefield's framework for systematic colonisation but in doing so turned Wakefield's prescription upside down. Alongside proposals for road construction and agricultural societies to secure agricultural improvement, Chauveau emphasised that the land price should be driven down, through a state subsidy, to induce the entry of putative settlers into colonisation schemes. The state subsidy was essentially a means of welfare for a surplus population in Quebec. This is the significance of the agrarian doctrine of development, a doctrine which was pursued through the colonisation schemes of the mid- and later nineteenth century.

Colonisation

Colonisation schemes owed much to the 1849 Report. As far as the Eastern townships were concerned, the original proponents of francophone colonisation had been clerics, such as Fr B. O'Reilly of Sherbrooke who, over the harsh economic winter of 1847–8, made the link between poverty, emigration to the United States and the need for French Canadians to colonise land in the townships. The proposals were taken up by the influential Bishop Ignatius Bourget of Montreal as well as by both the radical *Rouges*, who were centred on the *Institut Canadien*, and Cartier and the *Bleus*, who were in the coalition government. With this wide range of influence and support, and especially in the wake of the formation of colonisation societies, the Legislative Assembly debate on emigration and the Chauveau Report, state grants were allocated for roads in colonisation schemes. Typically, in these schemes, settlers were granted free grants of up to 500 acres alongside roads with options to purchase adjacent lots up to 2,000 acres.[271]

If the object of these schemes was to settle French Canadians on timbered but uncultivated Crown land, or land held by land companies, and at the expense of British immigrants whom the land companies were meant to have settled on their land, then the schemes were a success. Little, in his impressive case study of St Francis district, where a scheme was mooted in 1848, found that the population doubled between 1851 and 1861 with the proportion of francophones rising from less than 10 per cent in 1841 to more than a half in 1851. The number of very small-scale French landowners tripled during the 1860s in what had been the anglophone Compton township. Likewise, to take another example, the colonisation movement established twenty new parishes during the second half of the century in the Maurice region, north of the St Lawrence, while in Berthier county, the population increased by 20 per cent during the 1850s. In the Eastern Townships, generally, the proportion of francophones rose from less than a quarter of the total population in 1844 to nearly 60 per cent in 1871. Drapeau's 1863 study *On the Developments of Colonisation in Lower Canada* reported that the number of landed proprietors increased by 10 per cent between 1851 and 1861.[272]

But if the object of colonisation was to secure the establishment and improvement of agriculture on an extensive margin of land and to commercialise smallholding production to the extent that households might realise revenues which would be sufficient to withstand the prospect of emigration, then the conclusion must be that the schemes failed. Fowke, writing in 1947, gave a blunt appraisal: 'Canadian attempts to bait developmental schemes with offers of settlement on timber lands, on pine lands, were rewarded by the ultimate problem that settlers came to be timber-sellers, bogus settlers, interested only in stripping the timber and then moving on.'[273] Later, more detailed studies of the colonisation schemes are wary of making this kind of generalisation, but it is clear, as Ramirez has put it, that the 'settler more likely to succeed was the "*colon défricheur*" (i.e., one primarily engaged in forest clearing) as distinguished from the "*colon cultivateur*"'.[274] Little, in 1989, amplified what the 'ultimate problem' might have been. His argument is that the purpose of agrarian development was 'dynamic' or positive but that the means of colonisation was misconceived.

In short, colonisation established a dual economy whereby francophone colonists, officially settled for the activity of agriculture according to the prevalent homestead 'agrarian myth', faced the immediate presence of anglophone capital, whose sole purpose was to make profit from stripping timbered land under their possession and who secured the legal right and force to officially exclude colonists from the activity of forestry. To repeat, what Little meant by 'ultimate' was that a different policy, namely the mixed agro-forestry regime according to the Nordic model, could and should have been adopted to achieve the purpose of colonisation.[275]

In posing the problem as one which stemmed from the state bias for anglophone capital against francophone small farmers, Little has effectively turned back the clock to the debate over the 'agricultural crisis' in Lower Canada in the first half of the century.[276] To ask whether there was a crisis of agriculture and if, indeed, there was a 'crisis', to argue over the causes of the crisis is no mere academic quibble. As we have mentioned, the agrarian doctrine of development was driven by the awareness of emigration, which was adduced to the failure of existing agricultural capacity to maintain a population which had risen relatively rapidly in the first decades of the nineteenth century. By turning to the extensive margin of cultivation as a substitute for an increase in land-based productivity that would have to have been paralleled by a far more extensive commercialisation of agriculture, colonisation schemes only reproduced the agrarian problem of the old *seigneuries*, the very condition that had created the agrarian doctrine of development. As such, the very causes which were controversially alleged to lie behind the agricultural crisis before 1849 are the same causes which arise in the evaluation of the colonisation schemes after 1849.

Norman Macdonald's comment on the dire poverty of colonists in the Eastern Township schemes has been quoted by Little:

> In the back townships . . . settlers were not much better than squatters, employed no hired labour, cultivated only so much land as they required for mere subsistence, and consumed what they produced. They amassed little or no capital and contributed little or nothing to the immediate development of the country.[277]

Accounts of the schemes confirm this impression. To meet subsistence, households depended on periodic wage-work, which was offered by road construction. Road construction consistently failed to meet targets, thus reducing the rate at which land could be brought under cultivation and effective markets established for produce. Above all, as Fowkes had generalised, the main source of household revenue and indeed preoccupation of male household members was either self-employed and informal enterprise in timber-stripping and/or wage-employment offered by the land companies in the forests they continued to possess and control. However, where Macdonald is amiss, as indeed is Little more obliquely, is that there was no objective purpose for colonists to amass capital or contribute towards 'the immediate development of the country'. Colonists were part of a surplus population that had to be maintained on welfare grounds. Welfare maintains a population in poverty.

Thus when Macdonald criticised policy for settling impoverished families on relatively low potential land without the means for agricultural improvement, he mistook the premise of the policy. When Little questioned an assumption that colonisation difficulties 'could be attributed to failings on the part of the settlers themselves',[278] he was criticising the postulate Ouellet had made when he effectively opened up the agricultural crisis debate, namely that francophone farmers, unlike their anglophone counterparts, had no commitment to agricultural innovation and were reluctant to engage in a commercial agriculture. If the same point is made about small-scale agriculture on the colonisation schemes after mid-century, the same criticism can be made of the Ouellet postulate regarding the significance of a perpetual agricultural crisis in nineteenth-century Quebec.

Take, for example, the estimate of the record showing that the net export of 35,000 bushels of wheat from Quebec in 1827 was turned into a net import of 2.2 million by 1851.[279] Ouellet argued that this trend was not to be explained by state bias against francophone farmers; nor by increased exertions of absolute rent by *seigneurs* during a period when the person:land ratio rose markedly; nor by market conditions, including a British decline in demand for Quebec-produced wheat. While Ouellet's subjectivist postulate has been reasonably criticised on statistical grounds, other criticisms point out that a registered decline in wheat output was compensated for by an increase in potato production. Others argued that the 'crisis' was due to cyclical and coincidental facts of climate and the prevalence of pestilence. Still others have argued that the evidence shows that francophone farmers allocated resources as efficiently as their other North American counterparts before mid-century. However, these countervailing arguments cannot mask the fact that francophone small-farm households were becoming poorer before mid-century and that colonisation did not reverse the trend.[280] It is difficult to deny, from the evidence and argument of Ouellet's most astringent critic, McInnis, that if farmers were relatively efficient, then they were also relatively poor.[281]

The extent of land under cultivation in 'the colonisation zones' rose by a third between 1851 and 1861 and more than doubled by the turn of the century.[282] Total farm output from the Eastern Townships increased in the 1850s and continued to rise during the 1860s. Output of cattle products, including dairy output, potatoes and non-wheat grains increased in the same proportion, with exports to the United States rising rapidly during the 1850s. Yet poverty was reproduced in the colonisation schemes. Whereas some larger small farms, of 200 acres under cultivation, benefited from the minimal state expenditure on agriculture after 1849, there was 'economic stagnation' on one-third of the holdings of less than 50 acres in the relatively favourable Berthier county land of the Maurice region. 'By mid-century, day labour and landlessness', it is suggested, 'had moved from a marginal to a central feature of the socioeconomy of many counties.' After 1870, 'proletarianization had not only become an irreversible process' but 'the rural day-labour market had become a permanent structural feature of economic development'.[283] This was a form of poverty that was more than a replication of the pre-1849 period because the cause that had predicated colonisation – emigration to the United States – became a standard to establish, for colonists themselves, their own relatively poor condition.

Ramirez has reported that, for as late as 1870, the majority of first-generation adult French-Canadian emigrants to the north-east of the United States were general labourers and construction workers rather than textile workers. Children of emigrant families were sent to work in textile factories as the family household, through a multiplicity of occupations, strove to secure subsistence.[284] Expected earnings from wage-work in industrial New England became the reference wage for establishing the degree to which francophones would migrate as an alternative to either settling on colonisation schemes or searching for wage-work within Quebec itself.[285]

We have not found consistent time series for comparative wage rates before 1870. However, published data sources, as shown in Table 4.1, show that both industrial and construction *money* wages in the eastern and north-east United States were generally at least twice the corresponding wage rates in central Canada for the period after 1870. Cross-border variation may have been less in the case of general labourers; the dispersion in comparative rates may have narrowed during the 1880s; the difference in *real* wages may have been less than *money* wage rates between 1862 and 1874.[286] However, the evidence is overpowering for the proposition that *both* employment opportunities and household renumeration from wage-work in the United States were large enough to induce emigration. Any prospect that colonisation schemes might improve material well-being for farm families was insufficient to act against the economic pull of wage-work in the United States.

As Table 4.2 indicates, emigration increased at an exponential rate in the 1860s and then again in the 1900s. In each of these two periods, colonisation schemes

Table 4.1 Comparison of daily money-wage rates for selected occupations and years, Canada and eastern United States, 1871–1910

	1870	*1871*	*1875*	*1880*	*1885*	*1890*	*1900*	*1910*
				(*$ per day*)				
Canada[a,1]								
All manufacturing	0.80			0.90		0.85	1.20	1.30
Eastern USA								
Weighted average manufacturing[b]	2.30	2.00	1.90	1.70	1.90	2.20	2.50	4.50 (2.50)[4]
Canada[2]								
Average carpenters[c]	1.25	1.50		1.80	1.90	2.00		
Eastern USA								
Carpenters[b]	3.00	3.00	2.70	2.60	2.90	2.90		
Weighted average construction[b]	3.10	3.05	2.90	2.60	3.00	3.10		
Canada[c,3]								
General labourers	1.00	1.25		1.30	1.25	1.25		
Eastern USA								
Labourers[b]	1.80	1.70	1.60	1.30	1.60	1.60		

Notes:
1 Annual average wages series normalised by 260 days per year; for the period 1884–9, cotton mill operatives worked an enumerated average of 250 days per year. Series D208–231 in Urquhart and Buckley, 1965.
2 Made up of an amalgam of Montreal/Ottawa/Toronto average daily rates.
3 Skilled farm labourers for 1871, 1875.
4 For all USA.

Sources
[a]Leacy, Urquhart and Buckley, 1983; for 1910, Logan, 1937, Table IX, p. 143.
[b]Long, 1960, Table A–4; for 1900 and 1910, 1899 data normalised at 290 days for three states (NJ, MA and PA), Rees, 1961, Tables 9, 10, pp. 32–3; for 1910, all US, Logan, 1937.
[c]Urquhart and Buckley, 1965, Series D188–195, D196–207.

multiplied and became more and more grandiose both in intent and in the amount of state money expended on the schemes.

Migrants may have adopted 'survival strategies', switching between different combinations of revenue from colonisation schemes and wage-work in the United States. Yet increased state expenditure on colonisation towards the end of the century did as much to accelerate as to ameliorate the compulsion to migrate. In the same way that mid-century railways induced emigration, state loans for agriculture a half-century later enabled households to finance the costs of

Table 4.2 Estimates of emigration from Canada to the United States, 1820–1929

Decade	*USA official record of Canadian immigrants*[a] ('000s)	*Estimates of Canada emigrants*[b] ('000s)	*Estimates of Quebec emigrants*[c] ('000s)	*Emigrants as per cent of Canadian population*[d]
1820s	2.3			
1830s	12.0			
1840s	34.3	75–	35–	
1850s	64.0	150–135	70–	6
1860s	120.0	300–305	n.a.	8
1870s	324.4	375–325	120–120	13
1880s	500.0	450–410	150–165	17
1890s	3.1	425–380	140–195	20
1900s	122.0	325–225	100–75	22
1910s	706.0	250–110	80–100	17
1920s	953.0	450	130–150	13

Note
There are wide disparities between the number of official immigrants into the United States from Canada and estimates that have been made of emigrants from Canada to the United States. Until the end of the nineteenth century, the agrarian bias of development in Quebec did not stem the outward flow of migrants across the porous land border with the United States. French Canadians were estimated to have comprised one-half the number of migrants until the turn of the century; by the first decade of the twentieth century they fell to 15 per cent of estimated Canadian emigrants to the United States.

Sources
[a]Lavoie, 1972, Table 1, p. 10.
[b]Lavoie, 1981, Table 7, p.53 (for first estimates); Jackson, 1923, Table 2, p. 28 (for second estimates).
[c]Armstrong, 1984, Table 10.2, p. 158 (from Lavoie's estimates, [a] above for first estimates); Paquet and Smith, 1983, Table 5, p. 446 (for second estimates).
[d]Vedder and Gallaway, 1970, Table 1, p. 477. Percentage is for each decennial census year.

migration to the United States.[287] Therefore, accentuated attempts at national development, to lock up a surplus population of francophones in Quebec, could not withstand the process of development itself.

Little has concluded:

> The cultural identity of the French Canadians seems to have been a handicap primarily because it discouraged them from leaving Quebec, except as a last resort. Scots and English Canadians, on the other hand, did not hesitate to emigrate to greener pastures south and west. This in essence was the dilemma faced by French-Canadian nationalists who deplored the exodus from the province at a time when conditions were not ripe for large-scale industrialization. In idealizing agrarian self-sufficiency, they were making a virtue out of what they perceived to be a necessity.[288]

However, during the nineteenth century, the process of capitalist development was a matter of necessity. As national differences were opened up by constructive doctrines of development, they were continuously closed by the process of capitalist development, a process to which national development could only be a reaction. The same applied to the idealised difference of town and countryside.

Wakefield may have hoped that systematic colonisation would be the means by which a surplus population could be established on the land and thereby exert upward pressure upon money wages in manufacturing industry. The same principle was adopted by radical reformers in Britain at the turn of the twentieth century. Proposals for land reform, including colonisation to establish farm workers on their own plots of land, were driven by the belief that a reversal in rural depopulation would cause the flow of labour into manufacturing industry to be restricted and thereby force an increase in money wages for an industrial labour force. In mid-century Canada, the agrarian doctrine of development may have had the same intent, but it was no more likely to succeed than the Wakefield-inspired schemes in North America or the later British experience at the end of the century.

CONCLUSION: AN IRONY OF HISTORY

Marx, on the final page of the first volume of *Capital*, was struck by how a relative surplus population in Europe had been reproduced in the European colonies of Canada and Australia. It was as if he had read Mayes when he wrote that:

> The shameless squandering of uncultivated colonial land on aristocrats and capitalists by the English government, so loudly denounced even by Wakefield, has, especially in Australia, in conjunction with the stream of men attracted by the gold-diggings, and the competition from imported English commodities which affects everyone down to the smallest artisan, produced an ample 'relative surplus population of workers', so that almost every mail-boat brings ill tidings of a 'glut of the Australian labour-market', and prostitution flourishes there in some places as exuberantly as in the Haymarket in London.[289]

It was only at the end of the nineteenth century that the doctrine of development, embodying the British expression of the desire to develop, was to come to Britain through Joseph Chamberlain, as a response to surplus population, one very notable phenomenon of immanent process of capitalist development wherever it happens. History moved from British colonies to Britain itself. In doing so, it moved through the idea of the national economy, an idea which List had attempted to formalise as the foundation for development policy. The idea of a national economy, whose early practice was to be found in mid-century Australia and Canada, came to Britain through Chamberlain's understanding that development and unemployment were inextricably part of each other. It is one of the ironies of history that at the end of the century, a half-century after development doctrine had been practised in a British imperial arena and after List had set out a programme of development for Germany, that both the practice and the key ideas were to be appropriated by Chamberlain, one of the most important political figures of the very nation against whose free trade domination both the colonials and List had railed.

5

DEVELOPMENT DOCTRINE IN 'UNDERDEVELOPED' BRITAIN

Modern economic historians have striven mightily to convince later generations that what has become known as the Great Depression of the late nineteenth century was, in reality, a minor recession during a period of generalised prosperity.[1] Yet any historian who seriously examines the writings and politics of the period cannot help but be struck by the profound effect which this supposedly minor economic blip had on those who lived through it. This was particularly true of the impact of the depression on British agriculture. From the novels of Thomas Hardy to the proceedings of Parliament, the 'agricultural question' was a prevailing theme from the 1870s to the eve of the First World War.[2]

Between 1872 and 1895, the area under wheat fell from nearly 3.6 million to less than 1.5 million acres, or from about one-half to one-sixth of all arable land under cultivation. Fewer acres of wheat required fewer agricultural workers. English agriculture adapted to this marked decline in wheat production – itself the result of a development process in Australia, Argentina and Canada – by increasing mechanisation and livestock production. Both more mechanisation and livestock production accelerated the reduction in the demand for agricultural labour power whose work was displaced by the growth in overseas wheat imports. Unemployed agricultural workers and tenant farmers were forced to migrate to English industrial cities. There, they joined an industrial working class already at the mercy of overcrowded housing, an unstable labour market and inadequate social amenities.[3]

Among others, Joseph Chamberlain, the then Radical Liberal, warned of what might happen if the marked increase in rural–urban migration were to continue unchecked. Chamberlain painted a picture of deprivation which was reminiscent of Mayhew and Engels:

We are compelled occasionally to turn aside from the contemplation of our own virtues, intelligence and wealth, to recognize the fact that we have in our midst a vast population more ignorant than the barbarians who we profess to convert, more miserable than the most wretched in other countries to whom we attempt from time to time to carry succour and relief.

Disorder, Chamberlain continued, was threatened by the perpetuation of urban destitution:

If our middle class, and the press which panders to its prejudices, cannot reconcile themselves to the altered situation . . . they may wake some day to find their terrors realized, and themselves in the face of an organization whose numbers will be irresistible and whose settled principles will be hostility to capital and distrust of the monied classes.[4]

The Joseph Chamberlain of the 1870s and early 1880s was the quintessential 'Radical'. By the time of the sitting of the British Parliament of 1868–74, the meaning of the term 'Radical' had been secured through a process of historical accretion. 'Having descended from the seventeenth-century Levellers through the Wilkite movement, English Jacobinism, Philosophic Radicalism, Chartism, the anti-Corn Law League and militant non-conformity, the Radical programme – admittedly in various formulations – aimed at establishing political democracy (for men), and at breaking the privileges and power of Anglicanism and the landed orders'; Radicals sought:

a society entirely free from traditional privilege, open to the rise of the prudent, self-made men, and operating in every aspect by competition. Radicals differed from Conservatives and Whigs in their hostility to the landed orders, their dislike of tradition, and their ardent impatience with existing institutions. They differed from moderate Liberals in their analysis of society and in their perception of what must be done to bring it to the ideal state. Liberals believed in preserving a balance of interests and forces in society, and in providing opportunities for individual development, but they thought that by the 1870s only relatively minor reforms needed to be made to achieve these ends. Radicals . . . saw British society as still dominated by a traditional landed elite.[5]

Behind the Radical analysis, 'unrestrained by reverence for tradition and custom, lay a demand for immediate and extensive legislation, not to establish a balance of social forces, but to bring about a predominance of middle-class values, styles and institutions'.[6] Of the 89 and 130 members of the Parliaments of 1874–80 and 1880–85 who have been identified as Radicals, roughly one-half were occupied in commerce and industry, while over 80 per cent were non-conformist in religion.[7]

Chamberlain fitted this profile well. By birth, he was both non-conformist in religion and middle class in position, the son of an industrial capitalist who was himself the product of five generations of craftsmen and manufacturers. Before pursuing a political career, Chamberlain headed the family screw-making enterprise in Birmingham and presided over a thirty-fold increase in its capital value. His business methods were characteristic of the new men of capital. He integrated coal mining, iron production and wire extrusion into screw production, cut costs and prices, and bankrupted the competition. As the author of one biography of Chamberlain wrote, 'Nothing is more important for an understanding of Chamberlain than to remember that in Britain he had been a pioneer of large-scale production by consolidated enterprise.'[8]

During the decade following his first election in 1876, Chamberlain sat in Parliament as member of the Radical faction of the Liberal Party and, as such, both reflected and help to shape that faction's programme. The following was Chamberlain's diagnosis of the source of agrarian distress during the process of concentrating land holdings: 'The present system was devised with the object of creating and increasing the large estates. Silently and for generations the process of absorption of small properties has gone on, and all the time there has been nothing working in the opposite direction.' Now, Chamberlain argued, that system 'had broken down. Farmers have no capital; landlords declare they are penniless.' The result was the degradation of 'the agricultural labourer . . . the most pathetic figure in our whole social system. He is condemned by apparently inexorable conditions to a life of unremitting and hopeless toil, with the prospect of the poor house as its only or probable termination.'[9] The only remedy was the multiplication of 'small owners and tenants' by way of an attack on large-scale landed property. For generations, Chamberlain recounted, 'the squire and the farmer, and sometimes the parson, have all been lying on the agricultural labourer, but . . . they will have to find some new position'.[10] Of those who claimed that a reconstruction of landed property would be confiscation or theft, Chamberlain asked,

> what language will fitly describe the operations of those who have wrongfully appropriated the common land, and have extended their boundaries at the expense of their poorer neighbours too weak and too ignorant to resist them? If it be plunder to require the restitution of this ill-gotten property, I should like to know what we are to say of those who perpetrated the original act of appropriation.[11]

LAND

When Chamberlain proposed that land redistribution should be a partial solution to the late nineteenth-century crisis, he harked back to an earlier generation of Liberals including John Stuart Mill. Radicals, such as Richard Cobden and John Bright, had regarded land reform as an extension of their commitment to free trade.[12] In calling for a free market in land they had hoped to complete the political attack on landed power; an attack which had been stymied, after the 1832 Reform Act, by the restriction of the right to vote to those who owned freehold land whose annual rental value was worth 40 shillings. Following the repeal of the Corn Laws in 1846, the Anti-Corn Law League turned to the promotion of the formation of Land Societies through the purchase of landed estates and their subdivision into 40-shilling freehold franchises. The aim of this 'land reform' was not simply to enfranchise the 'aristocracy of the working classes' but to turn them into rentiers who would be able to provide welfare for 'their poorer brethren'. Although this aim, unsurprisingly, was not realised, it did provide one version of what was to become a model for land settlement.

At the same time, in the late 1840s, the Chartist leader, Fergus O'Connor, created a Land Company whose object was also to purchase and subdivide large landed estates. However, unlike the Liberal project, the aim of this faction of the Chartist movement was to disengage labour power from the wage-labour of industrial capitalism through the settlement of urban workers on the land.[13] Like the Liberal land societies, the Chartist company also failed, not because of the absence of social responsibility on the part of the beneficiaries of the scheme, but because it rested upon the belief that property in land could be distributed and reorganised from below and at a distance from state power, as in the Liberal case.

Another faction of the Chartist movement, associated with Bronterre O'Brien, bitterly attacked O'Connor's land company on the grounds that the plan 'would be to continue the evil of land monopoly under a modified form':

> you would abolish an aristocracy of thirty thousand men, and substitute in its stead an aristocracy of many millions and as every owner could 'do as he pleased with his own' some would sell their land to others, until in the course of time it would become the exclusive property of the few.[14]

Rather, O'Brien argued, 'true' nationalisation of land should be the object of land reform. To achieve this end, O'Brien proposed:

> The gradual resumption by the state . . . of its ancient, undoubted, inalienable dominion, and sole proprietorship over all the lands, mines, turbaries, fisheries etc. of the United Kingdom and our

Colonies: The same to be held by the State as trustee, in perpetuity, for the entire people, and rented out to them in such quantities, and on such terms, as the law and local circumstances shall determine.[15]

Pivoting upon the claim that the nation was the only absolute landlord, and linked to the object of making waste land productive to support a surplus population which would otherwise emigrate, O'Brien's proposal of 1850 for true nationalisation harked back half a century to Thomas Spence and the movement he had inspired. Spence's creed was 'land is the people's farm' and as such belonged 'to the entire nation, not to individuals or classes'. Private property in land violated the right of each individual to subsistence, while the common ownership of land, with land rented out to each household for production, would maintain individual subsistence. The rent, as a single tax, would provide for collective consumption through state expenditure. In Spence's formulation, people would take land into common ownership by invoking the millennium of the Hebrew jubilee.[16] This action of appropriation and redistribution, whose locus was to be local and municipal, was regarded by the state, not surprisingly, as subversive of its authority.[17]

However, as appropriated by the land nationalisers of the 1880s, the Spencian project, which now incorporated the Chartist argument for 'true' nationalisation, lost its subversive character. The mid-nineteenth-century experiences of both the Liberal and Chartist land societies provided a model for late nineteenth- and early twentieth-century campaigns to establish smallholdings. These campaigns were meant to have beneficial effects in both city and country as land resettlement was not only intended to prevent rural depopulation and to provide a ladder for land ownership by agricultural workers, but to ameliorate urban unemployment and to provide a means whereby the excess labour supply could be absorbed in agriculture to the extent that an induced shortage of labour would make average wage rates rise for all wage-workers. For example, the Liberal C.R. Buxton, who later found his way into the Fabian colonial nexus, was involved in an early twentieth-century campaign for smallholdings, as was Joseph Chamberlain's long-time political ally Jesse Collings. Initially directed from within the Liberal Party, the campaign, which later gained adherents within the Labour Party, was based upon the premise that labour absorption in agriculture would induce a general increase in average wage rates. Ireland and Denmark were used as examples of how state power and finance could be employed to make smallholding agriculture possible in England.[18] However, as had been true half a century before, scepticism towards a smallholding solution was expressed by radicals, now inside the Liberal

Party, who regarded private land ownership as the essential cause of the poverty of labour and who saw nationalisation as the way to ameliorate the condition of labour.

Pre-eminent among these radical movements to nationalise land was the Land Nationalisation Society (LNS). The stated aims of the LNS were: 'To affirm that the STATE holds the LAND in trust for each generation. To restore to all their natural right to use and enjoy their native land. To obtain for the NATION the revenue derived from its land.'[19] Peasant proprietorship was rejected in favour of what was called 'state tenancy' on the grounds that state-owned land required no management whatsoever. State tenants would not be deluded, as would be peasant proprietors, by the 'free trade in land' which would culminate in a new order of private landed monopoly.[20] The LNS was inspired by biologist A.R. Wallace, who had formulated the theory of natural selection independently of Darwin in 1858, but who based the natural advantage of particular species on their capacity of cooperation in competition with other species. Wallace was heavily influenced by the writings of both Edward Bellamy, the 'American kissing cousin' of Fabian socialism[21] and by the 'single taxer' Henry George.

In his most influential work, *Progress and Poverty*, which appeared in 1879, George argued that private landowners reaped an 'unearned increment' in the form of that portion of rent or sale price of their land which was over and above the value of the improvements they themselves had made to it. The source of this 'unearned increment' was to be found in the exertions of others who built the roads, railways, factories and towns which gave adjacent land its value. While state ownership of land might appear to be the logical conclusion of his argument, George avoided land nationalisation, partly because of his hostility to socialism, partly because, like John Stuart Mill, he gave primacy to competition, and partly because he objected to the massive capital outlay in compensation that a 'just' policy of nationalisation would require. Instead, he argued for a 'single tax' which would replace all other levies and 'tax away' the 'unearned increment'.[22]

Taking George a step further, Wallace maintained that the basic needs of labour could not be satisfied as long as the purpose of production was to 'create wealth for the capitalist employer'[23] and as long as land monopoly prevented the realisation 'of the birthright of every British subject to have the use and enjoyment of a portion of his native land'.[24] Through land nationalisation, Wallace envisaged the withdrawal of 'superabundant workers from the labour market' who would be organised as communities of workers to become 'wholly self-supporting on the land'.[25]

From the 1880s to the First World War, all the followers of precepts for land reform and land nationalisation put forward a basis for a palliative doctrine of development. Searching for remedies against urban unemployment and poverty, rural colonisation was promoted as the means to associate practically the 'two phenomena of unused land and unused labour power' and so eliminate the urban decay and destitution of British underdevelopment.[26] In so doing, land reformers, among others, were attempting willy-nilly to rework liberalism in such a manner as to answer the perceived late nineteenth-century crisis of British society. Others sought to do the same in a more explicitly philosophical fashion.

COMMUNITY: T.H. GREEN AND PROGRESSIVE DEVELOPMENT

The agitation over land had its contemporary analogue in the philosophical challenge to the utilitarian principles – that each individual's pursuit of pleasure and avoidance of pain comprised the basis of social order and progress – which had undergirded much of nineteenth-century liberalism. One of the most influential of those who sought the creation of a new, 'positive' liberalism, capable of meeting the social and intellectual challenges of the late nineteenth century, was the Oxford philosopher Thomas Hill Green.

Inspired in different ways by Hegel, British idealists such as Green made German Idealism palatable for an English sensibility in much the same way that early Fabians had domesticated Comtean positivism. British Idealism's politically radical effect, Stefan Collini notes, arose by way of its provision of 'an alternative set of categories' about the state and collectivism with which to confront utilitarianism. Or, as Rodman put it, 'it carried British thought from one world to another' as far as theory was concerned.[27] Green, judged a 'nice normal British neo-Hegelian' with 'theological twist',[28]

> tried to inaugurate a new concept of citizenship which would link men of different classes. The concept was based upon the notion that there was a good common to members of all classes . . . which could be established from German Idealist metaphysics and which could be made visible in actual measures of educational reform and social welfare.[29]

A central concern of Green's philosophical inquiry was 'development'. During the 1870s and 1880s, when Green wrote, the idea of development, despite Joseph Chamberlain's repeated references to a doctrine of development for the British imperial 'estates', had not yet taken on its present-day connotation of a primarily economic activity associated with an imperial or post-imperial 'Third World.' Rather, for Green, development was the process by which both specific, separate individuals and humanity as a general whole, came to full self-consciousness. A 'coming to consciousness', synonymous with moral and intellectual development, was posited by Green to be the existence of consciousness or complete (self) knowledge. Consciousness was realised *both* as an already existing teleological end *and* as immanent in the process of development itself.

In setting out this idea of development, Green was self-consciously rejecting the utilitarian views which had come to dominate nineteenth-century British thought. Rodman, Collini and Richter have all argued that Green's 'actual measures' differed little from the reforms proposed by J.S. Mill in his last 'socialist' phase. For Green, the poverty of utilitarianism lay, first, in its inability to express an idea of the 'common good' in any but the most banal terms. An aggregate of the 'pleasures' of individuals, with 'pains' netted out, could never be synonymous with the common good because the idea of the common, argued Green, was intrinsically social. Aggregation of utility, in utilitarian procedures, started from the premise of the separate, self-interested individual subject and then ended in the common good of social welfare. Green argued that if the purpose was to realise the common good, then it was only appropriate to start from the presumption that both 'the individual' and 'the social' were simultaneous parts of the consciousness of human history. Human history began as 'human' history only with social existence. No prior nor independent individual human existence was to be presumed.[30]

To this point Green followed Hegel perfectly. Hegel's critique of social contract theory was repeated and Green's theory of property seemed to have been lifted out of Hegel's *Propaedeutic*. But, in the face of the fundamental contradiction of the roles of state and civil society in development, Green recoiled, and flattened out the dialectic between history and logic while continuing to use Hegelian concepts.[31]

In Hegel's triadic scheme of history, the 'Idea' or 'Spirit' or 'Mind' that developed in the movement from *family* to *civil society* was made concrete in the *state*, the end of history. However, these were not simply moments in a linear unfolding of history but logical concepts. In the family, persons related to one another by instinct, each being an object, as if a natural being for the other. True freedom only arose when the subject of the Idea was alienated from the world of objects and then reunited with the object in the form of the state, the concrete complex of the Absolute Idea. Thus real human history began with the antithesis of the family – civil society. Here, through conflict – 'the revolt of slave against lord' – developed the consciousness of human interdependence through social obligation and absolute obligation to the state. This contention, that history, before its end in the ideal of a democratic state, was characterised by war, strife and revolution was uncongenial for Green, the 'nice British neo-Hegelian'.

In reaction to this unpleasantness, Green abandoned Hegelian dialectics in favour of the postulation of a Kantian eternal mind, or 'intelligence' of history, through which the 'progressive development' of the common good occurred. The common good could only be realised, contended Green, by a process of progressive development through which an

> eternal intelligence realised in the related facts of the world, or as a system of related facts rendered possible by such an intelligence, partially and gradually reproduces itself in us, communicating piece-meal, but in inseparable correlation, understanding and the facts understood, experience and the experienced world.[32]

Being divine in nature, when 'eternal intelligence' communicated, it both depended upon and gave rise to a development of consciousness which comprised feeling and thought 'inseparable and mutually dependent in the constitution of the facts which form the object of that consciousness'.[33] The process by which 'eternal consciousness' could be communicated could 'only be explained by supposing that in the growth of our experience, in the process of our learning to know the world, an animal organism, which has its history in time, gradually becomes the vehicle of an eternally complete consciousness'.[34]

Against bald theories of a biological evolution of consciousness, Green's ideas of eternal intelligence and consciousness were to contest the belief that 'human action' could be simply explained through some natural history of the development of 'brain and nerve and tissue'. Rather, action was only explicable by the existence of an eternal consciousness which acted through changes in such biological matter 'as its organs and reproduces itself through them'. Through supposing that a link between consciousness and matter could be put in this way, Green hoped to reconcile religion, in its broadest sense, with contemporary scientific theories of evolution and psychology. Thus, while it might be admitted that

> countless generations should have passed during which a transmitted organism was progressively modified by reaction on its surroundings, by the struggle for existence, or otherwise, till its functions became such that an eternal consciousness could realise or reproduce itself through them – this might add to the wonder with which the considerations of what we do and what we are must always fill us, but it could not alter the results of that consideration.[35]

The question of 'the development of the human organism out of lower forms' was 'quite different' from that of the transmission and reproduction of eternal consciousness. Biological evolution, and its metaphors, could be admitted without vitiating Green's argument. There was, Green maintained, 'an absolute difference between change' in the physical form leading to man and 'the intelligent consciousness or knowledge of change which precludes us from tracing any development of the one into the other'. As with Newman, 'development' as distinct from change implied 'an identity of principle between the germ and the developed outcome'.

When we speak of a development of higher from lower forms of intelligence, there should be no mistake about what we mean and what we don't mean. We mean the development of an intelligence which, in the lowest form from which the higher can properly be said to have developed, is already a consciousness of change, and therefore cannot be developed out of any succession in the changes in the sensibility, contingent on the reactions of the 'psychoplasm' or nervous system, however that system may have been modified by accumulated effects of its reactions in the past.[36]

A nebulous 'consciousness of change' could be both evidence of the working of an eternal consciousness and the means through which the reproduction of eternal consciousness could take place.

As to the obvious question of why 'there should be this reproduction', Green argued that this was 'as unanswerable as every form of the question why the world as a whole should be what it is'.

We have to content ourselves with saying that, strange as it may seem, it is so. . . . The unification of the manifold of sense *in our consciousness* of a world implies a certain self-realisation of this mind in us through certain processes of the world which, as explained, only exists through it – in particular through the processes of life and feeling.[37]

Evidence for the existence of such an eternal consciousness was to be found in humanity's desire for 'self-improvement' to a better human condition. Thus,

so long as a man presents himself to himself as possibly existing in some better state than that in which he actually is – and that he does so is implied even in his denial that the possibility can be realised – there is something in him to respond to whatever moralising influences society in any of its forms or institutions, themselves the gradual outcome through the ages of man's free effort to better himself, may bring to bear on him.[38]

It was the 'will' to desire improvement that distinguished human development from that of plants or animals, either through natural or artificial selection or training with which it had 'nothing in common'. Nor did human improvement have anything in common with a pre-conscious stage of human existence. To the question of the origin of this consciousness, if it was not to be found in nature, Green replied that it had no origin. 'It never began because it never was not. It is the condition of there being such a thing as beginning or end. Whatever begins or ends does so for it in relation to it.'[39] Development was made possible by 'reason' or 'the capacity on the part of a subject . . . to conceive of a better state of itself to be attained by action' when united with the 'will' of a 'subject to satisfy itself'.[40] Thus 'Men' came 'to seek their satisfaction, their good, in objects conceived as desirable because contributing to the best state or perfection of man; and this change we describe by saying that their will becomes conformable to their reason'.[41]

However, any process of human change had to be 'social' because the 'divine idea of man' could only take place in and through society where 'each being aware that another presents his own self-satisfaction to himself as an object, finds satisfaction for himself in procuring or witnessing the self-satisfaction of the other'. Without such a foundation of 'mutual interests' with 'each recognising the other as an end in himself and having the will to treat him as such there [could] be no society'. Society itself, then, was in each of its successive forms both a necessity for and the evidence of development. History was more than a succession of events, each determining 'the next in an endless series'. For if this were so, 'there would be no progress or development in it'.[42]

The rules enjoined by Green's idea of development for human conduct could be succinctly summarised as those which fostered the process of development itself. The good was that which fostered the 'good'. Green was well aware of the potential criticism that his schema was logically circular and therefore turned to the utilitarian alternative. Although one of the 'attractions', Green argued, 'of Hedonistic Utilitarianism' was that it seemed to avoid 'this logical embarrassment', in practice it only avoided circularity by eliminating any possibility of 'goodness'. A utilitarian calculus valued any act 'only as a means to an end' and not as a part of the process by which a 'perfect society' would be attained, namely the movement toward which was the very defining characteristic of goodness. Goodness lay in the process of development itself.

> Having found his pleasures and pains dependent on the pleasures and pains of others, he must be able in the contemplation of a possible satisfaction of himself to include the satisfaction of those others, and that a satisfaction of them as ends to themselves and not as a means to his pleasure. He must, in short, be capable of conceiving and seeking a permanent well-being in which the permanent well-being of others is included.[43]

Since the 'ultimate goodness' of a 'perfect society' could not be known by those who developed toward it, it was 'therefore not an illogical procedure, because it is the only procedure suited to the matter at hand, to say that the goodness of man lies in devotion to the ideal of humanity, and then that the ideal of humanity consists in the goodness of man'.[44]

Powerful evidence for the progressive development of humanity lay in the elaboration of 'rights' which, Green asserted, were progressively recognised, albeit gradually and piecemeal, in the sentiments, customs, 'laws of opinion' and the law proper. Green eschewed the tradition whereby 'natural rights' were derived from some pre-social past for the simple reason that the derivation of natural right, if it existed at all, would be external to the development of consciousness. Right had its origin within a 'community' rather than as some product of a contract between it and an individual. 'A right', Green maintained, of the individual, 'is a power of acting for his own ends – for what he conceives to be his good, secured to an individual by the community, on the supposition that its exercise contributes to the good of the community'. Again, however, Green insisted that for a right to be such required social consciousness because 'the exercise of such a power' could not be 'so contributory, unless the individual, in acting for his own ends, is at least affected by the conception of a good as common to himself and others'.[45] Compulsion to do good could not create a right because if an act was coerced it was not an exercise of the 'will'. 'The capacity for rights' was a 'capacity for spontaneous action regulated by a conception of the common good', a capacity which could not 'be generated – which on the contrary is neutralised – by any influences that interfere[d] with the spontaneous action of social interests'. Thus,

> any direct enforcement of the outward conduct, which ought to flow from social interests, by means of threatened penalties – and a law requiring such conduct necessarily implies penalties for disobedience to it – does interfere with the spontaneous growth of those interests, and consequently checks the growth of the capacity which is the condition for the beneficial exercise of rights.

It was for this reason 'that the effectual action of the state, i.e. *the community as acting through law*, for the promotion of the habits of true citizenship, seems necessarily to be confined to the removal of obstacles'.[46]

This was no simple argument for the principles of *laissez-faire*. Green was positive about the obligation of the state as 'the community as acting through law' to remove obstacles against 'true citizenship'. For example, drunkenness was one such obstacle since citizens who were addicted to drink could not develop their 'capacities' and, as such, the state was obliged to act against addiction in a 'positive' manner. Ignorance was a second obstacle which tended 'to prevent the growth of the capacity for beneficially exercising rights on the part of those whose education is neglected'. Last but not least, the 'accumulation and distribution of landed property' was potentially a third instance in which state action was justified.[47]

In Green's view, arguments about the historical origin of property, especially in land, had tended to conflate the question of how land came to be appropriated with the right to appropriation as a principle in itself. It was through the exercise of will, in this case the will to appropriate, that the general right to property was justified. The exercise of this will, to the extent that it came to be recognised by the state – the community as acting through law – was a development of the good in so far as it contributed to the 'emancipation of the individual from all restrictions upon the free moral life, and his provision with means for it'.[48] But the will to appropriate was not an unmixed good, for while the

> development of the rights of property in Europe . . . has so far been a state of things in which individuals *may* have property, but great numbers in fact cannot have it in that sense in which alone it is of value, viz. as a permanent apparatus for carrying out a plan of life, for expressing ideas of what is beautiful, or giving effect to benevolent wishes.[49]

In other words, the property of some prevented others 'who had not the chance of providing means for a free moral life, of developing and giving reality or expression to a good will, an interest in social well being'. Thus, Green argued that 'A man who possesses nothing but his powers of labour and who has to sell these to a capitalist for bare daily maintenance, might as well, in respect of the ethical purposes which the possession of property should serve, be denied rights of property altogether.' He then asked whether,

> the existence of so many men in this position and the apparent liability of many more to be brought to it by a general fall of wages, if the increase of population goes along with decrease in the productiveness of the earth, [was] a necessary result of the emancipation of the individual and the free play given to the will of appropriation? Or is it an evil incident, which may yet be remedied, of that historical process by which the development of the rights of property have been brought about, but in which the agents have for the most part had not moral objects in view at all?[50]

Green's reply was that poverty was both – both the result of the principle of appropriation and the particular history of property that had not met the criteria for what good development might be. The appropriation of land was 'only not an evil' if the 'restraints which the public interest requires to be placed on the use of land if individual property, were obeyed'. Thus, the 'turning of fertile land into forest', which made land 'unserviceable to the wants of men', had to be constrained by the public interest.[51] Activities which destroyed the agricultural capacity of land were the result of the history of appropriation. In England, Green argued, 'feudalism had . . . passed in unrestrained landlordism, almost untouched', with its 'landless countrymen, whose ancestors were serfs . . . the parents of the proletariate of the great towns'. Because of this line of descent, many lacked a developed will, right or moral sense and had been largely unaided by their landholding betters.[52] Taking this history into account, Green wrote,

> we shall see the unfairness of laying on capitalism or the free development of individual wealth the blame which is really due to the arbitrary and violent manner in which rights over land have been acquired and exercised, and to the failure of the state to fulfil those functions which under a system of unlimited private ownership are necessary to maintain the conditions of a free life.[53]

Indeed, it was capitalism which provided a partial compensation for the unlimited appropriation of land. For while it was 'true that the accumulation of capital naturally leads to the employment of large masses of hired labourers', there was

> nothing in the nature of the case to keep these labourers in the condition of living hand to mouth, to exclude them from that education of the sense of responsibility which depends on the possibility of permanent ownership . . . their combination in work gives them every opportunity, if they have the needful education and self-discipline, for forming societies for the investment of savings. . . . It is not then to the accumulation of capital, but to the condition, due to antecedent circumstances unconnected with that accumulation, of the men with whom the capitalist deals and whose labour he buys on the cheapest terms, that we must ascribe the multiplication in recent times of an impoverished and reckless proletariate.[54]

At the end of his discussion of property, Green considered Henry George's proposal for the state appropriation of the 'unearned increment' as a partial solution to the problems outlined in his consideration of 'Right of the State in Regard to Property'. His response was ambiguous. In principle, he found it 'fair enough' but maintained that any attempt to separate the part of 'value' earned 'through the expenditure of labour and capital' from that 'unearned', but arising from land ownership, to be so complicated that it could not be put into practice without lessening the 'stimulus to the individual to make the most of the land and thus ultimately lessening its serviceableness to society'.[55]

Green's ideal society was one of small-scale capitalism in which freedom would mean both the right to individual economic independence and the political obligation to participate in the life of the political community.[56] But, as Greengarten has shown, the idea that the individual should rightly achieve self-realisation through the pursuit of wealth was incompatible on Idealism's own grounds with the idea that human capacity could only be developed through the eternal mind, as both means and end, to making contributions to the common good.[57] Green's idea of development could not transcend Hegel's problem of poverty in relation to the state and civil society – the central contradiction of Hegelian development.

For Hegel, civil society was not the essence of social life but a means to the end of the consciousness of freedom or Knowledge which created universal altruism and culminated in the state. This is central to the distinction between Hegelian idealism and Lockean social contract theory, a distinction which governed a different basis for the official recognition of private property in land. For John Locke, the foundation of private property was necessary to secure the fruits of labour and thus the subsistence needs of work. An implication of Hegel's idea of freedom was that the state could not simply safeguard the active self-interest of individual property, rather the positive purpose of property was to realise human personality through the creative potential of working and not merely through the individual right to secure the products of work. Property, for Hegel, was the '*embodiment* of personality' involving both '*appropriation*' and '*an act of will* which must be acknowledged as such'; it is the 'objectification of self' in society. Possession was the means to the end of property which expressed the freedom of personality in society. It was a fact pertaining to the individual but it could have no personal quality without official recognition by the 'universal overseer of a social consensus' – the state.[58]

Whatever the intention of the individual to secure independence through work, the person was objectively alienated through immediate dependence upon wage-work. This was the 'cunning of reason' which moved through civil society. A lower form of knowledge, that of 'understanding' which was limited to the concept of civil society, would make individuals understand that they were interdependent via market exchange in meeting need. Through the historical development of civil society, 'understanding' as passive thought, would be transcended by 'active thinking', or the ideal form of knowledge for freedom. Such was the reason which culminated, Hegel supposed, in the form of the state through which active thinking brought the '*manifold into unity*'.[59] Thus, altruism in the state was not derived from some natural origin nor founded upon the biological instincts of family affiliation. Rather, while the economic market of civil society created the consciousness of the individual self and self-interest, the state reintegrated atomised individuals through the abstract, but really true, community of absolute selflessness.

The positive potential of civil society, constructed according to contract, was negated by the actual prevalence of the moral and thus absolute problem of poverty. Poverty was not just a relative problem for the many who were poor. For however much material needs were met, the universal desire for human recognition, so expressed by the essence of property, could not be self-realised. A regime of private property without 'ethical life' or without the state to enforce the personal obligations of 'love and trust', which Hegel associated with the 'family', would founder in the material and moral impoverishment of the many who sold and few who bought labour solely on contract.

Solely contractual relations between individuals were contingent, arbitrary, self-centred and devoid of personal ties and obligations. Civil society, if founded solely on such abstract relations, would be perpetually chaotic. State administration of 'the police' or law as the 'oversight and care of public authority' was, therefore, a logical condition of civil society. Public authority should protect the 'concept' of civil society against 'civil society's unimpeded activity' of the 'amassing of wealth' and the fall 'of a large mass of people beneath subsistence level'.[60]

The contradiction of poverty, which Hegel posed but left unresolved was this. *If civil society was a community without a state, the poor would be no part of it; but if the state, identified as the community, solved the problem of poverty the distinction between the state and civil society – so necessary for development – would collapse.*[61] British idealists such as Green may have seen the contradiction but were loath to recognise it through Hegel's own scheme of logic.

Hegel had insisted on working through contradiction in the Idea itself to understand that for different concrete 'experiences' there might be a logic of difference, but that it was the potential of development which created the identity between one category, the family, and another, the state. Green and his followers, such as R.B. Haldane the jurist, expunged Hegel's logic of difference from their Hegelian Idealism. This was tantamount to rejecting the logical gut of Hegel, who made it clear that the *Philosophy of Right* was based upon the method of his *Science of Logic*, the same logical method which was laid out, far more simply, in the *Propaedeutic*.[62] Here, Hegel considered the key idealist 'act of judgement' whereby '"the act of taking possession" is the expression of the Judgement that a thing becomes mine'. Judgement was then defined as the logical relations between the individual and particular and the particular and universal. He also carefully set out how the *syllogism* contains the 'mediation' or 'ground' through which relations between the individual and the universal were connected by the particular. The logical concept of civil society was derived, syllogistically, as that which mediated between the family and the state in history. The 'organism' or 'substance' or 'universal' – the various words Hegel used loosely to refer to the community of the family[63] – was dissolved into a world of particulars and individual differences in civil society, the particular differences which, by the logic of reason, had to be integrated through the universal will of the state.

Green's Idealism, with eternal intelligence and consciousness as the trans-historical standard for development as both means and end of the perfect society, ended up without either the logic or history of Hegel. Collini has pointed out that a trait of British Idealism, including that of Green's followers, was that it recognised 'no standard by which the present may be criticised that is not already implicit in the present, and yet it has no way of arbitrating disputes over the extent to which the ideal is actually realised'. British neo-Hegelians, Collini confirmed, had 'ignored the historicity of phenomena' and taken out 'the historical dimension' of Hegel.[64] Experience, for Green, was therefore abstract, arbitrary and without the contradictory but concrete 'determinations' of Hegel's logic.

THE BRITISH DOCTRINE OF DEVELOPMENT: JOSEPH CHAMBERLAIN

If Green shied away from the implementation of Henry George's single-tax proposals on the grounds of complexity of their implementation, Fabian thinkers, from the late nineteenth century on, made George's nostrums the foundation of their economic theory. Arising from an explicit rejection of Marx's value theory of labour, Fabian rent theory was meant to modernise Comte's thought by bringing it into line with the emerging theories of marginal utility and productivity and simultaneously to show how Comtean social trusteeship could resolve the perceived late nineteenth-century social crisis through an orderly redistribution of profits in excess of what was necessary to reproduce the wealth of the community in the name of humanity.[65]

Against George, the Fabian rent theory was based on the idea that all 'participants' in production, and not just those who owned land, realised rents according to their ability to contribute to production. In this formulation, capitalists or skilled workers were seen as just another active and knowingly self-interested party in society who received 'rents' in the forms of profits or wages, based on their skills, in analogous fashion to the way landlords received rents from land. Bankers, who received their rents in the form of interest, could not be made, the Saint-Simonians and Comte notwithstanding, the 'moral' altruistic trustees of society's capital, any more than landlords, *pace* George, could be trusted to fulfil their social obligation to make land productive and at the same time keep labour employed without destitution. Rejecting bankers, as a self-interested party who could not be trusted to act impartially, the Fabians fell back upon state officials whose own interests were not sectional but whose advancement ideally depended upon their administrative success in increasing social productivity. It was state officialdom, having the potential to be positively enlightened or be 'permeated' by Fabian thought, which was to become the Fabians' own version of Comte's 'positivist priests', and the trustees of society's development.

Elements of Fabian thought on the role of government found their echo in the writings of many who were frightened, at the turn of the twentieth century, by the spectre of relative British decline, and who called for policies of 'national efficiency'. Leopold Amery, co-founder of the Oxford University Fabian Club and future Secretary of State for the Colonies, explicitly took List's *Theory of National Economy* to be the inspiration for his *Fundamental Fallacies of Free Trade* in which he railed against the baleful effects to Britain of free trade.[66] At the same time, F.S. Oliver rehabilitated the thought of Alexander Hamilton, the American proponent of manufacturing, through protection, which he taught to the imperial proconsul Milner and his South African kindergarten. Another voice was that of J.L. Garvin, associate of Joseph Chamberlain, one-time head of the London School of Economics and long-time editor of the *Observer.*

Garvin, in 1905, cited the case of pig-iron production, arguing that while Britain in 1880 had produced as much as the rest of the world put together, by 1904 the British total was less than that of either the United States or Germany, which employed the protectionist policies Britain had eschewed.[67] Academic studies of almost a century later lend credence to Garvin's assessment. During the five-year period 1880–84, for example, the United States produced a mere 83 per cent of Britain's steel production; two decades later, for the 1900–1904 period, United States steel production was more than two and a half times that of Britain, while a decade later, for the 1910–1914 period, it was four times as much as that in Britain.[68] Garvin explained the relative British decline by free trade policy and the *laissez-faire* system economic liberals had inaugurated in 1846 when they finally abolished the Corn Laws and inadvertently pulled away the ladder which had enabled Britain to become the first modern industrial power.

Classical economists, claimed Garvin, forgot the fundamental distinction between raw materials and manufactures. The free trade principle, advanced by

Tooke's *London Free Trade Position of 1820 for the City of London*, held that what was best for the individual merchant was best for the nation – buy in the cheapest and sell in the *dearest* market. But this could not apply for manufactures where, under large-scale production conditions, turnover was maximised on relatively small profit margins and where 'buy in the cheapest and sell in the *biggest* market' was the appropriate maxim. Economic liberals overlooked 'the ability of States pursuing a positive policy of economic development to create competitive power where it had not existed. *Created* aptitude in the modern world is as conspicuous a factor as natural aptitude.'[69] Constructive economics negated free trade, which was to be be regarded as an abstract conception which cannot exist and is as remote 'as the federation of the world'. '*Laissez-faire*', wrote Garvin echoing List, 'attempted political economy with the politics left out.'[70] Free trade economics was a theory of structure offering a passive conception of the state, whereas constructive economics offered *savoir faire* in place of *laissez-faire* as a theory of energy to underpin an active state.

No universal principle could underpin the realisation of economic need other than the construction of state policy which should be directed, Garvin advised, towards re-creating the 'greatness' of the British national system of commerce and maritime power. As if to remember the Saint-Simonians, Garvin argued that:

> There are no economic factors so potent, so creative, as national strength and the sense of it . . . economic progress, no less than the political presentation of it, must largely depend upon the conscious purpose and efficient action of the state itself. Government, in a word, should be the brain of the state, even in the sphere of commerce.[71]

For Garvin, as for many others of similar inclination, it was Joseph Chamberlain who embodied the political possibility, at the turn of the century, that 'the brain of the state' might be directed towards constructive economic ends. By this time, Chamberlain had broken with his Radical *confrères* over the question of Home Rule for Ireland. Landlordism had finally led to a breakdown of the old agrarian system in England and so too, Chamberlain argued, had landlordism made Ireland underdeveloped. In railing against the then Tory leader, he made his case:

> Lord Salisbury cares nothing for the bulk of the Irish nation. . . . He has no sympathy – or at least he expresses none – for the great mass of the population . . . who have been subjected to undeniable tyranny and oppression, and whose wrongs cry aloud for redress. He can express to you in eloquent terms his sympathy for the Irish landlords, who have had to submit to a reduction of 25 per cent in their rents, but I find nowhere any expression of sympathy for the poor tenants who, for years, under the threat of eviction, and the pressure of starvation have paid those unjust rents levied on

their own improvements, and extorted from their desperate toil and poverty.

Chamberlain continued:

I say that in this matter as in so many others, Lord Salisbury constitutes himself the spokesman of a class – the class to which he himself belongs – 'who toil not, neither do they spin', whose fortunes, as in his case, have originated in grants made long ago, for such services as courtiers render kings – and have since grown and increased while their owners slept, by the levy of an unearned share on all that other men have done by toil and labour to add to the general wealth and prosperity of the country . . .[72]

Chamberlain's proposed solution to the Irish question was state-sponsored economic development. Written in 1882, the following is a perfect example of an early 'development mission' Chamberlain hoped to implement through the British state:

I believe that sooner or later it will be found necessary to undertake some public works in Ireland. . . . State assistance in some form or another is afforded, especially to the provision of communications and to great works of reclamation and main drainage, in every country of Europe, and indirectly by grants of land and other privileges in the United States of America. I fully admit the dangers and difficulties of any such undertaking; the probability of jobbing and inefficiency; but having regard to the political exigencies of the situation as well as the character and poverty of the Irish people, I would at once appoint the strongest scientific and technical commission it would be possible to obtain to report on certain broad classes of undertakings, especially on railways, reclamation, main drainage, and harbours, with a view to some considerable scheme of public works.[73]

Public works, Chamberlain argued, would be 'reasonable insurance against greater evils'.[74] However, if Chamberlain was fully behind development in, and for, Ireland, he was not prepared to concede what Irish nationalists ultimately wanted – independence. National independence was what he understood the Gladstonian policy of Home Rule as leading toward and Chamberlain supposed that there was no necessary historic link between development and national independence. His bitter comment that 'Thirty-two millions of people must go without much needed legislation because three millions are disloyal' revealed the extent to which Chamberlain's version of the British doctrine of development was made to fit a different, imperial link. It was doctrine in which social reform at home and the condition of Empire had become increasingly interwoven.[75] To understand the logic behind this linkage, Chamberlain's career has to be re-examined.

On becoming Mayor of Birmingham in 1874, Chamberlain carried through a campaign of meeting civic needs, later misnamed and misunderstood as 'municipal socialism'. Private gas and water supply were purchased by the city and run as monopolies. In turn, the municipal utilities were to help finance slum clearance and reconstruction. Municipal undertakings would provide the foundation, Chamberlain had envisaged, for a political pact between large-scale capitalist enterprise and a well-paid, well-housed and educated industrial workforce of high productivity which would guarantee the efficiency and competitiveness of British industry. For Chamberlain, his own lived experience made his vision of a rational capitalism more than a political expedient. When head of the family business, he had driven out competition, as we have noted, and had been deliberate in his destruction of the small manufacturers who had hitherto been predominant in Birmingham industry. The extinction of small-scale production was described by Chamberlain as 'an almost unmixed good' because the simultaneous destruction of the 'secret moderation' of artisan-based politics permitted a pact between large-scale capitalist enterprise and a higher-wage, more productive labour force.[76]

Social insurance and the state provision of labour exchanges, supported through the Birmingham Chamber of Commerce, would regulate the surplus population and fend off labour unrest. If the Birmingham schemes could be extended nationally, British national industrial and military supremacy might be regained. If the purpose of Chamberlain's support of land reform was to stem the influx of surplus rural labour, it would be to no avail if it were not accompanied by another strut of policy to strengthen British industrial competitiveness. For Chamberlain, a policy of the constructive development of empire was central in an increasingly protectionist world. Amery later recounted the argument made by Chamberlain in 1903:

> British export trade had been practically stagnant for thirty years, in spite of an increase of 30 percent in population; . . . the export of manufactures to our industrial rivals had steadily declined from £116,000,000 to £73,500,000, and that loss had only been made good by an increase of £40,000,000 in the Imperial trade. Meanwhile foreign manufactures imports into the United Kingdom had gone up from £63,000,000 to £149,000,000. The whole character . . . of British trade was changing, and Cobden's dream of an England importing foodstuffs and raw materials, and in return supplying the world with its manufactures, had vanished.[77]

By 1895, when he became colonial secretary, Chamberlain proclaimed why it was necessary for development to encompass the colonial empire as a whole. He offered his well-quoted words in a speech to the annual banquet of Birmingham Jewellers and Silversmiths:

[It is] not enough to occupy certain great spaces of the world's surface unless you can make the best of them, unless you are willing to develop them. We are the landlords of a great estate; it is the duty of the landlord to develop his estate. . . . In my opinion, it would be the wisest course for the Government of this country to use British capital and British credit to create an instrument of trade [i.e. railways] in all . . . new important countries. I firmly believe that not only would they in so doing give an immediate impetus to British trade and industry in the manufacture of machinery that is necessary for the purpose, but in the long run . . . they would sooner or later earn a large reward either directly or indirectly.[78]

British colonies, in Chamberlain's view, had to be constructively developed in order to provide both outlets for British investment and markets for British goods in order to meet the British 'national needs' of industrial expansion and employment.

For Chamberlain, as for Garvin, there were 'no economic factors so potent, so creative, as national strength and the sense of it', and no economic weaknesses so debilitating as the lack of the sense of this strength. Chamberlain, goading his countrymen, foretold a bleak future without a policy of constructive development in an era of protectionism:

Your once great trade in sugar refining is gone; all right, try jam. Your iron trade is going; never mind you can make mousetraps. The cotton trade is threatened, well what does that matter to you? Suppose you try doll's eyes. . . . But how long can this go on? Why on earth are you to suppose that the same process which ruined sugar refining will not in the course of time be applied to jam? And when jam is gone? . . . And believe me, that although the industries of this nation are very various, you cannot go on forever. You cannot go on watching with indifference the disappearance of your principal industries.[79]

By the turn of the century, then, it was the Briton, Chamberlain, who sought to reconcile List's 'head' and 'arm' of the nation through the economic rationalisation of Empire:

Great Britain, the little centre of a vaster Empire than the world has ever seen, owns great possessions in every part of the globe, and many of these possessions are still almost unexplored, entirely underdeveloped. What would a great landlord do in a similar case with a great state? If he had the money he would expend some of it at any rate in improving the property.[80]

As Colonial Secretary in the Tory Unionist government of the 1890s, Chamberlain advocated the allocation of funds for the improvement of roads, bridges, public buildings, railway and irrigation works throughout the Empire. Cyprus was an early priority for which he ultimately was able to secure an annual grant of £50,000 in grants to repair neglected public works and £100,000 in loans for roads, wharves and railway construction was sought by Chamberlain for the island of Dominica. The grant was made but the loan was blocked by the Treasury which saw little chance of it ever being repaid. Railway surveys were begun in Sierra Leone and what was to become Nigeria. Only the competing demands of other government departments limited Chamberlain's proclivity for spending on projects. For example, it was a Foreign Office concern, the Uganda railway, and the demands of the Navy for more and better equipment and ships which absorbed finance the Colonial Secretary claimed for his development projects.

Chamberlain attempted to circumvent inter-departmental competition by devising various schemes to create new sources of colonial development finance. His proposals for the use of the dividends from Britain's Suez Canal shares and to dun the 'white dominions' for a part of the cost of the navy were respectively rejected as being fiscally and politically unsound. More, though still limited, success was had from the Chamberlain-sponsored Colonial Loans Act of 1899. The Act was approved at a time when British possessions in the West Indies were undergoing one of their sharper periods of economic decline in what was a long downward trajectory. Under provisions of the 1899 Act the construction of railways was proposed, begun, or advanced in Cyprus, the Gold Coast, the Niger Coast Protectorate, Lagos, Sierra Leone, Trinidad, the Malay States and Jamaica, as well as irrigation in the first, harbour works in the first three, roads and surveys in the Seychelles and hurricane relief in the West Indies. Although it accomplished much, the 1899 Act was largely hindered by the outbreak and subsequent cost of the Boer War, which was regarded by the Treasury as a cause for stringency for civil expenditure, just as the outbreak of the Second World War would also delay the implementation of the 1939 Colonial Development and Welfare Act.[81]

The Boer War unleashed more ominous forces against which future plans for colonial development would have to contend. Liberal 'little England' anti-imperialism, historically more sentiment than practice, was given greater substance through opposition to the war. From the conflict's onset, war opponents were strident and vocal. As Stephen Koss notes, 'with remarkably few exceptions, the key figures in each of the various committees', through which much of the opposition to war was carried out, 'belonged to the Radical Wing of the Liberal Party'.[82] Chamberlain's name was now an anathema to those Liberals and Radicals who had remained with him when he parted company over Irish home rule and who now hated him as a result of his being implicated in the Jameson Raid. Aside from a hatred of Chamberlain and a new-found fondness for the Boers, whom David Lloyd George at one point glowingly referred to as 'Liberal Forwards', the 'pro-Boer' Radicals developed other, more distinctly unsavoury crotchets.[83] Rampant anti-Semitism should be recognised, not least because it is John A. Hobson, one of the most rabid anti-Semites of the period, who is the inspiration, alongside Schumpeter and Veblen, for Cain and Hopkins' 1993 magisterial treatise, *British Imperialism*.

Hobson's vaunted position arises, as it did for Keynes, because of his distinction 'between "Producers England", where industry set the tone of life and "Consumers England"', of the south where the 'moneyed class' reigned supreme as a 'class of "ostentatious leisure" and "conspicuous waste"'.[84] Cain and Hopkins owe much to Hobson for their thesis of an imperialism governed by what they call 'gentlemanly capitalism'. A brief perusal of the wartime writings of John A. Hobson who was employed by the *Manchester Guardian* and who was soon to gain fame as the author of *Imperialism: A Study*, provides ample illustration of this prejudice. Johannesburg, he wrote,

> is in some respects dominantly and even aggressively British, but British with a difference which it takes some little time to understand. That difference is mostly due to the Jewish factor. If one takes the recent figures of the census, there appears to be less than seven thousand Jews in Johannesburg, but the experience of the streets rapidly exposes this fallacy of figures. The shop fronts and business houses, the market place, the saloons, the 'stoeps' of the smart suburban houses are sufficient to convince one of the large presence of the chosen people. If any doubt remains, a walk outside the Exchange, where in the streets, 'between the chains', the financial side of the gold business is transacted, will dispel it.

In case there is any illusion that Hobson's anti-Semitism was the happenstance of a particular conjuncture of Johannesburg merchants, he made the same connection Hilaire Belloc, another anti-Semite, had more satirically contrived when he tied City of London finance to that of Jewish control and both to the cause of 'development'.[85] Hobson continued:

> So far as wealth and power and even numbers are concerned Johannesburg is essentially a Jewish town. . . . The rich, vigorous, and energetic financial and commercial families are chiefly English Jews, not a few of whom here, as elsewhere, have Anglicized their names after true parasitic fashion. I lay stress upon this fact because, though everyone knows the Jews are strong, their real strength is much underestimated. Though figures are so misleading, it is worth while noting that the directory of Johannesburg shows 68 Cowens against 21 Jones and 53 Browns. The Jews take little active part in the Outlander agitation; they let others do that sort of work. But since half the land and nine-tenths of the wealth of the Transvaal claimed for the Outlander are chiefly theirs, they will be the chief gainers by any settlement advantageous to the Outlander.[86]

Examples of pro-Boer propaganda which demonstrate that Hobson was not alone in holding such views are easily found in the materials produced by the anti-war committees and press. The point about anti-Semitism was that it was the motif for a fierce hostility towards all capitalists in South Africa. 'Jew' was perfectly understood to be 'capitalist' and Chamberlain's doctrine of development easily assimilated to the 'development projects' of financial speculators. Cain and Hopkins, for instance, are far from the mark when they claim the following:

> Since Chamberlain was making a direct assault on gentlemanly culture, it was inevitable that he would be condemned, in similar terms, by both sides of the political divide. Almost instinctively, his attempt to place industrial wealth creation and its problems at the head of the political agenda was condemned as 'utterly sordid' because it catered for the 'ignoble passions' of 'vulgarity and cupidity'.[87]

'Sordid', 'ignoble', 'vulgarity and cupidity', are the very epithets of 'Jew' which were used to mark Chamberlain with the taint of finance and its association with 'development'. What was so instinctive for Hobson was that if he shared the attack on 'gentlemanly culture', he did so perversely because the gentlemanly values were precisely presented as being the antithesis of what was implied by being 'Jew' and, therefore, of 'development'. When Cain and Hopkins plausibly suggest that Chamberlain 'linked the fortunes of City and industry together and then claimed that both depended on the future development of the empire under a regime of tariffs',[88] they reinforce the very impression that Hobson and the other anti-Semites used against Chamberlainite doctrine. The evidence is unmistakable.

The Radical Whig Philip Stanhope, in moving a motion to censure the committee appointed by Parliament to investigate the origins of the Boer War stated that 'Mr. Rhodes and his associates – generally of German-Jewish extraction – found money in thousands' and used finance to 'poison the wells of public knowledge.' The publisher and pamphleteer W.T. Stead, who did so much to influence the colonisation projects of General Booth in his Salvation Army's endeavour to settle London's unemployed on the land, compared the Boer leader Kruger to Christ and made a none too subtle appeal to his readership's prejudices by comparing the British government to the 'Jewish mob which cried out "Crucify him! Crucify him!"'. The admittedly eccentric Edward Carpenter described Johannesburg, since the discovery of gold, as 'hell full of Jews, financiers, greedy speculators, prostitutes, bars, banks, gaming saloons, and every invention of the devil'.[89] John Burns, another Radical MP, related to the House of Commons how 'the British Army, which used to be for all good causes, the Sir Galahad of history, has become the janissary of the Jews', alleged to be in league with members of Parliament who held stock in Rhodes' Chartered Company.[90]

Pro-Boer agitation was also often linked to a general critique of Britain's treatment of Africans. A common theme was that the Boer, though rough hewn, was a better master. Others went farther and launched a wholesale criticism of Britain's imperial record. One such critique, which appeared in the Radical pro-Boer *Morning Leader*, attempted to illustrate the hypocrisy of the pro-war claim that England 'was now fighting against the slavery of the African race as truly as the Northern States fought against it in the Civil War'. This critique itemised British misdeeds as follows:

1. In the strip of East African Coast – a British Protectorate – which faces Zanzibar, the full 'legal status of slavery' is maintained, and fugitive slaves have been handed back to their owners by British officials.
2. In Zanzibar and Pemba the manumission of slaves, presided over by Sir Arthur Hardinge, is proceeding slowly, and many thousands are still in bondage.
3. In Natal the *corvée* system prevails, and all natives not employed by whites may be imprisoned to labour for six months of the year on the roads.
4. In Bechuanaland, after a recent minor rebellion, natives were parcelled out among Cape farmers and indentured to them as virtual slaves for a term of years.
5. Under the Chartered Company in Rhodesia the chiefs are required, under compulsion, to furnish batches of young natives to work in the mines, and the ingenious plan of taxing the Kaffir in money rather than in kind has been adopted so that he may be forced to earn the pittance which the prospectors are willing to pay him.

And last, but not least:

6. In Kimberley what is known as the 'compound' system prevails. All natives who work in the diamond fields are required to 'reside' under lock and key, day and night, in certain compounds, which resemble spacious prisons. So stringent is the system that even the sick are treated only within the prison yard. On no pretext whatever is a native allowed to leave his compound. After all, the hands of the Northern States were a little cleaner than our own when they championed the cause of the slave.[91]

The last criticisms, in particular, were directed at the implantation of wage-labour in Southern Africa, an implantation associated with Chamberlain's doctrine of development. They were fuelled not only by the Boer War but by Britain's involvement in other colonial wars in Africa such as the brief though bitter conflict in the Sierra Leone Protectorate. The critique was fired as well by more general circumstances, including the revelations of British official forbearance over Belgian atrocities in the Congo.[92]

Such was the climate in the run-up to the general election of January 1906, when Chamberlain advanced his campaign for 'tariff reform' or 'imperial free trade', the campaign partly designed to resolve the problem of unemployment at home through the creation of an interlocking imperial economy. Looking back, Amery characterised the rationale of tariff reform, or the campaign which has come to be known as 'social imperialism', by saying that he could not

see any solution of our domestic social problems except in the framework of an expanding Empire economy. Convinced as I was of the necessity of social reform, it always seemed to me that it should be supplementing and not displacing a man's individual effort on behalf of his family or of his own old age; as meeting the need of the less fortunate for a decent minimum standard of living, not as providing a comfortable standard of living for everybody regardless of their own effort.

Amery then spelt out his conviction in British development doctrine:

But that was only possible in a framework of real opportunity and incentive. In a stationary economy, and still more in a regressive economy with heavy and continuous unemployment, an increasing social conscience and a democratic vote would inevitably lead to an evergrowing burden of taxation which such an economy could not afford, and to a dependence on public support which would still further weaken the mainspring of our productive effort. The continuance of Free Trade seemed to me ineluctably to involve the growth of confiscatory socialism and a consequent division of parties on class lines.[93]

But the same trajectory, the 'evergrowing burden of taxation' to confront unemployment, was what tariff reform implied, with the burden of protective duties falling on the real wages of producers at the cost of their consumption and productive effort.

Chamberlain and the doctrine of constuctive development he espoused met defeat in the election of 1906, which brought to power a government heavily stacked with anti-imperialist Radical Liberals committed to free trade. The Liberal Prime Minister, Campbell-Bannerman, played to this element when, in the campaign of 1905, he responded to Chamberlain's injunction to develop the imperial 'estate' with a development design of his own:

We desire to develop our undeveloped estates in this country; to colonise our own country; to give the farmer greater freedom and greater security in the exercise of his business; to secure a home and a career for the labourer, who is now in many cases cut off from the soil. We wish to make the land less of a pleasure ground for the rich and more of a treasure house for the nation.[94]

Campbell-Bannerman's renewed call for the reform of land was emblematic of the call for a host of remedial measures designed to combat the evils of poverty and unemployment in the wake of the defeat of Chamberlain's doctrine of 'constructive' development. Among the remedies was a reworking of the Poor Law, one weapon which had been employed to deal with the existence of a surplus population since Elizabethan times. As a recent historian of the Poor Law has emphasised, the Elizabethan legislation

> made a division between the 'impotent' poor (the lame, blind, old, invalids, feeble-minded and other persons not able to work) and those who were capable of working. . . . [t]he impotent poor . . . were to be given 'necessary relief', and the able-bodied poor were to be 'set to work' on 'a convenient stock of flax, hemp, wool, thread, iron and other necessary ware and stuff'. Able-bodied persons refusing to work . . . were to be punished with Draconian severity.[95]

In keeping with the rise of Benthamite utilitarianism, the system was 'reformed' by the Poor Law Amendment Act of 1834. Previous practice, it was argued by the proponents of the new law, had become lax in that many of the poor were said have registered for 'necessary relief' when they did not qualify as being 'impotent'. To resolve this administrative difficulty, 'outdoor' relief was to be all but abolished while the so-called able-bodied were to be confined to 'indoor' relief which was distributed in prison-like workhouses. Normal assistance was not to exceed the standard of life of that of an 'independent labourer of the lowest class'.[96]

From the beginning of its implementation, and however modified in practice if not in principle, the 1834 Act had been met with riot and resistance. By the 1880s, it was increasingly recognised to be untenable. Ironically, it was the then Radical Joseph Chamberlain who was responsible, in the midst of large-scale unrest of the unemployed during 1886, for the most important post-1834 changes to the Act. As president of the Local Government Board, he issued the 1886 circular which proposed the following 'emergency measure':

> What is required to relieve artisans and others who have hitherto avoided Poor Law assistance, and who are temporarily deprived of employment is – (1) work which will not involve the stigma of pauperism; (2) work which all can perform, whatever may have been their previous avocations; (3) work which does not compete with that of other labourers at present in employment; and lastly, work which is not likely to interfere with the resumption of regular employment in their own trades by those who seek it.[97]

It was this emergency measure as embodied in Chamberlain's 1886 circular which 'remained the policy of both Liberal and Conservative governments concerning the unemployed' until 1905 when a Royal Commission on the Poor Laws was set up to consider the problem. Conflict within the Royal Commission exemplified the way in which the problem of a surplus population was to be understood and to be dealt with in the aftermath of the defeat of the Chamberlainite doctrine of development. Here, the precepts of one strand of Greenian idealism, the Fabian version of positivist trusteeship, and the tenets of 'advanced' Edwardian radical liberalism were pitted each against the others.[98] One upshot of the Royal Commission was the 1909 Development Act, the first attempt to put the official practice of development to work within Britain itself.

TOWARDS THE 1909 DEVELOPMENT ACT

The Poor Laws, from their inception, had never been intended to be the sole or chief means of relieving poverty. Rather, the Poor Law provision of relief was intended to supplement the endeavour of private charity. In 1869 many of these charities came together in the Charity Organisation Society to begin to coordinate their efforts. It was, in part, from this body that the Royal Commission drew its members. Among these were Helen Bosanquet, the political economist, and her philosopher husband Bernard. Both had deeply imbibed Green's idealism. Thus the Bosanquets believed that the 'poor' were those who lacked a capacity to fulfil their obligation to contribute to the realisation of the common good through their own self-development. Undeveloped capacity, especially among the 'able-bodied' or 'helpable' poor, was believed to be due to an ethical failure of the individual and the community. Specific reasons for such ethical failure were to be uncovered by the method of social case-work which the COS had pioneered. In the view of the COS, the state could and should not undertake the task of social work. State responsibility for the poor should be confined to providing assistance to the 'impotent' or 'unhelpable'.[99]

Any Benthamite utilitarian would have approved of the Bosanquets' conservative interpretation of Green and the way in which Green's version of British Hegelian Idealism was set out in COS policy:

> The working man does not need to be told that temporary sickness is likely now and then to visit his household; that times of slackness will occasionally come; that if he marries early and has a large family, his resources will be taxed to the uttermost; that if he lives long enough, old age will render him more or less incapable of toil – all these are the ordinary contingencies of the labourer's life,

and if he is taught that as they arise they will be met by private relief or state charity, he will assuredly make no effort to meet them himself.

Then came the old crunch:

A spirit of dependence, fatal to all progress, will be engendered in him, he will not concern himself with the causes of his distress, or consider at all how the conditions of his class may be improved; the road to idleness and drunkenness will be made easy to him, and it involves no prophesying to say that the last state of a population influenced after such a fashion will certainly be worse than the first.[100]

Poverty, to repeat the old nostrum that had long been associated with the failure of progress itself, was the result of an individual subjective failing. If one part of the Bosanquets' gloss was new, it was their claim that this failing could not be put right by force, including that of market force, but by investigation into failure and then the case-work method to correct the reasons for poverty. Unlike Chamberlain's constructive doctrine of development, no attempt was made to identify non-subjective reasons for the failure to match capacities of work to social need and then provide the basis for state direction to make the idle productive.

Among those who objected to the Bosanquet view were the Fabians who, in the person of Beatrice Webb, were represented in the deliberations of the Royal Commission. As Fabian orthodoxy came to be congealed around the views of Beatrice and Sidney Webb, the ultimate source of the failure of capitalism to provide employment for the 'able-bodied' was traced to competition. COS investigative studies revealed the existence of the 'able-bodied poor' but the case-work method in itself, so Beatrice Webb came to believe after her own experience at such work, could not eliminate the problem. If production, together with consumption, was to remain money-mediated or commercialised, then the Fabian task was to ensure that production was capitalised on a moral basis and not according to the temper of competitive capitalist enterprise. Nor could labour, solely by virtue of its productive force and potential command over production, be trusted to control enterprise since the resulting competition between groups of producers would not realise social need. Both capital and labour, following the Saint-Simonians and Comte and when brought into line with 'modern' economic thought by the Fabian theory of rent, were to be administered 'in trust for the community' by scientific experts in the employ of the state.[101]

In the first decade of the twentieth century Empire had no place in the Fabian prospectus for development. State activity was variously to take the form of education and training designed to make the young more employable. Draconian measures were to remain in force to compel shirkers into work, but pensions for the old, labour exchanges and, after much argument with William Beveridge who was sceptical of the Webbs' preference for labour colonies in which the idle would be matched with work, counter-cyclical government spending was advised as well. This last was to be carried out in times of slump provided that it was undertaken to meet real, as opposed to invented, social needs. Although such work was to conform by and large to the criteria of Chamberlain's 1886 circular, it was also seen to be a basis for how state experts could increasingly direct the employment of labour and capital:

There are slums to clear, houses to build, land to redeem, and waste places to afforest. To get this work done there is need of armies of workers, engaged not temporarily to tide over a depression, but permanently to complete an undertaking, the amount of undertaking swelling or diminishing each year according to the state of trade.[102]

Early in 1909, the members of the Royal Commission issued two reports, a majority report representing, in large measure, the views of the Bosanquets and a minority report written by Beatrice Webb. Coincidentally, the 'most acute commercial depression since 1879' was experienced over the short period of 1908–9.[103] Responding more to evidence of depression than the Royal Commission reports, Winston Churchill pressed upon the Liberal government, of which he was a part, the need to adopt a 'scientific solution' to the unemployment problem. Some elements of the solution, which were also proposed by others, included old age pensions and unemployment insurance. Another significant proposal, which had the full support of the Webbs, was that a commission should be established to be in charge of 'national development'. Without much hesitation, the government complied and the Development Act of 1909 created a Development Commission. Sidney Webb became one of the commissioners while Keir Hardie, for the Labour Party, called the Development Bill, as it passed through Parliament, 'the most "revolutionary" measure ever introduced by a government and an implicit recognition of the "right to work"'.[104]

Under the authority of the 1909 Act, the Development Commission was mandated to: 'give financial assistance to agriculture and "rural industries"; to forestry, land reclamation, and road transport; and to the "development and improvement" of fisheries and the "construction and improvement" of harbours and inland navigations'.[105] The rural bias in this mandate is unmistakable. Moreover, permissible 'development projects' were to be in industries and infrastructure which remained, as Lloyd George made clear in his budget speech of 1909, 'outside the legitimate sphere of individual enterprise' and which had little expectation of profit.[106] Further, it was accepted that the use of the unemployed would be 'unlikely to be as efficient' as those 'hired through the normal labour market'.[107] At the outset of official development practice within Britain itself, therefore, the intention of development was palliative, to put the unemployed to work. Development work was at a tangent to work deemed to be efficient and productive according to the norms of accounting profitability and all that is entailed by the pursuit of capitalist rationale for productive activity.

Between 1907 and 1909, some elements of big business had called for constructive state policy. For example, Sir John Brunner, of the Brunner-Mond chemicals complex of companies, which later became ICI, organised a group of Liberal MPs who appealed to Prime Minister Campbell-Bannerman: 'We regard it as sound policy to spend national funds on the promotion of local interests provided that such expenditure contributes to the national welfare.' They called for policy 'to assist trade by developing the internal resources of the Kingdom', approvingly citing the German example of state expenditure for scientific research, especially for metal industries. Likewise, they wanted state spending on infrastructure – ports, canals and roads – to reduce the transport costs of big business. Campbell-Bannerman's response was to call these proposals 'wild' and to suggest that they would 'scare every railway shareholder in the country'. Tories 'would raise the cry of a new plunder, and all the quiet people who live on railway dividends would vote against us'.[108] It was the fear that the development implied nationalisation which deliberately made the Liberal government confine its 1909 Act to activities which would not provoke the anticipated reaction of plunder.

Under the terms of the 1909 Development Act, then, development was to remain confined to primarily rural or agrarian activities which would not greatly interfere with commodity and finance markets and threaten rentiers and antipathetic business interests. In the aftermath of the defeat of Chamberlain's constructive doctrine of imperial development, official British development doctrine, remaining within the contours of 'individual enterprise', was almost exclusively palliative in intent. Until such time that a programme for the nationalisation of productive assets was actually elaborated, some thirty years later, the Fabian version of the doctrine of development, which although understood to be the scientific administration of land, labour and capital in 'trust' for the 'community' of Britain, also remained trapped, despite itself, in palliative intent. The equivocations of Fabianism were mirrored by the provisions of the 1909 Act which were, in turn, matched by those arising from more theoretical musings on the idea of development. Within Britain itself, and for a British purpose, L.T. Hobhouse best exemplifies how the idea of development continued to be equivocal when tied into the 'new liberalism' of the early twentieth century.

THE PROBLEM OF INTENTIONAL DEVELOPMENT: L.T. HOBHOUSE

If, as we have argued in Chapter 1, much of the work of J.S. Mill should be seen as an attempt to reconcile utilitarian and positivist thought, then that of L.T. Hobhouse was an attempt to marry versions of late nineteenth-century liberalism. Hobhouse's purpose was to reconcile Herbert Spencer's liberalism with the idealism of T.H. Green. In the introduction to his *Development and Purpose*, first published in 1913, Hobhouse set out the difficulties he had in accepting both Spencer's rendition of liberalism and Green's idealism and why he was prompted to set out his own views on the problem of development.

Hobhouse's dissatisfaction with Spencerian sociology was rooted, as it was for many of the new liberals, in Spencer's 'uncompromising' evolutionary individualism. According to Hobhouse,

> This assertion of individualism coincided with the beginnings of a new demand for the extension of collective responsibility and the social control of industrial life. Economically the old individualism was dying, and apart from the evolutionist school, it was clear to thinking men that the idea of liberty required a new definition.[109]

Hobhouse was equally dissatisfied with Green's idealism, largely on the grounds that it was incompatible with modern – read evolutionary – science. However, Green's idea of eternal or 'permanent' consciousness, which Hobhouse mistakenly supposed that Green had got from Hegel, could be rescued. Supposing Green's problem to be one of the 'spirit' in a scientific age, Hobhouse thought he could interpret Green thus:

> I conceived that if the mental or spiritual side of evolution were treated quite dispassionately, without any attempt to minimise differences in kind, but setting them out impartially and using them to measure the length of the line which by whatever means evolution had somehow traced out, a very different interpretation of the whole process could be reached. As I followed this line of thought, it seemed to me, details apart, the Hegelian conception of development possessed a certain rough empirical value.

Hobhouse then explained what the 'empirical' might mean: 'There were grades or degrees of self-consciousness, and as personal self-consciousness was distinctive of man, so there was a higher self-consciousness of the human spirit, which would represent the term of the present stage in development.' He elaborated what the 'term' or end of development might be with regard to the process or 'course' of development:

Further, if this conception was interpreted in terms of experience, it indicated a point of union, where one would not expect to find it, between the Idealistic and the Positive philosophy. This higher self-consciousness would be the Humanity of Positivism, regulating its own life and controlling its own development. But further, if this was the true empirical account of Evolution, our interpretation of that process would be fundamentally changed. The factor of consciousness, as the late Professor Ritchie was already insisting, would influence the course of development. If my view was right it would turn out even to be the central point in development.[110]

Moreover, it was necessary for the future of humanity, Hobhouse contended, that self-consciousness be rescued and set upon a new and more secure footing. While Spencer had insisted upon the 'biological conception' of development when he argued that the struggle for survival was essential for 'increasing the fullness of vitality' of life, Hobhouse alleged that the biological metaphor of development failed 'to differentiate the aims of man from those of the tiger and the wolf'.[111] A 'strict spirit of biology' had prevented the constructive intent of development. Development, as if it were nothing other than an immanent struggle for survival, 'waged war for a couple of generations on such schemes of social and political amelioration as tend to peace and equity between nations, co-operation between classes, and mercy and tenderness for the weaker brethren'. A Spencerian view of development, complained Hobhouse, had become so pervasive that the biological metaphor made it possible to envisage 'artificial selection of types for reproduction as a civilised substitute' for the process whereby 'unfit' types in human society were naturally eliminated. Were such a 'ruthless doctrine' to become the harbinger of the future, Hobhouse thought it better that the 'process of evolution should cease'. But there was an alternative which was to be found in the study of the evolution of 'mind'.[112]

While mental and biological attributes of evolution were not separate but interacted in the same process, or so Hobhouse argued, the development of mind could not be reduced to mere biological change. In order to understand the evolution of mind, it was necessary to 'distinguish the successive phases' of its development as well as to estimate the overall 'direction' of that 'development as a whole'. There could be no understanding of mental evolution without a conception of the 'rational and the good', but this human conception was 'intimately connected' to the process of the evolution of mind.[113] How could such an obvious logical problem, namely that human values were necessary to understand mental evolution but the values were part of what was to be grasped, be resolved? Hobhouse thought he found a solution in the 'element of heredity' which formed the 'substructure of all our thought, feeling and action'. Thus 'reason and will' were as much 'hereditary as any capacity to feel or any tendency to a physical or mental response to special stimulus'.[114] If this was so, what was so special about the human values of the rational and the good? The short answer was that human instinct and habit, through which values of the rational and good were inherited, responded in a special way to their environment through the development of consciousness.

Mental evolution happened through a 'process of a psycho-physical structure' which grew up 'in interaction with the environment', and which acquired 'through the medium of correlations of which consciousness is the essential organ, the power of directing its own fortunes'. The first stage of mental evolution was that of 'Instinct', a stage in which the 'response to the environment, at first wholly random and useless', was 'gradually directed in paths which are normally suitable to vital needs by the action of heredity'. Following the stage of instinct, mental evolution passed on to 'correlation based on Individual and Social experience' which yielded 'habits or trained skill' as 'particular experiences' were 'articulately or directly correlated'. This second stage, the correlation of particular experiences, was then followed by the 'Correlation of Universals' in which experience was 'organised into bodies of thought and action subordinated to wide and permanent ends'. Up to this third stage, 'purpose', which gave direction to human action, was directed to the satisfaction of 'personal' ends. Now, in the third stage, 'social' purpose was made possible.

Hobhouse had now arrived at the final stage of the mind's evolution. This last stage was one in which the 'deficiencies and contradictions of the thought-order force on a process of reconstruction by which the underlying factors of heredity, of personal experience, and of social growth which go to the building of consciousness', were 'themselves brought within consciousness'. Such was 'the correlation of results with processes or principles'. To summarise, Hobhouse believed that it was only after the consciousness of social purpose had been developed that it was possible to be conscious of what had brought about development itself. Indeed, he stressed that it was only in the last stage of the evolution of mind that it now became possible to take a comprehensive survey of human development, tracing our life backwards to its ultimate conditions, and carrying its aims and efforts forward to their ultimate meaning and goal, to correlate human purpose as a whole with the conditions of development as a whole.[115]

As if to indicate that he wanted to remain faithful to Hegelian idealism, Hobhouse called the last stage of evolution that of 'reason'.

Although reason was a part of 'reality', which was deemed to be infinite and although reason could only comprehend discrete parts of an infinite reality, Hobhouse nonetheless claimed that 'the demand of reason and knowledge for wholeness and completeness' was compulsive. To grasp reality in its entirety was a compulsion of reason, characterised by 'the distinctive modern view of the world of human endeavour as relative and yet capable through self-criticism of transcending its own relativity, and relating itself to the vaster whole of which it is only one facet'. To understand experience as being relative but to reason that there was some purpose in the 'wholeness and completeness' of endeavour was a view which found 'its justification in the idea of development'. When 'applied to knowledge', Hobhouse wrote, 'the theory of development explains the actual limitations of the mind by the conditions of its genesis'. But development also revealed 'an indubitable growth of faculty, and what is more important, the emergence of powers and interests unconnected with mere survival and concerned with the expansion and improvement of life'.[116] Such might be the intent of development because the 'growth' of the faculty concerned with human improvement was synonymous with the 'Will' of development. The will to develop was to achieve an 'evolving Harmony' of 'feeling and experience'. Harmony, as Hobhouse understood it, would achieve 'Happiness' or what he also called 'the general character of the good'. Necessarily, the achievement of happiness through harmony involved a deliberate social intention, because

for the rational mind there can be no satisfaction in a harmony that anywhere involves fundamental discord. The rational impulse is to harmonise all that is susceptible of harmony, and that is the whole world of sentient mind. Hence for rational man there is no harmony within the self unless as a basis of harmony with other centres of experience and feeling, and the realisation of any one self is regarded as only an item in the development of society, that is in a Common Good.

Hobhouse continued, in an attempt to reconcile the social with the person:

This development implies an ideal of Personality in which the moral virtues as well as the intellectual and physical excellences are constituent conditions, and the promotion of which, when it conflicts with any warring impulse or interest, is felt by the individual as a duty. Finally, the instinctive or quasi-instinctive promptings that urge us without reflection to the action generally necessary to such a harmony, form the content of moral sense and the summed up judgement of present duty, in which the elements of direct feeling and rational reflection blend in a final deliverance which *in foro interno* is felt to be supreme, is the reality to which the name conscience has been given.[117]

Conscience was the key, the internal force of development, that opened the gate on the path, created by the desire for harmony, which led to justice. It was justice which was so conducive to 'social order'. Social order, so necessary for even 'the bare existence of human beings',[118] had been derived through the development of conscience but conscience, it should be noted, was both the instinct of feeling and that of 'rational reflection'.

If development was to explicate the implicit conscience of feeling through the mental evolution of the mind, then Hobhouse had not moved the theory of development beyond Newman, who had verged on the modern view of which Hobhouse was so conscious. Unlike Newman, however, Hobhouse was committed to a belief in progress, a belief which carried not only the mark of positivism but one which put the primary force of development upon social construction. But Hobhouse, like Spencer before him, eschewed such constuctivism. To repeat, Hobhouse wanted a reconciliation between the personal and the social.

Since he was a committed Liberal, and because of his later acknowledgement that he was obliged to Green for his analysis of how development happened, Hobhouse retained a commitment to the ideal of personal development. The problem, as he put it, 'was to conceive the heightened claims of personality as to make them not disruptive of social order but working constituents of social harmony'. It was the proper understanding of development which provided the 'keystone' of the 'synthesis' between the person and social order, a synthesis which required harmony 'not merely of the individual with social interests, but [of] a many-sided freedom, social and personal, with orderly and disciplined co-operation'. In working out what synthesis implied, Hobhouse pointed to the risk of disorder entailed by the will to develop when he wrote that

the implication of liberty is that error, the wrong and the discord which it renders possible are the price of truth, character and co-operation. In the end we get nearer to the truth by letting error develop its fallacies than by stifling it at birth. From the beginning to end we develop character not by sheer coercion, but by self conquest and the knowledge – or rather the full imaginative realisation – of the meaning of good and evil. We approach assured social co-operation not by compelling obedience, but by winning assent.

We have mentioned that Hobhouse had brought the 'deficiencies and contradictions of the thought-order' within the compass of development when he spelt out the stages of mental evolution. Here, likewise, he recognised that the 'error, the wrong and the discord' of personal development could not merely be avoided but was to be regarded as a positive antithesis to the 'true and right' for the synthesis of the common good. Hobhouse continued:

> In fine, those things which we ourselves hold true and right and socially just we know for partial truths which will gain in the end by the contest with their rivals in the open. But these considerations have weight only when we conceive the social order as a stage or a process of development, and that a development of a spiritual or rational kind. If it was merely a question of realising immediate good as it appears to us, coercion would always be in place. Liberty has its value only in a far longer game.[119]

The 'far longer game' was development itself.

Given his views of development, the qualified support which Hobhouse gave to the Webbs in their battle against the Bosanquets over Poor Law reform becomes explicable. Equally, the same views also explain why Hobhouse took the Fabians to task for an unjustified bias, so he thought, towards state authority in their plans for reform. Hobhouse, who attempted to reconcile liberalism and idealism, ended up by attempting to place positivism on a firmer scientific footing and, in so doing, to steer a middle course between developmental ideas rooted in state development, community development and the personal development of liberalism.[120]

CONCLUSION

However much it may be possible to dispute the extent to which the ideas of community, state development and the liberal reaction were to find fertile ground in official policy towards Britain itself, there is little question that the ideas have had a major impact upon official colonial and post-colonial policy in Africa during the course of the twentieth century. Each idea has come from outside Africa but has been a pivot upon which policy has been expressed to determine the direction in which property of Africans should be held and regulated. While each idea cannot be independent of the others, the three ideas have provided an overarching principle of policy for three distinct periods of history.

It was after the initial colonial occupation of Africa, and finally in the 1910s, that these ideas were in various permutations implanted in official practice in Africa. Elsewhere, we have shown how there was a complex of Radical Liberals and Fabians and idealists, who from different philosophies, missions and practical interests arrived at a common presumption that there was a natural African community, of persons and producers, who had to be protected from the historical degradation of industrial capital.[121] For West Africa, in particular, the figures who made up this complex provide a rich and well-known mosaic of influence and power over colonial administration.

Evangelicals, such as the Kingsleys, were avid in asserting a static and specific social order for Africans. Lugard, who had come from India with ideas for the capitalist development of Africa, turned a somersault when he postulated indirect rule and then, after his return to Britain from Nigeria, entered the inter-war Fabian colonial nexus. E.D. Morel, the radical campaigning journalist, started out from Manchester free trade liberalism to arrive at a Liverpool merchant capital conception of a national but communitarian African polity centred upon indigenous tribal and ethnic authority. Sydney Olivier, like the Webbs, brought Comtean positivism to bear on socialism and then used a Fabian version of socialist positivism to campaign against the presence of 'white capital' in black Africa. These colonial governors, missionaries and campaigners had permeated the Colonial Office in London and their associates were thick on the ground in West Africa.[122]

For Liberals and Fabians alike, it was Kenya, the case which follows in the next chapter, which came to epitomise the failure of colonial policy. Kenya, with its white landed settlement on agricultural estates, was synonymous with Chamberlainite development. It was where Africans had become poor, according to Radical Liberals and Fabians, because private estates had been created to meet the speculative propensities of European landlords and where producers had been induced into wage-labour to meet subsistence. As early as 1906, Lord Elgin, the Colonial Secretary took steps to constrain 'the evils of unrestricted speculation in land' in Kenya. His newly appointed Commissioner of Lands for Kenya was asked to report and justify why land should be concentrated in the hands of a relatively small number of large settler landholders: 'A new country can be developed in only one of two ways. Either you must allow speculation and leave it to the pioneers to develop the country. Or the Government must be prepared to do a great deal more than it has done in helping the settler.' This second alternative, to develop agriculture on the basis of a relatively large number of small farmers, whether European or African, would mean that the government 'must lay out much more capital'.[123]

As the twentieth century progressed, the thrust of the Kenya argument, albeit in somewhat different terms, was to be extended to Nigeria and the rest of non-settler British colonial Africa where production had been succoured through trade, under the slogan of free trade in goods but not free trade in land. This extension of the argument from Kenya to Nigeria invalidated, in practice, the distinction which, at least on a rhetorical level, continued to be made between East and West Africa, between the two African policies of trusteeship and development. For what had been required of the state to support white landholders in Kenya was also that which was now demanded for Africans.

A policy of state development appeared, out of Africa, in the immediate prelude to the Second World War, as a reaction against the idea of community which had now become associated with *laissez-faire* economic doctrine and thus the causes of the emergence of a relative surplus population. During and after the war, the idea of state development was brought into Africa by a British Labour government as the means to recharge the dual mandate of new imperial rule. Both British immediate material need and African subsistence, it was supposed, would be guaranteed by state involvement in production and trade. State development schemes, from groundnuts in Tanganyika to eggs in the Gambia over the fleeting period of 1947 to 1950, provided an unsuccessful experiment in direct state management and control over production. Yet, in the aftermath of this high-point of Fabianism and well into the post-colonial period of the 1970s, state marketing boards, the nationalisation of property, state-directed industrialisation, and state-regulated smallholding schemes of production remained the hallmarks of state development.

By the early 1980s, a reaction against state development had set in. Virtually no African country has escaped the injunction to de-nationalise property and deregulate trade. Under the aegis of the IMF, the World Bank and general panoply of agencies making conditions to enable aid for foreign exchange and internal welfare, African enterprise has been deregulated. The idea of liberalisation, so swift and sudden in its implantation in official practice, was conveyed to Africa from abroad. By way of example, the idea of liberalisation had been developed in Europe as a critique of state development and socialism as early as the 1940s. If the idea of state development had been criticised for 'not meeting African material conditions', then there is little to suggest that the proposed antidote to general African economic regress, the idea of liberalisation, is itself any more in keeping with those conditions.[124] The idea of liberalisation has been implanted in Africa, in the same way that the idea of community was implanted at the beginning of the century.

6

DEVELOPMENT DOCTRINE IN AFRICA: THE CASE OF KENYA

INTRODUCTION

In the previous two chapters we have shown how the concern over the loss of productive force and the recognition of the emergence of a relative surplus population were the stimuli for development thought and practice. We now turn our attention to the case of Kenya. As was noted in Chapter 5, development doctrine originally came to the fore in an African setting as a part of Joseph Chamberlain's abortive project of constructive imperial development. Ultimately, it was the Lugardian 'dual mandate' which replaced Chamberlain's constructivism. Although at the level of rhetoric the intent of this policy was the mutual benefit of Britain and its African colonies, by the late 1930s it had become increasingly clear to the higher orders of colonial officialdom that this policy of non-development had faltered. Bernard Bourdillon, the then Governor of Nigeria, wrote to the Colonial Office in April 1939 in terms which made these failures clear:

> Our duty to the people themselves is to promote their social and economic welfare, to stimulate their desire for, and facilitate the[ir] attainment of a higher standard of living. In so far as we have fallen short in the performance of our duty to the world by allowing the natural resources to deteriorate, and in that of our duty to the British taxpayer by failing to expand with sufficient rapidity the market for British goods, we have also failed in our duty to promote the economic welfare of the people.[1]

Earlier, speaking to the Royal Empire Society in London in January 1937, Bourdillon had signalled the need for an advent of a positive doctrine of development for colonial state practice:

> It is the fashion nowadays to describe the Colonial Empire, and the African part of it in particular, as a vast estate which it is our duty to develop. This description, like most metaphorical descriptions, is capable of being used to support quite erroneous arguments from analogy, but on the whole it is an improvement on the old view of Africa, as a sort of mine from which the contents, mineral, vegetable and in old times human, should be extracted as quickly and as cheaply as possible. The exploitation theory . . . is dead and the development theory has taken its place.[2]

Bourdillon's comments testified to an increasing awareness among colonial functionaries of both a deterioration in the conditions of production and/or the emergence of a relative surplus population throughout the British empire in Africa. In reacting against what he called 'the exploitation theory', Bourdillon presented development as the positive means to counter what had been recognised as the depredations of primary capitalist accumulation in Africa.

It is important for the discussion which follows to observe that the negative consequences of 'exploitation' were noted both in territories such as Northern Nigeria, where radical liberal land nationalisation precepts had been enshrined into law, and in those such as Kenya, where Crown land had been alienated to white settlers.[3] However, for Liberals and Fabians, Kenya had come to epitomise the failure of colonial policy. Kenya, with its white settlers, was synonymous in the minds of Radical Liberals and Fabians with the supposedly discredited Chamberlainite project of constructive development. It was in Kenya, about which it was believed, that Africans had become poor because private estates of land had been created to meet the speculative propensities of European landlords and where producers had been forced into wage-labour to meet subsistence.

Nevertheless, by 1931 it was possible for the imperial gadfly C.R. Buxton to observe that the 'problem of Kenya' was 'in a greater and lesser degree, and under many different forms, the problem of Africa' as a whole.[4] Buxton's view finds implicit support in another of Bourdillon's writings. The Governor of Nigeria had explained his desire and plans for encouraging Africans to move into large-scale oil-palm estate agriculture so that Nigerian exports could regain their competitiveness *vis-à-vis* Malayan and Sumatran plantations in the world market. In doing so he pointed out that: 'The main difficulty here is that in most of the palm belt the land is tribally owned, and the individual will not invest capital in a permanent crop on land to which he has no individual right.'[5]

An old conundrum had re-emerged. It was one thing to use an argument about the natural status of African communality, as had been repeatedly done in British West Africa, to exclude private landlordship and thus permit the state to become landlord and rentier. It was quite another to maintain labour productivity under the conditions of global competition. The liberal belief in progress had failed. No longer was it possible to suppose that free trade in commodities, without free trade in land, could make production grow without state organisation of production, without taking up the positive side of trusteeship, without, in a word, development.

Colonial policy in Kenya had been criticised by liberals and Fabians as a betrayal of trusteeship because the development, in Chamberlain's parlance, of 'our estates' had come to be interpreted literally as meaning 'our European-owned estates'. The thrust of colonial policy in Kenya had been to help European settlers through the idea that development was synonymous with state-assisted land colonisation. However, when Bourdillon restated Chamberlain's doctrine in 1937 for West Africa, the duty to develop 'our estate' was reworked to mean the development of African-owned 'estates'. Through this redefinition, prompted by the 1930s crisis, it was proposed that development as the positive side of trusteeship was to be extended to Nigeria and much of the rest of non-settler British colonial Africa. What had been required of the state to support white landholders in Kenya was also that which was now demanded for African farmers in Nigeria and elsewhere. The effect of this demand was to invalidate the distinction that continued to be made between East and West Africa, between the two African policies of development and trusteeship.

Yet if Bourdillon himself had placed his emphasis on the enhancement of production as the means of increasing the welfare of Britain's African colonial subjects, the Colonial Development and Welfare Acts of 1939 and 1945 came to enunciate an overwhelmingly welfarist agenda. Development was grounded in agrarian doctrine and directed towards small-scale production. Liberals remained loath to carry out direct state intervention in production. The implementation of the positive side of trusteeship would await the arrival of the Fabians to power in the context of Britain's post-war economic crisis. It was only then that a fully blown Chamberlain-like doctrine of development to serve the reconstruction of British 'national interest' and so promote the post-war colonial offensive was to be mounted.

Development doctrine for British Africa was laid out by the 1945-51 Labour government as part of its response to the 1947 sterling crisis and the final withdrawal from India. This was a late-imperial doctrine to maximise production in African colonies to meet British national material need. The intention to develop schemes for the large-scale production of primary products – from eggs in the Gambia to groundnuts in Tanganyika – was guided by a national need to swiftly expand exports to Britain and so save the need to spend dollars on imports of food and other immediate needs. As if inspired by Chamberlain's doctrine of development, the 1947-50 colonial offensive was the only occasion on which there was concerted British state effort to make colonial populations do the work of generally maintaining subsistence, and therefore productive capacity, in Britain itself.

When, in 1947, the development offensive was mounted, the most important question that arose was the same as that which had arisen over Chamberlain's initiative at the turn of the century. This question was whether the British empire could negate both the autonomy of individual British capitalists and bear the burden of becoming a fully-fledged, centralised imperial state capable of carrying out its developmental mission. In part, the answer to this question turned on a second which was, at least as far as the Fabians were concerned, of equal vintage. This was whether agricultural production could be made to meet what were established as social needs once land had been nationalised and vested under peasant proprietorship.

The Labour government in 1947 attempted to circumvent the first of these questions by substituting the state for private capital enterprise. State enterprises were directed as if they were part of a military campaign and resources were allocated according to the traditional Whitehall pattern of *ad hoc*, inter-departmental committees. Deeply informed by Fabian precepts in the colonial sphere, Labour attempted to answer the second question by establishing their own large-scale enterprises in areas which had been shunned by African proprietors. Large-scale enterprises were located where the resource base of climate and soils were relatively poor; where existing rights to land need not be disturbed; and where crops were to be produced under conditions which had been long made unprofitable, in terms of capitalist endeavour, by the very conditions created by peasant proprietorship itself. Whatever the technical failures of the colonial offensive, and these were legion, it was not simply a lack of expertise that doomed large-scale state enterprise to failure. In the end, it was the simple reality that, given the conjuncture of 1947 to produce commodities fast under the condition of nationalised land, peasant proprietors, short of imposing a *corvée*, could not be made to work productively on large-scale enterprises which were directly managed by the state.

In the event, when the colonial offensive projects for agricultural production on estates managed by the would-be developmental and imperial state failed to meet British needs, that failure was to be negated by moving the logic which drove the practice of development onto another plane – that of Africa itself. In the aftermath of the failures of 1947, by virtue of this negation, the integral meaning of the post-war practice of development came into being. The concern over unemployment and poverty, the manifestations of the growth of a surplus population that had driven mid-nineteenth-century land reformers, Chamberlainites, twentieth-century Radical Liberals and Fabians alike, was now to become 'Africanised'. And for development's post-war practitioners, 'good development' was to become synonymous with 'rural development'. By way of this formulation, the primary rationale for the development of rural areas was not to be, as it had been for both the Chamberlainites and for the Fabians of 1947, one of fulfilling the positive side of trusteeship by meeting social need through production, but rather, as for Campbell-Bannerman and the Liberals of 1909, merely the one-sided attempt to deal with the unemployment of a growing surplus population.

THE SIGNIFICANCE OF 1960

As had been the case a century earlier in Victoria and Quebec, the agrarian doctrine of development was enunciated in Kenya amidst the transition from colonial to post-colonial government. It was in 1960, three years before the government of Kenya became formally independent, that the doctrine was clearly formulated by A.G. Dalgleish in a Kenya government report, *Sessional Paper No. 10 of 1959/60: Unemployment*.

During the late 1950s, unemployment had become an immediate problem for the provincial administration in Kenya's Central Province. As a result of the Mau Mau revolt nearly 90,000 people were detained between 1952 and 1959. The majority of surviving detainees were released into the province. Over one-half of the 30,000 Kikuyu population rounded-up from Nairobi in 1952 and 1954 were detained; an estimated 100,000 of the Kikuyu population, mainly farm workers and resident labourers (squatters), were repatriated to the Central Province from the European-owned large-farm areas in the Rift Valley Province.[6]

'A growing unemployment problem' was reported from Kiambu District in 1956 where the District Commissioner, backed by the Special Commissioner for the province, urged his peers in the Rift Valley to permit the permanent return of Kikuyu families to Nakuru and Naivasha districts of the Rift Valley. 'It is our policy', wrote the Special Commissioner, 'to endeavour to place, as far as is consistent with security, as many Kikuyu as possible in work so that they do not become a charge on the state e.g. Kikuyu Relief Works.'[7] As the emergency regulations were wound down during 1958 and 1959, the Nyeri district commissioner reported that the 'unemployment position still left much to be desired'. 'It is', he wrote, 'to be hoped that with land consolidation there will soon be sufficient [Kikuyu] land-holders wealthy enough to absorb a large proportion of the unemployed on their farms.'[8] With the end of the emergency in 1959, it was reported that 'a large flood of unemployed' had moved out of the district to Nairobi and the Rift Valley.[9] As we shall see, neither the consolidated farms nor the estates were to absorb the unemployed as wage-workers on farms.

The growth of a relative surplus population in the 'African reserves' (or areas containing a preponderance of small family farms) of the Central Province during the 1950s was accompanied by an absolute labour shortage in the large-farm areas. When European farmers found their estates bereft of Kikuyu labour skills after the 1952 emergency, they implored the administration, as early as 1954, for labour to be returned.[10] However, the administrative fear was that an unregulated movement of people would rebuild the pre-revolt surplus population in the Rift Valley. More significant was the appreciation that former workers would be willing to return for work plus land but not for work alone. In 1956, it was reported that repatriates 'have exploited their scarcity value by doing very little work and demanding very high wages'. It was noted from Kiambu that repatriates were reluctant to go back to the Rift Valley or do unskilled work on local coffee farms at 24 shillings per month when the daily casual labour rate in the contiguous Nairobi district was up to 4 shillings per day.[11] Yet, as Dalgleish was writing his report, the existing mixed arable-pastoral large-farm demand for labour was about to dry up.

It was from February 1960 – immediately after the Lancaster House conference agreement to self-government and the beginning of the exodus of European mixed-farm owners – that African small-farm settlement began in the Rift Valley. The preference of the majority of repatriates as far as agricultural activity was concerned now changed, with them either preferring to return to the Rift Valley for land settlement or to remain in place and fully engage in small-farm household commodity production. This change in perspective was rapid. The Nyeri District Commissioner wrote of a serious deterioration in employment for 1960 and 1961 as large numbers of people, who had recently left the district for work, returned 'as redundant labour':

> Even skilled men found it increasingly difficult to get work. It was not unknown for artisans in possession of Trade Test Certificates . . . to accept work in a lower grade, with less renumeration. Literate youths, who in the past would only have considered clerical employment, accepted manual and domestic work.[12]

Map 2 Kenya *c.* 1960

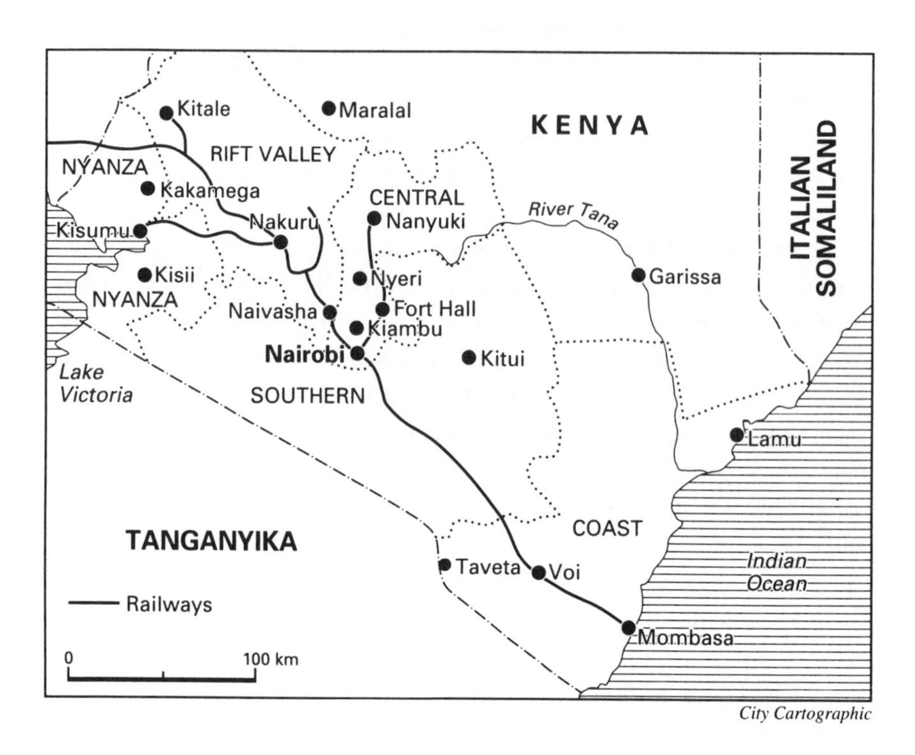

City Cartographic

Such was the immediate background to the 1960 report on unemployment, a report which Dalgleish wrote while there was an abrupt about-turn in the local evidence for the conventional view that aspirations for too high a money wage was the cause of unemployment.

Dalgleish rejected both the then current views that the unemployment problem was new and 'that it is basically a problem of the settled areas (or of the wage-earning sector of the economy)'. Nor did he believe that the problem would be resolved or alleviated 'by special measures designed to increase temporarily the numbers in wage-earning employment'. Unemployment, Dalgleish insisted was a problem of seasonal unemployment and/or underemployment in the African reserve areas.[13] Underemployment or 'redundance of labour on the land', Dalgleish made clear, was involuntary in that leisure was enforced on grounds Keynes had defined in the *General Theory.* In view of both the claims that Keynes' theory is irrelevant for the post-colonial world[14] and the anti-Keynesian disposition of anti-development ideologues such as Collier and Lal, who embrace Dalgleish for his prescience towards the policy and economic events of the Kenya post-colonial transition,[15] it might seem paradoxical to find Keynes' theory lurking behind a report which concluded that the solution for unemployment and underemployment might depend upon the rapid expansion of peasant household production. Yet the paradox is illusory.

Keynes may have had no general theory of colonial development but he had an abiding interest in India and Russia. In both cases, it has been argued, his prescriptions for policy followed the contours of agrarian bias. 'Keynes', writes Chandavakar, 'was always more solicitous about the interests of the Indian peasant as against those of urban or exploiting classes'. Rural cooperatives were to be 'a principal agency for the emancipation of the peasant from the village moneylender and for much-needed direction of new capital into agriculture'. His well-known proposals for government storage of agricultural products, together with a buffer stock scheme to stabilise producer revenues, found their way into the colonial African marketing boards of the late 1930s. Indeed, Chandavakar insists that Keynes had

> a strange, almost Physiocratic scepticism of India's industrial potential, believing that 'her future prosperity is to be sought almost entirely in the application of more capital to methods of agriculture, and, almost like a Gandhian, he did not believe that India will find in mills and factories the non-economic goods, which make up with wealth, the dignity of a nation.[16]

Likewise, Keynes' 1920s reflections on Soviet Russia were marked by a critique of what he called 'the official method of exploitation' based on, to quote Toye, a 'state monopoly of trade, manipulation of the internal terms of trade between agriculture and industry and the subsidisation of state-controlled industry'. More specifically, Keynes reckoned that the failure of the Soviet state 'to pay the peasant more nearly the real value of his produce' inhibited 'both the means and the incentive to a far higher output' in agriculture. The result was accelerated rural–urban migration. Toye notes that Keynes prefigured the Harris–Todaro 1970 migration theorem, itself formulated out of the experience of Kenya in the 1960s, and summarises Keynes thus:

> Keynes argues that the artificially high urban wages will draw too many people from the rural areas because they will fail to calculate correctly the probability of their finding employment in the towns. Hence the flow of rural migrants will be checked only 'after the towns have become overcrowded and unemployment has reached unheard-of proportions'. Because they are paid a dole 'from their trade', 'this vast army of unemployed is a heavy burden on the financial resources of state establishments'.[17]

It was far better, therefore, to ensure that would-be migrants remained committed to production in the countryside where their dole would be self-financed by their own productive effort.

While Dalgleish did not directly invoke Keynes, he rested his case upon a 1950 ILO report informed by Keynes' precepts.[18] Given that there was an essential continuity in Keynes' thought before and after the *General Theory*, it is possible to spell out its implications for late colonial policy towards the money-mediated economy of Kenya. It was the official desire to 'change underemployment into full employment', but households' revenue from the existing production potential of small-farm production could not satisfy the prospect of meeting 'economic, social, and other needs and aspirations of the people associated with' underemployment.[19] However, an increase in investment to heighten the production potential in small-farm production would, for any given population of producers, increase the effective labour effort of each producer. Therefore, such an increase in production potential would serve to eliminate underemployment in that the prospect of a higher average household income from production would permit each household member, who was capable of production, to work more. The total increase in actual production would then justify the initial increase in investment which made the expansion in production potential initially possible.

By the time that Dalgleish offered his diagnosis a major plan to improve small-farm production was already underway. The Swynnerton Plan, which was premised on a design for rapid agricultural improvement rather than on the direct promotion of employment, had been prompted by rural distress in Central Kenya. Such distress was, at the time, seen as a cause of the Mau Mau revolt. For Dalgleish, the Swynnerton Plan had to be supported, since, as he argued, 'the ultimate and only real solution to Kenya's unemployment problem' was to be found 'in the full economic development of the African land units'.[20] From this point on, all major reports on unemployment in Kenya have offered a similar prognosis.

REPORTS ON UNEMPLOYMENT

Political independence brought expectations that the post-colonial government would create jobs but there was no resolution of the unemployment problem. In May 1962, before independence and during the most intensive period of labour unrest and strike action in Kenya's history, the new Minister of Labour, Tom Mboya, met the British Colonial Secretary in London to discuss the 'gravity of unemployment' and secure further British economic aid. Two years later, and four months after a mutiny of some Kenya army units had been quashed with British military assistance, a crowd made up of the unemployed rioted in Nairobi: 'Their spokesmen said they wanted free housing and food until jobs were found, and unless jobs were forthcoming they wanted their voting cards returned so that they could put in power a government which would help them.'[21] Riot police dispersed the crowd of over a thousand on 13 July 1964. However, the perceived political threat was sufficient for the government to promulgate its first tripartite agreement between itself, private business and the trade unions. Large-scale enterprises and the state agreed to each expand the numbers on their payrolls by 10 per cent while the unions agreed to the establishment of an industrial court and wage guidelines to enforce wage restraint.[22] At the same time, it was rural development, based on further international economic assistance, which was given primary emphasis as the only authentic course of development to resolve unemployment.

A 1970 report on the 1966 Kericho conference *Education, Employment and Rural Development*, which led to the Special Development Rural Programme and its successors, emphasised the intractability of unemployment: 'The inability of any department or ministry to find realistic solutions to the school-leaver and rural employment problems is a major source of frustration to the field administration at the present time.'[23]

Also in 1970, a National Assembly Select Committee issued a report on unemployment. Chaired by G.N. Mwicigi, it was composed of *de jure* government MPs in the now one-party Kenyan state who could barely disguise their opposition to the government. The bulk of the 450,000 increase in the labour force expected between 1970 and 1974, the report concluded, would 'have to be absorbed in the traditional sector either as employees or as self-employed on farms and other enterprises or else be unemployed'. Structural unemployment that had been 'formerly prevalent in the countryside' was now 'open and explicit in the towns'. It was caused, the report postulated, by 'the very nature of the efforts to transform the traditional sector to the modern sector'. The report advocated the reversal of rural–urban migration through the application of an urban wages policy and subsidies for agriculture:

> When urban unemployment has decreased, then urban wage increases should be related to rises in rural incomes. Any increases of productivity in the modern sector in excess of that in the traditional sector should go to taxes for general development, or be transferred to the rural areas through lower prices for the products of the modern sector or higher prices for the products of the agricultural sector, or through additional provision of services in the rural areas.[24]

When the government responded to the 1972 ILO report *Employment, Incomes and Equality,*[25] it followed the policy which was advocated by the 1970 Mwicigi report. First, agriculture was emphasised:

Where the Development Plan 1964–70 focused upon rapid growth, the second Plan for the period 1970–74, attempts to shift the locale of growth toward the rural areas. In fact, rural development, the improvement in the quality of rural life, is the first priority of the Government and is so stated in the second Development Plan.[26]

Second, a wages policy was advocated and implemented through issuing state guidelines for wages and incomes to the industrial court.[27] Despite these injunctions, and a severe downturn in real wages after 1970, the extent of unemployment continued to grow during the 1970s and 1980s.

The extent of unemployment in 1970 and its increase during the 1970s can be gauged by the registration of job seekers who amounted to 46 per cent of those enumerated to be in wage-employment in 1969. A decade later, in 1979, the proportion had increased to 57 per cent of those enumerated in 1978; for Central and Western Kenya, the proportions were 85 per cent and 250 per cent respectively.[28] A 1986 Urban Labour Force Survey found a total unemployment rate of 16 per cent but 80 per cent of the estimated unemployed were less than 30 years old, while the unemployment rate for the 20–24 age cohort was 40 per cent.[29] At the same time the official index of real wages showed a 30 per cent decrease in the public sector wage and the official Nairobi minimum wage, with a 50 per cent decline in the private sector manufacturing wage, over the period between 1970 and 1989.[30] A sustained fall in real wages had accompanied the dramatic increase in unemployment. Yet, despite this evidence, the focus on rural development did not abate.

Hunt, for instance, called in 1984 for an 'egalitarian peasant-farm based development strategy' which, she claimed, offered 'much greater opportunities for employment generation' than 'unfettered capitalist development'.[31] In 1986, Livingstone reflected on the durability of the 1972 ILO report, claiming that its basic needs approach changed the focus of development policy from unemployment to poverty:

The problem was not one of finding jobs – formal employment – for those without jobs, but in finding *adequate* employment. This involved raising income levels in the sort of activities in which the majority of the population were *already engaged*, including particularly smallholder farming and activities in the so-called 'informal sector'.[32]

What had hitherto been cast as the aspiration for full employment, creating wage-labour jobs to eliminate unemployment, had become one of aspiring to the work of self-employment, in which non-wage-employment was seen as the key to subsistence for the poor without jobs. Non-employment was the corollary of the denial of state social security. Livingstone effectively made this point at the outset of his argument for non-employment: 'In a developing country which lacks a national system of social security, people must find some means of survival and eking out a livelihood, whether this is on a wholly inadequate piece of land or as a shoeshiner on the urban pavement.'[33] To make pieces of land or pavement 'adequate' for subsistence is the theme of non-employment. The means by which this might be done, through the supposedly benign 'sponge effect' of increasing 'the capacity of the rural economy to absorb'[34] a surplus population, is itself the malign effect of a present-day doctrine of development. It is a doctrine which has not yet even begun to comprehend its own past whether in Africa or the other continents, whether earlier in the present century or early in the previous one.

Tellingly, the rift between official policy and its critics has not been over the nature of the agrarian bias of development but the question of land reform and, thereby, the form of property which might make land and assets adequate for subsistence. The 1970 Mwicigi Report called for 'a ceiling on the amount of land which' could 'be owned by one single individual'; the 1972 ILO report urged 'a thorough reappraisal of the need to retain large-scale agriculture in the high-potential areas in its existing form' and proposed a progressive land tax to make land-use more intensive. Hunt's peasant-farm strategy entailed 'as its linchpin a redistributive land reform', while Livingstone claimed that 'employment and access to land may be synonymous; no land may mean no employment'.[35] Contrawise, with very few exceptions,[36] the thrust of official policy has been to ignore the clamour for such land reform by maintaining, without argument, the extensive margin of large-scale agriculture. For a post-colonial state founded on a national movement whose purpose had always been to secure African private landed proprietorship, including the large-scale occupied farm lands in the white highlands, it would have been economic and, most likely, political suicide for the politically established class to renounce its recent acquisition of large-scale farms in favour of land reform.

As in the colonial period prior to 1960, the problem of securing the productive base of land for the subsistence of the unemployed has been deflected onto the intensive margin of small-holding agriculture. Three examples of policy will suffice. In its 1973 response to the ILO report, we find that the Kenyan government accepted 'that programmes to achieve a more intensified use of land should be the central thrust of the country's policy for agriculture; this will be done in a framework which promotes co-operative farming in the large-scale farming areas'.[37]

Second, a Presidential Committee on Unemployment, chaired by Maina Wanjigi and with Professor Mbithi on board, reported in 1983 that,

> the migration of job seekers from rural to urban areas is now a very real and serious problem. We have therefore sought to recommend rural-based programmes which will not only benefit the greatest number of the unemployed/underemployed but will also stem the influx of people to the urban areas.[38]

Bemoaning a 'pattern of development' which had been 'heavily biased towards urban development', the Wanjigi Report followed its predecessors by invoking the imperatives of an agricultural subsidy and 'a national incomes policy' to promote 'a more equitable distribution of incomes' and thereby employment.[39]

Third and more significantly, three decades after Dalgleish and the official recognition of unemployment as the development problem of Kenya, the 1991 Ndegwa *Report of the Presidential Committee on Employment* shifted uneasily, in taking on board 153 written depositions and the advice of such luminaries of development economics as Livingstone, Killick and Singer, between the now familiar theme of rural non-wage-employment to absorb labour effort and an official admission that urban unemployment could not simply be resolved through rural development.

If the Ndegwa Report recommended the 'active promotion of an enabling environment for the development of smallholders, farmers, small-scale enterprises and the informal sector in general', it also admitted, for the first time, that 'it will be essential to take account of the existence of real urban poverty, more so as the numbers being absorbed into the urban labour force accelerate'. Indeed, the Report suggested that 'urbanisation and economic development are mutually reinforcing processes' in that economic transformation in Kenya would be one 'in which a predominantly rural-agricultural economy will progressively acquire an urban-industrial orientation'.[40] Yet, paralleling its focus upon smallholdings and the division of large mixed farms for labour absorption in agriculture, it envisaged that self-employed and small-scale enterprises would form the basis of labour absorption in the 'urban-industrial orientation'. Thus, in a report which represents the official advent of deregulation and liberalisation of enterprise in place of the state effort of development, we find recommendations for a subsidy to state-organised small industrial enterprises. Ndegwa followed the *Sessional Paper No. 1 of 1986* which decreed that, in making economic growth the 'primary concern of economic policy' to generate jobs, it was 'imperative' that 'the great majority of new jobs be created, not in the cities or in large industry, but on farms and in small-scale industries and services, both urban and rural'.[41] After Ndegwa, a 1992 sessional paper on *Small Enterprise and Jua Kali Development in Kenya* was issued to proclaim the official intent to develop small-scale manufacturing enterprise. 'Self-employment', the 1992 paper asserted, 'has to be viewed as a highly respectable occupation', echoing recent past reform in school syllabuses which emphasised business education by telling pupils that their career 'question should not be: WHO WILL EMPLOY ME BUT HOW WILL I EMPLOY MYSELF? In many cases, you will find that self-employment is more paying than being employed by another person.'[42] What had hitherto been the officially unregarded informal sector of unregulated enterprises, growing 'spontaneously' and facing much official harassment, was now to be the positively developed *jua kali* 'sector' of African-owned manufacturing and self-employed enterprise, working outdoors and under the 'strong sun'. This sector, the 1992 sessional paper confirmed, 'has already officially reached a stage at which further growth is not possible without strong and effective support from the Government, the aid agencies, the business community and the people at large . . . all acting in coordination'.[43] Aid agencies and academics had long called for a 'comprehensive national strategy for small enterprise', what Kenneth King and Charles Abuodha eulogised as an 'enabling environment' for small enterprise in which 'the most remarkable aspect of the collaborating strategy is the detailed specification of actions to be undertaken by government ministries and by financial institutions'.[44] Such is the latest practice of development doctrine.

From 1960, and immediately afterwards, agrarian development schemes were expanded to follow the spontaneous surge of former agricultural workers and those with land but without jobs towards new forms of household commodity production. Education and training, means of extending new methods of production, credit and marketing organisation, for instance, were all to be directed towards small-farm production on the same grounds that 1992 doctrine aspires to coordinate effort towards small-scale industrial enterprise. As for agrarian development thirty years earlier, so self-employment in industrial enterprise would expand, among other objectives, output for the economy, create jobs 'at relatively low capital cost' and 'promote rural–urban balance'.[45] The 1992 doctrine continues to be pervasive. For instance, in launching the government's 1994 economic survey, the vice-president and minister for planning and national development, George Saitoti, claimed that 'the majority of future non-farm job opportunities will be in the informal sector – small-scale manufacturing, marketing, repairs and other service activities – located mostly in market centres and towns throughout the country'. A month later, in July 1994, the minister for research, technical training and technology, Zachary Onyonka, expected that one-half of the half million of 'new entrants to the labour force' would have to find jobs created 'in the rural non-farm and urban micro-enterprise field, of which 75 per cent would be in the urban areas'.[46]

It is not surprising, therefore, that the Ndegwa Report, which represents the official advent of deregulation and liberalisation of enterprise in place of the state effort of development, should recommend a subsidy to state-organised small industrial enterprises.[47] Nor should we be surprised to find that the minister of commerce and industry in 1994 should implore that banks should be compelled to lend to small business. 'Banks', the minister claimed, 'must start lending to the risky informal sector enterprises, rather than operate as mere conduits for making money.'[48] As such, and as will become clearer later, and as Chapter 7 will reinforce, the intention to develop – in confronting unemployment – was here little other than the transposition of agrarian doctrine into a wider setting.

We have seen the effects of the agrarian doctrine of development in the nineteenth century. Doctrines of development, generally, have always reproduced the original positivist intent to reconcile progress and order. Systematic social security represented the culmination of the aspiration to maintain the effort of work, the condition of progress, while contriving to find the immediate means to make effort part of the productive force of labour. These 'means' were nothing other than the means of production which were denied to the unemployed through the process of capitalist development, a process which cannot assure full employment when it creates a surplus population relative to the population of productive force under the jurisdiction of a particular state.

Social security is the positive development which has historically accompanied capitalist development, the second-best alternative to the full socialisation of the means of production in establishing the basis for positive economic and social order. It is instructive to return to Dalgleish's report of 1960 and ask what he meant by 'full economic development'. This question is at the core of an understanding of what has made the process of capitalist development consistent with the positive doctrine of development in Kenya.

STATE AND COMMUNITY

Dalgleish emphatically proposed an immediate solution to the underemployment problem. He counselled that,

> for the individual worker, it takes little to change underemployment into full employment – no more for example, than that a smallholder should decide to farm his land by his own and his womenfolk's efforts than rely, as in the past, on the assistance of other male members of his family or tribe.[49]

Two significant points are raised in this important passage. The first is the presumption that it would take 'little' to create full employment were individual effort to be concentrated in the households of the family farm. More than a century earlier, in the Quebec case of land colonisation, we saw how the presumption that 'it takes little' to absorb a surplus population on the land was a major cause of the perpetuation of rural distress. Mid-twentieth-century development doctrine, in incorporating the experience of the doctrine of the past, took on board Chamberlain's injunction that, contra Dalgleish's premise, 'it takes much' to induce labour effort for full employment. Indeed, the cardinal feature of Chamberlainite doctrine, and the obvious cause of the rejection of the doctrine, was that the aspiration to develop had to be backed by the intent of the state to engage in investment spending if work effort was to be productive. We shall see why and how state investment, from 1960 on, was applied to family-farm production on a scale which had hitherto characterised Kenya government support of large-scale estate agriculture. A policy of 'full development', even when premised on the family farm, still implied the state-financed direction and organisation of production.

Second, in denying 'assistance of other male members of his family or tribe', Dalgleish rejected the projection of community, which was implicit in 'family or tribe', onto the indigenous organisation of landholding, production and welfare. The idea of community was inscribed in colonial policy from the early colonial period of *community* development, a period in which the liberal-radical injunction to develop communities in Africa was the justification for the official proscription of African individual landed proprietorship, to the mid-century policy of community *development*. By mid-century, community development was part of state development policy since community became the means for achieving the end of development.

Although the history of the intellectual formation of the 'community' in community development has roots at least as far back as the writings of Henry

Maine, which were discussed in relation to the modified positivism of John Stuart Mill in Chapter 1, for British colonial Africa, the immediate history of community development may be argued to have begun in 1925 with the publication of the White Paper 'Education Policy in Tropical Africa' later released as Cmd 2374. This paper, which was a response to the criticisms of British rule by the Phelps Stokes Commissions of 1921-23, was produced by the newly formed Advisory Committee on Education in the Colonies. It argued that education is 'intimately related to all other efforts, whether of Government or of citizens for the welfare of the community' but warned that the 'real difficulty' lay in 'imparting any kind of education' which had 'a disintegrating or unsettling effect' and argued for the necessity of conserving 'as far as possible all healthy and sound aspects of social life'.[50]

If the concern of 1925 was how to respond to criticism without fostering the 'de-nationalisation' or individualism of Africans, by 1935, the year of the publication of an important Memorandum on the Education of African Communities, new concerns had been added: 'The world wide depression of the 1930s made it necessary for each colony to review its resources. . . .'[51] Now community development was to be undertaken to 'direct' and 'stimulate' the production and 'efficient marketing of colonial products' in order 'to more efficiently respond' to the 'requirements of the world's markets'.

The need for a response based on economic efficiency was highlighted by the war-time production demands of 1943, the year of the publication by the now renamed Advisory Committee of the Colonies of 'Mass Education in African Society'. Here education was made the linchpin of community development, by arguing that 'the general health of the whole community, its general well-being and prosperity, can only be secured and maintained if the whole mass of the people has a real stake in education'. Warning of the dangers of 'partial development', again a euphemism for individualisation, the report argued that if government was 'to secure the improvement of the life of the community, we are brought face to face with the conclusion that improvement depends upon the training of the community as a whole . . .'.[52]

Community Development, enunciated in 1949 as official colonial policy by the British Labour Party government, was also referred to as 'social development' or 'mass education':

> We understand the term 'mass education' to mean a movement designed to promote better living for the whole community, with the active participation and, if possible, on the initiative of the community but if this initiative is not forthcoming, by the use of techniques for arousing and

stimulating it in order to secure the active and enthusiastic response to the movement.[53]

It was stressed that this was 'no new movement but the intensification of past plans for development by means of new techniques'. As far as colonial Kenya was concerned, 'new techniques' involved the inculcation of neighbourhood 'self-help' by reinventing local work parties, such as the Kamba *mwethya* or Kikuyu *ngwatio*, to temper the earlier colonial experience of communal labour as the forced levying of labour for land terracing and local road and dam construction, which had provoked so much nationalist and local resistance.[54] In other words, community or social development was an official strategy to make each household bear the private cost of the social investment necessary for agricultural improvement.

When *Harambee* was inscribed upon the flag of the post-colonial state, the official idea of 'let's pull together' as much expressed the motif of 'order and progress' as the actual inscription of these words on the Brazilian flag had done at the end of the last century. If harambee implied the official aspiration for a national community of the different ethnicities of tribe, then it was also to become the name for 'community co-operation' in local projects of self-help based on the superabundance of local initiative and enthusiasm after political independence in 1963.

Over two decades, from 1965 to 1984, 37,300 harambee projects were completed, contributing some 12 per cent of national gross capital formation. In 1976, it was officially estimated that harambee contributed to 'nearly 40 per cent of capital development in all the rural areas'.[55] Moreover, it has been claimed that the harambee movement 'developed rapidly throughout Kenya in response to people's actions and aspirations rather than simply as a creation of the government'.[56] 'Harambee', insisted Philip Mbithi, when writing earlier as an academic sociologist rather than later as a head of the Moi regime's civil service, was 'not a means by which peasants in Kenya are manipulated by the elite'. Indeed Mbithi argued that the 'concept' embodied ideas of 'mutual assistance, joint effort, mutual responsibility, community self-reliance'. These precepts, because rather than in spite of Mbithi's insistence, were the source of trusteeship for what the then sociologist called 'the more significant contribution of Harambee', namely 'its ability, as a development strategy, to increase significantly the mobilization of hitherto unavailable or immobile resources'.[57]

Harambee has covered a manifold of reciprocal activity, including the organisation of women's groups for the purpose of collective saving, the funding of individual house-improvement, the purchase of private landholdings, as well as province-wide, ethnic-based movements for the establishment colleges of technology or the construction of cathedrals. The manifestation of reciprocal self-help is vibrant in so far as a social project starts from the premise of an active self-interest in private benefit, realises benefit for the individual household and is based in a locality which has undergone agricultural improvement.[58] Since 1963, the most significant projects have been directed at secondary schools. It has been estimated that for the period 1965-72, the period when local self-help was most active and during which the number of secondary schools tripled, 60 per cent of both general education expenditure and the funds for new schools came from harambee contributions. These contributions were made for the clear purpose of making it possible to obtain the certificates of education necessary for wage-employment outside the rural household.[59] In other words, the provision of education was accelerated by the harambee movement in the expectation that each school-leaver would find non-agricultural wage-work to relieve underemployment on the farm and would be able to contribute to the school fees of each successive sibling seeking the necessary educational credentials. This, indeed, was a spontaneous movement. However, it was also one that was informed by the positivist precepts only in so far as it enforced mutual responsibility within the family. When it resulted in a flood of unemployed school-leavers in a labour market that could not satisfy the demand for wage-employment, the official response was to control the movement.

As we have shown above, the policy of community development was not rooted in some 'post-colonial' fantasy of an imagined community of the poor. Its origins were in official colonial doctrine. Thus, the original impetus behind local self-help projects for constructing schools, dispensaries, dams and cattle dips came from what had been called in the colonial period improving or progressive farmers who were so often seen as the catalysts of community development. This same driving force, with official support, had established African cooperative societies during the colonial period and accelerated cooperation after 1963.[60] One study of harambee concluded that 'the activists in the locations studied have not usually been poor and marginal citizens as Mbithi suggests'. Indeed, Mbithi's own data shows that more than 90 per cent of harambee project 'leaders' were businessmen, clerks, clerics, artisans, district and sub-district government officials, politicians, teachers and farmers.[61] Termed a 'rural petit bourgeoisie', this cadre of leaders, as Ngau points out, 'found it expedient to help the peasants organise and to create social amenity services, partly using funds extracted from the government and private sources, in exchange for political and economic gains'.[62] As such, they performed the role of 'political entrepreneurship' regarded as both a logical condition for social cooperation and the historical key to the political networks which give what Bayart has called rhizomorphic form to the state in Africa.[63]

If harambee is a cooperative alternative to lump-sum taxation as a means of providing public goods, then, as has been argued, it can only be effective if each contributing household expects that its contribution will elicit the same from others. Given this premise for cooperation, the degree of cooperation will depend upon the extent to which potential contributors are monitored so that the opportunity for defection from a project and free-riding will be minimised. Monitoring, according to this analysis, is done by the political entrepreneur.[64]

What is interesting about the Kenyan case is that the logical fiction of political 'entrepreneurship', with its implication of a purely mediating role of official altruism, elides into actual economic entrepreneurship. Project entrepreneurship is expedient for the leader when he acts both privately, as a contributor and beneficiary of the project, and politically, to secure channels for state finance and political donations to the project. The political gain a project secures for the local leader is the reward for handing over the particular constituency of locality to the wider and higher political domain of the state regime. Political competition, whether by political faction, religious denomination or any particular form of ethnic affiliation, is the impetus of harambee, but competition is subject to a system of political control which subsumes the local leadership of projects to the personal authority of the state presidency.

Under both the Kenyatta and Moi regimes, harambee was the means by which the financing of a local project was politically linked to presidential power in its injunction for development. Contributions by big business to Kenyatta's Gatundu Self-Help Hospital or the Forces Memorial Hospital, for instance, were regarded as obligatory when state favours were sought. These grand harambee projects were shells within which funds could be amassed and then distributed to a plethora of local harambee projects through political entrepreneurs in the name of the President. As such, a system was developed by which finance was raised and distributed. The system may be a patrimonial alternative to that of the rational-legal authority of a state tax-and-spend system, but it has been no less a hierarchical system through which the injunction to develop is induced and controlled for the purpose of asserting state power.

During the late 1980s, a state official explained to NGO staff who became involved in a faulty Coast Province settlement scheme that:

Politics in Kenya is the politics of small projects. In place of party politics we have project politics.

So in politics to succeed, you must design yourself, or buy yourself a project. So you, as the NGO, if you have a good project, some politician will find it, and that applies to the individual women's group you help as much as to the big project.[65]

Studies of women's groups confirm this rule by showing the extent to which it is the political patronage of politicians, again to realise private benefit of households, which accounts for the seeming permanence of spontaneous self-help.[66] Project politics, to repeat, is what connects the roots of locality to that of state regime but, contra the post-modern inclination, the possibilities of designing or buying projects may be endless but the rule is systematic – harambee inserts individual desires for improvement into the meaning of intentional development. We may agree with Widner that the harambee was used differently by Kenyatta and Moi. By meeting part of the costs of local projects, the argument runs, the Kenyatta regime provided compensation for political acquiescence, as in the Central Province, where state expenditure was directed towards development outside that Province. Moi abandoned complex 'extra-parliamentary bargaining', using harambee as a simple instrument to favour one political faction against another within each district or province separately.[67]

In both regimes, however, money was handed down according to a political intention which was tangential to the intrinsic economic worth of particular projects when that economic worth, which was normally to qualify the individual for employment, did not meet either the individual desire for improvement or the social problem of unemployment. Holmquist, in arguing that 'with Presidential encouragement self-help has become a very popular form of development', revealed the opportunism of political intention when he wrote that:

The point is that the intrusion of big money into self-help politics has not automatically coopted or defused self-help, or made it simply dependent on the bourgeoisie or the state. People will usually take whatever money they can get, but the money, more often than not, buys only transitory loyalty.[68]

Questions are raised here about how administrative control has attempted to force harambee into the 'rational' purpose of rural development and why money of the bourgeoisie should have been handed down in the transitory ways of harambee.

Political competition and patronage from the presidency make support for projects spontaneously snowball and confound administrative attempts to plan the disposition of effort. State-sponsored agrarian schemes of development do beget self-help, but largely in localities where political competition is relatively intense. Thomas' evidence suggests that it is in relatively well-developed locations that 'wealthier segments of the rural population' pay more 'for self-help projects than the poorer segments, both relatively and absolutely'. By contrast, it is in the poorest locations, in areas of marginal agricultural potential, that there is 'a comparatively high level of coercion' and a regressive system of contributions for harambee.[69] This explains why no projects may be forthcoming in some areas, while in others, projects are started to pre-empt would-be competitors and, if and when completed, they often replicate existing projects. It is this paradox, that a competitive impulse is necessary to secure the end of cooperation, that is resolved by the force of state control over the spontaneity of the harambee movement.

In 1964, barely a year after spontaneous self-help had been unleashed, community development committees were established to control the disposition of resources for harambee projects. By 1970, development committees had been established at the provincial, district and division levels, the very units of administration which defined the scope of obligation for the end of development. From 1982 on, a presidential policy of 'district focus' was implemented to further the cause of rational planning for rural development.[70]

Since the end of development had been registered as the solution of unemployment, a solution which had not been provided by direct state effort in the immediate transition to post-colonial rule, it is not surprising that self-help reappeared in the late 1960s as a prescription for this irresolvable problem of development. But while Dalgleish's ideal of self-help, with its narrowed scope of kinship relations and minimal affiliation of the family to landholding, aspired to the individuation of productive effort, self-help now took on a different meaning in the practices of state development. Self-help came to connote the widening of kin, beyond the scope of any defined lineage or clan, by fixing personal obligation for productive effort within an administrative unit of government. As such, self-help came to be regarded as a substitute for the direct state financing of production and welfare.

The tension between spontaneity and control, which is at the core of nearly all analyses of harambee,[71] is usually supposed to be that between the contending forces of 'the people' and the state. However, it is better comprehended through an understanding of the way in which these forces pivoted upon the 'rural petit bourgeoisie' of local leaders. No matter how much this term may be disclaimed, the existence of the phenomenon it describes cannot be denied. Local leaders, the big of the small locality, are the smallest of national notables. They constitute the intersection between the force of the state and the spontaneous movement of self-help, each demanding relatively small amounts of money for investment in their own single enterprise but large blocs of capital investment for the large number of small farms and other rural enterprises in any given locality. State control, if it were to eliminate the duplication, waste and corruption endemic in projects, would thus have to eliminate the entrepreneurship which fosters self-help. To eliminate such entrepreneurship would be to kill the impulse of harambee. To foster local leaderships, without the unstable and transitory complexities of political bargains, would be to make harambee a worthless political instrument. To do any or all of these things would mean that the wider meaning of harambee – the end of intentional development – would be lost as the motif of the post-colonial state. It is little wonder that G.K. Kariithi, who was the head of the civil service in 1978 during the transition between the Kenyatta and Moi regimes, exalted harambee as 'a total system which underlies our political, economic and social procedures' referring to it as a system of 'participative centralism'.[72]

TRUSTEESHIP

It is through an understanding of how harambee funds were commanded and distributed among a plethora of small projects and fewer large projects that we can begin to see how the practices of an indigenous form of trusteeship were elaborated in post-colonial Kenya. By the early 1970s, it was the urban-based bourgeoisie – from senior state officials to professionals, owners of big business to directors of multinational company subsidiaries – who had come to command the cause of self-help.[73] Whatever their links, through biography and genealogy, to an earlier generation of farmers, the members of this class, who now commanded state power, were to immerse themselves in the cause of rural development in a different manner than their antecedents. This would not be a practice of *immediate* development for one's self by way of an appreciation that the community effort of projects would realise a disproportionate private benefit for those who had more capacity, in land and assets, before the project was undertaken. Rather, in the new integument, these bourgeois aspired to meet the demand for technically qualified labour in the face of the threat posed by the mass unemployment of academically certified school-leavers. This was the development of the 'other', the person in non-employment. Such was the rapidity with which trusteeship came to be associated with the doctrine of development. However, this was a particular form of trusteeship and one in which the meaning of full development could be expressed.

An immediate origin of the positivist precepts which Mbithi, for instance, attached to the harambee movement and, thereby, the 'authentic' strategy of development is to be found in a signal 1965 state document, *African Socialism and its Application to Planning in Kenya*, written by the then Minister of Economic Planning, Tom Mboya, as a political document to confront the anti-capitalist opposition which had been created in the first years of the Kenyatta regime. Although it was set out formally as an argument for a 'third way' of development, between communism and capitalism, Kenyatta put a definitive imprint upon the document by decreeing that its purpose was to end official ideological debate in the state.[74] The socialism of the then extant anti-capitalist coalition, led by the vice-president Oginga Odinga, can best be described as a conservative communalism premised on the belief that capitalism would undermine the presumed natural community of Africans. In this context, *African socialism* can be seen as the astute means by which the President and Mboya employed state power to subvert that belief through the assertion that community, because it was natural and therefore amenable to another form of content, should become a necessary part of capitalist development. As such, Kenyatta and Mboya could claim, 'they were for progress while the others were antidevelopment'.[75] Thereafter, 'antidevelopment' and political opposition became synonymous and called down the repression of both the one-party and party-state regimes of Kenyatta and Moi.

African socialism had been generated by the pan-African movement and, in particular, that vital part of it which had assembled at the sixth Pan-African Congress of 1945. It was there that George Padmore's approach of 'democratic socialism' captured the movement. The Congress, whose credentials committee was chaired by none other than Jomo Kenyatta,[76] endorsed both nationalism and 'the doctrine of African socialism' as the twin elements of the 'struggle for political freedom and economic advancement'. Following Padmore, the link between nationalism and socialism was made on the basis that Africa had a 'single traditional culture', that of communalism. The common colonial experience, it was argued, subdued communalism and exploited resources in Africa for the benefit of non-Africans. To achieve post-colonial economic advancement or progress, rational planning of resources would be required. This commitment to planned and rapid progress had led, by the time of Senghor's address to the 1962 Dakar colloquium on African socialism, to a recognition of 'the importance of development and the need to choose the most efficient ways and means for this development'.[77]

The Fabian Margaret Roberts, in her 1964 essay on African socialism, argued that it was in the identification of socialism with a 'sense of community', and then the latter with the essence of African-ness, that prevented the link between socialism and nationalism from being construed to be one of national socialism. Roberts quoted President Senghor:

'Socialism for us is nothing but the rational organization of human society considered in its totality, according to the most scientific, the most modern, and the most efficient methods' – might easily be said to be indistinguishable from a fascist approach to development had not [Senghor] later added, 'More than the use of the most efficient techniques, it is the sense of community which is the return to African-ness'. But, at Dakar, it was clearly the developmental aspect of socialism that struck the common chord.[78]

This constructivist meaning of development followed from the earlier bearings taken by W.E.B. Du Bois, the late nineteenth-century student of the German Romanticism of Herder, Hegel, Weber and Schmoller, who had set out one theme of pan-Africanism from the vantage point of an American.[79] It has been pointed out by Gilroy that when Du Bois referred, in his early work of the first decade of this century, to the problem of the '"backwardness" of American blacks': his belief was that, 'their backwardness could be remedied by an elitist and paternalist political agenda that viewed racism as an expression of stupidity and implied that progress, rational social policy, and the Victorian moral virtues advocated by the talented tenth could uplift the black masses'.[80] If such a 'model of national development' had appeal,[81] it did so not because it was inflected by what has been called 'the dictatorship of developmentalism',[82] but by virtue of its appeal to the universality of the democratic virtues of local community. Padmore, for instance, quoted approvingly from Du Bois' address to the third 1921 Pan-African Congress in London:

The habit of democracy must be made to encircle the earth. Despite the attempts to prove that its practice is the secret and divine gift of a few, no habit is more natural or more widespread among primitive people, or more capable of development among masses.[83]

Surely, Du Bois claimed, by emphasising local government 'there can be found enough of altruism, learning and benevolence to develop native institutions'.[84] Sidney and Beatrice Webb strongly supported the 1921 Congress where their Fabian associates, Sydney Olivier and Norman Leys, were present. The Fabian connection was of long standing. Olivier and H.G. Wells had been at the fourth Congress in 1923.[85] Annie Besant, the early Fabian, had been at the 1911 Universal Races Congress in London, alongside Du Bois and Gandhi. Although there was mutual suspicion between Padmore and Fabians such as Arthur Creech Jones, the Labour government colonial secretary in 1947, and Leys and Rita

Hinden,[86] Padmore came to be impressed by the Fabian method of political permeation.

Padmore's political trajectory was in the opposite direction from that of Du Bois, away from affiliations with Communist Party organisation. Twenty years after Moscow accusations that his tendency represented 'petit bourgeois nationalism' and his abrupt expulsion from the Comintern in 1934, Padmore remained faithful to the Soviet model of resolving the Russian imperial problem through industrialising the eastern regions and 'raising the cultural level of the former subject races of Central Asia' relatively rapidly through 'co-operation between the various races and peoples'. His long-standing defence of Soviet policy against British colonial experience was what irked his generation of Fabians. Equally, however, he reflected that pan-Africanism had 'adopted more Fabian-like methods' than either those of 1930s communists, who took Africans to be 'backward, unsophisticated tribesmen', or of Marcus Garvey, whose 1920s platforms of 'black zionism' and 'demagogic racialism' were castigated as nothing other than an attempt to colonise Africa from the United States. Thus, Padmore wrote in 1941: 'I have always considered it my special duty to expose and denounce the misrule of the black governing classes in Haiti, Liberia and Abyssinia, while at the same time defending these semi-colonial countries against imperialistic agression.'[87] Padmore's principle, he insisted, was that pan-Africanism had to find expression by Africans in Africa.[88]

Despite much historiographical dispute over the priority to be given to Du Bois and his 1919 Congress, what was arguably the first pan-African Congress of 1900 was conceived by the London-based West Indian barrister Henry Sylvester-Williams.[89] Williams made no reference to self-government but claimed that he wanted 'to start a movement looking forward to the securing to all African races living in *civilised countries* their full rights and to promote their business interests'.[90] The incubus of business interest may have been submerged by subsequent pan-African socialisms but it was ever-present and was well understood by early nationalist groups in Africa who just as easily could have expressed an extension of Williams' aim for Africans living in Africa. This was certainly the slant of the Kikuyu Central Association whose representative, Jomo Kenyatta, arrived in London during 1929 to become an immediate protégé of George Padmore. Thirty years later, another protégé of Padmore arrived in Accra to be the chairman of the 1958 All African Peoples Conference, the first pan-African Congress in Africa. Tom Mboya was to be fully blooded in African socialism.[91]

Partly schooled within a Fabian nexus in Britain, but with a public face which was succoured by the Cold War extension of American imperial interests to Africa, Mboya, who was assassinated in 1969, seemed to be an enigma. His perceptive biographer describes him as being one who was '"of" the ruling group yet stood somewhat out of it'. Being for the development of private property and enterprise, Mboya stood against the 'massive and rapid capital accumulation through the opportunistic fusing of political, administrative and business roles'. He was opposed to straddling, the process whereby the primary accumulation of private capital starts from individual employment in the state and international corporations.[92]

In proposing the prospect of an ideal capitalism made up of moralised capitalists and in projecting a 'capitalism as it might be' rather than deferring to the course of capitalism as it has been in Kenya, Mboya's African socialism was a particular vestige of nineteenth-century Millian positivism. As 'the single and most powerful man in government capable of explaining and rationalizing Kenya's economic policies', Mboya's credo has been explained succinctly:

> He had a clear ideological line, that of a social democrat, and he believed in capitalist development in Kenya in which the state would play a leading role in accumulation, manned by competent civil servants. He could not, however, stand 'socialist adventurists' . . . who wanted to build socialism by redistributing the poverty of the country.[93]

It is clear that it is the 'competence' of a civil service through which Mboya sought to connect social democracy to the relative poverty of a predominantly agrarian country. Leaving aside the question of the need to tie the historical emergence of social democracy to the maturity of industrial capitalism, we cannot help but note the stringent hope voiced here that what was represented by the experience of social democracy, with all its positivist roots, could be impressed upon the minds of state servants within, rather than in opposition to, the body of early indigenous capitalism in Africa.

Indeed, it has been argued that just this aspiration has been achieved. David Leonard has explained the relative success of post-colonial agriculture in Kenya, namely the growth of production within the expanded framework of colonial infrastructure, as being due to efforts of senior civil servants who have pursued a collective interest for the capitalist class as a whole. Generalised capital accumulation in Kenya, where 'the largest private business entrepreneurial class in sub-Saharan Africa' has allegedly developed, has been also used as a prescriptive model for what has been called a 'positive capitalist state' for Africa: 'By organizing private accumulation of capital, state elites would feel sufficiently "in command" of the development process and would be so recognized by those who came to feel that their personal prosperity depended on the political leadership of their rulers.'[94] 'Kenyatta', Leonard has written, 'seemed to see himself first as the trustee of an emergent African propertied class and only second as a patron dispensing individual favours.' Senior civil servants, instead of having shown a predatory pursuit of self-interest, are seen as having pursued 'enlightened self-interest'. This self-interest was enlightened because it was

> the pursuit of the greater gain of the larger number of members [of the class], rather than the ruthless pursuit of advantage by some members at the expense of others in the group; a concentration on long-term interests, rather than on immediate, short-term benefit; and sufficient care for the welfare of those outside the group so as to maintain the stability of the system, rather than nonmembers being so deprived that they challenge the order's continued existence.[95]

Nevertheless, it was because a state salariat were engaged in the individual pursuit of private property that an active self-interest undermined the social democratic basis of Mboya's credo. Civil servants own and control capital in large farms, real estate and a gamut of enterprises in trade, tourism and some sectors of manufacturing industry. As such, trusteeship is abstracted from the ideal of public service but embodied in the very nexus through which service and acquisition were intertwined.

When large farms produce the same range of products as smallholdings, then Leonard argues that self-interest of owners of large farms coincides with that of household producers. It is seen as in the interest of a straddling salariat 'to maintain healthy terms of trade for agriculture and to support services that will keep it expanding'.[96] This influential argument follows that of Michael Lofchie, who argues that 'so far as agricultural matters are concerned, Kenya's government is one of farmers, by farmers and for farmers'.[97] Albeit, from a different standpoint, the same kind of argument has been made regarding how private land proprietorship became the common interest of producers and capitalists. Freehold title for the smallholding was the same title, Apollo Njonjo has argued, which afforded the large-scale acquisition of land.[98] Yet, there is a world of difference between 'patches of land' for immediate subsistence and landed estates for speculation, production and accumulation. The argument of coincidence only makes sense in explaining the immediate and historical basis of trusteeship and not the other way around. It is not coincidence which explains trusteeship but the source of trusteeship which explains why there should be coincidence between smallholdings and large farms. During the colonial period, as we shall see, the similarity between products of settler estates and smallholdings was a source of contrariety for settlers in a regime of labour shortage because of the fear that smallholdings would drain wage-labour away from estates. Under the late-colonial and post-colonial regime of surplus populations, smallholdings absorb what the farms and enterprises of the salariat cannot employ.

A senior state salariat, reared by the subaltern ranks of colonial servants, socialised in the schooling and professions of the late-colonial period, and parachuted into administration, inflected the ideal of service onto the terrain of enterprise. One of Leonard's informants explained what his teachers at Alliance High School had expected:

> They had in their minds stereotyped professions for which they were preparing us – teaching, medicine, [the] civil service, [the] ministry, agriculture, and veterinary [medicine]. It never occurred to them that some of us could decide to go and do law or join a private firm. . . . That set of values was being ingrained in us. . . . A lot of emphasis [was placed] on *public* service.[99]

Charles Karanja, one of the four state servants whose cases form the bulk of Leonard's studies, was general manager of the Kenya Tea Development Authority during the period 1970 to 1981 when the volume of smallholding tea production nearly doubled. Having established a range of private enterprises including involvement in a private tea factory, Karanja typified how trusteeship of state authority was expressed:

> in order to enhance faith and confidence in government [among the large number of unemployed], I have always deemed it my duty to see job seekers and to help where possible. The process is straining and time-consuming in a busy office, and . . . I have only been able to help a small fraction.[100]

Karanja's own enterprises were constrained by profit-maximising criteria, thereby minimising his political desire to absorb the unemployed in wage-work. Equally, it was Karanja's influence within the KTDA, as much as that of external agencies such as the Commonwealth Development Corporation and the World Bank that financed the authority's investment plans, which forced the KTDA to adopt cost-minimising practices. Cost-minimisation constrained off-farm wage-employment for the sake of on-farm tea producers' payout from the authority. As such, agrarian doctrine, in its abstract form of rural development, had to bear the burden of absorbing the unemployed.

It is not surprising to find that Karanja was a contributor and organiser of harambee projects, usually in education, from 1964. It is reported that 'many times he was giving more than his salary at the KTDA'. He was able to do so precisely because his own business income when he started at the authority was as large as his official salary.[101] However, trusteeship, through the form of straddling, is inherently unstable. For each case of the 'positive values' of professional and collegial accountability, which Leonard enumerates, we can find the converse. A typical 1993 case is a report from the auditor-general on the state-owned Kenya Post and Telecommunications Company:

> The report shows that KPTC had accumulated arrears on its own foreign debt, on its contributions to the National Social Security Fund and on a special telecommunications tax. The monies owed are said by western officials to total more than $125m, an amount equivalent to about half the company's annual turnover.
>
> KPTC deposited hundreds of millions of shillings in scandal-tainted domestic banks. According to a western official investigating the KPTC management, these funds were onlent to government supporters. . . . A former manager at KPTC has also accused company executives of faking the theft of millions of dollars of equipment so that it could be resold to KPTC for instant self-enrichment.[102]

Self-enrichment of this kind, so suggestive of the 'negative values' was, as Leonard notes, a pervasive threat to the positivism of the early post-colonial period. They are the extreme attributes of straddling which find historical expression as part of the logical extension of indigenous capitalism to all regions and ethnic blocs of Kenyan society.

The Kenyatta regime, so the familiar argument runs, spawned an expansion in the scale of indigenous capital in so far as it was identified with an ethnic bloc of Kikuyu capital. State power was used to create spaces for indigenous capital accumulation and the spaces were filled by the senior state officials, professionals and/or senior staff of multinationals' subsidiaries who were promoted and sponsored by the regime. In the immediate post-colonial period 'affirmative action' carried out in pursuit of African-ownership was used to break down the European domination of large-farm agriculture as well as the domination by Kenyans of Asian origin of indigenous-owned trade and manufacturing industry.[103] Through various forms of straddling and joint ownership of enterprises, African accumulation was fuelled by both managerial control and rent-seeking by those who had acquired state power or generated access to it through ethnic affiliations and the cross-ethnic coalitions which formed part of 'project politics' of harambee.

If the Moi regime replicates the experience of accumulation in the earlier post-colonial period, through the phenomenal form of straddling, it does so differently from the earlier period of accumulation and the difference is not epiphenomenal. Once racial or ethnic space has been opened up and then occupied for accumulation, both the economic and political task of re-creating space for different ethnic blocs without violating the core of African socialism becomes more difficult precisely because the existing space has been occupied according to an indigenous trajectory of African accumulation. When the Moi regime self-consciously claims to represent the rights and capacities for non-Kikuyu property acquisition, thereby extending the indigenous trajectory of accumulation to a Kalenjin-Masai bloc of capital, it does so in the manner in which the original trajectory struck at colonial and racial domination. But when the claims are expressed through an attack on 'tribalism' as a surrogate for 'Kikuyu domination' with the aim of dislodging the pre-eminent identity between the force of indigenous capital and capitalists of Kikuyu origin, then the original basis of African socialism, of 'positive capitalism', is severely undermined.

Two examples of this post-1978 process suffice. The first is the searing experience of ethnic cleansing which has occurred from 1991 on in the Rift Valley from where an estimated 300,000 people of Kikuyu origin have been forced to find refuge mainly in Kiambu district in the Central Province.[104] Forty years after the Mau Mau repatriation of population, the 1962 threat of Martin Shikuku, then a subordinate politician in the Kenya Africa Democratic Union of which Arap Moi was a major leader, was realised. Shikuku had warned 'the Kikuyu' that 'if they come to take our land by force they will meet our spears'.[105] Thirty years after the threat, spears, as well as bows and arrows, were met.

White Highlands land of the Rift Valley had been occupied by the settlement of small farms on subdivided estates and through the acquisition of existing large farms, both of which, in the majority, went to those of Kikuyu origin.[106] Likewise, a multiplicity of trading and other enterprises had been bought or started by both small-scale and big Kikuyu business. Kikuyu smallholders, in the majority of cases, were forced to sell their landholdings, when they fled from the violence, at less than a quarter of the then market price.[107] Contra Bayart's thesis of African manoeuvre within the confines of state power,[108] there is little room for hiding in this central field of Kenya polity and economy. To use state power to force 'the Kikuyu' out of the Rift Valley was nothing other than to re-create the space in which a new basis could be created for a different ethnic bloc of putative capital.

Second, the KPTC example, in which a state corporation is looted for individual self-enrichment, is one of many which arises because, through the prior closures of space for primary accumulation, the state corporation or parastatal authority is the one obvious singular shell of activity for the pursuit of individual acquisition. Virtually all agrarian-based corporations and authorities have been disorganised, according to the positivist premise of organisation, on the grounds that by dislodging Kikuyu management and control, the purpose of small property ownership can be served.[109] Case after case can show how no such general purpose is achieved and how the interest of self-enrichment, again as in the KPTC example, forces the emergence not of an indigenous but of an international force of trusteeship in the World Bank and IMF.

Mboya, acting in the pan-Africanist tradition, sought to graft the positivist basis of trusteeship onto African roots, thus disconnecting it from colonial and neo-colonial sources of influence and control. He wanted to give trusteeship historical depth in Africa. In a widely quoted passage that has been called 'almost Burkean' but which cannot help but strike the late twentieth-century reader as 'Afro-centric', Mboya wrote:

> When I talk of 'African Socialism' I refer to those proved codes of conduct in the African societies which have, over the ages, conferred dignity on our people and afforded them security regardless of their station in life. I refer to universal charity which characterises our societies and I refer to the African's thought processes and cosmological ideas which regard man, not as a social means, but as an end and entity in society.

And, after referring to society as the interdependence of individuals in a social organism, Mboya reflected: 'I think it worthwhile emphasizing the fact that these ideals and attitudes are indigenous, and that they spring from the basic experience of our people here in Africa and even in Kenya.'[110] The trusteeship which Mboya sought to re-create in Africa was to be the positive side of the dismantling of racial barriers against African aspirations in business and enterprise.

In 1963, when he was minister of labour, Mboya referred to the need for Africans in business 'to share in the task of development as industrialists and in commerce both in rural and urban areas'. Four years later, when he was minister for economic planning and development, Mboya led the drive behind the 1967 Trade Licensing Act with its attendant citizenship provisions that acted to affirm African business and led to the 1968 Asian exodus from Kenya.[111] Africanisation, however, had to be informed by two principles which accorded to the fundamental nationalist stance of what had made socialism African. The two principles which, according to Mboya, informed African socialism were laid out in the 1965 document.

The first principle, that of political democracy, was the source of the criticism of the concentration of economic and political power. Concentration, for the African socialist Mboya, meant the fusion between the active self-interested, desire for property and the clasping of state power by those so self-interested resulting in the indigenous capital accumulation which characterised the immediate post-colonial regime of Kenyatta. 'Political rights,' the author of *African Socialism* wrote of 'traditional' African society, 'did not derive from or relate to economic wealth or status.' This contention is highly disputable in that it can be shown that, in the nineteenth-century prelude to colonial rule, political notables were indeed so through wealth garnered through the trading and raiding of cattle. As if to admit this objection, Mboya claimed that: 'Even where traditional leaders appeared to have greater wealth and hold disproportionate influence over their tribal or clan community, there were traditional checks and balances including sanctions against any possible abuse of such power.'[112] Such 'checks and balances' and 'sanctions' are not described, but it was the competition for personal political power, buttressed by capacities to acquire wealth and the opportunity to colonise land, which made the threat of a shift in allegiance from a despotic notable act as a check and sanction.[113] Mboya's central point was that 'traditional leaders were regarded as trustees whose influence was circumscribed both in customary law and religion'. Religion 'provided a strict moral code for the community'. The moral code, and this is the constructivist, second principle of African socialism, was one of 'mutual social responsibility'.

Mutual social responsibility, as defined in *African Socialism*, was the obligation of each person, endowed with the equal political rights of citizenship, to contribute according to ability 'to the rapid development of the economy and society'. If 'every member of African traditional society had a duty to work' then, Mboya continued, the 'society' had 'the power and duty to impose sanctions on those who refused to contribute their fair share of hard work to the common endeavour'. For post-colonial modern society, it is the state which rewards effort and takes 'measures against those who refuse to participate in the nation's efforts to grow': 'Sending needed capital abroad, allowing land to lie idle and undeveloped, misusing the nation's limited resources, and conspicuous consumption when the nation needs savings are examples of anti-social behaviour that African socialism will not countenance.'[114] Here was an allusion to both Kenya Asian business and what was to become the source of disillusion with self-enriching straddling as the phenomenal form of indigenous accumulation. Back in 1962, Mboya had asked of African businessmen:

> will they evolve business ethics suitable to this situation while at the same time adhering to the standards of efficiency, initiative, and thrift which will enable them as a group to help Kenya make the maximum use of its resources in the context of economic planning?[115]

As if to give a doubtful reply to the proposition contained in this question,

Mboya, in the 1965 document, made the state legitimate on the well-trodden constructivist grounds:

The idea of mutual social responsibility presupposes a relation between society, its members and the State. It suggests that the State is a means by which people collectively impose on individual members behaviour that is more socially constructive than that which each would impose on himself.

State control and planning, it was concluded, were 'to ensure that productive assets are used for the benefit of society'.[116] This precept, the positivist kernel of Mboya's doctrine of development, is the key to the understanding of the predicament of trusteeship in the Kenya case.

Particular policies enunciated in *African Socialism* held implications for later unemployment policy. Mboya stressed both 'the rapid development of agricultural land' and a 'wages and incomes policy that recognises the need for differential incentives as well as an equitable distribution of income'. Self-help, as it was put in the document, 'has strong roots in African traditions and has therefore important potential for development. *But it, too, must be planned and controlled.*' A spontaneous mushrooming of harambee schools and clinics, where rural development was relatively advanced, made pre-emptive claims upon state expenditure for teaching and nursing staff and, therefore, violated Mboya's injunction on 'the need to plan and control how resources are used'.[117]

An obvious question to ask is 'to plan and control by and for whom?' The answer, again to be searched for through the focus upon particular policies of both *African Socialism* and what followed after it, was to be found in Africanisation. Mohiddin, for instance, claims that the answer is to be found in 'Africanization of the colonial ideology' of the post-war period; Leys, that it lies in the 'formulation of comprador ideology' or, in other words, the Africanisation of land ownership and business but on the basis of private property and in collaboration with international capital.[118] If, as is claimed, colonial ideology was perpetuated, then trusteeship was now invested in an African administration. But the Africanisation of property gave vent to indigenous capitalism whose course, as capitalism, was such that it was the object of what trusteeship was meant to subdue. Such has been the predicament of African trusteeship in Kenya.

The characterisations of African socialism of Mohiddin and Leys are equally one-sided and miss a vital point of trusteeship. A community of interest, centred on a class of capital, whether characterised as the rich or *matijiri*, or 'the family', as of Kenyatta, or bourgeoisie, finds its collective expression through the state. The state is forced to promote another community of interest, that of the poor, as in 'small-farmers' or 'self-employed' for which policy is designed to support a relative surplus population. These two communities only find a coincidence of interest when the suppositions of class are expressly denied or, as we have emphasised, where space can be created to further the accumulation propensities of one without undermining the conditions of existence for the other.

There is no formulation of African socialism, nor its perpetuation in the various forms of dependency theses, which does not project possibilities for socialism without supposing that, as in Roberts' 1964 words, 'the African "haves" – the potential "exploiters" – are so insignificant in number as to be scarcely worth bothering about'.[119] A paucity of number, in one period, can become a significant source of bother in another. When the expansion of middle peasant production happened in Central Kenya, the barrier it imposed on the acquisition of land for accumulation, rather than on money-mediated subsistence, appeared to have been transcended by the acquisition of land and property in the Rift Valley, which provided the opportunity of economic space. Likewise, expanded household commodity production provided an internal market for the agricultural and non-agricultural enterprises of the indigenous class of capital. However, from 1978 this national space began to be fully filled because the potential 'exploiters', no less than the numbers of the community of the poor, had become, in relative terms, so many. At this point the burden of trusteeship returned to its primordial task, that of the state administration aware of its obligation to secure the potential productive force of labour power.

What is striking is the way the 1965 precepts of *African Socialism* have been offically restated time and again in seemingly perpetual tangent to the indigenous trajectory of capitalism. Thus the 1970 Mwicigi Report insisted that 'Government should exercise greater control of the economy' through the 'systematic' and 'properly planned' nationalisation 'of all major enterprises' in the economy. Like Africanisation, this could only be achieved 'if the degree of dedication and competence of the Civil Service' was 'maintained at a high level'. Pointing out that nearly one half of allocated development funds were returned to the Treasury unspent in 1969–70, the Report complained 'that some of the civil servants are spending their official time doing other side business'.[120] A decade later, the 1983 Wanjigi Report claimed that 'most Kenyans do not acknowledge their role and duty in the alleviation of unemployment' and 'this is in direct conflict with the national philosophy of mutual social responsibility as articulated in *African Socialism*'. This 1983 Report devoted its second chapter to restating the 1965 precepts and recommended that 'a concerted effort should be made to teach and popularize the concept of Democratic African Socialism and its application to Kenya'. Finally, a restatement of 'mutual social responsibility' was made once again in the opening of the 1991 Ndegwa Report.[121]

Such views were not confined to the pages of these reports. Harris Mule was a one-time protégé of Tom Mboya who became a permanent secretary in the Kenyan Ministry of Finance in 1978. In the course of his career in that ministry, he influenced the Mwicigi Committee Report, was largely responsible for *Sessional Paper No. 10 of 1973 on Employment*, and had a hand in the 1983 committee. He was also the driving force behind the 1982 plans for the 'District Focus for Rural Development'. His belief, following Mboya, in the assertion of national policy over 'sectional interest' was inflected by a friendship 'with O. Norby, a Ford Foundation adviser in Planning. Norby's Pakistani experience helped him to persuade Mule that the only way to keep people out of the cities was to raise rural incomes.'[122] The ready acceptance of such advice when combined with the unmistakably positivist tone of each report on unemployment provides evidence of the extent to which the official desperation to find a solution to the intractable problem of a relative surplus population could not be dissociated from the presumptions of the nineteenth-century idea of development. At the same time, the evasion of the twentieth-century experience of capitalist development in Kenya made it difficult to find a social source for trusteeship and thus still more difficult to discover a solution to the development problem which did not make it more, rather than less, likely that the problem would be accentuated by the implementation of that solution.

Dalgleish evaded an understanding of the course of capitalist development in Kenya when he urged that assistance of members of family and tribe be denied to the individual household workers. It was also evaded by all the later reports which held to the positivist precepts of *African Socialism*. If there were the development of wage-employment and/or rural development which realised revenue for each individual worker, then there would be no need for the family to subsidise the costs of education and non-employment of family members. If a cadre of positively minded civil servants were paid according to their subsistence need, they would not need to engage in private business. In other words, it is the end of official development which, yet again, is being presumed to already exist as the conditions for an ideal course of capitalist development. The predicament of trusteeship was no less present in post-colonial than in colonial Kenya. It arose out of the idea that any propensity for indigenous capitalist accumulation had to be disciplined in the name of a community. Indigenous capitalists were meant to act according to 'social responsibility' rather than private profitability in which the productive force of labour was left to find its own means of employment for the most basic needs of subsistence.

AGRICULTURE AND INDUSTRY

Presumptions of nineteenth-century development doctrine, from official trusteeship to the practice of the agrarian doctrine of development and its urban analogue of informal sector self-employment, have been the repeated recourse of official reaction to the explosive growth of urban unemployment. However, the phenomenon whereby a marked rate of urban population growth is less than the rate of growth of industrial employment has been regarded as a 'new kind of unemployment'. Walter Elkan proclaimed in a 1970 overview of the problem: 'We now know how to avoid mass unemployment in the industrial countries. What had not occurred to us was that we would soon be facing a new kind of unemployment in the non-industrial world.'[123]

If, two and a half decades later, 'we', Keynes notwithstanding, no longer 'know' how to avoid mass unemployment in the 'industrial countries', then we also know that the growth of a relatively surplus urban population is no new phenomenon. Positivism addressed the old nineteenth-century problem of the relative surplus population through an idea of development which put a premium on official policy. Elkan, a fervent anti-development apostle, explained the origin of the problem in much the same way as did the early Keynes. Remarkably, anti-Keynesians such as Collier and Lal, as well as early proponents of underdevelopment theory for East Africa such as Arrighi and Saul, located the same source of the problem.[124] It was to be found – for all of these writers – in the application of a 'misguided' relatively high, urban, and, particularly, industrial, wage policy that fostered the creation of a labour aristocracy composed of 'a small but prosperous proletariat' at the expense of maximising 'the growth of employment opportunities'.

Elkan's point in 1970, following Todaro's formulation, was that if employers' demand for industrial wage-employment were to increase as a result of the implementation of 'appropriate policies' which abolished official minimum wage rates and other interference in the 'market', then the number of urban job-seekers would rise disproportionately to the number of available jobs because each putative rural–urban migrant would have a heightened expectation of finding a job.[125] This paradox, that unemployment would increase simultaneously with an increase in the availability of wage jobs, rested upon the assumption that urban wages were 'artificially high' during the 1960s. Such was the evaluation which lay at the core of the elaboration of development doctrine in Kenya.

As far as policy was concerned, not all antagonists of the labour aristocracy, an ideal source of trusteeship if ever there was one, were agrarian-biased. For instance, while Elkan did not dismiss 'rural regeneration' out of hand, he did argue that '*in addition* it is a worthwhile endeavour to increase urban employment, not because it will necessarily reduce urban unemployment but simply because it would increase the choice open to people'.[126] House and Rempel directly questioned state support for agrarian development and argued for state support of the 'intermediate' artisans of the informal sector who have come to be seen as the urban analogue of peasant household production.[127] In addition, the unemployment reports, like the Development Plans, all paid lip-service to industrial development.

For example, Dalgleish, quoting from the East African Royal Commission Report of 1955, thought it was clear that neither industry or irrigation, for intensive agriculture on less fertile land, would 'be able, in any short period of time, to absorb many people on an economic basis, and that the real need in East Africa is the economic development of fertile land'. Given this assumption, Dalgleish supposed that 'possibly, as much as one-third of the number employed in agriculture itself, would be engaged in the future in non-agricultural services and secondary industry'.[128] Similarly, the 1970 Mwicigi Report asserted that 'While industrial development is very important for long-term economic development and employment creation . . . agriculture is also very important and its development must be accelerated.' The 1991 Ndegwa Report repeated much of the recommendations of the 1983 Wanjigi Report to the effect that the elimination of protection on imported inputs for export-led manufacturing industry was necessary to secure 'a strong industrial development base', while the promotion of linkages to small-scale enterprise would be the source of employment creation.[129]

These words had a hollow ring for the simple reason that there has been no systematic intention to advance manufacturing. As we shall see, the argument that a subsidy existed for manufacturing arose, *ex post facto*, out of the argument that the aim of tariff protection was the pursuit of an industrial policy of development rather than out of an analysis of the real, largely non-manufacturing, purposes to which these funds were put. Despite all the verbiage, the intention to develop for, and the emphasis of policy in, both post-colonial and colonial Kenya, remained resolutely agrarian.

AGRARIAN POLICY

Everything grows from agriculture. Look after agriculture and economic and industrial growth will take care of themselves.[130]

Such was the motif of 1982 World Bank policy which was quoted in the 1983 Wanjigi Report and which has been the historical theme of Kenya government policy. In the colonial period, agricultural growth was that of large-scale estate agriculture with which European settlement was identified. To 'look after' agriculture was to look after settlers in Kenya. If there was an agrarian doctrine of development, to promote the employment of wage-labour in agriculture, it was in support of the extension of large-scale agriculture for the purpose of production and not to absorb a surplus population in agriculture. The latter was the task of small-scale agriculture, whose growing intensity of production was made necessary by the extensive margin of estate agriculture. Indeed, the historically short trajectory of estate agriculture, from the 1920s to its apogee of 1959, was one in which wage-labour was increasingly incorporated as productive force in the white highlands, the 'scheduled' large-farm areas, while population was simultaneously expelled to the reserves of the non-scheduled 'African land units'.

Company-owned plantations engaged in the specialised production of tea or coffee may have been capitalist enterprises, but large-scale, mixed pastoral/arable agriculture in Kenya, whose productive force rested upon a labour process of wage-labour, was not motivated by profit for capital accumulation but by the securing of subsistence for settler farmers according to European norms of consumption.

It could not be decided, as Wrigley pointed out, whether settlers were to be a substitute for an indigenous aristocracy 'with supervisory and directive functions, set over the native population', or whether they constituted a 'distinct community' of poorer emigrants from Britain and the Dominions.[131] As such, white settlement was established according to a contradiction between the intention and process of development. Confined to settlers who had some money to purchase tracts of land and with access to a modicum of finance to establish conditions for production, estate agriculture depended on agricultural subsidies from the Kenya colonial government. These subsidies were justified by settler landowners, whose estates were carved out of Crown land, on the grounds that they were the active proponents of the intention to develop what they hoped would be a self-governing colony under their political force and aegis. When the counter-claim was made, that the subsidy was unjust and inefficient because it was financed by taxes on imports falling primarily on African wage-earning and revenue-producing consumers but spent on unprofitable settler estate agriculture, the intent to develop was confused with the process of development.[132]

It was the very process of development which estate agriculture set in train which was to undermine the premise of all that was involved in the claims for the intention to develop upon the basis of the association between race and large-scale land ownership. There was the development of a rail and road network, agricultural infrastructure, extension services and cooperatives which, however centred on the white highlands, made Kenya a focal point of colonial development, including that of early industrialisation in eastern Africa. The development of a wage-labour force in agriculture accentuated this process of development. An intent to develop is confused with the process of development when the one-sided focus on the claims of intentional development obliterates the course of capitalist development itself. To deny the particular claims of intentional development is not, in itself, sufficient to deny the course of capitalist development.

From the vantage point of the mid-1930s Depression, it may have appeared that the colonial association between race and land ownership was to be undermined by the economic propensities of estate agriculture. However, a quite different picture was painted in the period between 1939 and 1959 when the war-related demands for food and the post-war commodity shortage induced an increase in supply such that, despite the decline in commodity prices from the early 1950s and the threat of the Mau Mau revolt, it was seen as reasonable to conclude that the 1950s were 'the most prosperous phase of European farming in Kenya', the peak being reached in 1959.[133] It was the peak from which Dalgleish looked down in setting out what was to become the new agrarian doctrine of development designed to confront the threat of unemployment. It was this threat which hung over the 1960 Lancaster House conference for African self-government and thereby the final dissolution of racially exclusive estate agriculture.

Post-war growth in estate production induced a once and for all change in wage-employment between 1947 and 1950 when the enumerated number of agricultural workers doubled to over 200,000, representing nearly one-half of all wage-employment in the colony. Then, when a plateau of one-quarter of a million was reached in the early 1950s, the number employed in large-scale agriculture remained constant. Mechanisation, coupled with a lengthening of the working day for each worker, increased the average value of output per worker, during a period of falling farm-gate prices, from £112 in 1954 to £140 in 1962; expenditure on capital equipment per worker rose by 50 per cent between 1954 and 1958.[134] During the same period, the labour problem of estate agriculture, that of endemic labour shortage, became one of excess labour supply at the same real wage rate and thus a generalised problem of relative surplus population for the colonial administration and its post-colonial successor. In so far as settler trusteeship for development was guided by 'the assumption that Africans would be more productive working for European settlers than working in their own areas', the failure to provide productive work for Africans in the 1950s signalled the end of a local but settler-dominated and anti-colonial trusteeship.

LAND IN THE CENTRAL PROVINCE

The proponents of colonial trusteeship, especially those of Fabian inclination, came to regard Kenya as an aberration because of the claims European settlers enforced upon the Kenya government. From the early 1930s, when notice was taken of an emerging surplus population, the official response was to take European settlement as an historical fact and emphasise the improvement of agriculture within the African reserves.

Two examples of the official response to unemployment are sufficient. In 1930, Nairobi Municipal Council faced up to African unemployment by declaring that: 'the town is a non-native area in which there is no place for the redundant native, who neither works nor serves his or her people. The exclusion of these redundant natives is in the interests of natives and non-natives alike.'[135] Akin to this view of unemployment in the 'town', was the pre-1936 administrative view of petty trade in the countryside where a second generation of school-leavers, in the midst of the inter-war depression, had neither the offer of nor desire to engage in wage-work in large-farm agriculture. Such petty traders were 'the unemployed and mostly the unemployable'.

> As they emerge from the mission school their main ambition appears to be the achieving of a competency by means of petty cheating. . . . This finds expression in the incredible number of middlemen – traders who are found on almost every reserve road trading on a miserable capital, but persuading the native producer . . . to dispose of his produce at a figure much below the market prices. It further finds its expression in the shops . . . butcheries and 'hotels' which have sprung up like mushrooms and owe their existence to that misguided patriotism which is the refuge of the scoundrel.[136]

A second example is provided by the response to squatting. The 1932–4 Carter Land Commission[137] estimated that between 1902 – when the Central Province (Kikuyu) reserve boundary began to be demarcated for the purpose of European land settlement – and 1931 the population density in the province was constant. Land congestion, in the sense that average output per family would be higher were more total land available to a given population, was apparent but was due to the 'lack of skill' in farming rather than any total lack of land in the reserves. The 'method of disposing of the surplus population', the report continued, had been squatting on alienated, European-owned large-farm lands.[138] Squatting, until 1911 when the reserve boundary was finally demarcated, had provided a window of opportunity for livestock-rich migrants to get pasture on alienated land; thereafter it became a form of labour-rent as opportunities for poorer migrants to occupy land for household production diminished while obligations to engage in wage-labour on the 'squatted' estates increased. Thus, as European estates extended their margins of cultivation during the 1920s, and as they were later to do more emphatically after 1939, administrative and physical force were employed to push the means of family household livelihoods off the large-farm areas and into the reserves.

If the Carter Commission approved of the tendency of replacing a tenancy contract by a wage-labour only contract, then it was forced to acknowledge that there had to be agricultural improvement in the reserves. 'When Government fixed the boundary', Carter put it,

> it prevented the Kikuyu from expanding in the manner which was natural to them, and . . . we have the duty of providing for their needs in some other way – not necessarily as a tribe, and not necessarily by the provision of a block of land, but by providing scope for their development in the way which appears most suitable for them.[139]

The colonial 'duty' of trusteeship now underwent a significant change. As late as 1930, the Lands Trust Ordinance, following the 1929 Native Land Tenure Committee report, had emphasised the 'protection' of nationalised 'tribal land' as the cornerstone of indigenous land and agrarian policy.[140] This policy, Carter reported, was at the expense of 'internal management, administration and private control over land'. Lands were not only to be protected but also to be 'developed':

> We prefer to think of a dynamic condition in which natives are themselves the principal developing agents by cultivating their own crops on their own land. . . . Our proposals throughout will be found to give far more prominence to private right-holding, and they are based on a frank recognition that

the tenure of land is becoming more individual, and that the problem of land for natives certainly cannot be met simply by the expedient of reserving areas to specific 'tribes' of the colony for ever.[141]

In proposing that there be a gradual move towards freehold title, the Carter Commission returned to what governor Northcote had promised in 1919 when chief Kinyanjui was informed: 'His excellency has predominantly in mind the desirability of Individual Tenure in the Kikuyu Reserve whereby every garden owner would ultimately receive a certificate of title thus ensuring security of tenure.'[142] Carter also foreshadowed how the irresolvable problem of settler agriculture was about to be replicated in agrarian policy towards an aspirant African class of indigenous capital.

The motivating force behind agricultural improvement in Central Kenya was what were then called the *athomi*. Literally 'the educated', the *athomi* were the first generation of mission-educated junior government officials, clerks, teachers and artisans who, from the 1910s on, in being paid salaries three to four times the average unskilled agricultural worker wage, possessed revenue to invest in agriculture and small-scale business. This was the historical origin of straddling between employment and private enterprise, an economic phenomenon not be confused with the sociological and empirical category of class according to occupation. Straddlers were no petit bourgeoisie in so far as the endeavour of the first period of straddling was an aspiration to accumulate property and capital.[143] Their restricted scale of enterprise and farm was until 1959–60 part of the predicament of colonial Kenya which both confined the scheduled large-farm areas to white settlers and restricted the types of jobs and enterprises in which Africans could be engaged. Racial prohibition and exclusion made this indigenous class of capital both anti-settler and anti-colonial but never anti-enterprise. When the early African claims for freehold title to land were made from the 1910s on, including what Kinyanjui had demanded of the colonial government, the claim was both that land title afforded protection against European settlement and secured the basis for individual accumulation through private property. From the official colonial perspective, it was because freehold title threatened indigenous accumulation that it had to be denied and that when all landholdings were compulsorily registered for the purpose of making titles generally available in the 1950s, the policy could only be implemented in the state of emergency and direct colonial rule which was imposed on Central Kenya upon the 1952–59 occasion of the Mau Mau revolt.

Although the Carter Commission had admitted that there was 'no evidence whatever to show that even the word Githaka, which means bush, had acquired any technical significance when the protectorate was declared',[144] it had justified its contemplation of individual freehold title on the grounds that recognition had to be given to the extent to which, particularly in Kiambu district, the subdivision or 'partition' of the 'ephemeral' *githaka* system of landholding had occurred. *Githaka*, in referring to landholding in the widest possible extent, had come to represent the communal holding of land upon a 'tribal' basis. Thus, the *githaka* was both to be employed to justify the colonial nationalisation of indigenous landholdings *and* the assertion of a general Kikuyu claim for Crown land which been appropriated for European settlement.

The highly significant category, when Carter wrote, was the *mbari*, the landholding category which was identified by lineage or sub-clan and within which a claim to land ownership could be made by virtue of genealogical descent from the common ancestor who had first created an estate of land. It was by right of first clearance and occupation that estate was created and thereby acquired as a right of private possession. Generational descent, through kinship affiliation, incorporated more and more persons into this estate of land in different discontinuous places. This estate was then perpetually subdivided – usually upon a principle of equipartible, patrilineal inheritance – for the purpose of household production.

Carter was concerned with subdivision of *mbari* land in that partition meant fragmentation. The latter was assumed to be inimical to agricultural improvement due to diseconomies of scale at the level of the fragment of the field and including the time it took to move from one fragment to another. Carter wrote that:

> While it is obvious that this practice of subdivision is calculated to aggravate the problems of local congestion, attempts to govern it by rule might result in an undue dislocation of population at a time when other agricultural right-holdings are hard to get, and there are no derivative industries in the reserve capable of absorbing any considerable number of the population.[145]

Carter's solution was to propose the consolidation of fragments of land while forgoing the registration of title for landholdings of existing individual households for fear that 'a landless class might be created prematurely'.[146] In the absence of administrative 'rule' the burden of enforcing consolidation would fall to the *muramati*, the guardian, custodian or trustee of what Carter called the *githaka* but was, in effect, the *mbari*. Here was the nub of the problem of indigenous trusteeship. By the late 1920s, the *mbari* had become significant because its guardian was not the *muramati* of old but the *muthomi* of the new period of agricultural improvement.

It was in the inter-war period, and especially from the mid-1920s, that Kikuyu country was struck by a 'land crisis'. Straddlers competed with each other to acquire land by reassembling the *mbari* as a new vehicle for the acquisition of landed property. Through spates of land litigation, the guardian of a *mbari* sought to redefine the breadth and depth of kinship charts which mapped out the ties of kinship affiliation to fragments of estate land. Litigation was the means to exclude individual households from a particular estate and/or enlarge the extent of one estate at the expense of another. *Athomi* became the guardians of estate land because they had the skills and money, through education and work, to fight court cases. The reward, which went to the guardian for protecting or extending an estate in the struggle for land, was such that an after-case settlement of land multiplied the share of a *muthomi* landholding within any given estate of land. Thus, intra-*mbari* litigation accelerated the process by which landed property was concentrated in the hands of acquisitive guardians.

There were two dimensions to guardianship. Warrior-raiders of the pre-colonial period, through the defence and acquisition of land and livestock in any given territory, engaged in primary accumulation on their own account. Indeed, it was from their ranks that the *athomi* had emerged. In the colonial integument, the new generation were regarded as warriors for land according to the older principle. The second dimension was trusteeship. Godfrey Muriuki has given the ideal form of control over acquisition:

> The *mbari* land tenure was a safeguard against exploitation by any one member of the clan, however strong or influential he might have been. Moreover, in a society in which communal solidarity was essential for survival, the welfare of the less fortunate members was ensured by the rest of the community. Anyone without land, for example, became a *muhoi* (tenant-at-will) on someone else's land, with the assurance that save for misconduct his tenancy would be secure.[147]

Tenancy-at-will, however, was an elementary part of the kinship affiliation which created the pattern of claims to right over the possession of land. In a typical land case 'A' would claim that his lineal descent from 'Y' gave him an exclusive right to the estate against the claim of 'B' whose descent from 'Z', which defined B's claim to part of the estate, was based on some evidential fact that Y had given tenancy to Z as a reward for his labour effort in clearing forest land. It would then be claimed that B's descent from Z was either unjustified and/or that B's labour effort was redundant as far as land clearance was concerned. The *ahoi* phenomenon, as it was called, was accentuated by the land crisis, itself the immediate cause of the threat of landlessness.

Athomi propensities for land acquisition were minimally speculative. Land was wanted for the expansion of commercial agriculture on landholdings which had been farmed, individually, for money-profit through the employment of wage-labour. The scale of such mixed-farm agriculture was necessarily limited by the extent of land within the 'reserve' and the intra-reserve competition for land which was revealed by the land crisis. Early agricultural improvement was spontaneous and devoid of any plan by colonial administration for development. Wattle exemplified the process whereby an activity, in this case a tree yielding the joint product of bark and a log, was first introduced by improving farmers for the express purpose of making money-profit from the sale of products. Tree planting accompanied the land crisis to put land, subject to litigation, under immediate cultivation. The original principle of individual land possession, namely that labour effort was fixed in land, was thereby reproduced. There was little which would have forestalled the process of land acquisition for profit-motivated improving agriculture, and thus the development of capitalism in Kikuyu country, other than the development plans of the colonial administration.[148]

State intervention had followed when, as in the case of wattle, a spontaneous expansion of production increased the supply of an unregulated Kenya product and thus threatened the confines of an internationally regulated market. By inducing the expansion of wattle production on all landholdings, the colonial administration attempted to exercise control over the amount, quality and price of the output which would be realised for sale. As each landholding, in principle, was ensured of a money-income from the sale of a product, landholdings within each *mbari* were given an assurance of survival as economic units of production. The engrossing of land, which might have attained the scale reached during the land crisis, with the majority of landholding households dispossessed of land, was restrained. However, before 1945, state schemes of production, despite their ambiguous intention, acted to sequester indigenous capitalism in the labour 'reserves' of small family farms.[149]

The same process followed the post-1945 removals of racial prohibitions on the African production of tea, coffee and improved dairy cattle enterprise. In each case, commodity production started as an autonomous drive to make money-profit out of agricultural improvement and resulted, for the preponderance of small farms in Kikuyu country, in development schemes which purposefully served to reproduce subsistence for middle-peasant producers. As in the case of wattle, the control behind the schemes met conflict from the political organisation of those who aspired to accumulate capital from landed property. As before 1945, there was ambiguity in the intention to develop the agrarian capacity of relatively high potential land. After 1960, if not before, and in spite of the Mau Mau revolt, official ambiguity was aggravated by both the growth of a relative surplus population and the official inability to confine improvement to an ideal minority of yeoman or 'progressive farmers'. The Central Province senior agricultural officer wrote in 1948 that while

> a landless class is springing up, it is not really to be regretted as it is the poor and ignorant who ruin the lands and are the least amenable to instruction. Better that these people should seek a living elsewhere or learn to work for the bigger land owner for wages.[150]

Reflecting on the attitude with which he undertook the 1954–1959 Swynnerton Plan for agricultural improvement, a plan which accompanied the enforced consolidation and registration of freehold title for all Central Province land-holdings during the emergency, the Nyeri District agricultural officer G. Yates wrote:

> I took the attitude that there was going to be inequality, there was going to be a certain amount of skulduggery, for want of a better word.... Then the benefits which would accrue to virtually the majority of the population were so great that I adopted the attitude that, let us get it done, let us get on with the farm planning and let's see the thing to a situation where we could see cash crops and coffee and pyrethrum and tea flowing out of these districts.[151]

Yates' adoption of a resigned attitude may be easily understood. When former Rift Valley farm workers, now unemployed on their holdings or in emergency villages, wanted the new enterprises on their consolidated farms to reproduce subsistence there was little that could be done to achieve the intention of agrarian reform which had been to:

> create a prosperous class of contented and cooperative African landlord farmers who will have everything to lose in the face of political disturbance. It is really the policy of developing a contented and loyal middle class of African[s] as a barrier against the African extremists such as that of S. Rhodesia. In the case of Kenya the land reforms are the colony's insurance against a repetition of Mau Mau.[152]

What was developed instead, until the 1980s at any rate, was a majority of middle-peasant family farms. This was achieved through aid programmes carried out on land of high quality. As the availability of good land for these schemes lessened, they became increasingly difficult to replicate on land of lower agricultural potential. These programmes involved a deliberate subsidy for small-farm agriculture, a subsidy which would be repaid in foreign exchange earnings through the export of relatively high-quality commodities. To achieve high-quality export produce, output was controlled through state schemes. Thus family farms were 'independent' only in so far as the family farm regulated its own disposition of labour effort.

We have seen that the acquisitive propensities for land, on the part of the aspiring class of capital, neither 'loyal' nor 'content', were displaced into the large-farm space of the Rift Valley. Through the conglomeration of enterprises, itself an attribute of straddling, small-farm occupation of high-potential land was accommodated by the internal market which was expanded by the agrarian schemes. Disposable incomes of farm families specialising in relatively labour-intensive but high-value activities was expended on food and services produced by indigenous enterprise. Food, private health, school and transport expenses made up the bulk of household expenditure.[153] This internal market spawned the multiplicity of enterprises – from micro-buses to private health centres – owned by a new generation of straddlers amidst spirals of competitive conflict between different layers of concentrated property.[154] Such enterprises may have generated wage-employment, as indeed did self-employment in trade, but they were a far cry from the high-wage policy of early African socialism.

Given that agrarian innovation has not been a continuous process of replacing lower by higher land-value activities to absorb the rising population within each family farm, there was no illusion on the part of the family that underemployment on the farm could be eliminated by more intensive labour effort. Without an increase in agrarian subsidies on a scale which would have to, but which could not, correspond to the conjuncture of the 1950s and 1960s, underemployment on the farm has been displaced off the farm.

Dalgleish hoped that the pre-1960 experience of unemployment would not be repeated, but certified 'job-seekers' who could not find employment still made their way into the residuum of the informal sector. Such was the experience of the 1930s to the 1950s, as recounted by Kenneth King, of artisanship. It is an experience which was made more emphatic, as Kitching is right to emphasise, after 1960.[155] Underemployment off the farm was only at one remove from the grinding life of poverty of on-farm employment. No imagined distinction between a benign 'intermediate sector' of relatively prosperous artisans or traders and a 'community of the poor'[156] can disguise the fallacy inherent in the belief that full development could be achieved upon the premise of doctrines of development which aspired to put the burden of subsistence directly upon 'independence of the poor'.

MANUFACTURING INDUSTRY, COMMERCE AND SMALL ENTERPRISE

Full development, whether for Dalgleish in 1960 or for Carter, who in 1933 referred to 'derivative industries' in the 'reserve', encompassed the development of manufacturing industry. Yet, it is a remarkable fact of post-colonial policy that the thrust of national development has not pivoted upon industrial development. Despite the failure to articulate an industrial policy, it has been ever present as a derivative of what have appeared to be, at any given time, more pressing official priorities. Thus, industry has been subjugated to priority of agricultural growth; or the trajectory of indigenous accumulation; or the need for state revenue; or the compulsion to absorb the relative surplus population. Nor was it different during the colonial period. Immediately post-war, when unofficials in the Legislative Council pressed for industrial development and when there was little challenge to the official desire to foster industrial development, 'positive pressures to shape and direct' industrial development, as Michael Mcwilliam observed, 'were lacking'. Even 'unofficial opinion', Mcwilliam continued, 'tended to welcome industrial development not so much for its own sake, but as a palliative to absorb surplus rural populations'.[157] As such, when policies for industrial development have been promulgated, the stress has consistently been on the hoped for derivatives of rural industries and not industry in general.

This argument may seem surprising in view of the repeated claim that it has been the magnitude of the post-colonial subsidy to manufacturing industry which has been at the core of the inefficiency, distortion and incapacity of the Kenya economy. Manufacturing industry, so it is argued, has passed through the consumer-good phase of import substituting industrialisation (ISI) but cannot compete in international or even regional export markets.[158] It has been estimated, from 1964 to 1984, that 70 per cent of manufacturing output growth came from the increase in internal demand within the economy and one-third of this increase was attributed to the three sectors of food, tobacco and beverages. Since only 5 per cent of output growth came from exports and the remaining quarter of output growth was explained by ISI upon the assumption that were there to have been no growth in internal demand, local industrial production would have increased to substitute for imports. Overall, industrial production has expanded by meeting the growth in domestic demand within the incubus of protection which, the argument concludes, afforded an implicit subsidy to capital invested in manufacturing at the expense of agriculture.[159]

An implicit subsidy, arising out of protection, is not necessarily determined, as we saw in the Canadian case of Chapter 4, by an industrial strategy of development. In the Kenya case, so the Sharpley–Lewis evidence shows, the state revenue-raising element of effective protection is more significant than that of nominal industrial protection. Tariffs have cascaded, until the late 1980s at any rate, as the source of protection has moved from final consumption to intermediate and capital goods within the production process. In other words, the state recoups revenue lost from tariffs on final goods, which are no longer imported as ISI proceeds, to intermediates which continue to be imported by industrial firms. Protection is thereby deemed to be inefficient for the economy because firms recoup increased costs of production, due to tariffs on imported inputs, by increasing the prices of final goods output. Effective protection may be positively related to nominal protection when, as the Sharpley–Lewis evidence indicates for clothing, textiles and steel re-rolling, 'tariffs on intermediate inputs raise the costs to protected industries by less than the tariffs on their output raise the final selling prices of their goods'.[160] However, for agro-processing industries which are afforded little nominal protection, tariffs on intermediates has resulted in negative effective protection.

Proponents of industrial protection, from Friedrich List onwards, argued that higher productivity, and therefore more national capacity for international competition, would compensate for protection's initial costs of higher output prices and lower real wages than would otherwise be the case. Kenya is the case where List's premise has not been fulfilled because there has been neither the policy nor explicit subsidy of the magnitude which could have been expected for large-scale industrial development.

David Himbara, in his recent thesis on the development of capitalism in Kenya, has concluded that it was in the late colonial period that industrial policy was evident: 'In the period from the Second World War to 1963 when the Kenya state pursued more consistent policies that encouraged both the domestic and foreign capital to invest in Kenya, a spectacular industrial and commercial development took place in a remarkably short time.'[161] Industrialisation in Kenya started as a pre-war phenomenon, partly of the associated control, through cooperation, of estate producers over the processing of agricultural products and partly that of indigenous capital of Asian origin which moved from trade into industry to manufacture consumer and intermediate goods. It was from the 1950s on, as was true elsewhere in sub-Saharan Africa, that the subsidiaries of multinational companies came to share a place in import substituting manufacturing with local Asian manufacturers. This process of industrial investment, however, was not directed by state policy and certainly not of the 'late industrialisation' kind, towards which Himbara beckons as being the proper state developmental model and which we examine in Chapter 7. Indeed, what Himbara refers to as 'industrialisation taking place by default'[162] after 1963 was equally what made industrial development a spontaneous process before political independence.

Himbara quotes Sir Ernest Vasey, the Kenyan minister of finance who attempted to construct industrial development during the 1950s:

> Kenya's real history of economic development . . . seems to have been, for many of those early years, a development through accident rather than design. And by that I mean that the channels . . . of our social and economic development were not consciously planned, but evolved from the circumstances of the day.[163]

Despite his intention and effort, Vasey's own words might well have applied to his own period of development. 'The truth', Mcwilliam explained, 'was that during the period up to 1960 private capital for large-scale developments was generally forthcoming':

> There was no need either for a special development agency supported by official funds, or for special industrial promotion policies beyond keeping the rate of company taxation moderate and giving generous tax treatment to capital expenditure. A more lively development corporation might indeed have found interesting opportunities; but its presence was not a precondition of industrial development.[164]

In so far as there was development effort at work, and the effort which Himbara emphasises as being perpetuated into the post-colonial period after 1963, it was directed at what Mcwilliam described as being 'behind this first wave of Kenya tycoons', namely 'a vigorous jostling' which 'could already be discerned amongst emergent African businessmen, in trade, transport, building'.[165] Himbara's thesis is that the failure of colonial government effort to develop African entrepreneurship in industrial production was replicated by the experience of post-colonial state effort to promote indigenous African enterprise, especially after 1966. As such, the enervation of African enterprise in industry only highlighted Asian capital as the authentic, indigenous and vigorous source of industrial development.

When the claim is made, usually but not always by innuendo, that manufacturing constitutes a 'foreign' enclave in the economy in opposition to the 'authentic' core of agriculture and that it is this alien command over the enclave which has distorted the development of the economy, some well-worn tropes are elided in such a fashion that the intention to develop is conflated with the process of development itself. Himbara is right to insist that it is only by a racial sleight of hand that 'Asian capital' is deemed to be foreign.[166] Industrialisation 'by default' occurred through the investment of money-profit, gained from trade accompanying the expansion of commercial agriculture, in manufacturing industry. Commercial banks, which made over 60 per cent of all loans during the 1970s and 1980s[167] and whose lending practices had been oriented to trade as Table 6.1b shows, facilitated the transformation of money into industrial capital. After 1970, the profile of commercial bank lending increasingly shifted away from trade, and towards other sectors of the economy, including manufacturing. Absorbing a larger share of domestic credit expansion than was warranted by its share of measured value added in the economy, manufacturing was where the banks' foremost clients were Kenya Asian firms who could both be expected to invest profitably and who relied most on the commercial banks for financing investment.

This was a process of capitalist development which continued in spite of official discrimination against 'Asians' in trade and the myriad of instruments to promote the development of African business. It was not only that the intent to develop African business accelerated the movement of local Asian capital from trade to manufacturing. During the 1980s, disinvestment by international firms, largely through the sale of subsidiaries to predominantly Asian-owned firms, increased the extent of Kenyan Asian ownership, as Himbara's 1990 data shows, to three-quarters of all large-scale manufacturing firms.[168] It is the paucity of purely African-owned enterprise in large-scale manufacturing, little more than 5 per cent in 1990, three decades after official policy was directed towards Africanisation of business, which Himbara attributes to the perverse practice of development doctrine in Kenya. In support of his thesis, Himbara quotes D.J. Penwill who, when a district officer in Nyeri, wrote in 1950 of the Kikuyu 'vocation' and 'enthusiasm for trade' which had 'in no way abated' despite 'innumerable failures':

> The question now arises as to whether the Government, in its capacity as trustee for the emergent communities, should attempt to temper the wind of normal economics to give Africans some protection and assistance while they gain experience and understanding of economics and commerce. We can already see the results of a policy of laisser-faire. Much African capital and endeavour will be wasted . . .[169]

To give 'some protection and assistance' to African business, on the basis of trusteeship, was a two-pronged policy. African business, as business to secure means of production, was to be protected against Kenyan Asian business

Table 6.1a Government recurrent and development expenditure on agriculture, manufacturing and roads, 1963–90

Year[a]	*Agriculture*[b]	*Manufacturing*[c]	*Roads*	*Total* K£m *Current*	*Total* K£m *Constant*[d]
	(*per cent*)			(*1972 = 100*)	
1963–66	18.5	0.3	5.8	70	77
1967–70	12.5	3.6	7.8	102	117
1971–74	9.0	4.1	12.0	192	181
1975–78	11.2	2.4	6.9	420	276
1979–82	7.3	2.2	6.0	948	460
1983–86	7.8	2.6	4.4	1396	393
1987–90	8.6	2.3	2.5	2663	467
1963–90	10.7	2.5	6.5		

Notes
[a] Annual average of financial years within each period.
[b] Includes forestry; excludes land.
[c] Commerce and industry until 1974/5; mines, manufacturing and construction thereafter.
[d] We have deflated total government expenditure by a labour cost index derived from splicing the Cowen–Newman real wage index for artisans, 1963–71, into the official 1972–90 labour cost index for the construction sector. See Cowen and Newman, 1975; Kenya Republic, 1975, Table 118, 1978, Table 149, 1983, Table 115, 1986, Table 116, 1990, Table 116, 1991, Table 116.

Source: Kenya Republic, 1968, Table 172, 1971, Table 188, 1974, Table 207, 1979, Table 230, 1987, Table 199, 1991, Table 198.

enterprise which developed according to the 'normal' process of development. Equally, African business, as the means to secure subsistence through trade and non-agricultural production, was to be protected from itself. The 'vigorous jostling' for trade created, so ran the official complaint, too many traders for any given volume and velocity of business. It was the surfeit and not the deficiency of competition which wasted both capital and productive effort. This was the waste of unemployment which the doctrine of development aspired to ameliorate and which accompanied the trusteeship for African business as business. Development doctrine was not so much perverse as overburdened.

Agrarian schemes of household production, as we saw from the case of the Central Province, were developed by parastatal authorities to both secure the basis for employment, as an alternative to direct wage-employment, and the productive base of the economy. Central government expenditure, as Table 6.1a indicates, was biased towards agriculture, in that it absorbed four times more than what was directly allocated to manufacturing by way of state services. Given that security,

Table 6.1b Commercial bank loans and advances to agriculture, manufacturing and trade, 1963–90

Year[a]	*Agriculture*[b]	*Manufacturing*	*Trade*	*Total*[c] K£m	
		(per cent)		*Current*	*Constant*[d]
				(1972 = 100)	
1963–66	13.8	13.2	47.8	47	52
1967–70	13.3	20.8	43.2	66	76
1971–74	12.5	23.4	34.1	126	120
1975–78	14.6	18.5	22.4	324	213
1979–82	18.0	24.6	21.2	577	280
1983–86	15.3	22.0	23.2	928	282
1987–90	15.8	23.0	18.0	1670	293
1963–90	15.0	22.0	28.0		

Notes
[a] Annual average profile as at December of each year.
[b] Includes forestry, fishing, wildlife.
[c] Includes government; excludes private households and deflated as for Table 6.1a.
[d] 1963 and 1964 estimated since data included bills in exchange.

Source: Kenya Republic, 1967, Table 115, 1968, Table 118, 1970, Table 122, 1975, Table 138, 1979, Table 160, 1984, Table 131, 1986, Table 132, 1991, Table 132.

defence and general administrative expenditure typically absorbed one-third of the state budget and education and health another third over the 1963–90 period, government spending on agriculture, as it is so often claimed, was neither deficient nor overstated.[170] Indeed, much multi-purpose expenditure, such as on roads, served the purpose of production and accumulation from agriculture. Apart from the 1967–74 period, when the Kenyatta government made its most direct effort to promote African enterprise, the direct and explicit subsidy to manufacturing was overshadowed by agriculture. Subsidies to manufacturing came from a different source, the dual burden of trusteeship which was inherited from the colonial government and directed by the premise that African business meant both securing means of production for African enterprise and employment in parastatal enterprise.

Development banks, rather than commercial banks and external finance, through aid and loans, rather than government spending, were the means by which productive enterprise was to be developed to secure African business and wage-employment. To take the first of three examples, the *Industrial and Commercial*

Development Corporation, started modestly by the colonial government in 1954 as the Industrial Development Corporation, was turned into the instrument to finance indigenous African business to acquire firms from non-citizen Asian entrepreneurs and compete with Kenyan Asian enterprise. Jointly established and owned by the ICDC and external agencies, such as the Commonwealth Development Corporation, the *Development Finance Corporation of Kenya* was set up in 1963 to extend investment in manufacturing and other sectors. Although established in 1973 and owned by the Kenya government in association with a state bank and insurance company, the *Industrial Development Bank* advanced medium- and long-term loans, and also underwrote securities and owned equity in large-scale new enterprise upon lines of credit from the World Bank, external development corporations, and merchant banks.[171] By 1990, these three development banks were involved in the indirect ownership of 100 out of 255 parastatal enterprises. In accounting for 10 per cent of total wage-employment in the economy and 11.5 per cent of the value added share of national income,[172] parastatal enterprises have borne the brunt of the criticism that wage-employment and African-owned business have been implicitly subsidised and that state development doctrine can be deemed to be perverse.

Criticism of parastatal enterprise is twofold. For Himbara, the management of state enterprise represented the transfer of 'public service' mentality into business but this mentality was transmuted, as we saw above, into the private opportunity for regarding state enterprise as a public source of primary accumulation. Enterprise was denuded of productive function, the World Bank has asserted, 'because the state proved to be an uninspired entrepreneur and a bad manager'.[173] When, as in Himbara's thesis, the first is elided into the second, the fundamental weight of criticism is that African entrepreneurship, despite four decades of state assistance after 1950, continued to lack 'vision, management and the ability to conceive and execute investment plans'; or, as one of Himbara's informants told him, 'African business men do not grasp the concept of money and how to transform it into capital.' Conversely, the 'commercial skills' of Kenyan Asian entrepreneurs were evidenced by 'sheer determination and hard work; their vision of the potential mass market; their general efficiency and competitive edge' all buttressed by community organisation and family 'in providing mechanisms to engender discipline and cohesion'.[174] This dichotomy, whereby the Schumpeterian 'factors' of African and Asian entrepreneurship are the inverse of each other, is nothing other than a barely disguised inversion of the intent and the process of development.

What is normally attributed to the intent to develop, its positive purpose in confronting the negative of the process of development, has been inverted so that the intent to develop African business negates the potential to develop further what Kenyan Asian capital has achieved through a normal process of development. In other words, it is the intent to develop what cannot be developed artificially, the recurrent reconstructions of African business, which arrests the natural course of capitalist development. There is neither a normal nor natural course of capitalist development that can be treated, and predicted to be, a sequence of effects which are determined by the causes of phenomena. Virtuous factors of Asian entrepreneurship may be as much the effects of a process of development as their causes. It was the effect of colonial government policy, due to the development of European estate agriculture, which shut out Asian immigrants from land ownership during the colonial period. Likewise, and as Himbara points out, the exclusion of Kenya Asians from the high politics of state power during the colonial period reinforced commitment to business and the transformation of money into industrial capital.[175] Artificial exclusions of state policy were part of the course of development.

Contrawise, the premise behind the intent to development cannot be refuted on the ground that there was neither will nor vigour in African business including aspirations to engage in large-scale manufacturing. On the contrary, as we saw from 1950, the official complaint has been that there is too much vigour behind the pursuit of money-profit. If, as Himbara has argued, it is the capacity for business which has to be developed and that this capacity has been tested in large-scale industrial production and failed to develop as part of the normal process of capitalist competition, then we must be clear what criteria are involved in the tests of development. A relentless search for money-profit, and not some ideal of national development, is what made indigenous capital develop in Kenya. National development, as we have suggested, has been the means to compensate for the consequences of the process of development.

We saw earlier that state power extended merchant and commercial capital opportunities to African enterprise on the basis of private agency. State agencies of development, from development banks to parastatal enterprise, were funded to pursue a course of national development but the social 'waste' of such external funding was the means by which money could be accumulated for the pursuit of private enterprise without any compulsion to engage in manufacturing industry. As such, there has been no state interest in trusteeship for large-scale manufacturing. On the contrary, Himbara reinforces the evidence showing how the protective apparatus for ISI enterprise is perpetually undermined by propensities for state agents to make private gains from enabling the imports of consumer and intermediate goods which compete with the very industries for which they are entrusted. National development becomes perverse only according to its own ideal. State agency of the intent to develop, as we have suggested, derives from the results of a process of capitalist development and not the other way around.

Development doctrine was not the impetus behind the trajectory of indigenous industrial development. Absorption of labour power into manufacturing industry was neither driven by need for international competitiveness nor, apart from limited industrialisation during the Second World War, from planned postulates of autarchy. Rather, spontaneous change in the structure of industry, following change and growth in internal and regional demand, has been the main source of evident increases in average labour productivity.[176] Over the two decades between 1965 and 1985 the share of manufacturing in the gross value of GDP increased from 11 to 13 per cent. Large-scale manufacturing's share of the enumerated wage-labour force increased from 5 to 7 per cent over the same period and since 1985 these two proportions have remained constant.[177] It is this gap between the proportionate growth of output and that of measured labour input in production which gives rise to the argument that the course of industrial development in Kenya has aggravated the employment gap and, with few exceptions, is responsible for general bias against large-scale manufacturing industry.[178] There may be dispute over the precise source of the cause of evident increases in average labour productivity, which accounts for the employment gap, but comparisons between subsidiaries of multinationals and local predominantly Kenya Asian-owned industrial firms fail to find any significant systematic difference in industrial performance. Nor, to make the same point, is systematic difference to be found between the industrial performance between expatriate and Kenya African management of parastatal enterprises when the enterprises in question conform to the same set of commercial objectives in the same industrial sector.[179] From the standpoint of commercial criteria for profit-making enterprises, the intention to create employment-maximising conditions of production belongs to development doctrine and not to the intention to develop capitalist enterprise.

SMALL ENTERPRISE

It is for the reason of promoting employment that a chorus of opinion, from contrary sources, directs the trusteeship of development towards the informal sector of 'micro-enterprise' embracing the gamut of urban-based *jua kali* sector of small-scale manufacture and artisanship, street trading and rural non-agricultural activity of the same kind. Government ministers and official reports, as we have reported, find no solution to their problem of the exponentially explosive employment gap other than to find recourse in micro-enterprise. Radical-liberal economists and other academic advisers, who might otherwise have no truck with authoritarian state regimes, either motivate and/or find a marked degree of consensus with official opinion. Again, it is in *jua kali* that the dual burden of trusteeship rears its head.

Jua kali, King and Abuodha asserted in their panegyric for small enterprise trusteeship, has 'broadened to stand not just for a particular form of microenterprise, but for a Kenyan African version of capital accumulation to be contrasted with that of the multinationals or Kenya Asians'. To drive the point home, Himbara's prognosis was that the potential for the development of African business enterprise is to be found in the *jua kali* sector.[180] Any ideal of national development is reduced, therefore, towards directing African enterprise towards the arena of the economy in which development doctrine can fulfil its role of ameliorating unemployment. This arena of self-employed enterprise and casual work is one of unremitting poverty. From the standpoint of constructivist development doctrine, it is tantamount to a system of outdoor relief.

That there has been a phenomenal growth in the number of self-employed artisans is not open to dispute.[181] Nor would we contest the evidence, as presented by King and Abuodha, about the extent to which artisan production has become increasingly complex and the quality of output 'improved tremendously' for metalwork and woodwork since the mid-1980s.[182] What we do dispute, however, is that there is some essential virtue that can be constructed out of the dire necessity for a relative surplus population to find subsistence out of self-employed enterprise. All the constraints, from markets for products to assets for production, that inhibited the absorption of potential productive force in the full development of commercialised agriculture during the high age of state development, are likely to be reproduced in the more 'competitive environment' of small-scale enterprise. A particular focus of development doctrine is the manifold of small enterprise belonging to the category of rural non-farm activity (RNA).

Evidence that doctrine makes a virtue out of dire necessity is not difficult to find. A 1977 Central Province survey of 850 enterprises showed that virtually none of the presumed virtues of RNA were achieved in practice. Land and cash constraints which prevented the desired expansion of farm production by family farms were replicated as equipment and cash constraints upon RNA enterprise. There was a marked differentiation in RNA activity, a difference between a small number of growing enterprises and a large number of owner-operated enterprises which followed patterns of differentiation in agriculture, namely the differences between small- and intermediate- and large-scale farming:

> While it appears that the rural non-farm sector provides low paying employment opportunities, and produces low cost goods and services to the poor, it does not offer universal opportunities for economic advancement. On the contrary, the more profitable activities have quite sizeable barriers to entry, require fairly substantial capital inputs, are less than perfectly competitive, and do not rely heavily on indigenous resources.[183]

Despite this warning from the province where conditions were most favourable for RNA, small-enterprise continued to be advocated as the means to absorb population in productive activity in the countryside.

Livingstone had commented on the 1977 survey by repeating that there was less than an average of two workers per enterprise, and then suggested that 'very small informal enterprise – as opposed to small factories – should be seen firstly as a major form of *employment* rather than as a type of enterprise which will develop and grow over time'. When King and Abuodha criticised Livingstone for later reinforcing the impression that the generality of *jua kali* enterprise was necessarily confined to the simple machine production of '"appropriate" goods' demanded by the poor at 'a very low supply price', their argument was directed at the extent to which potential for the technological development of enterprise was achieved during the 1980s.[184] Yet, the King–Abuodha evidence, from their two sites of production in Nairobi and Kiambu district, qualifies the potential which was evident in the official reason why development doctrine was to be oriented towards *jua kali*. 'One indication that spontaneous growth cannot adequately develop the sector', the 1992 sessional paper complained, 'is that only very few of the small enterprises have actually been able to graduate into the formal sector'. Or, in suggesting that 'the number of Kenyan manufacturing firms employing 10–50 persons is relatively small',[185] the official argument was that very few enterprises had become intermediate or large-scale industrial forms of capitalist enterprise.

Not only does the King–Abuodha evidence confirm the official complaint but the weight of their argument is directed towards showing the extent to which *jua kali* activity is highly differentiated. Two of 100 enterprises, enumerated in the 1990 King–Abuodha survey replicating an earlier 1977 one by King, had become small factories employing less than 80 and 50 workers in woodwork and metal work production. Since tailoring and candlework were overwhelmingly self-employed activities, the average level of wage-employment was four workers per enterprise.[186] Differentiation is about the kind of activity, the technology to which it is attached, and the extent to which *jua kali* is a means of survival for subsistence as much as a means for the development of productive force directed by African indigenous-owned capital. After all, the state and quasi-official effort to promote self-employment, especially through RNA, is directed towards the claim that self-employed enterprise, 'in many cases' realises more income than wage-employment. To examine this claim, and given that there is no differentiated concept of profit in such surveys to account for the phenomenon of capitalist enterprise, we have to rely upon the undifferentiated concept of income.

During 1985, at the alleged onset of the explosion in *jua kali* and when the official focus of development was on RNA, an official national survey of RNA was undertaken. In giving an average of three workers per enterprise, the 1985 survey confirmed that the bulk of RNA manufacturing enterprise was very small. Bearing in mind the caveat about the concept of income, income for the average enterprise, *including* 'operating surplus' and 'labour costs', accounting for less than 10 per cent of average gross output value, for all 10,000 enumerated RNA manufacturing and artisan enterprises, was K£54 per month. A different quasi-official 1985 survey on urban-based small manufacturing enterprise in Nairobi, Mombasa and Kisumu estimated *median* earnings at K£66 per month.[187] In the same year, average, undifferentiated wage and salary earnings in 'formal sector' large-scale manufacturing enterprises, without any concept of profit included in the estimate, was double the RNA average at K£100 per month.[188]

A later survey of 130 mainly manufacturing enterprises in western Kenya, reported in 1989, showed a bi-modal distribution of income, confirming what King and Abuodha reported in their 1989–90 survey.[189] After the 1985–90 period of hyperinflation, the King–Abuodha survey revealed an average monthly income of K£600 for the 'rural' Kiambu area and K£820 for all enterprises, heavily weighted towards Nairobi. Since the then average, undifferentiated large-scale manufacturing money wage was K£190, it is clear that small enterprise, as an enterprise, gained an appreciable advantage over the average industrial wage but there was marked variation in the King–Abuodha enterprise accounts. Metal and woodwork sectors realised an average of K£1,120, with a marked variation in the range of enterprise incomes. Candlework, at the other extreme, with far less variation in the range, realised an average of just more than half the industrial wage at K£100 per month.[190] This income, representing the margins of hardship for self-employed artisans in the setting of Nairobi, is what corresponds to the mainstream poverty of RNA. It is in the particular, of the 'many cases' of self-employment, as evidenced by King and Abuodha, and not in its overall scope, that *jua kali* represents potential for improvement in standards of living.

Systematic state assistance to small-scale industry, to develop 'African entrepeneurship through the establishment of nursery-type industrial estates', started in 1967 with the formation of Kenya Industrial Estates (KIE), a wholly-owned subsidiary of the ICDC. By 1994, 6,000 enterprises were supported through the provision of factory space, technological and economic extension work and loan finance. In 1983, a World Bank report repeated KIE's reports in bemoaning the failed development of industrial entrepreneurship: 'Most "entrepreneurs" normally come from the ranks of salaried employees in public or private concerns or are self-employed in some successful venture and are not willing to take in a new activity as a full-time undertaking.' Industrial enterprise, the report continued, was mainly 'the outcome of a process of project identification and development by KIE staff'.[191] In other words, the industrial scheme was an analogue to the agricultural scheme which was developed by state management and control.

Jua kali, it is supposed by its academic proponents, differs from industrial enterprise in the industrial scheme because enterprise is not an element in the process of straddling but a full-time preoccupation. If the presupposition for authentic industrial development is the unequivocal commitment to manufacturing activity which follows from the autonomous entrepreneurial development of enterprise, then the proponents of trusteeship for *jua kali* face a predicament.

Evaluation of KIE enterprise in 1983 is markedly similar to what the 1990 King–Abuodha survey revealed for *jua kali*: 'High profit margins coexist with high rates of business failure.'[192] A decade later, and after the general policy of industrial liberalisation, while KIE reported that 70 per cent of its clients were 'doing well, industry sources contend that nearly 50 per cent have collapsed already or are on the verge of doing so'. It was also reported that manufacturing commitment was undermined by 'governmental corruption' and 'an element of political favouritism in the granting of loans, and hence the half-hearted attempt to make the enterprises really viable'.[193] Contrawise, it can be argued, self-employed *jua kali* enterprises do not collapse because they do not face a profitability constraint while wage-employing entreprises, devoid of credit liability, political favouritism and general industrial supervision, survive both because they are not supervised and because they are committed to manufacturing activity. *Jua kali* may substitute for both large-scale and KIE enterprise during the period of liberalisation which followed the policy of eliminating industrial protectionism but it cannot be autonomous of supervision if it is to be the authentic source of industrial development.

Small enterprise, with the capacity for technological innovation to compete with manufactured imports, must be supervised. Sources of supervision may be disputed but the principle of supervision is nothing other than what was always entailed by the trusteeship of development. The World Bank 1983 report praised KIE supervisors who 'exhibited a degree of energy and attention to practical detail unique to public agencies' but then denounced their focus 'on a relatively small clientele', 'administrative overstaffing' and a 'dearth' of indigenous expertise in 'critical technical areas'.[194] To widen the extent of supervision, reform the system of loan finance and, especially, 'upgrade technology' formed the basis of the World Bank's 1983 policy recommendation. All this was equally emphasised by King and Abuodha, as we mention below, for their trusteeship of *jua kali*. KIE's managing director proposed in 1994: 'We need to link the small manufacturer to the big manufacturer through subcontracting. This is the only way the small manufacturer will be assured of a market and constant advice on technology.' What was proposed was what had always been the hallmark of small enterprise, namely its subcontracting to, and supervision by, the most concentrated form of property in the centralised capital of big business.

Irrespective of whether it is based in city, town or countryside, *jua kali* is a conceptual subset of RNA activity. In 1985, according to the official national survey, manufacturing accounted for 30 per cent of the 100,000 enterprises enumerated in all RNA activity. Moreover, the number engaged in mainly small-enterprise production was then less than one-fifth the number of workers employed in large-scale manufacturing. Self-employment generally, like RNA itself, is predominantly trading activity. Distribution and services enterprise, and one-half of 70,000 so engaged in 30,000 enterprises were in the three activities of wholesale, retail and restaurant trade, realised an average monthly income of K£130 in 1985 and far more than the corresponding income from manufacturing enterprise.[195] Enumerated trading enterprise, albeit not wage-workers employed in the enterprise, appeared to realise relatively higher average revenues than did artisan work.

Official national surveys of household activity confirm what was apparent from the enterprise-based surveys. Data from Integrated Rural Surveys (IRS) of the 1970s show that the contribution of RNA to total household income was then relatively slight. One-half of households had no RNA; 40 per cent had one or two non-farm activities. A later 1988–89 *Rural Labour Force Survey* indicated that only one-fifth of all average labour time was spent in RNA; the proportion for men was no more than one-third of total labour time expenditure. IRS data showed some variation between provinces, as did the 1985 enterprise-based survey, but it is clear that the intra-provincial variation in RNA participation is far greater than the degree of variation between provinces.[196] RNA is poorly renumerated work for households which do not have access to relatively stable per capita cash revenues from farming activity and which do not have access to remittances from individual members of their households who are engaged in relatively well paid wage-work.

A number of local studies, which were undertaken in different ecological zones of western Kenya during the 1980s, reported that off-farm income was 'significant' or 'highly significant' for total household income. Hebnick's survey of Nandi district, relatively well-endowed for farming activity, showed that more than one-half of mean total household income came from off-farm sources; the same proportions were reported for two out of the three villages surveyed in Lavrijsen's Bungoma study; the third village showed RNA realising one-quarter of mean total income.[197] Other surveys in western Kenya, by Francis and Hoddincott, show that off-farm proportions of household income are far higher than one-half. Orvis' detailed sample, for Kisii district, showed that off-farm income amounted to 80 per cent of the household income of the wealthier small-farm owners, the 'pioneers' of tea and coffee production in the area; the same proportion applied to a group of local wage-labourers' households. It was only for 'farmers', or middle-peasant households, that agriculture provided over one-half of total household income. Last, but not least, the Paterson survey, for Kakamega district, revealed a similar trend: agriculture provided about one-half of household income for the smallest (up to 2 hectares) and largest (over 5 hectares) holdings; for holdings in the middle of the distribution (2–5 hectares), agriculture contributed as much as 70 per cent of total household income.[198]

However, a closer look at the western Kenya household studies reveals that a substantial proportion of off-farm income was in the form of wages from public sector employment and remittances from labour migration and circulation. In Bungoma, the proportion was one-half; in Nandi, only 40 per cent of off-farm income was from RNA.[199] Other surveys give similar results. Francis reported that the average two-thirds of household accounted for by off-farm income was shared roughly equally between RNA and remittances. For the poorest households, remittances and RNA were substituted for each other as combinatory but unreliable sources of income. Where the development of commercial agriculture had regressed in recent decades, as in the case of Koguta, Francis reported that returns from RNA enterprise had 'been held down, *inter alia*, by low spending power, competition, lack of working capital and the demands of relatives'.

Hoddicutt's case showed that wage income alone accounted for half of average household income and difference in wage income was the foremost source of generating income inequality between farm households.[200]

There is a world of difference between small-farms maintained by executives, managers and professionals, at one extreme, and households which are dependent upon remittances from wages paid by absent members in unskilled and semi-skilled wage-work at the other. Whereas average wages/salaries (excluding non-salaried renumeration) of the executives were typically up to twentyfold greater than the average wage paid for unskilled and semi-skilled work, the corresponding ratio for teaching and clerical work, accounting for one-quarter of the national wage-labour force during the 1970s and 1980s, was of the order of 2–4:1.[201] Differences in wage-earning potential, which have widened since the early 1970s, determine differences in the capacities to make over remittances or contributions to the operations of small farms and, even without investment in agriculture, to finance the education which enables sibling household members get the wage-earning potential for family off-farm income. As Francis pointed out, remittances and wages from employment make any given distribution of rural household incomes more unequal.[202]

What might determine the initial capacity to fund the commercialisation of small-farm production, the ability to save out of wages from employment to finance investment in production, is also a marked fact about the expansion of *jua kali* enterprise. King and Abuodha reported from their survey that 'no single entrepreneur' got 'a formal loan to start business'.[203] Unfunded external loans played a major part in financing infrastructure and credit to farmers within agrarian development schemes to maintain the expansion of production. Official promotion of small-scale enterprise follows the experience of agrarian development doctrine and makes the provision of credit the major plank of policy. The 1992 sessional paper pointed towards 'problems' involved in commercial bank lending to small enterprises. To direct privileged borrowing by enterprise is the last resort of a policy which recognised that borrowing from commercial banks 'will be achieved only if banks perceive lending to the sector as a financially attractive part of their lending portfolio'. Thus, 'to ensure increased flow of funds', the 1992 paper continued, 'the Government will intensify its efforts to acquire supplementary soft foreign loans for on-lending to public and private financial institutions for lending to the SSE [small-scale enterprise] and Jua Kali sector'.[204] Devoid of profitability criteria for capitalist enterprise, this particular purpose of credit was to promote small enterprise as a source of employment.

That the doctrine of RNA enterprise, as for *jua kali* in general, has the potential to close the employment gap is axiomatic in so far as activity has to be found by those who cannot find no other recourse to subsistence. Livingstone's assertion, for instance, that the expansion of self-employing activity is independent of 'the formal sector' and does not necessarily *reduce* 'average earnings' is consistent with the tendency for such activity to perpetuate the extent of any given rural, and urban, poverty. This is not the idea of development expressing improvement but the application of development doctrine, to make the expanding population of self-employed labour bear its own burden of subsistence within the enclosed social arena of the community of the poor. Livingstone concluded:

> One should not expect the sector to absorb additions of labour without problems, any more than one would expect this of agricultural land, other things being the same: in each country it is likely to require a balance between these additions and the capacity of the informal sector to absorb more without reducing incomes.[205]

Such 'problems' are twofold since capacities of small enterprise to develop as sources of employment, namely the intent to develop, are not independent of what is alleged to ground *jua kali* in the autonomous process of the development of Kenya African-owned industrial enterprise.

The advance of factory *jua kali* production from the mid-1980s occurred in the period in which large-scale manufacturing faced market and profitability 'problems' of the kind which we earlier recounted. Between 1983 and 1991, for instance, overall investment in the manufacturing sector actually declined.[206] It has often been pointed out, not least by King and Abuodha, that the capacity of *jua kali* factory and artisan production to develop is dependent upon the external technology of manufacturing. 'All its crucial technology has come from the formal sector', wrote King and Abuodha adding that, in view of a debilitating large-scale 'industrial base', the 'phase when the formal sector could provide technological pushes to the informal sector is actually rapidly dwindling'. To move to 'higher technological echelons', there 'will always be need for another external source' of technology to give 'a whiff of new life' to *jua kali* capacity. Another external source of technology, as for credit itself, is the 'help' provided by state trusteeship for development.[207]

Likewise, as we have also recounted, the capacities for the full gamut of RNA to develop is the extent to which commercial agriculture has developed. Small-farm revenues are expended on buying goods and services from RNA enterprise. And, according to Livingstone, it is the 'development and *level of technology* in agriculture' which determines the extent to which *jua kali* develops to supply intermediate and capital good inputs into agriculture. Such backward linkages might well be generally limited in the Kenya case and thereby provide the kind of 'problem' of small enterprise development to which Livingstone referred.[208] However, Kenya is a case in which agrarian schemes of production have been developed on the basis of state trusteeship and the failures to maintain, improve and extend such schemes forces individuals to engage in RNA. Where average incomes remain constant, let alone decline, the external of trusteeship reappears to develop the spontaneous outgrowth of non-farm activity, the very conditions which are created by the failures to develop agriculture and industry according to the intention that development is the means to generate employment without reducing 'average incomes'. If this is the general premise of development doctrine, that it acts to ameliorate, then the twist of intentional development moves in the Kenya case through a dual burden of trusteeship.

Penwill's law of African enterprise in the post-war period, one which was mistakenly represented by Himbara to be an ideal age of development, was that competition between African-owned and Kenyan Asian capital aggravated intra-African-owned business rivalry and caused business failure because there was too much activity in trading. One driving force behind agrarian schemes of development was to induce traders into small farming. Now, fifty years later, when development is both a palliative, and the means to develop African business, the same logic is at work in a different form. An arena of self-employed enterprise and casual wage-work, where there is competition for work and in production and where livelihoods can barely be secured, is one in which African-owned capitalist enterprise can find the labour power, at lowest possible cost, to compete with other layers of capital. Official policy, not for the first time, is far more frank than the academic adherents of doctrine. The third objective of 1992 small enterprise policy was the 'development of a pool of skilled and semi-skilled workers who are the base for future industrial expansion'.[209] A system of small-scale enterprise, funded externally, is one which is intended to make artisans survive in enterprise as an alternative to wage-labour.

Intentional development, as we have seen elsewhere, moves through a cyclical path of history, thereby moving according to the original metaphor for the process of development as physis. Chamberlain's doctrine of development, towards the end of the nineteenth century, originated in his Birmingham endeavour to promote his model of the modern large-scale industrial enterprise, a model which was both borrowed from Germany and created on the basis of his own industrial enterprise. A rationalised capitalism, to confront the plethora of small and artisan enterprise, was the purpose of doctrine. Municipal welfare schemes would eliminate rural and urban squalor while making labour effort more productive for capital and British industrial capital more able to compete internationally.

For late twentieth-century Kenya, development doctrine is perpetuated by its latest intent to develop small enterprise and give focus to the unravelling of public authority welfare schemes. When King and Abuodha suggest that 'it has been felt that it is the *jua kali* economy that provides the bulk of people with their work, health, law, housing, and their training',[210] then the implication is that it is private payment and contribution to private enterprise, for operating school, clinic and security force, and not centralised public authority welfare which can and should meet need. In so far as it had been extended Fabian-colonial-wise from Britain to Kenya, public authority welfare was regenerated by the late nineteenth-century idea, for Britain at any rate, that inchoate and part-private provision of such services was inimical to the social and national efficiency which the design of national development was intended to create.

While the purpose of intentional development has been turned on its head, the source of positive intention, that there be trusteeship for development, remains in the aspiration to help, organise and regulate and create a system of small enterprise. There is also historical irony at work in that the very radical-liberal complex, which successfully fought the extension of Chamberlainite doctrine to colonial Africa in the early twentieth century, based policy upon the precept that it was the development of indigenous capitalism in Africa which had to be countered. Radical-liberals based policy on the ground that 'development' was the interpenetration between Chamberlainite doctrine and the propensity of local would-be capitalists to undermine the natural course of development in Africa through small-scale production. Ambivalence about the trajectory of African-owned big business is still prevalent.

CONCLUSION: THE INTERNAL AND EXTERNAL OF DEVELOPMENT

Kenya is a case, as Himbara has observed, in which 'the state as an agent of development remained in question'. In arguing that 'there can be no substitute for the state in capitalist development' and a 'national interventionist state that conceives and implements a consistent program of development', Himbara ended his book thus: 'The fate of Kenya, therefore, hangs on the rise of this historically important agency.'[211] National development, from the nineteenth century, was founded upon state trusteeship to bridge the external and internal of development because it was supposed that the 'mind' of the state could be exercised to direct the external body of matter, of trade and industry, which moved through national territory.

And, as we have seen in earlier chapters, the interior of development in expressing the potential for self-determination as national self-determination was premised upon trusteeship for the mass of the body of population which resided in the nation. Development doctrine, we argued, was the foundation of policy made to keep population confined to territory, to fix and confine the potential of productive force to the local place, of field and factory, where the need of local capital for labour power might be constructed.

It is the question of what 'the local' might mean that provides two accounts for why the presumed association between the state and the agency of development has been unravelled. The first, as might be expected in Africa, comes from the place of tribe and community in relation to the state. John Lonsdale, in his essay 'The Moral Economy of Mau Mau: wealth, poverty and civic virtue in Kikuyu political thought', has argued that 'ethnicity was not reduced to a residual category by the formation of classes':

> Indeed tribe was the imagined community against which the morality of new inequality was to be tested. The moral language of class could not be freely chosen. What ancestors had taught, or were said to have taught, on the relation between labour and civilization was the only widely known measure of achievement or failure in man- or womanhood.[212]

Lonsdale invoked Benedict Anderson's *Imagined Communities* of nations to transpose nation by tribe. A post-colonial concept of 'nation', for Anderson, was constructed out of colonial administrative practice whereby native and subaltern official ranks, and traders, moved between the tribally defined provinces and regions of each colonial territory but not between territories.[213] However, Lonsdale has argued for the Kenya case, that 'teachers and traders . . . reshaped ethnic institutions in the mind, enlarged a known category of reputation in order to dignify self-interest, and created tribal nationalism. At the heart of their new language were the old moral equations of wealth and virtue, poverty and idleness.'[214] And it is not by coincidence that Tom Mboya, 'Kenya nationalism's chief tactician', should be the peg upon which Lonsdale hung his sermon on the ethnic source of trusteeship.

Mboya refused 'discussion' on what the difference between being Kikuyu or Luo and Kenyan might mean and therefore opened what Lonsdale, in post-modern parlance, calls 'the "Mboya gap" between argued ethnicity and undiscussed nation-statehood', a gap which 'still needs to be closed'.[215] We have seen how Mboya, in the general pan-African tempo of his time, projected positivist categories, of how self-interest and capital might be moralised, upon an undifferentiated African community to found a source for trusteeship on the part of the post-colonial state. He assumed, again in the tempo of his time, that state development would be the means by which the intent to develop might make the process of capitalist development, dominated by the 'foreign' non-local enclaves of Asian and European capital in Kenya, conform to the prerequisites of authentic national development.

For all the objection to 'evolution theory' and 'modernisation', the synonyms for 'progressive development', Lonsdale's resolute refusal of the concept of development and his use of 'civilisation' as a surrogate of progress, this interpretation of the disassociation between state and agency of development is immersed in the 'community' of development. In reacting against the claims for national economic development, Lonsdale's treatment of tribal nationalism represents a return to the early twentieth-century attack on Chamberlainite development doctrine. As we saw in Chapter 5, it was the British radical-liberal complex which played the major role in putting paid to Chamberlain's development schemes in Africa.

A leading light of radical-liberalism was E.D. Morel. In his influential prognosis for British colonial policy in West Africa, Morel expressed the fear that development doctrine would undermine 'African Nationalism'.[216] At issue for Morel, and the Liverpool-based merchant trading companies for whom he spoke, was the question of West African land tenure. During 1910–11, in various places, he made it clear that 'African' nationalism was nothing other than tribal nationalism. When explaining why land should belong to the people of West Africa 'whose trustees we are, they being our wards' Morel also referred to land as the capital of the whole people with chiefs as trustees for the respective communities to ensure that 'the native shall have the opportunity of expanding along his natural lines and conserving his economic independence'.[217] Accordingly, he suggested that hegemony of the Yoruba states should be restored 'as the surest method of reconstructing Yoruba "nationalism" upon a proper basis', by enabling chiefs to act in concert with provincial councils.[218] And, in his campaign to turn the Colonial Office from a non-policy of drift and shift over land tenure, Morel made a clear choice between, first, the preservation of native customary law, improvement of the 'natural line of development', being the revitalising of 'African nationalism' and, second, the adoption of European law and individual land tenure involving the rejection of both indigenous development and the principle of African nationalism.[219]

Colonial policy should be directed, Morel advised, at the 'conservation and vitalising' of African nationalism against the Chamberlainite protagonists of development who were accused of representing City of London-based finance houses and land companies. African nationalism had to be promoted for supporting the 'free development of West African commerce and industry' against finance capital which threatened to direct development in Africa: 'Development – it paralyses initiative, progress and production, causes servitude rather than free expansion.' Forced development, Morel continued, spelt dividends to the City of London 'imperialists' who 'play for their own hand while prating of "development" for the good of the state'.[220] Here, in a nutshell, was the liberal reaction to the Chamberlain–Garvin doctrine of development for a centralised imperial state.

Morel, whose aspiration was to found a Liverpool school of politics as an adjunct to the Manchester school of economics, suggested that nationality could be meshed with free trade to find a third way between development doctrine and philanthropy. Development doctrine, for Morel and his associates, was based on a form of Social Darwinism that damned the African to wage-work, as Nworah put it, forced the 'damned nigger' to work for European capitalist enterprise.[221] A wage-labour force, on the model of what Morel's patron John Holt, the merchant capitalist, called the South and East African 'craze' of development, would be created without the status of a proletariat – without the citizenship and the 'common sense of civilisation'. On the other hand, philanthropy was as objectionable as development doctrine. Church missionary traditions of philanthropy sought to make the African civil by 'divesting him of his natural world'. The 'educated' African was denaturalised and, in being artificially European and devoid of 'national pride', would only fulfil the material interest of European financial capital.[222]

Since both development doctrine and philanthropy 'dislocated the African polity', the third party's mission was to renew the polity to guarantee that 'the inevitable development of modern industrial effort should not be prosecuted at the expense of the future liberties, free expansion and progressive advance of the West African peoples'.[223] To succour a multiplicity of West African nationalities, and it is worth noting the Edwardian radical-liberals were equally concerned about the Balkans,[224] was a condition for incorporating West Africa in the economic 'vortex' of international trade and production while ensuring that liberalism could take root among non-European peoples. A polity which should develop along its natural line would be a West African condominium of chiefs and merchant capitalists with Holt, Mary Kingsley half-joked, as Bismarck.[225]

Morel's school was part of the more general complex which came to incorporate, as we have mentioned, different strands of British neo-Hegelian idealists and Fabian socialists in their common endeavour to promote the development of community against the individuation of active self-interest which they associated with the maleficent development of indigenous capitalism.[226] This endeavour permeated colonial officialdom whose problem, as expressed by Charles Strachey, at the Colonial Office in 1910, was how to get 'the un-Europeanised native represented' against 'the educated' who, in Holt's words, were a ' greedy selfish lot for the most part and care not for abstract native land rights and customs'. Morel's accusation that the 'class of educated natives' were 'engaged in battening upon the ignorance and cupidity of their illiterate brothers' was answered by one organ who marked him out as 'the dangerous adversary' for having adopted 'his welfare strategy with hypocrisy, mellifluous cant with stupendous ignorance'.[227] Indeed, what Lonsdale's account of Kenya reveals is the extent to which Morel's precepts were turned on their head.

If teachers and traders, in Kenya, invented tribal nationalism they did so along the lines which Morel and his merchant capitalist associates might hopefully have envisaged but not along the line of development they imagined. No community or tribe could contain, what Morel and company had always feared, the indigenous propensities for freehold land title, credit, trade and the accumulation of money capital. For all of this, and including what was involved in the movement of ethnic populations over territory to engage in wage-labour, it was not the territory of tribe but putative national territory over which the state was to preside. It was, paradoxically, development doctrine, with its agrarian development and rural enterprise schemes, which fixed population to its given ethnic territory and thereby provided the means by which tribal nationalism could be asserted. Development doctrine had always meant 'more' and not less of community and there was little that was naturally African about this particular history of development.

A general logic of development also lies behind Lonsdale's history. Ancestors of the dead, according to Lonsdale, impose conscience upon the living and provide the 'test' of achievement and failure of development upon the living. However much it is refashioned to renew 'dignity' for self-interest, the dead and the living are enclosed within a community. Such, as we saw in Chapter 2, was one logic of the interior force of development in the mid-nineteenth century. There is little basis for Lonsdale's suggestion 'that the chief fault of evolution theory was its belief that religious – or at least, moral projects would give way to merely material self-interest'. The interior of development conveyed moral belief. And, it was the internal/external of development which Lonsdale addressed towards the conclusion of his essay:

> External architectures of control have stimulated the rival solidarities of political tribalism; interior architectures of civic virtue have prompted the awkward questions of moral ethnicity. If we are all condemned to live with the first we need to enlarge the languages of the second. If states cannot eradicate tribalism, the arguments within ethnicity must somehow infuse states.[228]

While the interior of development is positioned here by the external condition of the state, then it is equally the case, for Kenya and more generally, that the external of state development is the international agency of finance and aid and without which national development could not be imagined. The Kenya case is one in which any prior national, and local, source of state trusteeship has been infused by the external architecture of control exercised by the international agencies of finance and aid.

The second account of why the state has been dissociated from the agency of development gives stress to the withering away of state capacity for development from the standpoint of national development. Himbara's argument, for instance, is that the post-colonial Kenya government inherited a development apparatus from the colonial government but pursued a course of policy which undermined the development functioning of the apparatus. Policy which favoured Africanisation of the state apparatus and the promotion of African business reduced the potential of the indigenous development of capitalism. Much emphasis is placed, on this account, on the extent to which the public administration for development has been desiccated by the looting of the state apparatus for the private pursuit of active self-interest.

Rivalries of political tribalism, for Himbara, are not simply the means for external control over the interior 'community' of development but prevent the state as acting as the agent of national development. As such there is a 'serious dilemma' for 'Kenyan capitalism' since ethnic contestation 'has not been accompanied by an emergence of a coherent national leadership that would use the state to formulate a steadfast national agenda, 'on the contrary, some key state institutions that have historically played a leading role have either collapsed or barely function. It is this vacuum that some international financial institutions have attempted to fill'.[229] Among the leading institutions, such as some government ministries and parastatals, which were handed over to effective managerial control by the World Bank and its associated agencies in 1990, were the development banks.[230]

There is abundant evidence about the extent to which the accelerating rate at which state expenditure and revenue has been creamed off by state agents for private gain has been a feature of an increasingly disorganised state administration. Himbara's evidence from Auditor-General reports shows that the proportion of unauthorised to allocated state expenditure doubled from an annual average rate of 7 per cent during the period of the Kenyatta government to 15 per cent in 1986/7.[231] External *funded* public debt, as a proportion of the Gross Domestic Product, increased from an annual average rate of a relatively stable 12 per cent between 1963 and 1978 to over 30 per cent and rose rapidly after 1979 during the period of the Moi government.[232] Indeed, the renowned complaint, which Himbara repeats, is that the rise in external debt corresponds to the private creaming of the state budget thereby causing an increasing proportion of state expenditure to be funded by external aid and borrowing.[233] While external loans and grants had financed just less than one-half of the overall annual average fiscal deficit in the four years before 1978, external financing accounted for nearly 90 per cent of the fiscal deficit in 1982.[234] There may well be correlation between the extent of disorganised administration and the rise in external debt, even if part of the explanation for the break between the Kenyatta and Moi periods was the coincidence between change in regime and the independent impact of the 1979 external oil shock on the economy.[235] However, it is not possible to simply infer descent from a developmental state to an internationally funded kleptocracy from the one-sided ideal of national development.

The national development ideals, to which Himbara adheres, were conditioned and contained by the development doctrine which was spelt out, by our example of Bourdillon in the introduction to this chapter, for colonial Africa before 1939. Colonial development and welfare became national development and welfare and according to the same condition which informed the post-1945 period of state development in Africa. The condition of development doctrine was that British official funding would be applied, in the form of the subsidy, to colonial development and welfare effort. We have emphasised, as in Chapter 1, that the short-sighted history of development, finding the 'beginnings' of development and underdevelopment between 1945 and 1949, arises out of the changed scope of intentional development effort in this period. However, neither principle nor practice of development was new. Rather, it was the more concentrated and extended international scope of the principle and practices which were developed in the high age of state development.

At the outset of the aspiration of national development for post-colonial Kenya, therefore, international finance for development was the essential condition for exercising development doctrine. Trusteeship, when extended into the post-colonial African arena, could not become the exclusive internal basis of national development but had to be fused with the 'given' external architecture of control, the international domain of development. To use the now fashionable Latin American term of development, the ground for trusteeship was neither international nor domestic but 'intermestic'. For a 'bourgeoisie-in-formation', the perpetual expression for the African and aspirant indigenous class of capital, there was as much positive as negative in the Kenya development dilemma.

International finance made the burden of development doctrine compatible with the primary accumulation of capital. Welfare could be funded through the work of development and as long as development did its work there need be no recourse to the indignities of actively pursuing private self-interest in the name of national development. The negative of the development dilemma, as Himbara concluded for the Kenya case, is that the failure of development to do its work brings the development functioning of international agency of development into play. An international agency of development thereby directs development by default, to fill the place of agency which should be occupied by the nation state. Such an eventuality, it is suggested, is a matter of conjuncture and circumstance which could be avoided were there a particular and different variation in policy on the part of the state and the international agency of finance.

Long-term loans raised mainly in London, and giving rise to externally funded debt, had long been the main source of finance for government development spending. Between 1921 and 1933, for example, more than half of Kenya colonial government spending was financed in this way. 'Whenever such borrowing was possible for them', John King commented, 'colonial governments such as Kenya's tended to regard the spending which it could finance as their duty, in the interests of the development of their territories.'[236] National development, with the continuing emphasis on *duty*, made no difference to the principle of borrowing. It was the opportunity to incur unfunded debt, increasing sixfold between 1961 and 1970, which was the apparent source of the Kenya development dilemma. British government aid accounted for 70 per cent of all Kenya government unfunded debt, amounting to 30 per cent of all government debt as late as 1970. By 1990, all bilateral aid contributed towards only 30 per cent of unfunded debt.[237] The main external agency of finance was the World Bank and its affiliate, the International Development Association, which accounted for nearly one-half of external unfunded debt in 1990. As such, the international agency of finance for development was the World Bank.

Commercial and investment banks lend money to make money-profit. Governments may have a duty to borrow but banks have no duty to lend. A principle of Saint-Simonian development doctrine, as we saw in Chapter 1, was that investment banking should be endowed with the duty to lend. Unfunded debt is founded on the principle that both the lending bank and the borrowing government possess a duty both to borrow and to repay debt. Duty is the obligation of trusteeship, vested in the agency of both bank and state, and it is the obligation to lend which gives the development bank the potential to direct development. Kenya national development is the practice whereby the government mediates in the process of distributing finance from the international agency of development to the range of enterprises which also borrow according to the precept of the duty to invest for development. An overpowering weight of the social science of development studies, and the attendant practices of consulting to account for the use of finance advanced for development, is directed towards the analysis and prognosis of the variation in policy. Variation in policy derives from the potential to direct, and mediate in, a process of development. The conceptual principle which creates the potential for the direction of development has become lost in the welter of policy and accountability.

If Kenya is a case in which the state mediation of the direction of development has corrupted the ideal of national development, then explanation for the failure to meet potential, as in Himbara's thesis, is attributed to factors of functionings of administration and capacities of entrepreneurship. Associated with both is the theoretical mainstay of Joseph Schumpeter. It was the same Schumpeter who addressed the question of direction over the process of capitalist development. His general theory of economic development was laid out more than thirty years before the World Bank appeared in 1944, partly due to Keynes' imagination finding official acceptance because he offered the means of making a national need to invest, that of Britain, appear to be consistent with an international duty to invest, that which was to be exercised by the United States. Behind this scheme, however, lay a more fundamental intellectual urge. In the process of capitalist development, it was not the duty to invest but the expectation of money-profit of enterprise, both of lending bank and borrowing entrepreneur, which governed investment decisions. Keynes aspired to make the trusteeship of development, as expressed by the duty to borrow and lend, consistent with the process of capitalist development.

Schumpeter's general theory of economic development aspired to Keynes' vision before Keynes, to find a source of trusteeship that would be consistent with capitalist propensities for investment. Against the background of the dying years of the Austro-Hungarian empire, and after both failed state policies for national development and during controversy over the place of banks in directing industrial and commercial investment, Schumpeter's focus was on the general process of development. For reasons which will become clear in the next chapter, Schumpeter held no brief for state mediation and direction of development and did not envisage any such bank as 'The Bank' of the World Bank, extending finance at the behest of government and especially that of the United States. Following one path of the positivist tradition, Schumpeter made the bank the source of trusteeship for development for the simple reason that state agency in general could not be trusted to direct the work of development. Banks, the institutions of the money market, were burdened with the duty for making investment socially productive for capitalist enterprise. The ephor of development, for Schumpeter, was the bank.

Part III

7

JOSEPH SCHUMPETER AND 'FAUSTIAN' DEVELOPMENT

Joseph Schumpeter's theory of economic development, partially embodied in his so-named volume of 1909, was self-consciously a general theory that attempted to comprehend economic development within a general account of how and why development proceeded and how limits to the process of capitalist development arose out of development itself.[1] While the significance of Schumpeter for post-1945 development aspirations in the colonial and post-colonial world has been questioned, there is little doubt about his presence in the early years of the academic discipline of development economics and the attendant political science and sociology which fostered what came to regarded, by the 1960s, as the orthodoxy of modernisation theory. Ragnar Nurkse, one of the 'founding fathers' of development economics would write in 1953 that 'it is scarcely possible to consider' the subject of development economics 'without finding one's mind turning to Schumpeter's great work'.[2] To understand why this should be so it is necessary to address the problem of Faustian development and Joseph Schumpeter's place within it.

FAUSTIAN DEVELOPMENT

Marshall Berman, in his extraordinary *All That is Solid Melts into Air*, characterises Robert McNamara, Jean Monnet and, in particular, Robert Moses as exemplars of the twentieth-century Faustian developer who, especially since 1945, has made 'contemporary capitalism far more imaginative and resilient than the capitalism of a century ago'.[3] McNamara, the President of the Ford Motor Company who became Kennedy's Secretary of Defence, was responsible, as head of the World

Bank in the 1970s, for that institution's preoccupation with the provision of the 'basic needs' of the poor of the Third World and Africa in particular. Monnet aspired to create, through the nexus of the European coal and steel industry, the idea of European Union which would override the boundaries of the nation state and national identity. Both, by Berman's reckoning, epitomise the shift between 'public and private power' which has characterised the high age of post-war development as well as the arrival of the Faustian developer who selects projects of development to fulfil a purpose which was not necessarily congruent with either that of money-profit or national development.

It is especially Robert Moses who is emblematic, for Berman, of the modern-day Faust. Moses, from the 1920s on, inspired the redevelopment of New York City. Amongst much else, he used US federal government money to employ the jobless of the inter-war depression on public works to open up, in his re-creation of Jones Beach on Long Island, a 'pastoral world just beyond the city limits'. However, because this 'vast park . . . was only to be accessible by the private car', its creation thus necessitated the destruction of Berman's own neighbourhoods of the Bronx through which massive highways were rammed with great social destruction.[4] In the process of realising his vision of a transformed New York, Moses, Berman informs us, created 'a network of enormous, interlocking "public authorities", capable of raising virtually unlimited sums of money to build with, and accountable to no executive, legislative or judicial power'.[5] It is this absence of an external source of accountability over the internal will to develop which is, in Berman's rendition, the nemesis of the Faustian developer. It is not simply that the vision of development becomes undermined, as in the case of Moses, by its antithesis of the corruption which makes the work of development possible, but that the individual vision of what modern New York might have been was overwhelmed by the quantitative scale of development work. Berman notes that by the time of the height of Moses' power in the 1950s and 1960s, his projects had ceased to have the 'beauty of design and human sensitivity that had distinguished his early works'. His development agencies now 'stamped' constructions 'on the landscape with a ferocious contempt for all natural and human life'. Berman argues that Moses became 'scornfully indifferent to the human quality of what he did; sheer quantity – of moving vehicles, tons of cement, dollars received and spent – seemed to be all that drove him now'.[6] Such sheer quantitative change is closely akin to the 'natural development' against which, as we have seen in Chapter 3, Hegel asserted his principle of human development. By this reckoning, the high age of twentieth-century development, as exemplified by Moses, was human development's negation.

One paradox of development now becomes clear. Twentieth-century development, as Berman sees it, conveys an affinity between the 'ideal of *self*-development and the real social movement toward *economic* development'.[7] To realise the development of the individual self, a self freed from the restraint of community and tradition, is the ideal of modernity. The ideal, and here Perry Anderson has underscored Berman's invocation of Goethe's *Faust*, was to free the self from 'the fixed social status and rigid role-hierarchy of the pre-capitalist past with its narrow morality and cramped imaginative range'.[8] This ideal, which could only be expressed as a possibility in the early nineteenth century, was made probable in the twentieth century through the modernisation of economy and society entailed in the development of capitalism and its dissolution of the old social world by, as for Marx, the 'constant revolutionising of production, uninterrupted disturbance, everlasting uncertainty and agitation'. Moreover, with its onset at the beginning of this century, modernity offered the prospect, recognised by Marx, that this unselfconscious destruction of the status, tradition and locality of the old social order by techniques of building and production, which were directed towards the achievement of an expansion of output and money-profit, could be given the conscious intent of the liberation of the individual self. Now, at the end of the twentieth century, we can see more clearly that the conjuncture between modernity and modernisation was not to survive into the period following the end of the Second World War. There was, as Berman puts it, 'a radical splitting-off of modernism from modernisation',[9] a split which developed from within the conjuncture itself and, as such, contained the potential for human development to be either cast aside from, or subsumed under, the non-development of modernisation with the consequence that the movement towards economic development would be neither real nor social.

This first paradox then reveals a second. Just at the moment that the possibilities of self-development were seen to become more probable, in so far as production was revolutionised, subsistence disrupted and population was disturbed by the development of capitalism, the *ideal* of development, which should be real social movement, increasingly has come to be the endeavour of locking the individual self into status and function. Social movement was judged not to be real but deemed authentic through the creating of fixed channels by which a pre-capitalist past might be preserved or re-created in order to halt the destruction which the historical development of capitalism brings in its wake. Through his recapitulation of *Faust*, Berman lays bare the tragedy of development, the tragedy which envelops one paradox within the other.

Berman recounts the story of Goethe's great allegorical tragedy of the three metamorphoses of Faust, the archetype of development. Faust, who had been a practitioner of public health for, and in solidarity with, the poor, desires development before he becomes the developer. Between dream and action, Faust discovers the altruism with which to motivate development: 'In his first phase', Faust 'lived alone and dreamed.' During his second, 'he intertwined his life with the life of another person, and learned to love'. It is 'in his last incarnation' that Faust 'connects his personal drives with the economic, political and social forces that drive the world; he learns to build and to destroy'. The tragedy of Faust is that while he learns how to build, he also learns that he cannot do so without also destroying. And he cannot so destroy without destroying himself.

Faust envisages and plans projects to reclaim land and sea for human purposes, to build cities, and to re-create a productive agriculture out of waste and unproductively used land:

> He expands the horizon of his being from private to public life, from intimacy to activism, from communion to organisation. He pits all his powers against nature and society; he strives to change not only his own life but everyone else's as well. Now he finds a way to act effectively against the feudal and patriarchal world; to construct a radically new social environment that will empty the old world out or break it down.[10]

Some of Goethe's favourite reading, as Berman notes, was contained in the Saint-Simonian publication *Le Globe* which proposed large-scale development projects to provide work and income for the unemployed and the poor. Goethe himself had reflected upon such projects which were to include the Suez and Panama canals as well as another, located in the then Austro-Hungarian Empire, linking the Danube to the Rhine rivers.[11] Thus, it comes as no surprise that Berman acutely appreciates that Faust the developer is no mere entrepreneur motivated by the relentless search for money-profit and the accumulation of money-capital. As Berman tells us, 'Mephisto is constantly pointing out money-making opportunities in Faust's development schemes; but Faust himself couldn't care less.'[12] Instead, Goethe's Faust is that archetypical positivist developer, the Saint-Simonian banker, played opposite Mephisto the individual capitalist, the entrepreneur.

Berman writes of how Mephisto shows Faust *how* to do the developing and how to do it fast. Mephisto's mediation of 'love' is not the Comtean altruism of 'universal altruistic love' but sex and power. Mephisto has technique and is motivated by the pursuit of money. Above all, it is 'Mephistopheles, the private freebooter and predator who executes much of the dirty work, and Faust, the public planner who conceives and directs the work as a whole'.[13] Faust orders

Mephisto to destroy an old couple who obstruct his plans by refusing to move from their home. The pair, Philemon and Baucis, own and live on a small piece of land by the sea from which they have saved the ship-wrecked according to the Christian virtues of 'innocent generosity, selfless devotion, humility, resignation'. They are the final symbolic remnants of the past and Faust orders Mephisto to destroy them. While not wishing to know the details of their end, Faust finds that no matter how much he attempts to distance himself from their destruction, he cannot assuage his guilt for their murder. He comes to realise that their world, now destroyed, was the origin of his own; its altruism was that which had motivated his own mission to develop. Faust comes to understand that once this old world has been eliminated there is no longer any purpose to development. Berman comments: 'It appears that the very process of development, even as it transforms a wasteland into a thriving physical and social space, re-creates the wasteland inside the developer himself. This is how the tragedy of development works.'[14]

Wasteland, the external of development, is here internalised by the developer; the external is re-created within the interior purpose and intent of economic development. What Berman says of wasteland echoes what was said in the nineteenth century by the Saint-Simonians about the wastage which occurred as the force of labour was made more productive. The tragedy of development is thus recounted by Berman in the way that Robert Moses combines the two dimensions of Faustian development – Faust the trusted banker and Mephisto the unscrupulous capitalist entrepreneur. One question posed by the 'Faustian bargain' was whether Mephisto, the anti-hero, could be disposed of in the modern work of development. A second was whether the entrepreneur, so closely associated with the will and intent to develop, was to be reinvigorated by the modern ideal of self-development. Still a third was whether Mephisto the entrepreneur could be brought under the control of Faust's banker.

During 1909, the year, it is to be recalled, in which a British Liberal government passed a Development Act signalling the advent of development practice for Britain if not for British colonies, Joseph Schumpeter formulated a modern view of economic development in Austria. He did so by recasting the Faustian developer in a modern light. For Schumpeter, development was a process whose driving force, the entrepreneur, was one which was internally and immanently restrained by the supervisor of development, the heroic banker. Schumpeter's idea of development was one in which the self-destructive propensity of the unrestrained will to develop of the entrepreneur was to be restrained by the pivotal role of the banker as mediator of the manner in which new techniques and new combinations of techniques of production arose out of the destruction of the old. In setting out his theory in this way Schumpeter aspired to make the ideal of self-development, embodied in the entrepreneur, conform to what he took to be the real social movement towards development represented by the banker. As such Schumpeter's theory is a crucial moment in the transformation of nineteenth-century ideas of development into mid-twentieth-century economic theory.

The intellectual context of Schumpeter's work, Vienna in the last years of the Austro-Hungarian empire, is a prime example of what Berman refers to as the conjuncture between the modern and modernisation and it is, perhaps, unsurprising that a modern theory of economic development should have been formalised amidst its controversial modernist currents. These currents, which ranged across Freud's psychology, Schoenberg's music, the painting of Klimt, Schiele and Kokoschka, and Wittgenstein's philosophy, were also manifest in what has come to be known as the Austrian school of economics.[15] Conceptual controversy was the source of Austrian theory. Inspired by Carl Menger's method of pure theory, the Austrian school took its roots from what Menger called 'exact' laws as opposed to the 'empirical' laws, derived on the basis of naive inductivism, which were attributed to the historical method of economics of Wilhelm Roscher and Gustav von Schmoller. Exact laws, Menger determined, were to be used to establish the basis of value through the subjective, rationally constructed, 'typical' individual evaluation of goods and services premised upon essential human desire rather than any social construction of what the common good might be.

Menger, as T.W. Hutchinson has pointed out, did not object in principle to the theory and laws of development which had become so closely associated with the historical method. He merely regarded development as being a 'secondary' and one-sided endeavour of economics:

> Theoretical economics is the science of the general nature or forms of phenomena (*Erscheinungsformen*), and general connections (the laws) of economy. In contrast to this comprehensive and significant task of our science the establishment of 'laws of development' of the economy . . . must, though by no means unjustified, seem still to be quite secondary.[16]

Second, Menger, and the faithful adherents to the Austrian tradition such as von Mises and Hayek, regarded themselves as being moderns who were engaged in an internal critique of modernity, criticising 'both the degeneration of modernity as represented by *constructivist scientism*, and also the degeneration of democracy in the welfare state'.[17] Their criticism provoked its own, immediate reaction.

Coeval with the flowering of the Austrian school was the formation of what is now called Austro-Marxism. In reaction to what Joseph Schumpeter called the 'methodological individualism' of the Austrian School, Austro-Marxists such as Max Adler, Otto Bauer, Karl Renner and Rudolf Hilferding brought Marx to contend against the Austrian School by renewing the emphasis upon 'society' and the social relationships of commodity exchange as the foundation of value in capitalist society.[18] However, while the purpose of Austro-Marxians was to be one which excluded 'all the individualistic, psychologistic theories of social life' from their construction of a social science of marxism, their method was to treat 'motives as causes' of phenomena. As Tom Bottomore notes in his interpretation of Adler, 'motives' here 'were analysed not as individual psychological phenomena but as forces at work in "socialized humanity" and thus having their effect as "social forces"'.[19] What was significant about Adler, for our purpose here, is that he brought positivism both into the attack against Menger's individual-psychologism and into the Austro-Marxist endeavour to purge Marx of the 'teleological' and 'metaphysical' and thereby claim marxism as an empirically oriented social science. Schumpeter, whether independently of the Austro-Marxists or not, followed the same current of thought.

Although giving primacy to Kant over Hegel, Adler admitted that it was Hegel who had 'first magnificently stated' the 'idea of *development*'. As well, in both his attack on Menger's individual-psychologism and in the Austro-Marxist endeavour to reclaim marxism as an empirically oriented social science by purging Marx of the 'telelogical' and 'metaphysical', Adler invoked positivism. Claiming in 1904 that Comte's purpose was '*savoir pour prévoir*, the culmination of all knowledge in a comprehensive politics', Adler pronounced that Marx's object had not only been the same, but rested upon the same 'great doctrine of the primacy of practical over theoretical reason'.[20] Later, Adler referred to Comte's 'correct methodology' of science, and then, separately, suggested that this method, which had been further developed by the Austrian physicist Ernest Mach, was that of Marx, whose work Adler characterised as 'only a form of natural science positivism, more or less in the manner of Ernest Mach'.[21] Mach's neo-Kantian philosophy played a key role in generating the Vienna Circle of logical positivism. His conception of theory was instrumentalist. Instrumentalism, as a variant of positivism, supposed that the purpose of theory was that of an instrument for the practical application of knowledge, a tool which would be used to deduce the prediction of future events. Theory's purpose was not to establish the truth or falsity of phenomena, through description, but the means of systematically organising the potentially infinite 'facts' of experience.[22] It should be clear from our discussion in Chapter 3 that this interpretation smacks of the 'true socialism' of which Marx was so disdainful. Adler sought to press Marx into the service of trusteeship for the 'social force' of the working class, class being here treated as if it was a simple empirical correlate derived from 'practical reason'. Despite the fact that such trusteeship was emphatically not Schumpeter's motivation, the same currents of thought, and Mach in particular, that played upon Adler and the Austro-Marxists, influenced what was distinctive about Schumpeter's method of economic theory.

In 1908 Schumpeter set out a method in his *Das Wesen* which, according to Shionoya, carried the writ of instrumentalism into pure economic theory. Since instrumentalism was a variant of positivism, there is 'little doubt', Bottomore also concluded, 'that Schumpeter was a positivist in several important senses of that protean term'.[23] Holding that facts existed independently of theory which constructed 'a scheme for facts',[24] Schumpeter was able to use Mach to contest Menger's theory because its causal essentialism put primacy upon describing the eternal, typical facts of economic life. Both the Austro-Marxists and Schumpeter held to a basic instrumental tenet, here voiced by Otto Bauer in reference to the virtues of socially planned activity,[25] that rather than a mechanical conception of cause and effect there could be only a purposeful evaluation of means for 'goal directed aspirations'. In 'exact reasoning', Schumpeter implored, 'we avoid the concepts of "cause" and "effect" whenever practicable and replace them by the more satisfactory concept of function'.[26] Function, it will become clear, was what the 'ephor' of development was all about.

THE BANKER AS THE 'EPHOR' OF DEVELOPMENT

Despite *Faust* and the many Saint-Simonian injunctions recounted in Chapter 1, it may still seem strange to a generation ingrained with the mystique of the entrepreneur that Schumpeter should have cast the banker in the heroic role of the ideal 'developer' of modernity. What was characterised as the will and thrust of energy for development of Doctor Faust has been largely transposed, not just in readings of Faust but of Schumpeter as well, into the figure of the entrepreneur, the practical man of action, whose self-development is set out as the guiding beacon for how development can be achieved by all. Yet, the Saint-Simonians and Goethe were not alone in their characterisation of the banker as development's hero. Francis William Newman, of whom it was said that he diverged from his brother John Henry Newman as if they lived in 'different worlds', echoed this aspect of the Saint-Simonian view.[27] A mid-nineteenth-century theist, social reformer, and proponent of colonial self-government and of land nationalisation, Francis Newman argued that while the usurer was a despised figure of history, the credit-advancing banker was a respected person portending progress.

In his 1851 *Lectures on Political Economy*, F.W. Newman distinguished between 'two different conditions, – the *savage* and the *civilised*' and then, to contest socialist doctrines, declared his intention to show through 'the natural history of economic progress' that 'the civilized state is one in which markets are perpetual; – that markets imply Competition, and that Competition has been most erroneously and causelessly vituperated'. The 'savage' state was one of 'individual self-sufficiency' of household production, a condition in which 'no perceptible *progress*' was possible, due to no specialisation of function, no change in technique and unremitting poverty of the producer.[28]

Marx thought F.W. Newman's distinction between usury and bank credit, being founded upon the different effects of each upon the working poor and aristocratic rich, was inane. Usury, Marx contended for both the ancient and the feudal world, had impoverished both 'the poor petty producer' and the 'rich landed proprietor' alike. In the transition to capitalism, Marx argued, the 'credit system develop[ed] as a reaction against usury' with usury being subsumed, not extinguished, by interest-bearing capital. This interest-bearing capital was that of the banker who searched for money-profit and whose potential for the monopoly profit of usury was thwarted by the competitive creation of credit.[29]

The ideal of bank as 'developer' could only arise, Marx suggested of Saint-Simon, where large-scale industrial production and the credit system of money markets was relatively undeveloped.[30] Such was arguably the case of mid-century France where, as we have seen in Chapter 1, the Saint-Simonian vision was of a bank which would stand between the 'idle capitalist' and the '*travailleurs*'. Here, *travailleur* meant both worker and industrious capitalist, the particular capitalist who was meant to be moralised by socialist doctrine.

Marx quoted Saint-Simon's fear that the possible 'advantage' of the banker's intervention was

> often outweighed and even destroyed by the opportunity that our disorganised society offers for egoism to hold sway, in the various forms of fraud and charlatanry; the bankers often intervene between *travailleurs* and idle capitalists simply to exploit both sides to the detriment of society.[31]

Marx's understanding of the Saint-Simonian dilemma was that, within the 'disorganised society' of capitalism, bank credit was interest-bearing capital because it contained the speculative propensity for making money out of money inside the function of systematically making money-capital available for production. Marx emphasised that the credit system had this 'dual character immanent within it':

> on the one hand it develops the motive of capitalist production, enrichment by the exploitation of others' labour, into the purest and most colossal system of gambling and swindling, and restricts ever more the already small number of exploiters of wealth; on the other hand it constitutes the form of transition towards a new mode of production.[32]

For Marx, it was through 'the full development of the credit and banking system' that the 'social character' of capital was to be realised. By 'social', Marx meant that the potential of capital to command profit was determined by its general or total character. Private property was generally established by the state and capital circulated, in the general forms of money and commodity, among particular individual capitalists whose command of the means of production was predicated upon the social form of property. It was, for Marx, 'by the total surplus labour that total capital appropriates, from which each particular capital simply draws its dividends as a proportional part of the total capital'. Credit institutions, and banks, in particular, held 'the available' and 'potential money capital that is not already actively committed at the disposal of the industrial and commercial capitalists'. The development of the credit system 'thereby abolishes the private character of capital and thus inherently bears within it, though only inherently, the abolition of capital itself'.[33] However, this trust was placed under stress by the potential of bank holdings to be used by the bank to further its own pursuit of money-profit through speculative ventures as well as by the propensity to centralise the ownership of capital. Marx concluded that it was 'this dual character' that gave 'the principal spokesmen for credit, from [John] Law through to Isaac Pereire, their nicely mixed character of swindler and prophet'.[34] In other words, the doubled archetype of the banker as despised swindler or respected prophet was inherent in the tension which characterised the role of trust as it was developed through the credit system.

The credit system made the application of finance to production more systematic. For instance John Law, the English founder of a Paris bank in 1716 which was nationalised in 1718 but which collapsed under the weight of its speculative activity two years later, was characterised by Marx as 'Law the First' in relation to William Patterson, the founder of the Bank of England and the Bank of Scotland. Behind the founding vision of these banks was the belief that the monopoly of precious metal could be broken by the creation of credit money with which to finance commerce and industry.[35] Notwithstanding his scepticism about the ability of the credit system to break free from its foundation in precious metals through inconvertible note issues, Marx referred to the Bank of England's 'tremendous power' over trade and industry. Despite the passivity of the Bank towards commercial and industrial capital, the movement of which remained 'completely outside its orbit', it was such a bank, Marx argued, which supplied 'the form of a general book-keeping and distribution of the means of production on a social scale, even if only the form'.[36] Those who in the nineteenth century hoped to create a socialist bank sought to fill the social form of accounting with the content of the distribution of means of production between agents of productive enterprise.

Marx quoted the radical Saint-Simonian, Constantin Pecqueur, who in 1842 proposed that the general banks should 'govern the entire movement of national production' in a planned, socialised economy in which borrowers were to be bound 'together compulsorily in a close solidarity in production and consumption'. Without such a planned economy in which credit would be advanced 'to people of talent and merit but no property' and 'in such a way that they themselves determine what they exchange and produce', Pecqueur advised that what would be achieved would be only,

> what private banks already do achieve, anarchy, a disproportion between production and consumption, the sudden ruin of some and the sudden enrichment of others; so that your [national credit] institution will never do more than produce a sum of benefit equally balanced by the sum of misfortune borne by others . . . you will simply have provided the wage-labourers whom you assist with the means to compete with one another, just like their capitalist masters do now.[37]

In Pecqueur's view, any real social movement of development would have to forestall the immanent tendency towards destruction which was inherent in the credit system.

Pecqueur's vision of a socialist bank, as this came to be understood at the turn of the twentieth century, rested upon the supposition that the basis of competitive enterprise had been destroyed by the development of capitalism itself through the centralisation and concentration of capital. Rudolf Hilferding's *Finance Capital*, itself based upon a close reading of volume 3 of *Capital*, proclaimed the reality of the fusion of banking and industrial capital. This work was, in turn, central in the formation of Lenin's belief that capitalist development had already made the application of finance to production systematic. Paradoxically, it was the Saint-Simonian type of general bank, on the style of the *Crédit Mobilier*, which was seen as the most developed capitalist institution in that it had unified commercial and investment banking functions, broken the remnants of the boundaries between private and public property, and had thereby created the basis of a socialist bank to govern the distribution of the means of production between enterprises of nationalised blocs of monopoly capital.

Goethe's Faust, Berman notes, had envisaged development to be free from the restraint of the nation-state and national boundaries. Pecqueur, however, interpreted the Saint-Simonian system of general banks as one which was to govern 'national production'. It was this national vision of trusteeship – but as the 'trustification' of monopoly capitalism – which was to have such an important influence on Lenin's understanding. In his *Imperialism*, such nationally trustified blocs of monopoly capital, succoured by state regimes, competed across the world.

If the credit system was characterised by the tension between the propensity for speculation and swindling and the accompanying prophetic vision of systematic, planned production, then this tension was compounded by that created by the choice between the taking up the trust for the purpose of national development and the mission to make capital free of national restraint. It might have seemed to those of a Listian bent that the modern idea of self-development could be displaced and then personified in the body of the nation-state, which would then act in the name of the self-determination both for, and as if it actually was, the individual being of the person. However, as far as modernity was concerned, the ideal of self-development was emphatically the ideal of, for, and to be realised by, the individual person whose desire and capacity for development was to be made free of the restraint of state and nation.

AUSTRIA

If the immediate intellectual turmoil of Vienna formed one context for the formation of Schumpeter's theory, then the economic transformation of the Austro-Hungarian empire formed another. Between 1909 and 1911 the Austro-Hungarian empire, and especially the Austrian-governed provinces of Cisleithania, were in the midst of a relatively rapid spurt of industrial expansion after a long period of slow but steady industrialisation.

During this short period of rapid industrialisation, the Austro-Hungarian imperial state was a highly centralised and hierarchical apparatus lording order over an empire rent by national, ethnic and religious division and, as far as Austria

Map 3 Austria-Hungary 1890–1914

City Cartographic

was concerned, racked by a marked degree of labour insubordination and insurrection. Industrialisation within the integument of an old aristocratic order was typical of the 'modernist conjuncture' in which the aspiration for development was expressed within an old social order of the *ancien régime*. As Perry Anderson, among others, suggested of Europe, it was only in 1945 that 'the old semi-aristocratic or agrarian order and its appurtenances was finished in every country'.[38] An old social order of *ancien régime*, whether or not associated with monarchy, aristocracy and imperial aspiration, was extant until 1914. As Anderson also suggests, the Russian revolution of 1905–7 was 'emblematic' of upheaval facing the European state, a general 'ambiguity' in which the prospect of a new order could be either 'more unalloyedly and radically capitalist' or socialist.[39] This is the point that should be remembered as we proceed with our discussion of Austrian development and Schumpeterian theory.

Between 1904 and 1913, Austrian iron ore production nearly doubled from 1.7 to 3 million tons. Pig and cast iron production, as well as zinc and lead output, also doubled over the same decade. Between 1907 and 1914, the number of power stations doubled while energy output tripled. Together with textile production, metal-making and machine-building this largely accounted for the rapid industrial expansion during the pre-war decade. Thus, during the peak boom years of 1903–7, at the height of the 1895–1914 wave of industrial expansion, known as the second *Grunderzeit*, metal-making and machine-building output grew at an average compounded rate of little less than 10 per cent per year while the composite rate for all industrial production was 6.3 per cent.[40] Even though questions have been raised concerning the significance of the turn of century *Grunderzeit* in relation to longer periods of industrial growth in the nineteenth century, it is difficult to imagine that Schumpeter, despite his absence during 1906–7 in England and then Egypt, was unmoved by the unmistakable and immediate thrust of industrial growth in Austria.

From mid-century, rural–urban migration increased rapidly. Vienna's population, for instance, quadrupled from half a million to 2 million between 1850 and 1910. From 1880 to 1910, well over a half of the increase in the city's population was accounted for by a net gain in migration of half a million; by 1910, only half of Vienna's population had been born within its boundaries. Also, between 1880 and 1910, the number of Austrian crownlands' towns whose population exceeded 10,000 increased from 90 to 150. Only one-sixth of the total population of 24 million in 1890 lived in such towns; by 1910 well over one-fifth of 28 million did so largely as a result of the tendency for the rate of urbanisation to exceed the rate at which industrialisation spread through the crownlands.

While wage-employment in Austrian crownlands' industry increased during the period 1900–10 from nearly 3 million to 3.5 million workers, the rate of increase was less than that of industrial production. In 1913, only one-quarter of Austria's 'gainfully employed persons' were in industrial production. Despite the extent of rural–urban migration, agricultural work still provided the major source of income for subsistence of more than one-half of the population, even in the more industrialised Austrian provinces, during this first decade of the twentieth century.[41] This is why the Austria of 1913, at the onset of world war, has been described as 'an insufficiently developed agrarian state' or as being 'typical of an agrarian country in the process of industrialisation'.[42]

If the industrialising Austria of this period can be cast as a developing 'centre' in relation to its underdeveloped Hungarian 'periphery' within the Dual Monarchy, then Austria can also be depicted as a 'typical' case of Central and Eastern European underdevelopment in relation to Western Europe, and to the pivotal economic centre of Germany in particular. Moreover, it was during the first *Grunderzeit*, the wave of industrial growth during the 1850s and 1860s, that Austria's general economic performance significantly lagged behind that of Germany. It is estimated, for example, that real per capita income in Austria fell from 70 per cent of that of Germany in 1850-60 to 55 per cent in 1890–1913, while per capita consumption of anthracite coal, iron and petroleum was only one-third of German consumption in 1907.[43]

Adding to the image of turn-of-century Austria as a typical case of under-development is the fact that industrial production developed within a protectionist regime which favoured relatively small business. Moreover, a disproportionate number of Austria's nineteenth-century industrial entrepreneurs were immigrants from Western Europe, and Germany and Britain in particular. From the late eighteenth to the mid-nineteenth century, the Austro-Hungarian monarchy had encouraged the immigration of German industrialists, British master mechanics and Jewish bankers and wholesalers.[44] The structure of agriculture, as well as the policies applied to it, both serve to reinforce this perception of underdevelopment as well. Family household farms, of less than 50 hectares, represented 99 per cent of landholders but commanded 60 per cent of cultivated land, while the 1 per cent of landed feudal estates controlled the remaining 40 per cent.[45] Agricultural productivity was generally low and remained lower, at the turn of the century, than in Western Europe.[46] During the 1890s, and following rural agitation, an agrarian policy 'of small steps' was instituted by the Austrian Parliament in a belated attempt to increase small-farm productivity. Made up of 'favourable credit regulations, expansion of agricultural education, and encouragement of cooper-atives' as well as a doubling of the Ministry of Agriculture budget in the decade after 1898,[47] the agrarian bias of this policy was directed towards making agriculture a secure form of employment in the face of rapid urbanisation.

Yet, one singular feature of the Austrian economy, the development of the banking system and development of a system of general banks in particular, stands out against this imagery of underdevelopment. By 1913, during the second *Grunderzeit*, the ten 'great' or general or universal banks centred on Vienna had come to command the heights of industrial capital in the Austrian crownlands. They accounted for nearly 70 per cent of all banking capital in Cisleithania and it was roughly estimated that by early 1914 over one-half of the capital of Austrian limited liability, joint-stock companies was held by ten banks.[48] Incorporated during the *Grunderzeit*, and through the agency of the banks, such companies accounted for one-half of industrial production. Both Rudolph and Marz have shown the extent to which control over industrial companies was exercised by interlocking bank directorships. The *Creditanstalt*, for instance, exercised majority control over 121 companies in Austria in 1909. Marz has provided the detail to show how the control, over 91 companies and their 37 subsidiaries in 1913, was spread over twelve industrial and commercial sectors and, thereby, virtually the whole gamut of the economy.[49] The question of whether or not this banking sector was, in a broadly Saint-Simonian sense, developmental was to be at the heart of much theoretical controversy.

There is little dispute over the origin of *Crédit Mobilier* style general banking in Austria. Rothschilds in Paris copied the idea of *crédit mobilier* from its founders, the Pereire Brothers, and then competed against them to win the favour of the Habsburgs in 1855 when the monarchy sought finance for railway construction and industrial promotion at the onset of the first *Grunderzeit. Creditanstalt*, the Rothchilds' bank in Austria-Hungary, was thus established as a general or universal bank whose orientation, until the financial crisis of 1873, was investment banking.[50]

It was the example of *Creditanstalt*, much emulated in Austria in the same way that *Crédit Mobilier* had been copied, that led Berend and Ranki to conclude that 'banks may be said to have been the primary force in financing economic development and the modern economic transformation. The famous Austrian *haute finance* actually became the master of the economy'.[51] Marz, a student of Schumpeter, made the high finance thesis – that the bank was 'the master' of development – more emphatic, especially for the second *Grunderzeit*:

> [Banks] promoted the development of modern large-scale enterprise by financing new companies and mergers, by encouraging the moneyed classes to invest more and more in the capital market, and by ensuring the liquidity of the firms with which they were associated. They acquired a position of power and influence whose outward manifestation was the close collaboration among the firms concerned.[52]

Rudolf Hilferding shed light on how this power and influence had been attained in Austria: 'For historical reasons a genuine capitalist wholesale trade has not developed fully here. In the mass consumer good industries, particularly where speculation plays a role, as in the sugar trade, the bank has assumed the functions of the wholesale dealer.' With a relatively small outlay of money capital banks could command a far larger degree of control over capital invested in trade.

Hilferding continued: 'bank capital acquires an interest in cartelisation both as a dealer and supplier of credit. Austria, therefore, provides the clearest example of the direct and deliberate influence of bank capital upon cartelisation'.[53] Having established cartels of companies involved in trading, across a range of commodities from sugar to coal, lumber and steel products, the bank's money-profit would be realised from a fixed commission which it could command as sales agent to the cartel. If this was to be one, relatively risk-free, function of commercial banking, it was to be interwoven within the different function of investment banking. The investment bank's function was inherently speculative in that it advanced credit to industrial enterprise with the uncertain expectation that this enterprise would realise money-profit, or what Hilferding called industrial profit, and would thereby be able to pay interest.

Hilferding's concept of finance capital, the fusion and symbiosis between banking and industrial capital, was duly acknowledged by Marz who also referred to the advancement of this same proposition from a different quarter. Eugen Lopuszanski, a high-ranking official of the Austrian Ministry of Finance, claimed that the universal bank was 'a social force exercising great influence within the commonwealth'.[54]

However, the idea of the bank as master of development was to be contested upon the very ground on which Hilferding had elevated Austria to be a unique case of development – that the universal bank's motive was that of a commercial agent as much as a missionary for investment. This was what Marx had suspected – as much for later as early capitalism – about the domination of commercial agency of banking over the intention to develop industrial production.

Industrial production was financed through the current account business of the banks. Rudolph has explained why 'debtors on current account' were at 'the very heart' of the 'industry bank relationship':

> The core of the current account had two sides. On the one hand, it consisted of deposits by commercial and industrial firms, and on the other, of loans granted to these firms. A given firm could have a credit at one time and a debit at another, while interest was granted or taken according to periodic balances struck on the account.

Such was the basis of the bank credit multiplier, through which an initial deposit of money on the current account created the opportunity for the bank to expand the basis of credit creation. An advance of credit returned to the bank as a new deposit which was used, through successive rounds of advancing loans, to multiply the original assets which the bank employed to yield money-profit. Regulation may make bank assets, its loans and investments, correspond to its liabilities, the deposits held on current account, by enforcing the bank to hold a proportion of its assets in the form of reserves. But, and this is what Hilferding had noticed about Austria, the general banks secured assets through commission banking. Rudolph continued to explain:

> Further, the bank rendered various banking services to its commercial and industrial clientele and the charges and commissions found their way into the accounts. The current account business provided the banks with the greatest source of commissions and profit. More important, the current account provided the means by which firm bonds grew between the banks and given enterprise.[55]

It was through the general bank being 'maid of all work' for industrial enterprise, as Riesser described the German 'great banks',[56] that the general banks in Austria came to be characterised as 'master' of the economy and development.

The overwhelming bulk of money-capital which was advanced to Austrian industrial enterprise at the turn-of-century was granted to existing firms which had become indebted and bonded to banks, was short-term in nature, and rolled over from year to year. Thus, in the *Grunderzeit* period, 80–90 per cent of bank assets were made up of short-term credit: bills of exchange, 'debtors on current account' and advances of securities and goods. Industrial firms used retained profits to finance investment in fixed capital of plant and equipment while short-term credit was used to finance the working capital expended on raw materials, wages and the transaction costs of incorporation and merger. It was estimated that short-term credit advanced to all industrial enterprise, excluding mining and timber mills, amounted to nearly 80 per cent of firms' working capital.[57] Seen in this light, what made the Austrian case unusual was not simply the fusion of the commercial and investment functions of banking in the single institution of a general bank as part of a system of general banks much like that which had been advocated by the Saint-Simonians. Austria was unique because the weight of commercial banking seemed, until the turn of the century, to have come to dominate the original investment function of the *crédit mobilier*-type general bank, the bank which was supposed to be both master of, and handmaid to the economy.

On the basis of this evidence Rudolph concluded that the universal, *crédit mobilier* banks were not primarily and deliberately investment banks: 'It is also clear that the idea of banks as entrepreneurs, initiating industrial development and tiding nascent firms through the dangerous years of youth and adolescence must be largely discarded.' Being risk-averse and cautious, Rudolph continued, banks could not be cast as 'leaders of industrialisation in Austria'.[58]

In other words, the question was whether the historic mission of the development bank, whose function was to advance credit to new industrial enterprise as a deliberate act of policy, had been corrupted by the commercial functioning of the general bank, or whether the very intent to develop – the historic mission of the development bank – could ever be embodied in the general bank as an institution searching for money-profit. This was the question which Schumpeter faced when he cast the bank as the ephor of development and it was the puzzle which was handed down to his doctoral student, Marz, who struggled to find the historic mission of development banking in the unique case of Austria.

Schumpeter had told Marz that *crédit mobilier* was 'one of the nineteenth-century's most consequential innovations' but then Marz, as if stung both by Marx and the twentieth-century academic orthodoxies of economic history, hastened to add that 'due to misapprehension in economic history', the *crédit mobilier* 'were not committed, even in their own explicit view to any "missionary" tasks'. And, 'While either economic historians or politicians may be inclined to impute lofty motives to the sober business of banking, bankers themselves usually see their function in a more prosaic light.' For *Creditanstalt* and the Viennese banks, any 'historic mission' to develop industries was 'defunct'; banks generally promoted, on their own account, 'commerce and industry in a general way' to further their own money-profit.[59] Marz's solution to the Schumpeterian puzzle was that banks became the 'master' of development by default, insinuated into the role of being the 'maid of all work' to industrial enterprise but without the Faustian will of effort which made intent to develop prior to the work of development itself. Who, therefore, was one of the misapprehending economic historians? Marz pointed at Alexander Gerschenkron, the twentieth-century theorist and historian of nineteenth-century late industrialisation.

Gerschenkron, in his essays on European late industrialisation, had made much play of the discontinuities of development and the 'essentially different instruments' of industrialisation. He clearly followed Schumpeter in explaining why the general bank was the model of 'continental practices in the field of investment banking' which 'must be conceived as specific instruments of industrialization in a backward country' and that it was 'essentially' in continental Europe that 'the historical and geographical locus of theories of economic development' assigned 'a central role to processes of forced saving by the money-creating activities of banks'.[60]

This is why Gerschenkron had made much of the fact that French industrialisation under Napoleon III was sponsored by industrial banks of Saint-Simonian inspiration such as *Crédit Mobilier.* Through the Pereires, who forced the 'old wealth' of banking, such as the Rothschilds, to adapt to investment banking, the general bank model was universalised. Although the extent of the bank as an instrument of industrialisation was limited by the extent of backwardness, Gerschenkron's argument was that it was belief in intended development which formed the pervasive ideology of a common policy of late industrialisation.

Although the state substituted for the bank, Gerschenkron emphasised that the policies pursued in Russia during the pre-revolutionary industrial spurt of the 1890s 'resembled closely those of the banks in Central Europe'. In both their 'origins' and 'effects', Gerschenkron emphasised, Russian policies were imbued with the Saint-Simonian precepts of the scientific organisation of industrial society, with the state's function being 'to guarantee workers from the unproductive action of idlers, to maintain security and freedom in production'.[61]

But why was it, Gerschenkron paused to ask himself, that the garments of socialism, including concern for the 'most numerous and most suffering classes', the abolition of inheritance and privilege, planning, had been draped around the industrial ideals of capitalists? Industrialisation in a backward economy, he answered, demanded faith, a break with routine and 'a stronger medicine' than 'the promise of better allocation of resources or even of the lower price of bread'. The Saint-Simonians offered a golden age that 'lies not behind but ahead of mankind': 'This, no doubt, greatly appealed to the creators of *Crédit Mobilier*, who liked to think of their institution as of a "bank to a higher power" and of themselves as "missionaries" rather than bankers.'[62] Such was the mission which Gerschenkron had generalised from France and, eastwards, to Germany. For instance, Friedrich List, 'a man whose personal ties to Saint-Simonians had been very strong', attempted 'to translate the inspirational message of Saint-Simonism into a language that would be accepted in the German environment, where the lack of both a preceding political revolution and an early national unification rendered nationalist sentiment a much more suitable ideology of industrialization.' Further east, Gerschenkron contended, 'orthodox Marxism' fulfilled 'a very similar function'.[63] As we suggested in Chapter 3, the Saint-Simonian mission did enter into Marxism but it was tied up, confusingly, between the palliative and constructivist components of trusteeship. When Marz criticised Gerschenkron on the ground of whether the bank had a mission or not, he forgot about the instrument of the state and the way that the state, despite Schumpeter's scepticism, assumed the investment function of the bank.

Indeed, Gerschenkron took Austria in his stride, writing a book on the one episode in which the state, from 1900 to 1904, committed itself to a spurt of industrialisation but failed. Prime Minister Ernest von Koerber's plan failed because it lacked any 'mission', whether or not of the Saint-Simonian kind, to pursue an industrial strategy of development.[64] In his *Economic Backwardness*, Gerschenkron represented the Austria-Hungary dual monarchy as the two 'models' of trusteeship. Austria, 'backward' in relation to Germany but 'advanced' in relation to Hungary, was where banks promoted industrial activity; in Hungary, the state had to do so.[65] Koerber's developmental plan, Gerschenkron argued in *An Economic Spurt that Failed*, had been primarily motivated to politically sublimate intense Czech–German ethnic conflict which was a part of the political turmoil that had occasioned the collapse of six Austrian governments in the five years after 1895.[66] There is a twist here, however, and one which Gerschenkron noticed in passing. In Austria, the failed spurt also involved the constructivist and palliative components of trusteeship.

Rudolph, for instance, explained the second *Grunderzeit* by way of the same explanation for the first, of the mid-nineteenth century. Expansion of infrastructure and industrial production in Germany generated effective demand for Austrian exports. Second, after state spending on railways, construction and armaments stimulated the wave of industrial expansion through the investment and employment multipliers, the universal banks pursued profit-making opportunities through commercial, current account banking business.[67] Gerschenkron addressed the question of the second, if not leading, role of industrial expansion. If it was not the bank, in what sense was the state the leader of development? After all, Gerschenkron's general thesis was that the state substituted for the universal bank, or vice versa, when, for whatever political and economic conjuncture, there was a lack of intent or capability for the bank, or state in the obverse case, to play a developmental role.[68]

The general theme of the intent to develop, to forestall the emigration of population bearing the potential of productive force, was also at work here. Significantly, it was the Liberal Deputy Max Menger, brother of Carl, who forsook his liberal disposition in 1901 to argue for state development expenditure on the ground that canal construction, as Gerschenkron translated, 'would stop the emigration from Cisleithania to overseas, which had been draining the country of its labour power'.[69] Although it might seem ironic, given that Carl Menger's pure method was directed at expunging from economics the historical method to which Friedrich List was so attached, the Listian overtones of Max Menger's complaint about emigration were commonplace in Austria at the turn of the century. Table 7.1 shows that the annual average number of emigrants doubled in the period from 1896 to 1901, and then more than doubled in the period to 1907, after which the 1908 annexation of Bosnia and the Balkan wars of 1912 inhibited

Table 7.1 Emigration from Austria-Hungary 1893–1913

	Annual average emigrants from Cisleithania[a]	*Net receipts from emigrants' remittances to Austria-Hungary (annual average million crowns)*[b]
1893–95	32,200	24.3
1896–98	33,000	31.6
1899–1901	60,000	70.0
1902–04	91,700	123.6
1905–07	149,300	215.5
1908–10	111,400	288.0
1911–13	125,000	442.0

Sources:
[a]Estimates compiled by averaging two series of annual data from Bolognese–Leuchtenmuller, 1978, Table 45, p. 132.
[b]Rudolph, 1976, fn. 58, p. 269.

emigration because would-be emigrating young men were pressed into military service by conscription.

Equally, however, emigration was deemed to be a virtue compelled by necessity for an economy which faced, between 1908 and 1913, year-on-year deficits on its current balance of payments account due to the sudden growth in net raw material and semi-fabricated goods imports which accompanied the wave of industrial expansion. Remittances from emigrants, as also shown in Table 7.1, proved to be the largest source of foreign exchange receipts at the end of the second *Grunderzeit*; Pasvolsky has estimated that the value of remittances, 1500 million crowns over the 1909–13 period, amounted to nearly half of total net foreign exchange receipts.[70] Since such remittances also found their way into savings bank accounts, and since the *crédit mobilier* style banks agglomerated such saving for their current account loans business, emigration served doubly to finance industrial production. To forestall emigration, on national grounds, was the intent of development, but one result of emigration was to provide an international source of the potential financial means for a process of development. Such was the ambiguity of the Austrian case, an ambiguity which had marked the Quebec case in the nineteenth century and one which came to be a hallmark of twentieth-century cases of underdevelopment.

Koerber had enunciated his 1900 programme in the familiar medium of state-sponsored national economic development:

> Despite the presence of abundant preconditions the development of our productive activities has been greatly impeded and has suffered grievously from the consequences of the continuing nationality strife.
>
> At a time when in the whole world the industrial upswing means intensification of effort and unification of [productive] forces, with us such forces are rendered lame by nationalist strife. To set them free and to place them in the service of welfare and social progress in the State as a whole is a thought that must warm every patriotic heart. Our task is to create for our State a period of repose.[71]

To intend to make development work for a 'road' which 'may remain free for the spiritual and economic development of the State',[72] as Koerber later put it in the Austrian parliament, was sufficient for the regime to garner the support of a parliamentary majority, officialdom, the big banks and business. But the intent was not sufficient to make the programme work as a strategy for industrial development. Minimal in extent, the 1900–4 programme eventually provided for less than the 250 million crowns anticipated development expenditure on plans for canals, especially the much vaunted but never to be completed Danube–Oder link, port construction at Trieste and a new railway through trans-Leithania to the port. The actual expenditure on construction came to less than the sum of emigrants' remittances during this period and was hardly more than one-tenth of the total annual budget of the Austrian state in 1901.[73]

According to Gerschenkron, Koerber's programme was not Saint-Simonian. It was neither akin to Louis Napoleon's version of the mid-century second French Empire nor that of the contemporaneous Count Witte's programme for the industrialisation of Russia. Witte, as Gerschenkron pointed out, controlled the Ministry of Finance and was politically able to increase taxation on the peasantry and cut non-development state expenditure.[74]

Above all, Gerschenkron maintained that the programme could not work because it was squeezed by the two forces of theory, namely the Austrian School and Austro-Marxism, whose contending theorists were also the foremost political practitioners in the arena of the state. Eugen von Bohm-Bawerk, successor to Menger as the leading light of the Austrian School and early theoretical scourge of Marx's economics, was Minister of Finance during the Koerber regime. According to Gerschenkron, it was Bohm-Bawerk who sabotaged the programme when he refused to countenance an increase in state borrowing for state-sponsored development work. Ill-disposed towards the constructivist effort of development, a disposition much accentuated later by Ludwig von Mises and his student Friedrich Hayek, Bohm-Bawerk pointed to the damage which further issues of state bonds, mainly purchased by 'foreign' banks would do to 'our credit'.[75] Indeed, Cisleithania public debt did double between 1902 and 1912; while towards the end of the *Grunderzeit*, bank note issues, the bank rate and the general price level all rose rapidly.[76] Whatever the cause of these aggregate movements, and they had much to do with rail nationalisation, military spending and industrial expansion itself, it was in this period that the long-lasting general idea of an inherent correlation between the intent of development and process of inflation entered into the firmament of Austrian economics.

For the Austro-Marxists, on the other hand, such as Bauer and Adler, Koerber's programme represented a retrogression to the Habsburgs and aristocratic atavism. Despite the fact that he was the first Austrian Prime Minister of non-aristocratic origin, with all this implied for a new, 'progressive' direction of the state apparatus, Koerber's programme was castigated by the leadership of the Social Democratic Party as an attempt to subordinate the interest of Austro-German labour to the nexus of agrarian, and thereby multinational, sources of imperial power.[77] Any development programme which was founded upon the productive force and industrial power of the Austro-German working class would of necessity be, in Austro-Marxist eyes, socialist. Agrarian biased, and seeking to incorporate a 'periphery' of Hungarian, Slav and Czech lands within the German core of Cisleithania, Koerber's 'bureaucratic programme' would, by Austro-Marxist reckoning, only serve to dilute the true internal development of productive force. This Austro-Marxist view spread far beyond the boundaries of Austria-Hungary, expressing the attempt to fend off the external impression of development upon its authentic interior and thereby generalising a condition of underdevelopment from periphery to core. Schumpeter, the 'Austrian patriot', it should be noted, parted company with those Austro-Marxists who had sought a socialist *Anschluss*, the unification of German-Austria with Germany.[78]

SCHUMPETER'S PURPOSE OF DEVELOPMENT

Joseph Schumpeter's theory of economic development can be read as an attempt to confront the underlying causes of Austrian underdevelopment. As such it was also a reaction against the theoretical pincers of the Austrian school and Austro-Marxists who were accused, as in the latterday Gerschenkron, of having made Koerber's development programme fail. Along with the economists of the Austrian school, Schumpeter stood against any state-sponsored effort of economic development. For Schumpeter, development was initiated by the private agency of the entrepreneur. Yet from Marx, Schumpeter had come to understand the process of capitalist development as one of creative destruction. The question then was not one of supplanting entrepreneurial activity but of supervising the process through which creative destruction took place. It was with regard to the question of supervision over the process of development that Schumpeter stood apart from the Austrian school, the Austro-Marxists and, indeed, from Marx himself.

If it was from Marx that Schumpeter captured the process of capitalist development as one of creative destruction, then his key to the general process of development, unlike Marx, was that there had to be intent to supervise the economic system. Marx's interior of development, the active capacity of labour, was alien to and absent from Schumpeter's scheme. It might be obvious, as Suzanne Helburn puts it, that Schumpeter 'could hardly envision the working class becoming a revolutionary class, that is becoming the subjects of history, the major actors and forces for change'. However, and notwithstanding the positivist, Austro-Marxist import of the 'social force' that appears here, we must be careful of the corollary, that Schumpeter's 'evolutionary' theory was one which 'substituted his own theory of class and class relations based upon his ideas of leaders and followership in which entrepreneurs carry out the "new combinations" that promote capitalist development'.[79] Robert Heilbroner, for instance, also emphasised that Schumpeter's scheme was underwritten by a 'vision of the fundamental nature of the body politic' but in then concluding that 'the mass provides social continuity and the elite provides leadership and social change',[80] he failed to notice that the ephor of development was at work to bridge change and continuity.

Schumpeter's purpose of development, within a restricted domain, did give the lead to innovation in production technique. However, 'leadership' was to be given direction through the process of change which was to be supervised. The premise for giving direction to the 'leaders' was that there should be both a conservative principle and 'technique of public life' at work within the general domain of development. Heilbroner was struck by Schumpeter's 'extraordinary conception of capitalism without capital', a conception which followed from making '*the dynamics characteristic of capitalism arise from non-capitalist sources*'.[81] This, however, is to put Schumpeter's purpose the other way around. His ephor of development is of capital and a capitalist source but it is meant to serve the principle of 'social responsibility' within the discontinuous process of capitalist development.

Schumpeter's early ideal of a polity, or what is now called a model of 'governability' for an ideal conception of political order based upon social responsibility, was that of what he called English 'Tory democracy'. Tory democracy was 'the technique of public life which England', Schumpeter reflected in 1916, 'has developed to its absolute perfection'. Systematic control of the media, 'loyal and intimate cooperation with the administrative apparatus' and 'cabinet ministers who reach out to the general public' were the hallmarks of the state regime which he used to confront the faults of development in Austria-Hungary. This was Schumpeter's 'perspective' of being 'conservative in basic questions' which coloured his general approach to development.[82]

Much has been made of Schumpeter's 1945 endorsement of the Catholic doctrine of corporatism when he declared, during a speech in Montreal, that the Papal *Quadragesimo Anno* of Pius XI 'recognises all the facts of the modern economy' and constitutes 'a *practical method* to solve a *practical* problem of immediate urgency'. The immediate problem, for Schumpeter, was the 'social disintegration' spelling class conflict 'between worker and owner' which thereby foretold another post-war, renewed prospect of 'centralist and authoritarian statism'. What Schumpeter emphasised, however, was that the practical problem was a historical culmination of the baneful inheritance of utilitarianism. 'This system of ideas', Schumpeter said, recognised 'no regulatory principle other than that of individual egoism' and expressed 'only too well the spirit of social irresponsibility' which had been perpetuated, as a philosophy and in practice, for a century and more. It created the destructive propensities of 'economic liberalism'. Corporatism, Schumpeter suggested, offered a prospect of putting 'the functions of private initiative in a new framework', securing the good of self-development within a responsible social order and one in which 'peaceful cooperation between worker and owner' would be founded upon the basic income of the 'annual wage'.[83]

Catholic corporatist association in 1945, like the 1916 idealisation of English Tory democracy, was a variation on the theme which gave purpose to the general domain of development. Schumpeter's search was for a means to assert trusteeship in a modern world which encompassed the core of the economic system. Trusteeship, so named, may not have been part of the Schumpeterian lexicon but its burden was ingrained within the role of the ideal bank which, according to Schumpeter's conceptual scheme, gave balanced order to the functioning of the economic system. Development, according to Schumpeter, was to be supervised by the ideal bank which, though private in constitution, was deemed to be a public agency of social order.

An example of the agency by which this trusteeship arose can be seen in

Schumpeter's reversal of the Austrian School's equation between constructivist development and inflation. Schumpeter argued that if inflation, the result of credit creation by banks, acted to stimulate a process of development, then the question of the functioning of the economic system in a manner beneficial for development was conditional upon the trust which was invested by the bank in the entrepreneur. In so far as it was the bank that selected proposals for investment projects, the bank's constructivist purpose had the potential to determine the course of the economy. Such potential, as Michio Morishima has recently noticed in his account of Schumpeter, involves the bank in a role of trust:

> In making decisions on loans, financiers will not only examine the technological aspects of the proposals but also carefully test personal qualities (trustability, leadership, etc.) of the entrepreneurs, especially in applications from small businesses. Decisions are not entirely economic, but sociological and philosophical too.[84]

The question which arises here, as it arose before, was why and how Schumpeter should have presumed that the bank, the institution of capital engaged in the pursuit of money-profit, could assume the role of trustee as if it were a public and social form of authority. To answer this question we must again turn to the Austrian experience.

A singular feature of general banking in Austria was that bankers were 'part of the administrative elite', the 'second society' ranked below the upper aristocracy of counts. Streissler notes:

> . . . the banks had always been considered as organs of a semi-governmental kind of economic policy. Supervision of banks by the Ministry of Finance meant that this Cabinet Minister could, if he wished, more or less direct the general direction of the business policies of the banks. Banking was seen as above all a *public utility* and not as mere business.

And,

> The upper Civil Service prided itself on the indirect ways of achieving its policy aims. Such an indirect policy is the stimulation of innovative activity via bank credit. Banks were public utilities.

At least a part of the origin of Schumpeter's presumption about the role of the bank is clear. If, as we shall see, innovation defined development, then the social authority of the general bank stemmed from its public utility in promoting innovation according to administrative precept. Entrepreneurs carried out particular innovations but bankers supervised the general and social process of innovation. Streissler continued:

According to Schumpeter, innovation was of eminent public utility; and in this attitude he echoed the cant of Austria's top administrators though in fact they did little to turn their professed convictions into practice. Can one wonder then that Schumpeter assigned to the business elite, with which the class he aspired to was closely linked, that he assigned to the banks in other words, above all to the bank managers, a leading role in germinating and fostering the efforts of the heroic entrepreneur, the dynamic innovator?[85]

Such an interpretation accentuates the view that Schumpeter's invocation of the bank was an illusion of the immediate experience of general banking in Austria. If we add to this view the evidence provided above that *crédit mobilier* banking was devoid of a development mission in its commercial banking practices, then we might be led to see Schumpeter's presumption about the bank's role as trustee as 'echoed cant' arising purely from 'Austrian social conceptions'.[86]

It is, however, possible to see Schumpeter's theory in a different light, that of the nineteenth-century doctrine of trusteeship in general, within which he aspired to find an ideal type of trustee which would fit the general theory of development. In this reading of Schumpeter the development mission of the ideal investment banker as trustee was asserted against its subsumption *both* by commercial banking and against the administrative machinery of the state. In other words, and particularly in so far as he generalised from Austria, Schumpeter reacted against what he understood to be a failed case of development and, as such, offered a theoretical account of the possibility by which capitalist development could generally continue to proceed.

If Austria was cast as a corrupt case of development then the true model of development was England. Despite the seemingly perpetual complaint in Britain about the incapacity and/or disinclination of the City of London, so overwhelmingly internationally oriented in its functioning, to direct long-term finance to foster British industrial production, the argument is that Schumpeter, like Marx, was conscious that British banking was more developed than any on the continent. The entrepreneur in Britain, from a turn of century standpoint, was also regarded as a vital exemplar for an enervated Austrian counterpart. Streissler, for instance, concluded that Schumpeter was both 'describing what *should* happen in Austria' and 'presenting a model of development, not its description'.[87] In presenting such a model Schumpeter was very much the modern theorist of development, who associated self-development with the capacity of *personality* to fulfil the social role of trusteeship over the discontinuity and rapidity of economic change which made the development of capitalism possible.

Schumpeter's emphasis on personality followed from his critique of state management of the economy in the name of development. In the latter part of his 1918 essay, 'The crisis of the tax state', which is one of the rare instances in which he paid formal attention to the state, Schumpeter gave a homily on the 'automatism of the free economy' and 'the play of self-interest' in the conversion of Austrian industrial activity to meet the needs of the First World War. In proclaiming that 'nine-tenths of all industrial experience and all industrial talent are at the disposal of private industry and not the government bureaucracy', Schumpeter followed the anti-constructivist path of the Austrian School in order to politically confront the 'socialisation' programmes of the Austro-Marxists. Marx, Schumpeter contended, 'would laugh grimly at those of his disciples who welcome the present administrative economy as the dawn of socialism':

> that administrative economy which is the most undemocratic thing there is, that step back to what preceded the competitive economy which alone can create the preconditions for true socialism and finally evolve socialism itself. The social form of the society of the future cannot grow out of an impoverished economy thrown backward in its development, nor out of instincts run wild.[88]

If it is such comments that have led to Schumpeter being labelled as a 'bourgeois marxist',[89] then we should be clear as to what 'bourgeois' meant for Schumpeter himself.

The recurrent theme of personality in Schumpeter's writings, when coupled with the 'methodological individualism' which Schumpeter invoked to construct the ideal types of banker and entrepreneur, might seem to imply that he followed the Austrian School's reduction of social phenomena to elementary individual subjects of economic action thus confirming the ideal bourgeois as the subject of history. Such is the view that Catephores, for instance, draws from Schumpeter's 'bourgeois' tendency to award 'victory' to 'individual creativity' and to be 'so impressed with the achievements of the entrepreneur's individuality'.[90] Yet, Schumpeter himself proclaimed that he was 'not in the habit of crowning our bourgeoisie with laurel wreaths'. Time and again, in the pages of his 1918–19 essay on 'The sociology of imperialisms' and his 1927 work 'Social classes' Schumpeter implored that 'nothing' was further from his mind 'than to explain a historical process simply by the actions of individuals'. Rather, the purpose of the essays was to explain why, given that 'aptitude' or 'disposition' should be empirically distinguished from 'success' or 'decline', it was possible to explain why one class position, that of the bourgeoisie included, might not succeed and could decline relative to another. In response to those who thought otherwise, he vented his frustration. 'We cannot help those who are unable to see that the individual is a *social* fact, the psychological an *objective* fact, who cannot give up toying with the empty contrasts of the individual *vs* social, the subjective *vs* the objective.'[91] Schumpeter mocked the figure of capital who was moulded by the autocratic state of monarchical power, the European sovereign who 'disciplined the nobility, installed loyalty into it' and 'statized' it:[92]

> The feudal master class was once – and the bourgeoisie was never – the supreme pinnacle of a uniformly constructed social pyramid. The feudal nobility was once lord and master in every sphere of life – which constitutes a difference in prestige that can never be made up. Moreover, the feudal nobility was once – and the bourgeoisie was never – not only the sole possessor of physical power; it was *physical* power incarnate. . . . The nobility *conquered* the material complement to its position, while the bourgeoisie *created this complement for itself*.[93]

Schumpeter sought to assert that the bourgeoisie, by virtue of its command over industrial production, *should* have had the disposition to create its material basis of class power. However, whereas '"feudal" elements' possessed 'a definite character and cast of mind as a class', the bourgeois figure, for Schumpeter, had neither independent conviction nor the 'solidity of social and spiritual position'. Thus, 'while the bourgeoisie can assert its interests everywhere, it "rules" only in exceptional circumstances, and then only briefly. The bourgeois outside his office and the professional man of capitalism outside his profession cut a very sorry figure.'[94]

Schumpeter castigated the captains of business, and here the parallel with Marx's criticisms of List should be noted, because of the way big business sought protection from and of the state at the very moment when the logic of capitalist development impelled the extinction of the military spirit and activity from the economy and society. Such was the 'dichotomy' and ambivalence of the bourgeois mind.

Imperialism, defined to be the 'objectless disposition on the part of the state to unlimited forcible expansion', derived from the political predicates of pre-capitalist society. But national aggression was 'capitalised' in the capitalist epoch of capitalism when it became associated with national policies of protecting large-scale industries and making industrial profit subservient to demands of military expenditure.[95] Schumpeter's contemporary 'bourgeoisie' in continental Europe, when engaged in the pursuit of an imperial policy, therefore acted as if it belonged to the era of early capitalism in which the trade monopolies of mercantilism were the political result of the autocratic state. This was why imperialism was 'atavistic' and why in the peculiar case of England, where the bourgeoisie had assimilated the ideal virtues of a social class, the imperial policy of Joseph Chamberlain foundered in 1906.[96] If the object of capitalist enterprise was to engage effort and energy in the private pursuit of industrial profit, then the historical heritage of the bourgeois, which governed its personality, was disjoined from its function which the 'inner logic' of capitalism demanded. Capitalist development, when *'fused'* with imperialism, deviated 'from the course it might have followed alone'.[97] In the face of this deviation, Schumpeter sought to mark out a course for capitalism as it might be, a course which would culminate in the development of 'true socialism'.

True socialism, being the goal of development for Schumpeter, was the 'liberation of life from the economy and alienation from the economy'. It was 'the mentality' of the 'bourgeois businessman', 'precisely his experiences and methods', which was necessary for the process of economic development to make 'true socialism' possible.[98] This was why Schumpeter pointed to the 1917 Russian Revolution as *the example* of premature socialism. Here we find the basis of an argument that has continued to run to the end of our century, heightened in the wake of the 1989 demise of the administrative economy, that private enterprise is the unavoidable means to the goal of true socialism.

In pointing to the social meaning of private enterprise, Schumpeter referred to the economy which 'is satiated with capital and thoroughly rationalised by entrepreneurial brains'.[99] To urge that economic decisions could be made rational was to address the heart of the ambivalence which surrounded his appraisal of capitalist development. For Schumpeter, as for Keynes, this was theory's main purpose. What, after all, could be the social meaning of private enterprise other than the restricted domain of the economic system and all that is implied by the endlessness and vacuity of value of the economistic?

Between the 1912 *Theory of Economic Development* and his 1942 *Capitalism, Socialism and Democracy*, Schumpeter shifted the locus of enterprise decision-making from the person of the entrepreneur to the body of management. This shift was part of the justification for his seeing an historical break between competitive and 'trustified' managerial or monopoly capitalism. Schumpeter's student Paul Sweezy was later, with Paul Baran, to import this break into the neo-Marxist foundation of underdevelopment theory. However, for Schumpeter, the historical break was part of the logic necessary to establish the source of rationalism in his earlier thought. When he glorified the entrepreneur, Schumpeter accentuated the will and drive to invest of the person who was driven by the rational end of investment but who, in contrast to administrative strata who rationally disposed of economic resources for irrational ends, made decisions irrationally. In other words, the banker was the ideal agent of development because it was through the supervision of banks that investment could be decided upon rationally. As such, economic development was the process through which enterprise was made rational by the bank which bore the burden of development.

Austria, for Schumpeter, was a case in which the potential for 'enormous entrepreneurial achievements' had been throttled by imperial administration. The task of the post-imperial Austrian state, after the First World War, was to

> raise that enormous treasure of energy which in Austria is wasted in the fight against the chains into which irrational legislation, administration, and politics have thrown the personality, which take the entrepreneur away from his organisational, technical, and commercial tasks and which leave him merely the backstairs of politics and administration as the only path to success.

Schumpeter complained about 'widespread misuse of credit' in pre-war Austria:

> This misuse of credit was intensified by the further fact that political pressure repeatedly forced the banking system into an uneconomic compliance with demands for credit for consumption purposes or for agricultural projects of very low productivity, so that forced saving hindered rather than advanced economic development.[100]

Waste and misuse of resources were to become the hallmark of hindered, arrested development, the irrationality of which was then associated with the administered direction of the economy to coordinate industrial effort.

In his 1939 *Business Cycles*, Schumpeter pointed to the Gosplan, the planning authority of the Soviet Union, as an example of the administered coordination of effort to direct development. If the principle of economic development rested upon the destruction of old 'combinations' of techniques of production in favour of the new, through 'the shifting of existing factors from old to new uses', then 'in the case of the socialist community the new order to those in charge of the factors cancels the old one'.[101] If the investment in the new was financed out of past saving, then to cancel the 'old' was to destroy existing means of production in the present without the assurance that the new would produce the same or greater value of real output to compensate for the destruction of the old. In this regard Schumpeter also referred to the financing of enterprise by 'government fiat' as, for example, in the coffee plantations of Brazil during the 1870s: 'More frequently, however, this method was advocated without being actually resorted to. Friedrich List for instance – proving thereby how well he knew how to generalize from American experience – wished to see railroad construction (*sic!*) financed in this way.'[102] Schumpeter's sarcasm was no cheaper than the money which was used to finance railways, money which came from credit-creating banks. Banks, which fulfilled the analogous role of Gosplan or government fiat by allocating funds as part of the capitalist process of accumulation and credit creation, were a distinctive feature of the development of capitalism.

Schumpeter's theory rested upon the assumption that the bank was an agent which had the capacity to discipline the will and drive of the entrepreneur for only by doing so could banks fulfil what Schumpeter regarded as the banks' proper role, to advance money for the purpose of investment rather than 'consumptive expenditure'. A latterday venture capital banker, William Janeway, when interpreting Schumpeter, has formulated what Hyman Minsky has called a law: 'Venture capitalists do not, by and large, plan to be the pawns of rapacious entrepreneurs. Sooner or later, nonetheless, the venture capitalist learns this law of life: entrepreneurs lie.' Entrepreneurs do not simply lie because they use investment money for personal consumption. Rather they lie, as both Janeway and Minsky point out, because there is a systemic tendency for the flexible *income statement* of the entrepreneur to outstrip the inflexibly defined *cash flow* account of the bank statement. 'Countless are the ventures', Janeway states, 'that have run out of money while reporting record profits as they accelerate reported revenues by hook and by crook': 'It is a tribute to Marx that M–C–M, the flow of cash *out* – to purchase labour services and materials; *in*, from the collection of receivables for the sale of products – should remain the basis for policing the financial integrity of the capitalist venture.'[103] Without trammelling entrepreneurial activity, the banker was the authority which sanctioned the belief that risk-incurring investment plans would realise cash-earning payments from projects stemming from the entrepreneurial imagination. Despite a surfeit of buccaneering bluster, the entrepreneur, in Heilbroner's more perceptive words, was Schumpeter's 'tragicomic figure', the anti-hero of development.[104]

In so far as the bank manufactured money and possessed limitless ideas about how different kinds of money-instruments could be created, credit-creation was equally an entrepreneurial function. But it was the very ability to generate money instruments which led to the bank's speculative propensity for creating money for its own account and it was the bank's entrepreneurial function that made the theory ambivalent.[105] Speculative propensities, of the bank no less than the entrepreneur, had to be internally subordinated to the function of trusteeship for the general body of entrepreneurs whose only account for the use of money might seem to be the external authority of the bank. It was the bank's external authority that rationalised entrepreneurial investment plans but the money which the bank created was an essential part of the internal process of capitalist development and it was this that made capitalism a unique domain of development. Whether desired by bankers or not, the logic of trusteeship had to be thrust upon the bank.

The banker, Schumpeter wrote, 'should be an independent agent. To realise this is to understand what banking means.' Not only must the banker rationally understand the technical basis of a financial transaction 'but he must also get to know the customer, his business and even his private habits' when screening investment plans. Such surveillance required that banks be independent of industrial enterprise in the manner of his much vaunted model of English banking and City of London money markets. Equally, banks were to 'be independent of politics': 'Subservience to government or to public opinion would obviously paralyze the function of that socialist board. It also paralyzes a banking system. This fact is so serious because the banker's function is essentially a critical, checking, admonitory one.' Economic 'catastrophes' of capitalism, Schumpeter asserted, happened when 'the banking community' failed to act '*corporatively*', failed 'to function in the way required by the structure of the capitalist machine' and were in 'times of decadent capitalism' coerced into unpopular action by state legislation.[106] Previously, Schumpeter had commented:

> The banking world constitutes a central authority of the economy whose directives put the necessary means of production at the disposal of innovators in the productive organisms. A monetary process, the creation of money which is only a 'claim ticket' and not also a 'receipt voucher', and the rise of prices to which it leads, becomes a powerful lever of economic development.[107]

Schumpeter derived the concept of a 'claim ticket', for money of which 'no-one' guaranteed the 'commodity content', with all its implied positivist trustability from J.S. Mill.[108] Banks created money as a series of claims held in trust to title over parts of the sum of commodity output or 'social product' which might be expected to increase as money was used for productive purposes. Schumpeter continued: 'The essence of modern credit lies in the creation of such money. It is the specifically capitalistic method of effecting economic progress. It gives scope to the *capitalistic function* of money, as opposed to its market-economy function.'[109] Contained within this distinction between the *capitalist* and the *market-economy* functions of money was Schumpeter's claim that Marx had reduced social phenomena to the economic while he, Schumpeter, wanted to assert a fundamental difference between the economic system and social order of capitalism.

In drawing the distinction between the capitalist and market-economy functions of money, Schumpeter's endeavour was to distinguish between the static and dynamic processes of the economic *system*. The market economy corresponded to a continuous process, akin to Marx's M–C–M characterisation of the circuit of commercial or merchant capital, in which there was no immanent economic force of development at work within the system. In his 1928 essay on 'The instability of capitalism' Schumpeter, in setting out the logical basis for a distinction between the economic system and the social order, collapsed the capitalist and market-economy functions of money into his 'economic system':

> We mean an economic system characterised by private property (private initiative), by production for a market and by the phenomenon of credit, this phenomenon being the *differentia specifica* distinguishing the 'capitalist' system from other species, historical or possible, of the larger genus defined by the first two characteristics.

Schumpeter went on to set out why he wanted to distinguish system from order, writing of 'the capitalist *order* instead of the capitalist *system*' when the question was about 'the institutional survival of capitalism'.[110] The 'institutional' has come to be interpreted as meaning the hierarchical body of socially necessary functions and social 'positions' of class which Schumpeter employed to establish the social prerequisites of, for instance, entrepreneurship.[111]

Entrepreneurs, being motivated by the intent to raise their social position, embodied the function of investment within the economic system of 'competitive capitalism'. However, the 'organised', 'regulated' or 'managed' capitalism of the twentieth century shifted this investment function to a managerial cadre. For Schumpeter, it was the shift in the locus of investment decisions, and not the business cycle, which was the source of the instability in the social order of capitalism. There was an 'inherent tendency to destroy the "order" by undermining the social positions on which the "order" rests'.[112] Borrowing from Max Weber, Schumpeter used the concept of *patrimonialisation* to refer to the loss of social rank and prestige which was attendant upon the loss of social function in the economic system. Given his interest in the survival – as much as the arrival – of capitalism, Schumpeter emphasised the decline and rise of social positions as the basis of what can be called the general domain of development.[113] Indeed, for Schumpeter, it was precisely the means by which social positions rose and declined that constituted the crucial link between the economic system and social order of capitalism. As such, while the economic system was contained within a restricted domain of development, the general domain encompassed both the economy and the social order and included the link which bridged the restricted and general domain of development.

It was the bridge between the economic function of the bank within the restricted domain of the economy and the position of the banker in the social order of capitalism which was fundamental in Schumpeter's working out of a general domain of development. If Schumpeter eulogised the turn of century nexus of the City of London, it was because this nexus demonstrated how the ideal patrimonial of trusteeship might arise. Here, those of aristocratic rank, whose social function had formerly been that of trusteeship over land, had maintained their social position by becoming trustees over the gains of money-profit from banking. Yet, this instance of the possibility of ideal trusteeship, bridging the economic system and the social order, rested upon the presupposition of the development of the economic system itself. No general theory of the development of the economic order could be constructed out of such historically specific conjunctures as that of the rise of the City of London other than that of the appearance of the ideal social 'condition' which Schumpeter, as Dyer has pointed out,[114] assumed to be the prerequisite for development within the restricted domain of the economy.

Thus Schumpeter was trapped by the nineteenth-century conundrum of the positivist doctrine of development which assumed, but could not explain, the existence of that which it was argued made development possible. 'By and by', Schumpeter wrote, 'private enterprise will lose its social meaning through the development of the economy and consequent expansion of the sphere of social sympathy.' Social sympathy, which the Saint-Simonians had inscribed in their call for the realisation of the ideal of *crédit mobilier* banking, was transmuted, through the practices of policing and rationalising, into Schumpeter's conception of the bank as the supervisor of development. If the Saint-Simonian bank, as the amalgam of prophecy and swindling, was the negation of development for Marx, it became, for Schumpeter, the positive agency of development.

Schumpeter's return to the nineteenth-century idea of trusteeship did not mean a return to the positivist belief in the doctrine of intentional development as the means to reconcile order and progress. Nor did it mean that development could be assimilated into the 'progressive development' of nineteenth-century evolutionary thought. Yet, Maria Brouwer, in her introduction to Schumpeter states:

> Schumpeter's theory can be named evolutionary in two respects. First, he explained processes of economic evolution or productivity growth. Second, he dealt with changes in capitalism's institutions and its culture which, he so assumed, were paving the way towards socialism. His model of social evolution is uni-directional, whereas his purely economic model is cyclical.[115]

Given Schumpeter's recourse to the biological metaphors of environment, species, genus and organism, Brouwer's conflation between evolution and development is unsurprising. However, the distinction between what is *unidirectional* about the progressive development of the social order of capitalism and *cyclical* about the process of the economic system is vital for the understanding of why Schumpeter, as modernist, dissociated progress from development. Progress, either in the economic sense of a growth in productivity or in its expanded meaning of the 'liberation from the economy', was implied by development but the process of development was not unidirectional. Rather, what marked economic development was a conjuncture between the classical metaphor of cyclicality, regarded as both a conceptual and empirical phenomenon of capitalist development, and discontinuity. Discontinuity, unlike the classical conception of continuous sequences of growth, expansion and decay, involved simultaneous destruction and decline. Bearing in mind that the 'creative' stands in an adjectival relation to the substantive of 'destruction', Schumpeter's longstanding idea of development as a process of creative destruction should be used warily.

PROGRESS AND DEVELOPMENT

Development, for Schumpeter, was deliberately set in contrast to the norms of progress in order to highlight the significance of development as representing 'revolutionary' changes in production and consumption conditions which were independent of any prior set of conditions or preceding events. Schumpeter's purpose in referring to the model of the normal, circular flow of economic life (the *Kreislauf*) in which economic conduct was directed towards the satisfaction of need by 'habitual economic method'[116] was to associate what was necessarily slow, continuous and natural about progress with the simultaneous balance and ordered equilibrium of natural economic exchange. By this reckoning, the image of progress, as the adaptation and ordered evolution of improvement through linear stages of history, was consigned to what was habitual and non-developmental about economic life.

Economic development was part of the economic system but could not be reduced to the system itself. It must be borne in mind that it was from the original theory of Leon Walras, the utopian socialist whom he regarded as the greatest nineteenth-century economist, that Schumpeter constructed the non-developmental model of the *Kreislauf*. Schumpeter, in the *Wesen*, had borrowed the theological concept of 'catallactics' to designate an economic domain of pure exchange. By bringing in Walras' system, as a logical construct to establish the economic function of exchange, and thereby emphasising the functional as opposed to causal method which he associated with Menger, Schumpeter's purpose was to divest the normal economic process from any constructive intent, including the psychological, which he wanted to derive separately from the 'abnormalities' of the social order of capitalism.[117]

In contrast to the nineteenth-century laws of progressive development, development, for Schumpeter, was about the rapid and abrupt striking out, along tangents, from the normal circular flows of economic life. A trajectory of rapid movement was the modern idiom which expressed forms of change of development. However, Schumpeter simultaneously captured the essentially classical meaning of development, that the new was a moment in the destruction of the old. In doing so, Schumpeter attempted to transcend a metaphysical or evolutionary conception of development by insisting that, while development was about the causes of disturbance and destruction of existing material conditions of life, mere adaptation to disturbance from some exogenous or external source of change, such as colonialism, famine or war, was not development but part of normal economic life.

Thus, in his *Theory of Economic Development*, Schumpeter set out a model of a static economic system. This system was characterised by the timeless balance and harmony of equilibrium in which all individuals who sold services to meet subsistence adapted to any enforced changes in the circumstances of such transactions. Such changes forced a new calculation of any individual's decision-making but did not change the order within which decisions were made. As long as the source of change was exogenous and did not disturb the channels within which decisions were made, change only involved the movement from one equilibrium position to another within the same set of conditions of production and consumption.[118]

In Schumpeter's scheme, evolution was treated as the normal and normative order of adaptation to an environment and not as if it were a part of development.[119] Charles Darwin, in the *Origin of the Species*, had contributed to the attack on any idea of a pre-ordained order, including that of the design of Providence, and had arrived, as had A.R. Wallace, at the theorem of natural selection. This was the mechanism which offered possibilities of improvement that did not depend directly upon human will. Because variation in a population was random, it was possible that the 'unfit might be the progenitor of a new species'.[120] While Wallace came to renounce natural selection in favour of the spiritual force of 'overruling intelligence' as the explanation of human improvement, Darwin later fell foul of his temper by returning, in his 1871 *Descent of Man*, to the earlier and more positivist conception of intentional development. It proved to be too difficult, amid the peaks of the Victorian age of progress, to keep to the belief in progress while upholding that immanent evolution was predicted by the *random* properties of natural selection. In the wake of Darwin's retreat the door was opened for the Lamarckian version of progressive development to be rehabilitated through the philosophical idea of associationism.

Associationism rested upon the premise that the individual derived the sense of experience from an external environment and that natural selection was merely a means of formalising the adaptation to a change in a given external reality. When attached to utilitarianism it could be further supposed that the complexity of an adult mind was composed of simple childlike elements, such as those of pleasure and pain in the original utilitarian calculus, which unfolded in a particular manner as a result of being increasingly exposed to the experiences of reality.[121] Moreover, the differing capacities for association between child and adult, which were assumed to determine the development of the one into the other, could be transposed to those between the lower and higher order complexities of primates. Such differences also could and were transposed to the differences between non-adult and adult races, young and old nations and, then, undeveloped and developed countries.

This genesis, now referred to as developmentalism, was taken over by Herbert Spencer who, despite his adult inclination to reject all that he had youthfully learned from Comte, continued to argue from a Comtean biological analogy in taking development to be an increase in the complexity and differentiation of the 'social organism'. By the end of the nineteenth and well into the twentieth century, through the influence of Hobhouse in sociology and Marshall in economics,[122] development continued to be associated with the view that natural capacities could be organised to promote the development of progress. When political and ideological disputes ranged between those wedded to 'competition' and those to 'cooperation', the source of the dispute lay only in whether the random source of variation lay in the individuality or the collectivity of the person. And when capacities were deemed to be imperfect, as in the Webbs' perception that the English social organism was 'unfit' or 'diseased' because it showed signs of regression, the dispute was over the means by which the entity, of person, 'race' or 'nation', could be put back on the natural and correct path of progressive development.

Schumpeter, for all his express reference to the 'unfit' and 'untalented', expressly detached development from associationism. Rather, the associationist implications of developmentalism were carried across to the model of the circular flow of economic life. Economic conduct in a system devoid of development was assumed to take place in 'a mechanical organism' where individuals habitually adapt and act according to their 'experience' of life.[123] Adaptation was incremental, continuous and relatively slow whereas development, whose abiding attribute was discontinuity, carried the Faustian injunction to act immediately and fast. As such, the agency of development, consisting of the Faustian contract between the entrepreneur and the banker, was carried over from the social order to the economic system of capitalism. Schumpeter, acknowledging his debt to Marx, repeated time and again that while development was an *internal* process, there was no interiority, no deliberate intent to develop, within the circular flows of economic life. Economic development, therefore, was the process by which an external agency of development became internal to the economic system. It was the means by which the immanent spontaneity of economic growth was given direction by the phenomenon of credit-money, an agency which was external to the system of exchange but internal to the economic system of capitalism. Any intent to develop had to be conveyed to the economic, while being part of it. It was thus that Schumpeter gave new meaning to the trusteeship of progressive development.

THE INTERNAL AND EXTERNAL OF DEVELOPMENT

Schumpeter, in his 'Imperialisms' essay, argued that 'leading bankers are often leaders of the national economy' and then claimed that, in the bank, 'capitalism has found a central organ that supplants its automatism by conscious decisions'. Yet, earlier in the same essay, he had supposed that it was the development of capitalism, as a social order, that made the capacity for 'conscious decisions' possible. Capitalist entrepreneurs, 'industrial and financial leaders', professionals and intellectuals, workers, and 'even the peasant', 'were freed from the control of ancient patterns of thought, of the institutions and organs that taught and represented these outlooks in village, manor and guild':

> They were removed from the old world, engaged in building a new one for themselves – a specialized, mechanized world. Thus, they were all inevitably democratized, individualized and rationalized. . . . They were rationalized, because the instability of economic position made their survival hinge on continual, deliberately rationalistic decisions – a dependence that emerged with great sharpness.[124]

Here Schumpeter registered the immediate historical dilemma of the modern. The capacity for self-development was inherently thwarted by the dependence of the 'self' upon the instability of economic position. The prospect of development made economic position more unstable still. Given the instability of positions, Schumpeter sought recourse in the one source of 'rationalistic decisions' which was external to an ideal social world in which all 'new types' of individuals would have the economic capacities to act according to the precepts of self-development. That one source of 'rationalistic' decision-making was the bank.

Within the pure system of exchange, being the non-development model of circular economic life, individuals carried out transactions of land and labour, 'the original productive factors', to obtain consumption goods in circular, repetitive and simultaneous flows of exchange. Schumpeter assumed that there was no accumulated stock of goods, no money credit and no class whose characteristic is that it possesses a stock of either means of production or consumption. Workers made economic decisions with the command over labour being simply another kind of work. Thus there was no distinction to be made between 'directing and directed labour'. All commodities were 'transitory' in that they are immediately exchanged to meet need.[125] Equally, there was no rate of interest and no saving; nor could profit and surplus value arise both because no product of labour could command a surplus over the costs of what labour and land could yield and because the value of the 'productive service' of labour absorbed the value of the product. There was, then, no concept of exploitation in Schumpeter's non-development model.[126]

The marginal utility of each good was equal to that of any other in all possible uses. Thus, prices were proportional to these marginal utilities. Wages and rents, in being the prices for services of labour and land, were proportional to their marginal productivities. Given that all sellers of all commodities became buyers of consumption goods and means of production and all buyers sellers in order to maintain in any period the previous use-value of their wants for consumption, all commodities found a market and individuals' plans were realised in equilibrium. Since there were no stocks of commodities, productive forces or money, unemployment could not emerge from the properties of pure economic exchange.[127] All individuals drew conclusions from known circumstances and a change in those circumstances produced a new equilibrium which was of no different order than the old.

If it was supposed, Schumpeter theorised, that there was continuous change in the system of exchange, then it was through exchange transactions that individuals would adapt to such change in external circumstances of events or information through 'infinitesimal steps' or by 'variations at the margins'. Through a series of adaptations the gravitational centre of an equilibrium could itself change and this shift in the gravity of exchange could sometimes be regarded as being dynamic. If, for example, consumer wants changed, then the application of labour and land could be varied at the margin to meet a change in the relative demand for any two or more commodities. Continuous change in the use of resources would potentially balance, through the proportional equivalence of the marginal utilities of the commodities, the withdrawal of resources from producing one commodity to another. This change, as it came to be understood, was represented by the comparative statics of shifts in the gravity of exchange.

From Bohm-Bawerk, Schumpeter got the idea of 'friction', the idea that a shift in the gravity of exchange would be imperfect in the sense that the plans of economic agents could only be realised imperfectly. Friction was 'error, mishap or, indolence' or any other source of imperfection which kept exchange out of equilibrium and created profit and loss for transactors until such imperfection was corrected. For Schumpeter, 'profit' was a 'symptom of imperfection' which was eliminated by adaptations through the channels of exchange.[128] Yet despite these imperfections and adaptations the pattern of 'the economic organism' was deemed to remain unchanged. However much the order of the organism shifted from one position to another, from that of equilibrium to that of disequilibrium and back again, it did not follow that the system was dynamic or disordered.[129]

In questioning the contrast between the static but positive concept of social order and the dynamic concept of economic progress, Schumpeter later discarded the concepts of statics and dynamics on the grounds that the analogy with mechanics was misleading. Different and ambiguous meanings of the concepts, Schumpeter suggested, led 'back to John Stuart Mill, who owes the suggestion to Comte, who, in his turn, expressed indebtedness to the zoologist de Blainville'.[130] Expressing scepticism about the way in which Comte wished to make the change of progress fit the paramountcy of order, Schumpeter highlighted the discontinuity and disorder of dynamic change by counterposing the discontinuities of development against the recurrent routines of economic life through which the static order of equilibrium was established.

Equilibrium and development, Schumpeter had earlier made clear, but only in the original German edition of *Economic Development* 'are opposite phenomena excluding each other': 'It follows from our entire thought that *a dynamic equilibrium does not exist*. Development in its ultimate nature consists of disturbances of an existing static equilibrium and does not have a tendency to return to a previous or any other equilibrium.' Economic equilibrium, for Schumpeter, was the ideal of a 'static economy' and one which, to repeat, represented the order which was re-established through adaptation to the destructive forces of external events.[131] Development, however, brought its own internal sources of destruction and, again, it is the ephor of development, of the economic system but not simply 'the market', who has to be brought into play to counteract the fundamental change of development.

Development moved the economic system from one equilibrium to another by changing the channels of exchange. As such, development, for Schumpeter, was that fundamentally unstable and disequilibriating change whose source was endogenous and whose impact was to destroy existing production and consumption practices in the process of creating new ones. Channels of exchange, sources of information and data of production and consumption were all abruptly changed by development. In order to be development, improvement, which made industrial expansion possible, had to be of necessity expansion which was neither automatic nor induced by adaptation:

> Industrial expansion, automatically incident to, and moulded by, general social growth – of which the most important purely economic forces are growth of population and of savings – is the basic fact about economic change or evolution of 'progress'; wants and possibilities develop, industry expands in response, and this expansion, carrying automatically in its wake increasing specialisation and environmental facilities, accounts for the rest, changing continuously and organically its own *data*.[132]

By Schumpeter's reckoning, much of what is today taken to be the activity of 'development' was nothing other than an adaptation to externally imposed change by means of existing production conditions or, in economics jargon, to the information that is contained in an existing supply curve. Activities of this kind, including the backward and forward linkages which are transmitted from one firm or industry to another as, for example, firms internalise a reduction in costs made possible by changes external to the firm, were to be regarded as secondary phenomena because there was no 'break' in production practices.

Schumpeter explicitly questioned that which informs so much of the present-day practice of 'development':

> Industrial evolution inspires collective action in order to force improvement on lethargic strata. Of this kind was, and is, government action on the Continent [of Europe] for improving agricultural methods of production. This is not 'secondary' in the sense we mean it, but if it comes to creating external economies by non-economic influence, it has nevertheless been due so far mainly to some previous achievement in some private industry.[133]

The achievement of development was not to make more use of labour and land which were underemployed in a given use dictated by some past course of development but was the '*different* employment of *existing* services of labour and land'.[134] Improvement was first about, 'new combinations of existing factors of production, embodied in new plans and, typically new firms producing new commodities, or by a new, i.e. as yet untried, method, or for a new market, or by buying means of production in a new market'.[135] But, second, such improvement or innovation involved the destruction of old combinations, old plants, old firms, old commodities and old or existing markets: 'The carrying out of new combinations means, therefore, simply the different employment of the economic system's existing supplies of productive means – which might provide a second definition of development in our sense'.[136] The unremitting association of Schumpeter with the heroic archetype of the entrepreneur, who embodied the purely creative function of development by a will to innovate, reduced the meaning of development to subjective intention. As an embodiment of the thesis of discontinuous change, the entrepreneur became Schumpeter's unintended hostage of fortune because he presented development as the simple redeployment of 'the economic system's existing supplies of productive means'. His antithesis, that redeployment meant decay and destruction, was obliterated by the one-sided association of development with the subject who commences development at a tangent to the circular flow of economic life.

An ideal social order for a capitalist economic system was one which provided aspirant entrepreneurs with the means of production. This is why the bank was essential to supervise development. Following Marx, Schumpeter repeated:

> New men and new plans come to the forefront that would otherwise remain in the background. The obstacles are removed which private property places in the way of him who does not already have command over means of production. The banking world constitutes a central authority whose directives put the necessary means of production at the disposal of innovators in the productive organism.[137]

Schumpeter's fudge was to suppose that the redeployment of means of production could be rationalised through the Faustian contract within the economic system, the restricted domain of development, while the authority of development was founded upon the restraint which was exercised within the social order of capitalism.

INTENTIONAL DEVELOPMENT

It was no simple task for Schumpeter to explain how the productive force of labour might be 'simply' redeployed at a tangent to, but outside of, the circular flows of market exchange. In setting out his explanation of the fundamental internal process of development, Schumpeter was inspired by Marx's 'vision' of the creative potential of capitalism, a potential that contained within itself, and upon its realisation, its self-destruction. Vision here was a 'preanalytic cognitive act' of insight and imagination which provided the impetus for 'analytic effort'.[138] Marx's vision, Schumpeter reckoned, was far superior to that of the smug bourgeois Ricardo, the misanthropic pessimist Malthus who had been the immediate inspiration for both Darwin and Wallace, or of the one-sided optimist List who had hoped that national development could create an urban-industrial society for Germany. 'Marx's performance' was

> the most powerful of all. In his general scheme of thought, development was not what it was with all other economists of that period, an appendix to economic statics, but the central theme. And he concentrated his analytic powers on the task of showing how the economic process, changing itself by virtue of its own inherent logic, incessantly changes the social framework – the whole of society in fact.[139]

But while Schumpeter seized upon Marx's vision, which he took to be derived from Hegel's principle of development, he rejected the logic of Marx's analysis of capitalism. In his 1949 essay, 'The *Communist Manifesto*', Schumpeter expressed his repulsion for the logic of Marx's analysis of exploitation in the production process as the source of a purely economic conception of class and class struggle.[140] For Schumpeter, the creative potential of capitalism was not explained, as it was for Marx, by the tension between the difference between the value of labour power and that of labour. This was the difference which, through the development of capitalism itself, separated the capacities of individuals to do and create from the work that producers had to do to meet the needs of subsistence. In rejecting Marx's logic, Schumpeter eliminated the source of the exploitation whereby capital was formed and accumulated. There was then no source for the necessary investment that expanded the productive capacity that enhanced potential need while simultaneously denying any actual need that was not necessary for the domain of work to make labour productive of surplus value. Thus, while he made capital the central concept of development by drawing on Marx's recognition of how economic development was embedded in capitalism,

Schumpeter gutted Marx's account of why the object of the productive effort of labour should come to be a subject directing the deployment and work of labour.

There is much confusion over Schumpeter's logic of capital. Heilbroner, for example, had accused Schumpeter of gutting Marx's 'essential insight' by arriving at a version of 'capitalism without capital' when he appropriated Marx's vision by recognising why economic development was embedded in capitalism. Schumpeter, Heilbroner complained, had refused to acknowledge the 'function served by capital' in the *Kreislauf*, the function being to direct and control labour 'in the *normal* course of economic activity'. However, capital, for Schumpeter, was neither 'benign' nor 'powerless', as Heilbroner put it, but the very opposite. Schumpeter eschewed any interest in the immediate source of directed labour in the enterprise because, in displacing the logic of capital into the course of development, he derived the capacity to direct labour from the authoritative source of development. Such a source was beyond Heilbroner's 'fundamental conception of the unit of capital – the business firm'.[141] Capital was not about how work was deployed and directed in the firm but how the means of so doing was directed by the ephor of development.

Capital, for Schumpeter, was not to be found in the circular flows of exchange. Interposed between a world of commodities and the entrepreneur, capital was defined as the sum of the means of payment, the fund of purchasing power, which was represented by credit money and directed by the agency of the money market: 'The money market is always, as it were, the headquarters of the capitalist system from which orders go out to its individual divisions, and that which is debated and decided there is always in essence the settlement of plans for further development.'[142] As a fund of purchasing power, capital was the 'bridge' between the circular flow of exchange and the tangential trajectory of development – 'performing a task' before production could start. So emblematic of the bridge of development, capital was also the instrumental 'lever' which set development in motion: 'Capital is nothing but the lever by which the entrepreneur subjects to his control the concrete goods which he needs, nothing but a means of diverting the factors of production to new uses, or of dictating a new direction to production.'[143] Through the mixed metaphor of 'bridge' and 'lever', Schumpeter sought a way to express the idea that development was the means by which the routines of existing production conditions could be made consistent with the creative potential of production embodied in the principle of development. One could not administer the diversion of 'productive means' by 'dictating' a new direction of production. Direction had to happen according to the diachronic precept that the results of the redeployment of the means of production would be justified by gains in productivity and, thereby, the survival of capitalism. But it had to occur within the synchronic process of the economic system within which there was no intent to develop.

If it was the redeployment of productive means which gave development new direction through a symmetrical and simultaneous process of the destruction of the old in the creation of the new, then the adjustments and adaptations to new production and consumption conditions happened through sequences of the business cycle. As many commentators have pointed out, Schumpeter's gales of 'creative destruction', which made the economic system dynamic in the sense that capital moved the economy from one equilibrium to another, were uneasily situated within the stable framework of the circular flow model wherein the routines of continuous adaptation governed the organisation of production.[144]

In concentrating upon the single process of innovation, which was founded upon his specific meaning of capital, Schumpeter specified an imagined end-state of innovation in which old production conditions were eclipsed by the new. However, the process by which the old was destroyed was not envisaged either in the initial conditions of the circular-flow system nor in the intent which set capital at work to do the work of development. The upshot of this criticism of Schumpeter is that the intent and process of development were confused. In response, Allen Oakley, for example, has suggested that the bridge of development should have been founded upon the 'traverse', the 'set of sequential and parallel processes of adjustment' which moved the economic system from one steady state to another 'through 'real-historical time'.[145] Schumpeter's analytical genesis of development can be laid out simply. Unlike what was implied in the Gosplan model of saving for development, it is the innovative capacity of the bank to create money that equips new enterprises with the purchasing power of credit to innovate in production before increased real output is available for consumption. As a result of credit creation, prices for means of production and consumption rise and the increase in the set of output prices stimulates new production because, at old costs of production, profit is generated within the economic system. Given that real costs of production, including money wages, do not rise at the same rate as prices for 'new' output, and since production is dominated by old enterprise, the source of profit is forced saving. Forced saving, in turn, is the means to fund the repayment of credit extended to new enterprises by the bank. Credit is thus repaid out of profit which enterprises share with the bank. It is in this phase, of debt-deflation following that of credit-inflation, the period of depression following that of boom and prosperity, in which older enterprise is destroyed.[146]

Older enterprise, consisting of old techniques of production embodied in old firms, is destroyed during periods of depression to the extent that it cannot generate profit to repay debt. Thus, phases of depression are as necessary as those of prosperity if development is to take place. Moreover, it is during recessions that innovation is diffused throughout the economy. Through competition, 'new' output drives out old as new enterprise moves down a decreasing cost curve of production and the general price level, in reflecting the change in relative prices for old and new output, falls to stimulate consumption expenditure. A new equilibrium for the economy is established incorporating the qualitative change in production and consumption conditions. Thus, economic development creates new methods of production and new products which are deemed to be an improvement on the old. Schumpeter reflected that this analysis of development corresponded to what Marx had suggested of the 'constant revolutionising of production' and postulated that since 'capitalism is a process, stationary capitalism would be a *contradicto in adjecto*':

> But this process does not simply consist in increase of capital – let alone increase of capital by saving – as the classics had it. It does not consist in adding mailcoaches to the existing stock of mailcoaches, but in their elimination by railroads. Increase of capital is an incident in this process, but it is not its propeller.[147]

The 'propeller' which drove forward capitalist development, the bank–entrepreneur contract, might seem in Schumpeter's account to leave the old stranded on the shores of history outside the process of development itself. Alternatively, it might seem that Schumpeter intended that the limit of the Faustian contract was contained by some external authority of development, the authority which is internalised by the bank.

'How exactly the old firms are ousted', Brouwer wrote of Schumpeter's equivocation, 'is one of the puzzles created by the Schumpeterian scheme'.[148] At the outset of the process of development there is no intention on the part of the bank to destroy the old but it is the social function of the bank to account for the capital of development. It is the working of the accounting function of the bank, during the process of development, which destroys the old. When banks select entrepreneurial projects, including those of existing enterprise and transfers capital between enterprises, the purchasing power of old enterprise is squeezed as a result of a belief on the bank's part that new enterprise will generate greater money-profit for it. The extension of new credit to enterprise is limited by the bank's belief that the accounting value of output produced will not repay the value of credit and interest and that enterprise will fail to continue to realise money-profit from innovation.[149]

Interest which the enterprise pays to the bank is, as Bellofiore has pointed out, a tax on industrial profit and, as such, a brake on development.[150] And, as Schumpeter implied, trust was placed in the bank to ensure that this tax on industrial profit would be used by the bank to generate further rounds of industrial investment and not be exhausted in unproductive expenditure. Schumpeter's trusteeship of development is embedded in the confused tension between the intent and process of development. While the intent to develop may be rationalised on the part of the bank, it is the Faustian bargain – the relation of the borrowing enterprise to the lending bank – which makes the money market responsible for development. Through the money market, the entrepreneur is 'entrusted' with productive forces but their use for development cannot be predetermined on any basis other than what is necessarily emotive: the entrepreneurial function. The money market, Schumpeter proclaimed, 'becomes the heart, although it never becomes the brain, of the capitalist organism'.[151]

Finally, it is because the banker acts as the trustee for development that the hero of Schumpeter's developmental scheme is not the industrial innovator who demands credit-money but the banker. The development banker was not the commercial banker, the middle-man who bought and sold commodity-money, the exchange value of which depended upon the use-value of consumer goods, the starting point for the *Kreislauf*. In the circular flow system, money was a technical instrument of exchange – 'the cloak of economic things' – and whose aggregate was the total stock of money; credit had no relevance for economic life. Development banking dealt with the demand for new purchasing power and not the redistribution of existing purchasing power in the form of commodity-money.[152]

If Schumpeter supposed that entrepreneurs were drafted into the economic system from the social order of capitalism in order to fulfil the function of enterprise as a result of their searching for the means of success to make the family upwardly mobile within the structure of class society, then it is clear that the social source of the banker was both of the same order and had the additional function of advancing credit-money for investment to enterprise. While the bank was not ranged against industrial enterprise as a whole, in standing between new and old enterprise, neither the function nor the personality of the banker could be reduced to that of industrial enterprise.[153] The entrepreneur, according to Schumpeter, did not value the use of consumer goods but sought 'out difficulties, changes in order to change, delights in ventures'. Since ventures cost money to society, the entrepreneur became a 'debtor' to society 'in consequence of the logic of the process of development', or as Schumpeter put it, 'his becoming a debtor arises from the necessity of the case and is not something abnormal, an accidental event to be explained by particular circumstances'.[154] Schumpeter was clear that the advance of credit was to entrust the entrepreneur, the debtor, with the productive force of society 'because new combinations means a new use of productive forces whose product from the innovation may not repay the social stream' out of which old productive forces have been diverted.[155]

As we have seen, credit was necessary to permit the entrepreneur to lay a claim on inputs of commodities before the purpose of 'ventures' could realise improvement and as such it represents a withdrawal of existing commodities from the circular flow of exchange. This is why the limit on credit-creation is the fear that the entrepreneur will not fulfil the trust which is placed in the capacity of enterprise to produce commodities and repay the credit. In evaluating this capacity, the banker becomes society's trustee:

> He is essentially a phenomenon of development, though only when no central authority directs the social process. He makes possible the carrying out of new combinations, authorises people, in the name of society as it were, to form them. He is the *ephor* of the exchange economy.[156]

The banker, as the ephor or trustee of development, was the overseer of the stable properties of the economic system who ideally acted to ensure a balance between withdrawals of productive forces from the old combinations and their reconstitution by the entrepreneur. As the ephor acting within the domain of development, it was the banker in whom trusteeship was vested to order the spontaneous change of data which emerged endogenously within the economic system and to ensure that the purchasing power of money was used to fund the development of productive force.

BANK AND STATE POLICY

Schumpeter's ideal type of the bank, like that of the entrepreneur, rested upon the socially necessary function of organisation. While, as we have seen, Schumpeter's methodological individualism inclined him to embed functionality in the person, he realised that the phenomenal function of enterprise and banking could not be reduced to individuality. Long before his renowned 1942 text, *Capitalism, Socialism and Democracy*, Schumpeter had become exercised by the thesis that what had made the 'types' of capitalism rational and thereby functional for the development of the economic system had changed the social order of capitalism as the organisation of enterprise came to be vested in the managerial strata of a '"new" middle class'. Innovation in the case of 'trustified capitalism', Schumpeter wrote in his 1928 'Instability of capitalism', is 'not any more embodied *typically* in new firms, but goes on within the big units now existing, largely independently of individual persons'.[157] This is the 'Schumpeterian thesis' which has so powerfully exercised academic studies of innovation and fuelled the debate about the applicability of Schumpeter's work in the formation of industrial policy.

Schumpeter himself was not particularly exercised by the questions raised by the so-called Schumpeterian thesis, namely what specifically governed the diffusion of innovation and whether new, smaller or larger, managerially based, firms were to be officially preferred as the source of industrial innovation.[158] Nor was the question of whether Schumpeterian innovation was to be confined to private enterprise, as opposed to that of state apparatus, of major concern to Schumpeter. He reflected, in 1949, that the entrepreneurial function 'may be and often is filled cooperatively' and that the organising function varied according to the 'social environment'. One example of this was the way in which the United States' Department of Agriculture had 'revolutionised' farmers' production practices.[159] Rather, Schumpeter's major thesis was about the trust for development which was conveyed between the economic system and social order of capitalism and it is the change in the source of this trust which fundamentally exercised the general theorist of development.

What Schumpeter spelt out in his *Capitalism, Socialism and Democracy*, that the rationalisation of the capitalist economic system would *undermine* its social order, was hinted at in a 1928 article when he wrote that 'Progress becomes "automised", increasingly impersonal and decreasingly a matter of leadership and individual initiative.'[160] This was, to repeat, the idea that the capacity for self-development, which the social order of capitalism had conveyed to the economic system, was being throttled by rationalisation of the system itself. The moot point, however, was whether the undermined social order would dispose of the economic system within the domain of development. In 1928, Schumpeter was laconic. Organised capitalism's capacity for innovation was not diminished by the vesting of trust in the 'trustified' firm. Here, Schumpeter argued, innovation met 'with much less friction, as failure in any particular case loses its dangers, and tends to be carried out as a matter of course on the advice of specialists. Conscious policy towards demand and taking a long-time view towards investment becomes possible.' The banking function, as well, was increasingly embodied within the trust of the firm:

> Although credit creation still plays a role, both the power to accumulate reserves and the direct access to the money market tend to reduce the importance of this element in the life of a trust – which, incidentally, accounts for the phenomenon of prosperity coexisting with stable, or nearly stable prices which we have had the opportunity of witnessing in the United States 1923–1926.[161]

Given that this 'prosperity' should have been the necessary inflationary phase of his sequence of development, Schumpeter might well have added that such stability was 'incidental' to his analysis of development as well.

Schumpeter's optimism was shattered by the 1929 financial crash, which was, ironically enough, immediately precipitated by the crash of the *Creditanstalt*. This crisis heralded the period of depression and mass unemployment which, by the lights of Schumpeterian theory should have been, but was not, much tempered by the phenomenon of organised capitalism.

Unemployment, Schumpeter proposed in his *Theory of Economic Development*, was either the consequence of non-economic events such as war or drought 'or precisely of the development which we are investigating'.[162] Destruction of old enterprise during the developmental phase of depression, involving contraction, decay and death of firms, created the unemployment of labour power. Schumpeter implied, although he did not explicate this phenomenon of development, that the withdrawal of productive force from existing combinations in production would be compensated for within the economic system. His scheme suggested that displaced productive force would be compensated for by adaptations such that through the sequences of development, from industrial investment to innovation and then diffusion and imitation of new techniques, more and better productive force would be employed in new combinations of production. If such was the implication of the unemployment of development, then the corollary, that unemployment ought to be the object of the policy of development, was vehemently attacked by Schumpeter during his American period at Harvard University. In the course of the depression and its aftermath of the Second World War, when Swedberg admits that 'he was mentally out of balance', Schumpeter publicly railed against New Deal policy, including the National Reconstruction Agency and privately speculated about 'a ten-years's Roosevelt dictatorship' which 'will completely upset the social structure'. He folded his racial prejudice into his antagonism against mass culture in venting his spleen against the way that Keynes' theory had outbid his own: 'Just as the nigger dance [the jitterbug] is the dance of today, so is Keynesian economics the economics of today.'[163] The imbalance in Schumpeter's mind was matched by the imbalance in the scheme of development. Theoretical imbalance was part of the explanation of why change in the social order would undermine the economic system of capitalism.

We have shown that the immanent process of development consisted of sequential phases in the economic system and yet, within the restricted domain, the bridge of development made the process synchronic. The intent to develop, to synchronise the process without disturbing the stable consistency of the sequences through which adaptations of the circular flow model shifted the economy from one equilibrium to another, rested upon the belief in and of the ephor of development. Credit-banking was both the means of development and the source of what made the belief in development possible. Once belief was lost by the agency of the bank, when the bank was assumed to act as if it were an entrepreneur with expectations of return but without responsibility to bear the social risk of investment, then the ephor of development disappeared and with it the synchronic process of development.

When Schumpeter observed, as he did in 1928, that the banking function could be internalised within the firm but without impact upon development he chose to ignore the possibility that the function could be as much broadened beyond the firm as narrowed within it. After all, the state could assume the function of investment banking in the way that the entrepreneurial function could be assumed by the state apparatus. Indeed, in *Business Cycles*, Schumpeter wrote of the associated forms of banking. He distinguished between '*bankers' banks*' through which higher-order banks, including the central or reserve bank, replicated the function of lower-order banks, and '*member banks*' which keep accounts of, and 'manufacture balances for, firms and households' without disturbing the essential trusteeship of function.[164] It was the *policy* of development, policy which was vested in state power and which Schumpeter took to be the realities of politics, that fundamentally disturbed the function of the bank.

State action, whether through a state investment bank or not, which used purchasing power to maintain the means of consumption in the face of unemployment during the developmental phase of recession may have been intended to be a programme of 'national development'. However, such intention, according to Schumpeter, could not compensate for the withdrawal of productive force in the economic system, particularly when the system was assimilated into that of a national or regional arena, because it pre-empted the creative phase of the development sequence by denying the credit-creating function of the bank in selecting investment for enterprise, whether state or not, according to the principle of development. This is why, however much he was tied into the positivist tradition by method and concept, Schumpeter denied the plausibility of constructivist development. Indeed, he associated the received doctrine of development with the political economy of both idealism and utilitarianism. Once again, Schumpeter recalled Marx, 'the founder of modern political thought' to contest the foundation of policy:

> There is no scientific sense whatever in creating for one's self some metaphysical entity to be called 'The Common Good' and a not less metaphysical 'State' that, sailing high in the clouds and exempt from and above the human struggles and group interests, worships at the shrine of that Common Good. But the economists of all times have done precisely this.[165]

While excusing Adam Smith from this charge, Schumpeter was adamant that policy for economists treated public authority and government 'as a kind of deity that strives to realise the will of the people and the common good'. Marx was invoked against the economists because he did not abstract from 'the business process' as a matter of 'the businessman's interest'. Unlike the economists, Marx 'hauled down this state from the clouds and into the sphere of realistic analysis', the sphere which, according to Schumpeter, should include 'the behaviour of bureaucratic organisms, of political bosses and pressure groups, and the like'.[166] Such was the real matter of the social order of capitalism. It was meant to be part of, and not external to, the domain of development.

After understanding why he was antagonistic to economists' *policy*, it is easy also to see why Schumpeter provoked a reaction from development economists whose post-1945 commitment to development followed the contours of the constructivist intent. Henry Wallich objected to the 'production orientation' of Schumpeter's scheme which he saw as inapplicable to underdeveloped countries where the process of development ought to be oriented to raising consumption standards. Hans Singer agreed with Wallich. S. Dass considered Schumpeter 'completely inapplicable' as well arguing that the likely agency of development in 'an under-developed country' would be government. S. Pal concluded that Schumpeter was 'too much detached from the realities, and the welfare and policy implications of his analysis', concluding that 'This probably is the reason why we avoid him!'[167] Yet, as we shall now see, much of what has come to represent the ideal developmental state has been constructed, as it were, in the image of Schumpeter.

THE DEVELOPMENTAL STATE IN EAST ASIA

In so far as rapid industrialisation has followed from the intent to develop, it is the post-1945 example of East Asia which has come to represent an ideal model of the developmental state. This is the state which, according to Chalmers Johnson, 'gives its first priority to economic development'.[168] Japan, in particular, has come in the late twentieth century to represent what nineteenth-century Germany represented for Joseph Chamberlain and other proponents of the doctrine of development in Britain. Japan, and increasingly Korea, have become the models of state action to promote industrialisation. These are the exemplars of the case in which the state is purported to have had a programme of industrial development *before* development took place. From education and training, to the regulation of labour and trade, to the organisation of government, Germany was then that which the British state was intended to imitate. Japan is now that which is to be mimicked by all.[169]

It is important to note, then, that if there was one arena in which Schumpeter's theory was taken seriously it was Japan.[170] As F.M. Scherer has recounted, there was 'an intense debate over Japanese development strategy during the late-1940s':

> The terms of this debate were essentially the terms set out in Schumpeter's 1912 *Theory of Economic Development*, although the Japanese industrial 'motorcars' were also provided with brakes – crisis cartels, initial protection against imports, and similar measures if their investments were threatened by structural or macroeconomic setbacks.

This debate,

> was resolved in favour of MITI, which advocated a Schumpeterian emphasis on building high-technology industries over the contention of Bank of Japan staff that Japan should cultivate low-technology industries and pursue comparative advantage through its low labour costs.[171]

It is hardly surprising, given the abiding focus by economists on the so-called 'Schumpeterian thesis', that they have repeatedly shifted the focus of discussion away from Schumpeter's essential 'terms' of development, the bank–industry relation as exemplified by that between the Bank of Japan and the Ministry of International Trade and Industry, to the question of high versus low technology. Similarly, the brake upon development has been shifted away from the bank's trust in the belief of development to the 'terms' of state macroeconomic policy. In the process, the location of the ephor of development has been displaced from the general to the restricted domain of development.

Despite the shift in the terms of development, Schumpeter's ephor of development keeps creeping back into the general evaluation of post-war Japanese industrialisation. To take one example, G.C. Allen's 'conclusion is that the government's most important role has been first to see that finances were available to enable private firms to develop on the lines that they had persuaded officials to approve, and secondly to provide incentives to investment in the most up-to-date equipment'.[172] Until the 1970s, the means by which officials of the developmental state directed industrial investment consisted of direct and indirect systems of providing loan-finance and 'subsidies to cover costs of investment in new plant'. It was not the amount of finance that was significant but the capacity of the state to discipline industrial enterprise which determined the degree to which industrial innovation and expansion was successful.[173]

Schumpeter's ephor of development was meant to discipline industrial production, to make it act for the purpose of development. Were the investment banking function to be incorporated within state policy while the burden of discipline was vested in the technology of production, then it might seem that the 'terms' of Schumpeter's theory had been fulfilled in the outstanding case of late-industrialisation in the twentieth century. Yet, Schumpeter's injunction to develop, conveyed by the bridge between the economic system and social order of capitalism, did not suppose that the Faustian contract could be that of state and enterprise. Japan, and even more strikingly Korea, appear to be cases in which the precepts of a theory of development have been carried, perversely from the standpoint of the theory itself, into development practice.

Alice Amsden, in her account of South Korean industrialisation, has provided this significant general emendation to Schumpeter's ephor of development:

> Schumpeter analyzed a new basis for competition, a new mechanism to discipline firm behaviour. He recognized such a disciplinarian in technological change. It was the creative gales of new technological discoveries that uprooted old monopolies and increased production, not steadily but in spurts.[174]

Equally, however, she recognises that 'there is no neat mechanism in the market-augmenting paradigm that can be relied on to drive firms to be productive, because growth itself does not happen automatically'. Although without 'an automatic disciplinary device', the market system 'nonetheless has a premise on which industrial expansion depends'. The premise of late industrialisation is a reciprocal relation between the state and the firm. This does not mean close cooperation, which is sometimes the way business–government relations in Korea and Japan are simplistically depicted. Nor does it mean that sometimes the government wields the carrot and at other, unrelated times, the stick.

Rather,

It means that in direct exchange for subsidies, the state exacts certain performance standards from firms. The more reciprocity that characterises state–firm relations in these countries, the higher the speed of economic growth.[175]

In other words, the premise of late, but rapid, industrialisation is that the state substitutes for the bank as the ephor of development in so far as the state acts according to the precepts of Schumpeter's bank. State subsidies may be either substituted for, or become complementary to, credit-creation, but Schumpeter's principle of economic development remains. What Schumpeter regarded as perverse, namely the capacity of the state to direct productive force, became the norm in the East Asian case.[176] Here 'the state was transformed from speculator to investor in the course of economic development, just as the state transformed the course of economic development itself'.[177] It is in this way that Schumpeter's theory was emended. Amsden generalised further. Early eighteenth-century British industrialisation was based, she suggests, upon *invention* as the source of improved productivity. Here, enterprise, embodied in the small firms of the individual entrepreneur, was regulated through *laissez-faire*. A second period of nineteenth-century industrialisation in Germany and the United States, based upon *innovation* and enterprise embodied in the large-scale managerial firm, was regulated by 'infant industry protection'. East Asia is part of a third twentieth-century period of industrialisation based upon 'borrowing technology from more technologically advanced societies, or what may be called *learning*'. In the absence of novel technology, borrowed technology is incorporated into production to reach the highest equivalent levels of international productivity in this third period. The focus of enterprise, whether of small or large firms, is neither on the minds of entrepreneur nor the offices of a managerial elite but upon the production engineer and factory floor with, as we have reiterated, enterprise being regulated through the state *subsidy*.[178]

What, it may be asked is the content of such subsidy? Amsden explains generally:

The subsidy includes not just tariff protection of the home market but also incentives to export, subsidies on inputs, government investment to promote technical or economic linkages between industries, as well as the usual state support of social-overhead and big business diplomacy. To stimulate investment and trade, the state has used the subsidy to get relative prices deliberately 'wrong' – that is, different from what the forces of supply and demand would determine.[179]

More specifically, Amsden shows in *Asia's Next Giant* how the state subsidised selected industries through nationalised banks, and how, during 1965–72 when domestic saving was insufficient to meet investment demand, subsidies were made by arranging 'long-term international credit for favoured firms at rates far below those obtainable domestically'. Amsden emphasises that the developmental state is founded upon the capacity of the state to discipline enterprise: 'The state in Korea, Japan and Taiwan has been more effective than other late-industrialising countries because it has the power to *discipline big business*, and thereby to dispense subsidies to big business according to a more effective set of allocative principles.'[180]

There should be no misunderstanding about the means by which, and to whom, the subsidy is advanced. Amsden makes it clear that the Korean process whereby 'poor performers' were penalised by bankruptcy while 'good ones' were rewarded by a subsidy was 'highly politicised' in that political bosses favoured 'close friends'. For example:

In the cement industry, the largest producer in the 1970s went bankrupt because it tried to optimize an old technology rather than switch to a new one. Its production was transferred by the government to a *chaebol*, the Ssangyong group, owned by one of the ruling party's elders.

Yet, whatever the extent of political favour, the subsidy is advanced for 'good performance' which is 'evaluated in terms of production and operations management rather than financial indicators'.[181] Political bosses inclined to subsidise selected firms but why did they have the capacity to discipline the big business which they favoured?

In explaining why the political bosses of the state had the capacity to discipline business in Korea, Amsden presents a set of historical conditions which were present there by the beginning of the 1960s. As earlier in Japan, land reform had eliminated a landed interest which could have made a claim on the state subsidy. Merchant capital had been coerced into manufacturing by the inability to import commodities. Banks had been nationalised and, like the Japanese *zaibatsu*, industrial enterprise took the form of a diversified business group, the *chaebol*, whose immediate post-war origins lay in the windfall gains which were venally amassed by political bosses from United States aid programmes.[182]

Ronald Dore has attempted to reduce the issues involved here to the ability of the state to construct a particular national interest, commenting that, 'the unusual features of Japanese industrial policy derive not so much from specific imitable policy measures as from the ability of the bureaucracy, closely attuned to thinking in industrial circles to generate a consensus around particular interpretations of the national interest'.[183] Yet, however complex the myriad of historical factors governing the degree to which the state was able to act to make capital conform to a national programme of development, the question was never merely about the capacity of the state to act, but about the desire of enterprise to act as if it was part of the intent to develop. This desire cannot simply be reduced to the ability of the state to construct a national interest because, tautologically, the enterprise is then assumed to act because it has a desire to do so. For Amsden, it is clear that the question is to be answered not on the basis of regulation in the form of the subsidy, which perversely conforms to Schumpeter's scheme, but on the basis of the organisation of production. It is the attribute of learning as opposed to innovation which makes Amsden explicitly depart from Schumpeter.

Late-industrialisation, Amsden has claimed, 'is devoid of innovation and occurs on the basis of learning'. Learning, in turn, 'involves borrowing, adapting and improving upon foreign designs. The Schumpeterian model provides insights into the process of late industrialisation, but it cannot penetrate a process of industrial expansion in which the dynamic of new technical discoveries is missing.'[184] It should be noticed, first, that Amsden's objection to Schumpeter is based upon reversing his order of the dynamic sequences of technical change. Imitation, by way of the international diffusion of technology is the start, rather than the end, of the internal process of development. After borrowing technology, the enterprise engages in innovation and then innovation creates the basis for inventing new techniques of production. Learning, thereby, is made compulsive for the enterprise because without borrowing technology, the enterprise cannot compete internationally.

Second, whereas Schumpeter embodied enterprise and its capacity for innovation in the single-product firm and treated creative destruction as the process whereby new firms appeared and old firms either contracted or were destroyed, Amsden emphasises the diversified, multi-product enterprise for the Korean case. As such, creative destruction is as much an intra-firm as inter-firm phenomenon. Bearing in mind the first point of learning, it follows that Schumpeter's 'new combinations' do not arise from the will and mind of the entrepreneur. Thus, third, whereas the active productive force of labour was absent from Schumpeter, it is at the forefront of Amsden's account of late industrialisation. Labour has to be made to learn in order for borrowing to be the means by which technology is brought to the factory floor. Moreover, the person of labour power has to be educated for production before labour learns to adapt technique. In so far as training and 'transfer of skill' have become the shibboleths of development policy, developmentalism reduces the motif of self-development from that *of* the person, who develops through learning, to that of being *for* the individual whose only capacity is the power to develop skill for production. It is in this manner that our old friend trusteeship appears anew.

There may well be merit in making distinctions between the categories of invention, innovation and learning, all of which are held to be different attributes of the internal process of capitalist development through industrialisation. It is, however, *laissez-faire*, trade intervention and subsidy, which are implied to be different forms of regulation and thus the external sources of authority for a general theory bridging intent and the internal process of development. Seen thus, Amsden's theory is not essentially different from that of Schumpeter and, indeed, the nineteenth-century antecedents of the internal and the external of development.

Indeed Amsden herself recognises this nineteenth-century connection in her comparison of the German case of borrowing technology, as seen through the eyes of Veblen, with that of Korea. The question is one of a matter of degree. For Korea the 'acquisition of theoretical knowledge' was a major, rather than, as it appeared in Veblen's account of Germany, a minor problem. What is clear, however, is that there was in Korea, as there had been in Germany in the nineteenth century, an industrialising mission conveyed by the production engineer. This mission conveyed the meaning, even if it did not carry the name, of trusteeship. Production engineers in Korea, 'like their German counterparts', Amsden writes, 'were the gatekeepers of technology transfer' who came through the schools: 'And in a society hungry to catch up, with a steadfast faith in the values of education, the practical knowledge that these professionals wielded went a long way toward winning them influence and esteem.' Amsden continues:

> The industrial community in Korea, therefore, became 'surely and unresistingly' drawn in under the rule of, if not the expert, then the technological trainee. Once the entrepreneurs recognised that government subsidies could build ships that floated and steel that bore weight, they increasingly turned their attention away from speculating toward accumulating capital.[185]

Nineteenth-century European positivists would have had little problem in appreciating the extent to which their precepts found a source of practice in twentieth-century Asia.

AGRARIAN BIAS OF NON-DEVELOPMENT

The twentieth-century converse, rather than the inverse or perverse, of the developmentalism and developmental state is to be found in Africa. From List to Schumpeter, it has been the peasant farmer who has been regarded as the archetype of non-development. In symbolising the habitual routines and conserving propensities of economic life, peasant-like conservation has been routinely counterpoised to entrepreneurial change.[186] We need to be clear, however, that the associations between non-development, underdevelopment, mass poverty and economic regress cannot simply be reduced to some essential distinction between the development of industry as opposed to agriculture, or further reduced to a distinction between large- and small-farm production. During the course of the reaction against the Chamberlainite doctrine of development for Africa, the peasant farmer became an object of development precisely because the archetype of conservation was rejected in the course of the rejection of 'development' itself. What remained, in the British variant of positivism, was the trusteeship of development and all which was implied by learning and the subsidy. The peasant, for the early Fabians, no less than for liberals, had the subjective potential for improvement in productive effort. But that potential was seen as realisable (certainly after the Second World War) only through state administration.

In like fashion, before the experience of central planning in the Soviet Union or indicative planning in Western Europe, the idea of a national plan coupled the rationality of science with the mission to educate the farmer in science. Early twentieth-century radicalism in the United States, for example, put primacy upon learning and then made 'development' its stand-in. In 1914, the then influential American 'new radical', Walter Lippman, stressed the need 'to *educate* the industrial situation, to draw out its promise, discipline and strengthen it':

> You have to make a survey of the natural resources of the country. On the basis of that survey you must draw up a national plan for their development. You must eliminate waste in mining, you must conserve the forests so that their fertility is not impaired, so that the stream flow is regulated, and the waterpower of the country made available. You must bring to the farmer a knowledge of scientific agriculture, help him to organise cooperatively, use the taxing power to prevent land speculation and force land to the best use, coordinate markets, build up rural credits, and create in the country a life that shall really be interesting.[187]

Seeing how Lippman's words tie together agrarian bias, conservation and cooperation with the direction of science and technology, we can appreciate the way and extent to which the converse of development formed the radical basis for what was to become agrarian policy in late twentieth-century Africa. From the 1960s on, the key period of developmentalism in Korea, the principle of development was contested by those who regarded themselves as standing for the radical tradition of developing the small but poor farmer. To look at what is proposed as rural development practice is to find an orthodox view which abjures the role of the expert and the borrowing of technology through directed agency. Here, learning must be autonomous, directed by producers themselves. The role of the development practitioner is to enable the producer to learn. This position is well exemplified by Michael Lipton, the contemporary Fabian and neo-populist scourge of the supposed urban bias of development. In 1986, mindful of the then world-wide focus on famine in Africa, Lipton reviewed Paul Richard's *Indigenous Agricultural Revolution* thus:

> Even illiterate farmers can be scientific, and the work of West African farmers such as the Ikale with yams and the Meude with rice, is shown to be at once adaptive and innovative. This call for a populist approach to research and development, in which scientists would learn from the farmers, is not a rejection of modern science. But there is much to be learnt from African farmers about managing difficult environments. . . .

Lipton continued: 'African population growth may require major advances in research, leading to much more intensive farming – not through mechanisation, which Richards rightly dismisses, but through irrigation, fertilisers and new varieties of plant [seed].'[188] Robert Chambers, probably the most influential author of texts read by African administrators and expatriate development practitioners alike, provides a second example. He has written: 'For any urban-based outsider to state the priorities of poor rural people is yet another core-based act of paternal guesswork.'[189] The core here is the 'centre' or 'top-down' view of development from state bureaucracies and not simply that of the metropolitan or advanced capitalist part of the world. In opposition to the generalised core, Chambers has called for a 'prescriptive paradigm of reversals' for rural development against 'normal bureaucracy' which 'centralises, standardises and simplifies' from the vantage point of 'capital cities' where 'programmes are designed for whole countries and orders are issued for policy implementation, regardless of diverse conditions'.[190] This 'paradigm' of anti-bureaucratic 'reversals' is taken to be the view from the 'periphery', one of 'decentralised process and choice':

This is coming to stress not the transfer of technology in the form of packages of practices for the uniform, simple, controlled environments of the irrigated green revolution, but the provision of baskets of choice for the more diverse, complex and risk-prone farming systems of rainfed agriculture. Bureaucratic reversals are implied, with varied local requests passed up from farmers replacing pre-set technologies passed down to them. Approaches which put farmers' analysis and priorities first complement those which generate and transfer technology. In this mode, the state is not school but cafeteria, and development is decentralised . . .

And, then, this proponent of agrarian bias takes heed of the contrariety between the doctrine of national development, which he mistakenly calls 'neo-Fabian', and the liberal thesis to conclude that a 'paradigm of reversals' resolves the 'contradiction' between the neo-Fabian and liberal theses:

Here a new neo-liberal agenda can liberate the poor by abolishing the regulations used to exploit them. The task is to dismantle the disabling state. In parallel, there is more that the state can and should do. Here a new neo-Fabian agenda can decentralise while providing safety nets, secure rights and access to reliable information, and permitting and promoting more independence and choice for the poor. The task is to establish the enabling state.[191]

These examples, drawn from the two most influential proponents of agrarian bias, could be added to without difficulty since the neo-liberal doctrine herein espoused has become the latest doctrine of development. Nor, it must be stressed, is the charge against the trusteeship of developmentalism confined to that of the 'foreign' or expatriate expert.[192]

Superficially, these views repeat what was stated in the first decades of colonial rule in British Africa by those who adopted a liberal stance to oppose the implantation of the Chamberlainite doctrine of development in Africa. John Harris, of the Aboriginal Protection Society in England, giving evidence to the West African Land Committee in 1912, spoke of how an African interest might be represented in terms that could have been written by Lipton or Chambers:

I want to see that old farmer woman from Abeokuta to give the Committee a good deal of 'farmer' knowledge, which [we] don't find in the books; Pearce of Lagos who is above all things a native merchant; those hard-working fellows in the cocoa farms of the Gold Coast. I should *not* like P.A. Renner B.L., and 'legal adviser' to various Gold Coast concessions. . . . The natives in Britain are not the type to come before the Commission – mostly students seeking their education. I want the man who is already educated, artificially or by nature.[193]

The seeming continuity between this early twentieth-century liberal rejection of the doctrine of development and the latter-day refusal of Lipton and Chambers to contemplate the principle of development contained within developmentalism is separated by the mid-century historical experience of state development, an experience which was a part of the passage from colonial to post-colonial administration. We need to be clear, however, that while the renunciation of development doctrine implies a rejection of the theoretical bridge of development and thus the positive assertion of trusteeship, it does not imply its corollary of a rejection of the intent to develop. There is little evidence to indicate that the proponents of the agrarian bias of non-development are willing to assert that development is the continuous perpetuation of a timeless present without a future, with the present being only a replication of what has happened in the past.

CONCLUSION

The problem for the contemporary proponents of a pre-modern idea of development, then, is that it is not possible for them to find the bridge of development to span the chasm has been opened up between the external and internal of development. While there may well be criticism of the capacity of 'the bank', as of the World Bank and other associated agencies of development, to select projects of development, the external authority of development continues to be a crucial datum of development. The internal process of development, on the other side of the chasm, is to be literally confined to the indigenous potential of any given set of production conditions. Here the rejection of industrialisation or technology or science as a part of the argument against developmentalism in the name of an authentic case for development is made by their historical associations with what is the logical truism of the directed action of developmentalism. Meanwhile, destruction remains associated with what Schumpeter called the exogenous of development – colonialism, war, famine. There is no positive means to compensate for destruction other than the internal process itself.

In the face of the failure of the agrarian bias of non-development to find a source of trusteeship, the evasion of trusteeship re-creates the original but real dilemma of development. In the next and final chapter we will show why and how, in the face of this dilemma, 'choice' is fixed upon those who are to be developed. Yet, where the realisation of the 'values' of development are assumed to be the purpose of development, thus providing the authority for choice to be exercised, it must be presupposed that the 'subjects' of development are already 'developed' in that they have acquired the capacity to express choice before development proceeds to eliminate their material poverty. This is the tautology and jargon of development. In its wake an underworld of non-development is created. The dilemma is of Goethe's Mephisto without the Faust of development:

I am the spirit that negates all!
And rightly so, for all that comes to be
Deserves to perish wretchedly. . . .

[I am] part of the power that would
Do nothing but evil
And yet creates the good.[194]

8

CONCLUSION: THE JARGON OF DEVELOPMENT

Without judgement, without having been thought, the word is to leave its meaning behind. This is to institute the reality of the 'more'. It is to scoff, without reason, at that mystical language speculation which the jargon, proud of its simplicity, is careful not to remember.[1]

WHAT IS DEVELOPMENT?

Development defies definition, we have claimed, because of the difficulty of making the intent to develop consistent with immanent development. Intentional development, we have argued, consists of the means to compensate for the destructive propensities of immanent change. The difficulty arises because, while an immanent process of development encompasses the dimension of destruction, it is difficult to imagine why and how the intent to destroy should be made in the name of development. By way of conclusion, we will now explore both 'orthodox' and self-described 'alternative' views of development. We do so in order to show the ways in which recent texts merely recapitulate, and most often poorly at that, the older views which grappled with the difficulty of development. Recent texts, we suggest, evade the difficulty and thereby make it more difficult to understand what development means.

RECENT ORTHODOX TEXTS ON ECONOMIC DEVELOPMENT

THE INTENTION TO DEVELOP

It is a commonplace for textbooks on the academic discipline of development economics to begin by asking 'What is Development?' The authors of two widely used textbooks answer, with the precision of participants in some synchronized aquatic sporting event, by defining development as a '*process of improvement*'[2] with development economics being 'concerned' with '*rapid*' and '*large-scale improvements in levels of living* for the masses of poverty-stricken, malnourished, and illiterate peoples of Africa, Asia and Latin America'.[3] The implication of a process of improvement necessarily carries us back to the eighteenth-century idea of progress as a long, slow, immanent process. There is little problem in grasping what is meant here if improvement is taken to refer to the unfolding of human potential to improve material conditions of life.

Progress becomes a matter of economic growth, embedded in the degree of quantitative expansion in material indications of improvement. However, the stipulation that this improvement is to be 'rapid' and 'large-scale' denotes an idiom of development first expressed by early nineteenth-century positivists to denote a qualitative improvement in life. The idiom of development contained the deliberate *intention* of resolving immediate and pressing problems of poverty left unresolved by progress. Development rapidly emerged as state policy designed to deal with problems of productivity and unemployment associated with the growth of a surplus population, emigration and/or the threat of economic decline.

When progress is elided into development, the difficulty is not simply a semantic problem of language but one of history itself. Coleman and Nixson, for instance, understand that intention should be distinguished from historical process when they respond to the question of 'What is development?': 'The concept of economic development as an activity consciously engaged in mainly, but not solely, by governments with the intention of approaching or reaching certain specified objectives . . . complements the alternative concept of development as an historical process.'[4] The difficulty arises when it is asked why and how intention *complements* a process of development. To understand why intention should add to the process is to foreclose the difficulty since it is equally possible that the intention to develop may *substitute* for an incomplete process of development, a process which did not and could not conform to the original ideal of progress. In either case, we would want to know the source of the intention to develop.

Todaro, to take our other example of an orthodox text of development, unintentionally highlights this difficulty when he writes that '*development is both a physical reality and a state of mind*'.[5] If the 'physical reality' of development is what it does or has done, then the 'state of mind' is clearly meant to represent the intention or design of development. Todaro's twofold meaning of development implies that it is through a deliberate decision-making capacity of the mind that policies are chosen to pursue some stated goals of development. Development, then, as an intentional acitivity, does not merely happen but is first a state of thinking about acting in such a way as to make development happen. But the origin of the *intention* to develop is omitted from discussion.

The way in which intention is a part of the process of development is both a question of history and logic. We have provided historical examples, through the advent of self-government in Australia and Canada, to explain why and how development was officially formulated as mid-nineteenth-century doctrine. It was there and then practised either for the purpose of absorbing a surplus population into industrial production or agrarian colonisation. The creation of a surplus population, the hallmark of the advent of capitalism, was officially recognised in the same period for India, but the rise of development as doctrine was convoluted by the colonial integument. Similar complements between development, development doctrine and capitalist development were shown for twentieth-century Africa.

Given that the same complements between intention and process seemed to produce different outcomes in each of these cases, a history of development doctrine might seem to affirm the Coleman–Nixson homily that 'once the abstract objectives of development have been defined, the concrete, specific socio-economic and socio-political structures *within which it is believed that such objectives can best be realised* must also be made explicit'.[6] But simultaneously, the homily passes over, without comment, the logical origin behind the intention of the 'abstract objectives' of development. What is assumed is that there must be complementarity between intention and process so that once the process of development is made explicit, the intention to develop becomes less abstract and more real. A presumed complementarity between intention and process makes it appear that what is defined by intention, and then added to by practice, makes development happen. As Adorno warned, it is not the 'more' of historical reality, more of historical correlations between intent and process, which resolves what is the logical difficulty of development. There is precisely the logical difficulty of the 'more', of making explicit what is implicit in the history of development itself. And, this history involves making implicit intent explicit.

The modern pedigree of speculation on the origin of the intention to develop was reproduced, at the turn of the twentieth century, by L.T. Hobhouse in his effort to reconcile himself to British Hegelian Idealism in the formulation of his own brand of positive Liberalism. As we saw in Chapter 5, Hobhouse struggled to find the primacy of intent in the correlations of 'reality'. He asserted that progress, the 'extension of harmony', was a 'self-furthering process'. Except in its 'lowest stages' it was 'effected by conscious correlation, and its development depend[ed] on the extension of conscious control'. The first British academic sociologist went further when he argued that: 'In the undeveloped state forces are locked in conflict and cancel one another. There is mutual arrest and stagnation. In the developed state of Reality they co-operate in the maintenance of a harmonious system.' For Hobhouse, the 'operation of Mind' set free these 'locked forces', thus making possible the 'first act' of development. The second act was to ensure 'true development' rather than 'mere disruption' occurred, by ensuring that the freed forces were ordered in a structure.[7] Other acts of development followed, but they need not delay us here. Hobhouse's thesis was that it was purpose, conditioned by a developing reality, which directed the improvement of progress. Both the state of mind and physical reality developed.

The textbook accounts of economic development noted above are, in principle, no different from the general view of development which Hobhouse advanced through his biological metaphor. All that has occurred is that the prevalent nineteenth-century metaphor of the organism has been grafted onto the older mechanical metaphor of equilibrium. Economic analysis remains structured by the principle of balance – between market forces of supply and demand, the forces of capital and labour, labour and land, or the balance of trade and payments. Balance implies the desirable order of harmony and in the bulk of the textbook analysis of economic development, is treated as phenomena of 'objective' or physical reality.

What, then, does 'operation of Mind' convey for the specific purpose of economic development? 'Government, in a word,' J.L. Garvin wrote of his 1905 'constructive economics', 'should be the brain in the state, even in the sphere of commerce.'[8] Constructive economics was nothing other than the basis of the Chamberlainite doctrine of development for Britain, a doctrine which was ranged against *laissez-faire* economics. This doctrine, as we have shown, epitomised the doctrine of development in general, the same doctrine which continues to pervade the purpose of economic development. However skewed the presentation may be, the textbook premise of development is that 'the state of mind' is literally that of the mind of the state. Development goals are state goals and development values are what state goals for development should be. When normative economics is defined against the positive, the repeated claim is that the values of development should be embedded in the mind of the state and it is this mind which should work for true development.

The intellectual price our textbook authors pay for their omission of an inquiry into the origins of the intention to develop becomes evident when they turn their attention to the role of *values* in development.

THE VALUES OF DEVELOPMENT

Thirwall, in another much used textbook, asserts that 'A concept of development is required which embraces the major economic and social objectives and values that societies strive for.'[9] Coleman and Nixson maintain that: 'Not only are value judgments an inevitable part of deciding what concepts and relationships should be employed to answer questions such as 'what causes development?' or 'has development occurred in any specific instance?', but value judgments are also necessary in how to represent concepts empirically.'[10] Todaro, who pays far more attention to the meaning of development, insists that 'it is essential that value premises, especially in the field of development economics, be specified carefully and not hidden behind a smokescreen of "pseudo-scientific jargon" . . .'. Referring to value judgements as 'subjective values', he argues that:

> Once these subjective values have been agreed on by a nation or, more specifically by those who are responsible for national decision making, then specific development goals (e.g., greater income equality) and the corresponding policies (e.g., taxing higher incomes at higher rates) based on 'objective' theoretical and quantitative analyses can be pursued.[11]

We are being asked to accept by each of these 'matter of fact' pronouncements that development has a qualitative content that can only be defined with respect to its subjective 'values' which are, in the end, a matter of belief. While we may wish to agree that values, of one sort or another, have been integral to development as an historical practice, we are left with a few questions. Just how and from where do the values of development arise? Given the importance attached to the values as being prior to the goals and policies of development, is it not strange that there is so little attempt to recognise the difficulty that values are not independent of the process of development. For all authors of the orthodox textbooks, the values of development arrive from off-stage as independent variables, fully formed and ready to do their work for development before it begins.

Coleman and Nixson are queasy about the mid-twentieth-century 'increasingly abstract and utopian definitions of development', and rueful that,

> the now popular concept of development refers to an ideal world or state of affairs that is both ahistorical and apolitical – ahistorical because it postulates an idealised structure that does not and never has existed, and apolitical because development is defined in an abstract sense and is not related to any particular political/social/ institutional structure.[12]

Development has increasingly been abstracted from history, but banal references to social, political, institutional 'structures' cannot resolve the difficulty. They cannot do so because if the goals are made concrete in the way our writers suggest, there then arises the old fallacy of development consistently evaded by modern textbooks. This fallacy is revealed at the same moment as any goals of development are stated. To be able to decide upon policy, to have the desire, knowledge and capability to do so, is to have achieved what are routinely stated to be central goals of development. That which should be imagined as the aspiration for development is assumed to be actually present, in a set of necessarily real conditions, for development to be designed before it happens.

When official doctrines of development were defined for Quebec in 1849, or Victoria in 1860 or Kenya in 1960, they were anything but an abstract statement of values but rather defined by reference to the past course of progress and development in each case. Doctrine consisted of attempts to make possible either what had been lost in the past or what had been achieved elsewhere. To prevent the loss of productive force through emigration or compensate for loss of productivity in agriculture or catch up with an already advanced state of improvement by late-industrialisation are all part of an official desire for improvement which is grounded in the reaction to the deficiency of the past. Intention complements process through a set of values which consist of an evaluation of the shortcomings of immanent progress or from a judgement of the failure of previous instances of intentional development to remedy what progress had failed to achieve. Values may purport to be 'abstract objectives' to express a speculative prospect of improvement from the standpoint of the present. When made explicit, however, the values serve as a standard of the present to provide criteria for evaluating the past and are no less speculative because they are based upon, and modified by, the process of development in which the intention to develop is being made explicit.

While authors of the orthodox texts understand the difficulty, they are trapped by an understanding of development which is both devoid of history and the logical implication of a doctrine of development for its practice. Todaro, for instance, posits an ideal value of development to be that of Mohandas Gandhi when he referred to the 'realisation of human potential'; Gandhi's vision, significantly, was one of appropriating an ideal of the past to capture that of the future. Todaro also referred to the difference between 'aspiration and reality'.[13] However, it is the difference between the actual and the potential which is the true ideal of development. Values may represent an ideal as an aspiration to construct a future which is an improvement of the present and, equally, they can imply an absolute ideal to criticise the present. What matters here, however, is that the 'subjective values' are being employed to bear upon the 'reality' of a past and to correct for an objective course of development which has created the grounds for the 'abstract objectives' of development.

The failure to understand that the values of development are derivative from the past course of progress and development allows the question of 'What is development?' to be answered by skirting around the difficulty of how the means of development policy are meant to correspond to the ends of development goals. For if the goals are routinely viewed as being rooted in developmental values, the means to develop, we are repeatedly assured, are objective matters subject to the rigorous analysis without which the subject of development economics, we may observe, might collapse. The old trick being played is to list a self-enclosed set of subjective values and treat them as if they belong to intention, and derived from belief, whereas the process of development is to be subjected to the interrogation of analysis which rests upon the assumptions governed by the world of objective matter.

Development, when distinguished from the growth of quantitative expansion, consists of expansion and 'something else', usually referred to as the qualities of development. The 'something else' are regarded as if they belonged to the values of development. It is a logical truism that a process of qualitative change can only be evaluated through reference to those qualities, in this case values, which are asserted before the process gets under way. In practice, the aims of development have nearly always been set out in just this way. However, the 'something else' is part of what makes the idea of development different from that of progress. Development, in both its classical and modern expression, includes process of decay, decomposition and destruction as well as growth, expansion and improvement.

In so far as modern development consists of simultaneous processes of improvement and destruction, then it follows that the modern intent to develop is as much directed to the negative as positive qualities of development. This is why development doctrine was founded upon the antinomies of positive and negative as much as positive and normative. And, as such, intentional development is inseparable from the whole process of development, expressing the intent to compensate for what is regarded as being deficient at the same time that intent offers the prospect of an aspiration for an improved future. It has been long understood that values of development were inseparable from development itself and could in no way serve as some kind of autonomous standard by which development was to be independently measured. Such, in the nineteenth century, was the bridge of development, the hypothesis to account for the difficulty of whether there could be development of intent when development was the process of historical change encompassing destruction.

But an even larger diffculty looms. For once we come to see that the values of development, which are supposedly to guide and evaluate development practice, are the outcome of an evaluation of prior development itself, we are forced to appreciate that the *capacity* to carry out evaluations of development is also one of the results of development. We are next led to ask in whom evaluative capacity resides and, more, who has the ability to give it effective voice. The textbook answer to the effect that developmental values are the product of, and are articulated, by 'societies' and are then put to developmental work by governments will simply not serve in addressing the question of 'What is development?'.

TRUSTEESHIP

We can now see why the querying of the source of development's values should have been evaded by our textbook authors. The distinguishing feature of the late twentieth-century question of development is that trusteeship, the integral of the nineteenth-century doctrine of development, has been renounced as the source of action towards development. From the immediate historical experience of formal imperialism and its end, it is easy to see why trusteeship should be renounced when the goals of development for post-colonial Third World states are explicated by people who are neither of the state nor the Third World. Yet, in so far as trusteeship is rejected because of its colonial connotation, what implicitly reappears is the original mid-nineteenth-century version of trusteeship untainted by imperialism. This was the positivist version which married a conception of science with that of state direction to secure the basis of social harmony through national development.

If the authors of our textbooks are able to delude themselves into believing that distance mitigates accountability, they may still not evade the reality that, for 'the other' of Third World poverty, the definition and articulation of the values of development remain as issues of power and responsibility. Yet the renunciation of trusteeship is obvious. Todaro, as we saw, recommends glibly that the 'subjective' values of development should be taken from Third World state regimes, most of which happen to be dictatorships of one kind or another, and then mediates the goals through the absolute values of development. The Coleman–Nixson queasiness about this procedure is only settled when the development question is displaced onto the institutional structure of the regimes themselves.

It is only by conflating progress with development that Todaro is able to claim that the goals of development, when governed by values, become criteria for evaluating the 'economic progress' of the past. Thus he writes that 'Without sustained and continuous economic progress at the individual as well as societal level, the realisation of human potential would not be possible.'[14] We have seen that the eighteenth-century idea of progress rested upon an objective, self-furthering process of improvement in material conditions of life. Gradual linear movement through stages of economic life, from hunting to manufacturing, was potentially infinite and positive in that both material improvement and social order could be secured. This view was premised on the understanding that a social body of individuals were committed to work through reason and sociability by way of belief. We have also seen how the idea of development appeared when progress was perceived as faltering in the earlier nineteenth century. Regress, as in the growth of poverty or social disorder, could not be attributed simply to subjective failings in the capacity to reason.

The problem here becomes clearer when we look more closely at the values of development. Todaro, for instance, defers to Denis Goulet, who has set out the 'core values' of development as 'life-sustenance: the ability to provide basic human needs';'self-esteem: to be a person'; and 'freedom from servitude: to be able to choose'.[15] These goals are not only eclectic, but as Goulet admits, ambiguous and 'open-ended'. Self-esteem was Hegel's end of development, an absolute value about which Fukuyama has recently made much ado in his *End of*

History; 'to be able to choose' is the well-covered utilitarian nostrum which was resuscitated for the purpose of development by the Fabian Arthur Lewis in his *Theory of Economic Growth*.[16] We shall see, through the work of Amartya Sen, that the absolute value of capability, as in 'to be a person' or 'to be able to choose', is categorically different from the relative worth of 'basic needs'. Goulet warns that these goals coexist in 'creative tension' and that the relationship between them is 'dialectical'.[17] He might well have referred to Hegel's dialectic where the positive value of one is the negation of the other. Servitude, Hegel claimed, was an essential condition for the slave to develop self-esteem in revolt against the master; the master, with only the aristocratic value to choose, could never have self-esteem without the necessity to work.

In our development textbooks the tension between self-esteem and need is relaxed to enable development to become a set of state policy objectives. Thus, following his recapitulation of Goulet, Todaro concludes by insisting that 'development in all societies must have at least the following three objectives'. These are: to increase the availability of basic needs; to provide more jobs and education to generate self-esteem; and to expand choice in order to free the individual and nation from dependence.[18] In other words, the objectives of development merely make the values active in the present for state policy. What were absolute values now become mere administrative tasks for the intention of development. This is how the jargon of development arises.

Other textbook treatments differ little. Dudley Seers suggested that in order to evaluate development we must ask 'What has been happening to poverty, to unemployment; and to inequality?'

> If all three of these have become less severe, then beyond doubt this has been a period of development for the country concerned. If one or two of these central problems had been growing worse . . . it would be strange to call the result 'development', even if *per capita* income had soared.

Here, progress, with its quantitative dimension of economic growth, is subsumed by the values of development. Concerned with the intention to develop, these values serve as subjective criteria for evaluating both the intention and the objective process of development itself. Seers continues: 'This applies, of course, to the future too. A "plan" which conveys no targets for reducing poverty, unemployment and inequality can hardly be considered a "development plan".'[19]

Again, the future hoped for development is nothing other than a reflection of its past experience. To reduce poverty and unemployment and the relative dimensions of inequality are the hallmarks of past development doctrine. The future remains firmly trapped in the past.

ENTITLEMENTS

The sources for the absolute values of development are equally revealing. Goulet arrived at the conclusion that authentic development must be based on an interrogation of authentic need in a very modified trusteeship of expertise. Seers came to criticise his own 'meaning of development' because he was amiss, so he reflected, in not renouncing trusteeship for the Third World and stated that the missing essential condition, element and value of development – 'self-reliance' – applied as much to Europe as it did to the Third World. Emphasis upon the cultural independence of nationalism, he wrote, 'relieves us in "developed" countries . . . of paternalism'.[20]

It is an historical irony that the essential basis of intentional development is undermined by the concern for development's values. In Goulet's 'cruel choice', for example, 'authentic' development is a process whereby the 'freedom *from* the want' of basic needs is made possible by freedom *for* needs. But when people consume more than they need because they are induced to do so, rather than according to their autonomous dictate of need, then the result is the 'social destruction' of 'pseudo-development'. To make development authentic is to ensure that the self-esteem of 'being more' does not become the alienated freedom of 'having more'. People, according to Goulet, must be imbued with the quality of restraint. Austerity cannot be imposed but must be developed through a balanced political regime, a mean of 'elitism' and 'populism', in which the expert meets the people in a dialogue that makes the values of development adhere in policy.

We have seen that development, whether immanent or intentional, is active through imbalance. Needs are induced when production, as in the process of capitalist development, is governed by the dictate of money-profit for the accumulation of capital. Unbalanced growth is activated by the state when capital goods capacity is deliberately planned to outstrip current demand thereby creating the means of production which is meant to induce investment, whether private or not, in consumption goods and services. This, according to Goulet, is the 'vulnerability' of development. As Goulet would have it, 'men's desires are tampered with before they can realistically satisfy these desires'. The process of development creates stimuli which become autonomous of peoples' desires, desires which should express the values of development. It is this, in Goulet's account, that makes the objectives of development unrealistic. Between the desire for development and the goal of capability stand the means of development. The intention to develop, through policy, thus comes to constitute a barrier to development. Development, as Goulet understands, is like history – an open-ended adventure. But the journey ends where it begins.

It is perhaps not surprising that the textbook pantheon of development's supposed values makes much of the views of Goulet and Seers but neglects the far more astute account of Amartya Sen. A fervent critic of both utilitarianism and relativist approaches to poverty such as 'basic needs', Sen argues that 'the process of economic development is best seen as an expansion of people's "capabilities"'.[21] Sen writes of the enhancement of capabilities as an absolute value of development: 'Ultimately the process of economic development has to be concerned with what people can or cannot do. . . . It has to do, in Marx's words, with "replacing the domination of circumstances and chance over individuals by the domination of individuals over chance".'[22] Sen continues, quoting Marx's renowned image of 'free development', which 'makes it possible for me to do one thing today and another tomorrow, to hunt in the morning, fish in the afternoon, rear cattle in the evening, criticise after dinner, just as I have in mind, without ever becoming hunter, fisherman, shepherd or critic'.[23] Everyone, in this vision, is free to choose from a potentially infinite set of activities, or, for Sen, 'functionings', without being forced to make any activity emblematic of their labour power. If the conditions of free development of individuality are taken to include the elimination of a social division of labour, no constraint of society and, therefore, the absence of a state, there is no problem behind the vision. It is in imagining how such conditions might be established that the problem of free development is created.[24] Marx and Engels specified that it was to be in 'communist society', in which 'society regulates the general production', where free activity was to become possible. We can see here, in part, the way that regulation by 'society', a central tenet of positivism, was inflected into marxism by way of Saint-Simonian 'scientific socialism' and conflated with the free association between persons of society.[25]

Another problem arises when we consider what Marx meant when he referred to 'the conditions for the free development and activity of individuals under their own control'. We have seen, from Marx's domains of development, that the domination of circumstance over capability referred to the realm of necessity where primacy is put on the development of productive force. Development of productive force may be a condition for the absolute end of development. The question is whether productive force is to be addressed as an absolute or relative value of development.

During the course of the twentieth century the development of productive force has been treated – both with regard to the intention to develop and as a criterion for evaluating the process of development – as if it were such an absolute value. The obvious case is that of a 'corrupt' communism, where the outcome of development was a greater specialisation of function, an accentuated social division of labour, and the entrenchment of state control over the activity of individuals. Yet the historical experience of 'Stalinism' is just an extreme example of how the nineteenth-century positivist doctrine of development could be more generally inherited to become, in experience, the antithesis of free development. Yet if productive force is regarded relatively, again both with regard to intent and process of development, then there is no conviction that the absolute end of free development will be progressively assured.

Sen grapples with these problems in so far as he registers the end of free development to be an *intrinsic* value of development while viewing possible means to that end as *instrumental*. Since the capacity for personal choice is itself a functioning, the range of capabilities making choice possible together with the potential set of functionings, 'to do' and 'to be', will establish the intrinsic value of free development. The process of development, whereby the person is actually able to function and thereby exercise the capability of choice, is the means to free development. Intrinsic valuation, whereby one person has the freedom to choose to do something or be somebody different from another, Sen makes it clear, will be changed by the process of development itself. This is so because what he calls 'value endogeneity' will, in the actual course of development, change what is 'regarded as valuable and what weights are attached to these objects' of choice.[26]

As we have seen from Marx, the process of development establishes the conditions for the realm of freedom within that of necessity where material abundance exists as a potential to be realised in a non-capitalist world. The real difficulty of development arises out of the idea that free development may be contained within a world of necessity and actual relative scarcity. This is the impression conveyed by Sen, who focuses on the dire but relative poverty of populations who subsist on the margins of life and for whom absolute capability is living. It is here, in a world of necessity, that the freedom to choose is different from that of the freedom to do and in which the freedom to do, as if there was a world of free development, creates the possibility for enhancing the capability to choose. This is made clearer if we look at the value of development as a criterion for the process of development.

For Sen, the absolute criterion for an acceptable standard of living is capability. 'Development', Sen repeats, 'is not a matter, ultimately, of expanding supplies of commodities, but of enhancing the capabilities of people.'[27] The problem is that in eschewing the relativism of development goals which make needs available to particular populations, Sen's absolute value of development is made amenable to state policy because capability is made relative. If capability can be expanded or 'enhanced', then it has the quality of being 'more' and really more at that.

State policy acts on capacity through entitlements, 'the alternative commodity bundles that a person can command', given the limits of his or her endowment and exchange possibilities. Through entitlements, commodities are converted into capabilities and the question of policy becomes the way in which entitlements are to be mapped onto capability. Different commodity bundles may give ground to the same capability; the same commodity bundle may realise different capabilities. On the basis of entitlement, individuals can acquire some capabilities but fail to acquire others. 'The particular role of entitlements is *through* its effects on capabilities.'[28] It is the extent of capability, the absolute value of development, which serves as both a goal of state policy and criterion for the evaluation of the process of development itself.

Sen's approach to the values of development is an attempt to move beyond the utilitarian and postivist fallacies of development. From Marx, Sen is able to associate the development of capability with that of freedom rather than necessity. Poverty is not to be endured through the virtue of necessity as in the 'poor but happy' or 'poor but efficient' aphorisms of utilitarian nostrums. The discontent of material and social loss need not be tragic but 'a positive assertion of creative potentiality'.[29] In this way, the positive is arrayed against the negative and not simply the normative of development. However, it is in the tension between the freedom of a potential future and the necessity of the present that capability, the absolute value, becomes a doctrine of development and essentially that of doctrines of the past.

In the necessity of the present, Sen makes it perfectly clear that the basic determinant of entitlement is wage-employment since,

> For most of humanity, about the only commodity a person has to sell is labour power, so that the person's entitlements depend crucially on his or her ability to find a job, the wage rate for that job, and the prices of commodities that he or she wishes to buy.[30]

State activism, on Sen's own grounds, should be directed at employment because it is through employment that entitlement is being affected and to act on entitlement is to have an effect upon capability. Development doctrine in the past, as we have shown, was framed according to just this purpose. The extent of capability may have been diminished, as much as enhanced, by doctrinal practice. State activism may have done better or worse in enhancing capability but it was through the mediation of entitlements, entitlements of property directly for subsistence or for wage-labour in industrial production, that a change in capability was to be effected.

The encounters of intentional development, whether through systematic colonisation and the agrarian doctrine of development or through late-industrialisation, were aimed at enhancing the capabilities of a surplus population. Functioning, far from being made consistent with the conditions for free development, was assimilated further, in the person of labour power, to the functions of production. In countering the loss of productive force, purpose was restricted to enhancing the capability of production but it was an enhancement of capability nonetheless. The ultimate value of development, that of the freedom of the person to be capable 'to do', rather than to produce, was not the source of state action in the process of development. Where the idea of free development is imagined to be contained by the realm of necessity, as in Blackburn's *Vampire of Reason*,[31] then the corollary was that the state always acts relatively, to defend the loss of productive force from the threat of the present – whether that of famine, war or competition between nations – and not out of a vision of free development. Whatever the merit of public welfare in attempting to make good the deficiency between actual and potential well-being, state action stands between need and the capability of free development.

Values of development are shot through with ambiguity because they are presented as an abstraction of history. They are treated as if they were absolute standards. The values posit a definitive prospect of improvement through stating what improvement might mean but they also rest upon an historical condition of poverty which has made one population, or one part of it, relatively poorer than another. As such, these values, though necessarily relative, are presented as if they were an absolute standpoint from which to gauge the extent of the rapid, large-scale improvement of development.

RECENT ALTERNATIVE TEXTS ON DEVELOPMENT

DEVELOPMENT AND ITS 'MALCONTENTS'

Present-day writing on development has not only its textbook writers but its malcontents as well. For the alternative view of development, trusteeship is not merely to be renounced but denounced as the source of the domination of the Third World. Banuri, in *Dominating Knowledge*, a recent collected work on Asian development, recounts what one post-modernist, Deleuze, said of another, Foucault. 'You have taught us something absolutely fundamental: The indignity of speaking on someone's behalf.' For authors such as Banuri the question of development, if its meaning is taken to be what '"we" can do for "them"' is only a 'licence' for imperial intervention.[32] Yet, 'free development' is out of the question as well for Stephen Marglin, writing in the same volume. For a reformulation of the question from 'What can be *done for* the person?' into 'What can the person do?' can become 'licence'. 'Freedom to become one's own person' can become 'freedom to do one's own "thing"' and thus Western individualism.[33]

The model of authentic development, for Marglin and others, stands between the domination of Western imperialism and the licence of free development. For Banuri, the model is to be found in the community of people resisting domination because their needs are not satisfied. For Marglin it is the Japanese value of *amae*, the 'dependent love of the kind a child receives from his mother', developed into adult life in which the person need have no choice because his or her needs are anticipated as part of love. Authentic development, in forsaking images of the paternal for the maternal, the imperial for the national, free will for will-over-self, universal identity for that of self-identity and 'the other', is presented as the ultimate rejection of Eurocentric nineteenth-century doctrine. However, it is precisely because development is deemed to be 'authentic' that it is nothing more than another kind of trusteeship.

Mid-nineteenth-century development doctrine lay in a prevalent idea that there was a proper course of development which could be guided by intent. Positivists, in the Saint-Simonian tradition, reinvigorated older versions of trusteeship, lodged in the name of family or household and community or nation. These were the very sources of identity and affiliation which were deemed to have been corrupted by what they took to be a critical period of European history and which we know now to have been the onset of a period of capitalist development during which the productive force of industrial capital was unleashed. In deeming their own doctrine to be authentic, postivists postulated that the destructive and negative propensities of progress could be resisted either, in the case of Comte, by bringing knowledge to bear on the personal qualities of love and altruism, or, as in Mill's riposte, through enabling the person to knowingly realise these qualities through the associative experience of the self in the community.

Comte's positive knowledge of social science was inseparable from trusteeship in so far as science was to be conveyed by trustees who, by virtue of their discovery of the laws of progress, were to be entrusted to manage development. When Mill objected to Comte's 'religion' it was not because he rejected the belief that progress could be made positive but because he claimed that domination of knowledge over the self would negate the possibility for self-development through the active acquisition of knowledge in experience. In other words, the dispute was over the means by which authentic development was to be made possible. To discover social laws was, for Comte, to explain the possibility of an authentic course of development if the laws were induced from deductions of different social experiences. Mill, as we have seen, did not so much object to this procedure as to Comte's trust that the potential of authentic development could be made actual without a popular understanding of positive knowledge. The problem of trusteeship remained. How were conditions to be established to make the potential of self-development possible for people who did not possess the capacity to acquire the experience of development?

Mill's answer to the question was that the capacity for development could be found in the different deductions of social experience. Capacities for development would then be induced through trusteeship. Such, for instance, was the purpose of community, where potential capacity could be found and in which the self was associated but had yet to be developed. The example of village community, deduced from the knowledge of the particular experience of India, was presented as having the potential for authentic development. Authentic development would be made possible were the community not corrupted by an unleashing of progress and were the inherited imperial burden of trusteeship directed towards enhancing the capacity of the community. Nandy and Visvanathan, again in *Dominating Knowledge*, offer the following prospect of authentic development for present-day India through Gandhi's radicalism: 'In it not only does every man become a scientist and every village a science academy, but there is a demand for a cognitive resistance to the gross appetite of modern science.'[34]

For Mill, the Indian example was but a part of a general argument for authentic development, an argument which entered into development doctrine in Europe as much as anywhere else. Now, as then, the implication of trusteeship follows directly from the prospect of authentic or positive development. Adorno's idea of *aura*, which expressed the disjunction between the other and thing-in-itself is the key to his *Jargon of Authenticity*. The jargon of 'authentic' development arises from the way in which development doctrine is stated for people who cannot account for the source of the doctrine itself precisely because they are not developed. Development doctrine becomes jargon when there is both distance and disjunction between the intent to develop and the practice of development;

when there is an exercise of power in which the capacity to state the purpose of development is not accompanied by the responsibility of accountability.

When doctrines of development are now damned at the end of the twentieth century, the accusation is made that both the idea and practice of development is bound up with a European or Western system of thought whose purpose has been to impose modernity upon people who live according to 'tradition'. Within the incubus of capitalism, the argument runs, development is the means by which traditional sources of knowledge, learning, innovation and power have been deliberately undermined and then destroyed to secure the command over populations and labour for the benefit of those who live through the progress of the modern world. From the premise that it is development which destroys what is 'old' in favour of creating what is 'new', the malcontents of development seek to bring into question both the means and the end of development. They seek to demolish the power of trusteeship and reject the values of development used to establish the meaning of development. Thus, the orthodox textbook question of 'What is development?' becomes for them one of asking how it would be possible to construct the basis for what is called, for example, either 'an alternative development' or 'another development'.

Two presumptions enter into this general malcontent with development. First, malcontent is not seen as being confined to the 'Third World' of the South where mass unemployment and poverty are seen as persisting in spite or because of development practice. Since development is so closely identified with modernity, it would be strange indeed if the debunking of development did not seek to incorporate European 'counterpoints' or 'challenges' to the modern idea of development. Second, but not separately from the first, the historical focus of malcontent is upon post-1945 practice associated with the global aspiration for development.

Imperial in scope, and within the history of transition from the late colonial to post-colonial world, the systemic Western aspiration for the development of 'underdeveloped' peoples has perpetuated, it is presumed, a process of underdevelopment. In doing so, however, development has had the effect of attracting 'Third World elites' to the Western idea of development while re-creating the idea of popular resistance to the modern in the West since resistance, which is made in the name of tradition, has been conveyed from the non-Western world to the West itself. The positive design of development was supposed to create the modern by destroying the old but the malcontents now see it as terminating in the corruption of economic and political power. Now the persistence of the old is seen by them as the creative basis for a new but different design of development. It is in this inversion of the positive and negative dimensions of development that hope is given for, and belief made possible in, a different idea and practice of development.

Yet, as we have argued, development, in both its classical and modern expression, has always been haunted by an awareness of the process of decay and destruction as well as expansion and improvement. Whether expressed cyclically – as in its original classical form of reasoning where a phase of decay followed one of growth in a sequence of time – or sequentially – as in the modern form where growth and decay are simultaneous forces in both time and space – development cannot be reduced to a process whereby the desire for improvement is taken for granted. A desire for improvement may be stated and we may take it for granted, but the cause of the expression of desire remains a part of the process of development itself. This point, so seemingly obvious if only because of the historical course of development itself, is lost because it is inconvenient for a positive prospect of development. As we have tried to show, the official desire for improvement is a desire grounded as a reaction to the past deficiency of development with respect to the present; not merely a speculative prospect of a future from the present.

Our contention is that those malcontent with development are themselves a part of the problem of development. It is not so much that it is implausible, on speculative grounds, to think that a different development could be constructed in the general manner indicated here or that were it generally believed to be plausible, this different development would be a refusal of the absolute value of development, namely the realisation of human potential. Rather, it is on historical and logical grounds that the different development is wedded to the doctrine of development created in the nineteenth century to confront precisely the problem the post-modern malcontents portray as a problem of modern development. This, we repeat again, was the problem of how to contain the chaos of progress, with its economic disorder of unemployment, poverty and alienation from a social interest in production, within the confines of a traditional social order. In the injunction to make the change of development organic and continuous, to conform to a natural order of progress, wherein the new is encapsulated within the old, the apostles of a different development have displaced the twentieth-century idea of modern development back into its original nineteenth-century setting. It is in this sense that post-modernist malcontents re-create a very much pre-modernist prognosis for development.

'ALTERNATIVE DEVELOPMENT'

Stephen Marglin and his associates in *Dominating Knowledge* provide a very clear example of the extent to which the present-day malcontent with development is the source of logical and historical irony. In emphatically rejecting any absolute value of development, while strenuously avoiding any interest in the historical course of development through the view of imperial history, this left-radical appreciation of the idea and practice of development only ends up reproducing the case of the conservative counterpoint of Burke to the theses of development doctrine enunciated in the imperial integument of India.

A central premise of the authors found in *Dominating Knowledge* is that the absolute idea of development is founded upon the European tradition of knowing, a tradition which creates its own grounds for logically dominating, and thereby forcibly eliminating, non-European forms of knowledge. It is the culture of the West, and the movement of this culture, 'Westernisation', which 'in a word' is taken to mean development. This scheme of culture is set out by Marglin in his opposition of *episteme* to *techne*. Episteme, which he describes as the values and rules of the dominant European culture, is understood to be the cerebral analysis of Theory which passively rests upon the logical deduction from self-evident axioms. *Techne*, in opposition, is the indecomposable, implicit intuition of Practice which creatively expresses the emotional, tactile and personal values of authentic development. Other binary oppositions follow. Theory is transmitted formally through the articulation of codified knowledge; practice by the informal authority of apprenticeships. Innovation in theoretical knowledge proceeds by criticism whereas trial, error and emendation are the procedures of *techne*.[35] Likewise, in the same collection of essays, Banuri makes an opposition between the impersonal of the dominant West and personal of the subdued East.[36]

Behind these sets of oppositions stands the binary opposition between universal/absolute values of Theory and contextual/relative values of Practice. This is the fundamental opposition, Marglin alleges, which makes it impossible to reduce either *episteme* to *techne* or *techne* to *episteme*.[37] Development, when governed by the dominant form of knowledge, is the imposition of the absolute values of theory upon practice forcibly eliminating the relative but authentic values of indigenous culture.

Three points, by way of both criticism and commentary, can be made of this scheme which, despite conveying a plausible impression of how directed change may happen, is ultimately self-defeating as either a logical and/or historical basis for a new vision of alternative development. First, and most obviously, the malcontents themselves use theory to make the case for indigenous practice. Binary opposition, precisely the method used to construct the opposition between *episteme* and *techne*, is itself nothing other than a form of *episteme*. While the purpose of an alternative development is to reclaim the practice of techne for an ideal but relatively free form of development, *episteme* is being employed as an instrument for the intrinsic end of development. This is nothing other than an instance of trusteeship.

The second point, the context of *episteme* and *techne*, explains why the use of a relative means for a relative end of development cannot but imply trusteeship. One might easily gain the impression that *episteme* represents the whole culture of the North and West and is the relatively new, whereas *techne* is that of the East and South and relatively old. If this were so, there could be no 'development' in alternative development, since the antithesis of development would be simply a conservation of the old. However, this is not what Marglin conveys in his representation of either *episteme* or *techne*. In tracing the genesis of *episteme*, from the world of ancient Greece through Judeo-Christian civilisation and into the modern world, Marglin's purpose is to show that Theory is an old constructed form of knowledge which has been used in the West to make the work of producers subordinate to the managerial prerogative of capitalists. Producers, whose method of work is governed by *techne*, have had their control over work wrested from their grasp by the deliberate use of *episteme*. Whether from their forced movement from cottage to factory, from say the seventeenth century on, or by way of scientific management in the twentieth century for all enterprises, wage-workers have been subsumed and subdued by the domination of *episteme*. This, we are told, has occurred because of the belief, embedded in the culture generally, that the absolute value of work in its various senses of episteme is the instrumental means for the subsistence of necessity. Western workers' resistance against the domination of capital has been ineffectual, and will continue to be, as long as the belief in work is abstracted from culture.[38]

Westernisation, then, is nothing other than the means by which *techne* is subdued under conditions where the process of the development of capitalism is incomplete and where a belief in *techne* survives, not in the restricted domain of necessary work, but in that which governs the course of development itself. Where work is embedded in culture and not alienated from it the context is defined for the alternative development. Yet if the possibility of true development must be imagined, it can only be done so by those who are vested in the belief that they are, through the knowledge of Theory, relatively developed because they believe they are, again through the knowledge of Theory, relatively free from the cultural envelopment of *episteme*. Whether of East or West, only those who are conscious of being so free, and being relatively developed, can assume the burden of trusteeship to bring *techne* to balance out *episteme* for the purpose of the relative harmony of authentic development.

Third, the image of harmony of development takes us back to the modern origin of development in the light of the binary opposition between the 'modern' and the 'traditional', the other key motif of the development malcontent. The modern, modernism and modernisation are all elided in the writing of Marglin and his associates with this elision creating a kind of confusion akin to the more orthodox elision between Progress and Development addressed above. As we have stressed, the idea of progress could not have been imagined without the modern onset of Enlightenment reason. The modern idea of progress was a rejection of the ancient image of development, which contained sequences of expansion, decay and decomposition in recurrent cycles of nature. Robert Pippin, in his useful account of the problem of modernism, characterised the onset of the modern tradition thus: 'Nature was to be mastered, not contemplated; the distinction between theory and *techne* was collapsed and modern humanism, as a kind of technological self-assertion, was born.'[39] Nineteenth-century malcontent with progress led to the doctrine of development which attempted, through both the positivism of Saint-Simon to J.S. Mill and variants of Hegelian Idealism such as that of the British Hegelians, to make progress developmental. Various attempts were made to recapture both the organic metaphors of continuous change and the organic, as opposed to the 'critical' individual-atomistic, characterisation of economic and social organisation, for an industrial age. In doing so, the distinction between theory and *techne* was reassembled for the purpose of directing change on the basis of trusteeship. It has often been pointed out, as by Pippin, that the idea of tradition, along with religion and classical ideas, has been part of the confusing claims of progress. The same claims, and not least the classical idea of development, were made emphatic to become an ideal basis for history, whether this was Comte's 'religion of humanity' for the final positive age or Hegel's 'end of history'.

Modernism, far from embracing tradition, was a twentieth-century attempt to expunge the claims of tradition from modern life. Tradition, including the re-created 'true world' of nature and the organic conception of progressive development, imposed constraint upon the ideal of, and possibility for, self-development in a world offering the immediate prospect of material abundance.[40] Schumpeter, as we have shown in Chapter 7, expressed the modernist idea of economic development. The potential of development was a reaction against the traditional routines of economic life. Development was a discontinuous process, a rapid striking out, in the movements of capital and innovation, from the tangents to the circular flows of traditional activity. When decay and destruction of the 'old' returned to become an essential basis of the modernist idea of development, they did not do so as part of a sequence but as elements in a process which was simultaneously that of growth of the 'new'. Unemployment was not an object for development but part of the process of development. Yet, as we also have seen, Schumpeter ended up with nineteenth-century development doctrine. In so far as it enunciated the effective purpose of trusteeship to be the deliberate balance between forces of decay and growth, and attached the purpose to the bank as the supervisor of development, the modernist imprint of the theory was severely circumscribed.

Present-day malcontents of development reject the discontinuity upon which the modernist idea of development is, in part, premised. Thus Banuri writes that the 'alternative vision' of development 'must deny . . . the theory of discontinuous change which follows from the invocation of a crisis'.[41] Here as well, despite their professed contempt for trusteeship, the post-modern malcontents return to the progressive development of nineteenth-century doctrine.

The residues of positivism lie heavily in *Dominating Knowledge*. Banuri sees 'human actions deriving from a tension between conflicting obligations and commitments'. He argues that 'the project of modernization has been deleterious to the welfare of Third World populations' because 'the project has forced indigenous people to divert their energies from the *positive* pursuit of indigenously defined social change, to the *negative* goal of resisting cultural, political, and economic domination by the West'.[42] The burden of the core of the Comtean doctrine of development was precisely to define social change for the purpose of confronting the negative propensity of progress. It is unsurprising to find this doctrine reappearing in the claim that progress can be made consistent with an 'indigenous' conception of social order. What is surprising is where we find the 'bridge' of development, the missing link between progress and order, in the post-modernist rendition.

It is John Henry Newman's second domain of development, conceived as a reaction to the evolutionary metaphor of progressive change, that precisely fits the purpose of the 'alternative development'. Development, for Newman, was not only the value which established whether different practices of doctrine conformed to the principle of Christian revelation. It was also what made the illative knowledge of the face-to-face early Christian communities consistent with the impersonal Theory of scientific knowledge of the nineteenth century. Through development, *techne* has the potential to become *episteme* without losing the principle of belief in the revelation. What was true development for Newman is little other than the alternative development of the present-day malcontents.

Belief is central to Marglin's conception of development which rests upon the antithesis between the organic or holistic and the atomistic, mechanical values of culture and methods of knowledge. Marglin makes much of Keynes' organicist method in drawing out the distinction between the *organic* role of belief, 'for propositions the truth of which depends upon the belief of agents', and the *atomic*, for which the assessment of truth 'is independent of those beliefs'.[43] However, Marglin's insight into Keynes' method, and what makes Keynes's economics truly distinctive from the assimilation of his theory into utilitarian economics, abstracts from the immediate origin of the method and the purpose Keynes carried through into policy.

KEYNES

The genealogy of knowledge that links Newman's philosophy of belief to that of Keynes via A.N. Whitehead, the mathematician and philosopher of science who was later emblematic of a challenge to the mechanical materialism, remains hazy. However, a plausible account can be given of the theoretical vectors which conveyed, from the latter half of the nineteenth century, general discomfort with *episteme*. There was enough of discomfort with *episteme*, in the midst of European thought at the turn of the century, to generate a philosophical challenge to the claims of Theory and it is this challenge which Marglin effectively embraces. It was in the midst of the turn-of-century reaction against both utilitarianism and British Hegelianism, that Keynes came under the spell of the philosophical intuitionism of G.E. Moore, and it was through his emendation of Moore's idea of decomposable holistic propositions of truth, sometimes called 'Platonic entities', that the organicist method was conveyed to economics.[44] Of equal significance, if given far less emphasis in the literature, was Whitehead's interrogation of Keynes' theory of probability.

Whitehead, who had found it difficult to understand Keynes' original thesis on probability, later helped Keynes develop the idea of how actual entities of time and space were made divisible and how the dependence of one entity upon another was established through what he called prehension.[45] Whitehead's later philosophy can be baldly put as an attempt to recover the essence of experience from its eclipse by what he called the 'Subjectivist Doctrine' of the British empiricism of Locke and Hume, where experience was 'the bare subjective entertainment of the datum, devoid of any subjective form of reception', and that of Descartes and Kant, where it allegedly consisted entirely of conceptual knowledge. Subjectivist doctrine, which had come to command the basis of what Keynes called 'classical theory', expressed the dualism of 'two worlds', one of 'mere appearance' and the other 'of ultimate substantial fact'. Whitehead aspired to transcend this dualism through a 'reformed subjectivist principle'.[46] An 'actual entity' was formed, according to Whitehead, through 'the *potential* unity of many entities – actual and non-actual –' in a process. Thus 'in the becoming of an actual entity, novel prehensions, nexus, subjective forms, propositions, multiplicities, and contrasts, also become'. In other words, the real world of actual entities, or of 'creatures' or of 'events', is a process in which the actual, being the 'real concrescence of so many potentials', becomes active because it possesses subjectivity.[47] Such was the basis for cognitive experience of reality where subjective form gave meaning to reality. But all value could not be entirely subjective, and this is what Marglin emphasises to be 'the social construction of reality' in Keynes.

Controversy abounds over the extent to which Keynes' later account of the decision to invest was founded upon organicist method and its implication that separate, individual investment decisions could be unified through the organic entity of the 'community'. It is argued, at one extreme, that Keynes abandoned the logical postulates which he had critically derived from Moore and formulated in the *Treatise on Probability*. 'Rather than people making decisions based on a Platonic entity', Bateman argues of Keynes' *General Theory*, 'they now make their decisions on the basis of their personal beliefs (or in the case of an investor, what they believe others believe).' By assuming that there was an empirically rational basis for the possibility of individual decision-making, Keynes embraced logical positivism and entered 'into the pantheon of empiricists'.[48] Moreover, and Davis reinforces the point, Keynes used Moore's organic principle to affirm the contention that what was 'good' or socially desirable about a decision resided wholly in the individual state of mind. As such, there was no antinomy between the organic and atomic basis of method. Degrees of individual belief changed in the course of intersubjective action but there was little to suppose that individual capacities to make decisions were captured within a social complex which contained more than the sum of its constituent parts. In other words, what was socially developed during the course of action was the capacity for socially desirable decisions. Socially desirable decisions, which eliminated involuntary employment, were founded upon the development of the capacity for individual judgement. Individual autonomy and active self-interest was enhanced rather than subsumed by what is deemed to be the 'ethical subjectivism' of Keynes.[49]

This view of Keynes is implausible. It may be made plausible if it is buttressed by the argument that Keynes' premise of uncertainty was no principle but merely an opportunistic device to account for the instabilities of economic decision-making and especially investment demand.[50] However, uncertainty is inherent in the fundamental distinction between the present and the future, a distinction which Keynes founded upon the absence and incompleteness of knowledge. Keynes' uncertainty, Lawson has argued, is confined exclusively 'for the evaluations of *future* outcomes of *all currently possible* decisions or acts'.[51] What, however, does 'possible' mean here, other than the potentials which, by way of Whitehead's account of entitities, were to be prehended? While the future is an entity 'of which we know nothing',[52] Keynes wrote in his 1936 book on *The General Theory of Employment*, the present is a different entity because it consists of an incomplete 'datum', of which we know something. Investment, if we interpret Keynes as following Whitehead's idea of organicism, is about prehending the potentials of both entities to make the future become part of the present.

Keynes' argument was against the mechanical premise of utilitarian procedures which assume that present possibilities can be continuously projected into the future as part of a timeless present. Not only did these procedures deny the difference between the two different worlds, of present and future, but they irrationally compounded the 'uncertainty' of the present, the uncertainty which is controversially adduced to the incompleteness rather than compete absence of knowledge.[53] Moreover, instabilities were to be explained, Keynes suggested, 'due to the characteristic of human nature that a large proportion of our positive activities depend on spontaneous optimism rather than on a mathematical expectation, whether moral or hedonistic or economic'. Such were the animal spirits, 'of a spontaneous urge to action rather than inaction', 'the whim or sentiment or chance' which explained the motive for investment without any rationally governed belief that the future outcomes would be an improvement on the present.[54]

Discontinuities between present and future, Keynes stressed, made individual belief inert as a source of action on the rational ground that existing knowledge can only be made more complete by a different kind of belief:

Knowing that our own individual judgement is worthless, we endeavour to fall back on the judgement of the rest of the world which is perhaps better informed. That is, we endeavour to conform with the behaviour of the majority or the average. The psychology of a society of individuals each of whom is endeavouring to copy the others leads to what we may strictly term a *conventional* judgement.[55]

Convention for Keynes is constituted by social practice and the knowledge which is gained through convention is got, as Lawson puts it, by '*partaking* in social practice'. Second, convention is regulative, consisting of a set of rules which make it possible for action to happen. Belief in convention is made necessary because knowledge, or 'true belief' is incomplete.[56] It is the belief in convention, Keynes argued, which makes the decision to invest possible on rational grounds when he suggests that,

investment becomes reasonably 'safe' for the individual investor over short periods, and hence over a succession of shorter periods however many, if he can fairly rely on their being no breakdown in the convention and on his therefore having an opportunity to revise his judgement and change his investment, before there has been time for much to happen. Investments which are 'fixed' for the community are thus made 'liquid' for the individual.[57]

To release the productive force of investment for the community, and thereby achieve a socially desirable outcome for the employment of labour, was the stuff of policy for Keynes.[58]

Convention was so significant for Keynes because it was the means by which investors' expectations could be changed. As Carabelli has argued, it was through moulding belief that Keynes believed that contingent social practices, and the rules according to which investment decisions were made, could be managed.[59] Belief belonged to convention – the concept setting out particular conditions for prehension – and convention could be changed to establish rationally constructed sets of belief for individually-made decisions. It was according to this prescription, following from the premise of uncertain knowledge, that desirable social outcomes, such as the injunction to temper 'speculation' in favour of 'enterprise', could be achieved without thwarting the belief that individual judgements were autonomous.

Winslow corroborates Carabelli's interpretation by using Whitehead to show how a set of internal relations between knowing and believing individuals could be nested in a hierarchy. Since the hierarchy contains different layers, defined according to different attributes of social practice, the more general attributes, belonging to a wider layer, are less likely to change than a narrower layer of attributes. Winslow gives the following example of one possible set of internal relations: 'Clinically depressed entrepreneurs' are thus nested within a layer of entrepreneurs who are likewise nested within the general attributes of human nature. It is the 'particular occasion of experience', derived inductively from the set of internal relations, which allows Keynes to postulate that, by way of an inductive hypothesis, reality – as entities of real potentials – can be socially constructed.[60] Policy, Keynes implored, should be directed at moulding attributes which are less regular and stable and more amenable to constructive change. This was the principle which was both derived from Whitehead's interrogation of the *Treatise on Probability* and came back to Keynes through Whitehead's concept of prehension.

Policy, for Keynes, could work on the expectations of individual economic agents because the hypothesis of the economist's logic rested upon a common meaning of convention. Following from the uncertainty premise, both hypothesis and expectation substituted for incomplete knowledge. As if to recapitulate Newman's *illative*, Keynes wrote of the economists' logic as developing from the same premise that governed belief of 'ordinary men', 'modern mathematicians', 'Polish farmers' and 'savages'.[61] It was, therefore, through persuasion of policy that the social understanding of beliefs could be changed, and it was by virtue of public opinion that 'we can act as an organised community for common purposes' for 'social and economic justice' while protecting individual property, 'freedom of choice', 'faith' and the individual capacity for judgement.[62]

The same principle could be applied to entities of space. Keynes was less formally concerned about how the entity of one economy, that of Britain or India or 'the world economy', was constructed through the prehension of another. However, this was the stuff of his role in state policy, from his early period in India Office employ to his key involvement in collaborating with the United States Treasury in the setting up of the IMF and World Bank in the immediate post-1945 onset of the global aspiration of development.

The important point here is one of policy and method. What was for Keynes the subject of development – the dominant agencies of Third World modernisation – has become the object against which the malcontents of development rail. The purpose of Keynes' modified subjectivism was to make state officialdom the subject of prehension. In the name of 'the community' it was burden of the state to foreknow, through rational deliberation, what a socially desirable propensity for investment might be if the involuntary unemployment of labour was to be eliminated. To achieve what the animal spirits of capitalists, under conditions of chance and economic disorder, found wanting on the ground of pessimistic expectation was to make the future appear as if it were the present. 'It is by reason of the existence of durable capital equipment', wrote Keynes, 'that the economic future is linked to the present'.[63] If necessary, state investment was to be substituted for private investment to employ capital according to the premise that the link or bridge of development would create order. Keynes' doctrine of policy was suffused by a generalised trusteeship; the state was to become the social trustee for capital to conserve the process of capitalist development. When subjectivism was cemented into state policy, through the medium of Keynesian economics, consumption and investment came to be regarded as if they were objective forces denuded of 'subjective forms', which interacted to produce the economic order of equilibrium.

What had been renounced as belonging to the older mechanistic-atomic fashion of science, namely the treatment of economic forces as if they possessed the potential to be what Whitehead called 'eternal objects', reappeared in the constructions of policy. In so far as state economic management is criticised in the name of Keynes, it is because the mechanical purpose of management has brought back what he had hoped to discard in his vision of policy.

As Fitzgibbons and others have pointed out,[64] Keynes' vision of policy was guided by the version of trusteeship which had been enunciated by none other than Edmund Burke. We have shown in Chapter 1 that Burke was the eighteenth-century inspiration for the Conservative view that it was tradition which had to govern the course of progress. Tradition embodied knowledge and the conventions of custom. Burke's standpoint was what has been called an 'epistemological populism', the belief that 'if our feelings contradict our theories . . . the feelings are true and the theory is false'.[65] Yet popular feelings, the conventions of tradition, had to be adjudicated. Through the trusteeship of Prudence, old traditions were to be discarded in favour of the new which arose in the course of progress. The old was conserved through new methods, according to Newman's later adage, making it possible for everything to change so that all might remain the same.

The state official and the politician, to quote one interpretation, 'was the servant, in some ways, the agent of society and property'. Burke's conservatism can be seen as aristocratic populism: 'To understand that notion throws light on the simple fact that both Burke's early defence of parliament and people and his later defence of authority were simply different methods of resisting attacks upon aristocratic trusteeship.'[66] In the twentieth century, Keynes put new method into this old tradition. Fitzgibbons suggests:

Keynes knew that Burke was opposed to state intervention in the spheres of property or commerce, but he laid aside these circumstantial *applications* and drew upon Burke's *theory*, which says that ultimately the public wisdom must always be supreme over individual discretion.[67]

Keynes' radicalism reflects the extent to which the active agency of the state was inscribed in the organic order of society. While the method and its basis was different, the purpose of agency was the same – by making progress conform to ethics according to the judgements of public wisdom. Such purpose was trusteeship.

WHITEHEAD

For the malcontents of development the purpose of development cannot be different from how development happens. In *Dominating Knowledge*, the attack on the absolute, intrinsic values of development couples the old assertion of tradition over theory with a renunciation of trusteeship. For Marglin, there is neither objective truth nor falsehood in the realm of 'organic discourse apart from peoples' beliefs'. Marx's vision of 'free development' is attacked on the ground that it makes the individual emphatic, as in 'I have a mind', without any conception of the organic. Arthur Lewis is abjured for positing the absolute value of progress, namely that economic growth rests on the mastery of man over nature, in a purposive control which is destructive of freedom.[68] Banuri sees 'core values like freedom, justice, equality, fairness, universality, efficiency and growth through the prism of impersonality'. He writes that the *impersonality postulate* concentrates 'intellectual energies only on those aspects of social behaviour which can be encompassed within an objectivist matrix'.[69]

It is apposite here to return to Whitehead, whose concept of progress has affinity, it has been noticed by more recent commentators, with that of Burke in that he stressed the continuity of custom while he made 'feeling' the basis of his theory of prehension.[70] Whitehead claimed that,

> There are two principles inherent in the very nature of things, recurring in some particular embodiments whatever field we explore – the spirit of change, and the spirit of conservation. There can be nothing real without both. Mere change without conservation is a passage from nothing to nothing. Its final integration yields mere transient non-entity. Mere conservation without change cannot conserve. For after all there is a flux of circumstance, and the freshness of being evaporates under mere repetition. The character of existent reality is composed of organisms enduring during the flux of things.[71]

In searching for the inherent duality of change and conservation, which is what

Whitehead was after in explaining why the design of purpose cannot be made separate from what endures when change happens, destruction was not given the same significance as conservation. It may be strange to suppose that it is possible to imagine that development can be expressed by the design to destroy, in the same way that the current fashion for 'sustainable development', for instance, puts primacy upon a deliberate intention to conserve. Yet according to a relativist idea of development, if what happens in the course of events is not to be made separate from the purpose of what should happen, how is it possible not to make destruction appear as part of the purpose of development rather than that which merely happens as development proceeds?

As with all philosophy of organism, Whitehead's process of change was directed at a synthesis of progress and order, where 'order' is the becoming of the actual with a definite form or pattern; disorder, by implication, is the unbecoming and/or formless becoming of the actual. Disorder was a part of the process of change but so also was the possibility, and here Whitehead departed from Burke, that the theory of progress was the 'guiding' and driving force of ideas in history'. But 'successful progress' crept 'from point to point, testing each step'.[72] Progress had no end but the end of development, which we may infer, was to confront the disorder of progress. 'Successful organisms', Whitehead wrote, 'modify their environment.'[73] Nineteenth-century evolutionary theory, with its prescription for progressive development through cooperation, was the prescriptive message of Whitehead.

Science, according to Whitehead, was the method by which the relative attributes of the organism might be investigated. As we saw above, because the less developed attributes of the organism were contained within the set of more advanced attributes, it was possible for knowledge to be adopted by those who did not yet have it and so set their development in motion. When this view of science was spelt out by Whitehead it was accompanied by a belief that knowledge was transmitted through cooperation. He stated that 'Men require of their neighbours something sufficiently akin to be understood, something different to provoke attention, and something great enough to command attention.' Whitehead then went on to conclude that 'Modern science has imposed on humanity the necessity for wandering.'[74]

For Whitehead's contemporaries, in the then extant world of colonial administration, this was a congenial view of what the practice of development might mean. When 'wandering' comes to be associated with 'Westernisation', especially in a post-colonial world, the implications of organicism for trusteeship are abjured. The principle, however, remains.

Although they deny the concept, trusteeship for Marglin and his associates is about conserving what is wilfully destroyed by the planned change of development. What is destroyed is the absolute capacity to choose. From Sen's distinction between intrinsic and instrumental values, Marglin argues that the universal belief in intrinsic value may expand potential arrays of choice but free development is not achieved unless the actual sets of available choice include the old that have been eliminated by the new. Thus Marglin argues that,

> if growth subtracts choices as well as adds to them, we are in a position to argue that growth expands possibilities only if we are to assume that an individual could reverse the process at will, and in effect could choose between two choice sets, the modern and the traditional.[75]

It is now easy to see why, despite – or indeed because of – its disavowal, trusteeship is central to the present-day malcontent with modern development.

If the process of development, a process conveying both the expansion of growth and the destruction of decay, is irreversible, then it is the design of development which preserves what would otherwise decay. An individual would be enabled to choose between the modern and the traditional only if the process is socially reversed through actively constructing the traditional anew. The argument for doing so is nothing other than a restatement of Whitehead, who wrote of the gospels of 'Force' and 'Uniformity' as being incompatible with social life because 'differences between the nations and races of mankind are required to preserve the conditions under which higher development is possible'.[76] Marglin concludes: 'It is in our own self-interest as well as the global interest to promote culture diversity, and a corresponding diversity of development models. . . . Cultural diversity may be the key to the survival of the human species'.[77] The developmental model in which Marglin proposes to preserve '*space* for a relatively autonomous transformation of indigenous cultures' is that of artisan-based, small-scale production. Marglin points to the survival of *techne* in the work of artisanal producers in his case study of handloom workers in the Indian state of Orissa, where there 'a measure of artistic independence and creativity is present even today'.[78] Work continues to be embedded in culture with the religious and the economic, the mystical and the real, meeting each other in tasks of work. Producers involved in cottage industry choose the time, pace and intensity of their work and are relatively autonomous because the labour process is non-capitalist. However, they are not absolutely so because the control over production is vested in the capital, whether of the merchant or of the cooperative, which purchases the cloth. There is potential space for this enterprise because the 'flexibility' of artisanal production has survived successive changes, from the eighteenth century or earlier, in the exchange relations between household producers and the buyers of cloth.

What has recently come to occupy the space for autonomous development, however, is large-scale, state-financed and supported capitalist enterprise in the same area of production. In this entity wage-work is governed by the culture of episteme. 'The adoption of Western values by Westernised indigenous elites', writes Marglin, 'stacks the cards against tradition.' To stack cards for tradition and shift the imbalance between *techne* and *episteme* is the task of trusteeship. This is a task which cannot fall to that of the modern Indian state but which, for Marglin, should be vested in the tradition of Mohandas Gandhi.[79] Ironically, it was none other than Gandhi who did so much to politically deliver peasant and artisan producers to the twentieth-century national movement, a movement which had long incorporated a doctrine of development framed according to the fear that the loss of productive force would commit India, like Australia and Canada before it, to relative poverty. So much for Marglin's prognosis for India. A fleeting acquaintance with the imperial history of India is enough to show how much of it is a re-enactment of the nineteenth-century interplay of development. Bearce, an historian of early nineteenth-century imperial policy, commented that:

> To the Utilitarian Liberals a series of rapid reforms based on a rational consideration of utility and directed to a transformation of the social and political structure of India, was the only sound policy. To Conservatives, improvement lay in the genius of Indian society, and this improvement was the responsibility of Indians, through a process of organic growth, not to be tampered with by analysis, theory, and British legislation.[80]

What irony! What was once the part played by conservative doctrine becomes the script for a present-day, very self-conscious radicalism.

'ANOTHER DEVELOPMENT'

Bjorn Hettne's *Development Theory and the Three Worlds* is one of the handful of 'development studies' texts which pays some attention to what we have argued are the origins of development and its ambiguities in earlier nineteenth-century European thought. Indeed, after re-identifying the much vaunted 'crises of development theory' in the East, West and South, Hettne proceeds along the reasonable path of attempting to understand what he depicts as 'development ideologies' in Western thought. For Hettne these 'ideologies' and their progeny stand condemned as 'Eurocentric'. Yet Hettne never asks the obvious question of how they could be anything else given that the conditions which gave rise to intentional development as a redress to progress arose first in Europe.[81]

When he turns to the Third World, the dissemination of Eurocentric theories is condemned as an 'academic imperialism' whose counterpoint is taken to be the rise of the dependency school's analysis of underdevelopment. Here, however, the *European* intellectual origins of what is now called underdevelopment are not appreciated but are merely assumed to have arisen as a *sui generis* response to Eurocentric thought.[82]

This error is compounded when Hettne proposes 'another development' which, in going beyond the critiques of thinking about underdevelopment, will 'transcend the European model' and so create a new kind of development thinking. This new model of development will be 'egalitarian', 'self-reliant', 'eco-' and 'ethnodevelopmental'. However, when we are introduced to the precursors of the 'new thinking', we find ourselves face to face with the ghosts of the likes of Saint-Simon, Comte, Proudhon and the Norodniks – to name but a few. Further, it is not only admitted but emphasised by Hettne that the well-springs of 'another development' are to be found 'back where we started' in the 'developed world'.[83] Through the approvingly quoted words of Denis De Rougement, it is here in

> Europe the continent that gave birth to the nation-state that was the first to suffer its destructive effects upon all sense of community and balance between man and nature. . . . the continent which, therefore, has every reason to be the first to produce the antibodies to the virus it itself generated.[84]

So is it that, in Hettne's words, 'development theory returns to Europe'. From a base in the 'new Europe', the gospel of 'another development' is to be spread to the Third World. Hettne is aware of some of the obvious questions the idea of 'another development' raises.[85]

He raises two, in particular: why have the concepts of 'another development' – which imply 'small-scale solutions, ecological concerns, popular participation, and the establishment of community, etc.' – been met with relatively more enthusiasm in the rich countries, while they are to a large extent rejected by the poor?' Second, why is there such an interest in 'another development' in the North?

Hettne's answer to the second question is that the 'collective consciousness of the industrially advanced countries is going through a transformation' against which 'spokesmen for the Mainstream will have a hard time finding a way out of the present impasse; a solution consistent with a world-view of automatic growth and eternal progress'. Growth and progress are symptoms of De Rougement's 'virus'. Hettne's answer to his first question regarding the lack of enthusiasm for 'another development' in the South is that it is the result of the corruption of Third World leadership. Thus,

> Small may be beautiful, but it does not entail power (as far as the ruling elite is concerned). The masses in the Third World will never reach the material standard of living at present maintained in the West (and by Third World elites), but some urban middle classes in some areas may, at least theoretically, achieve this. Consequently those chosen to become 'modern' do not intend to be fooled into some populist cul-de-sac.[86]

Given this impasse, how is 'another development' to be brought to the 'masses'? The vehicle will be an amalgam of official and non-governmental aid organisations whose agency, in inheriting the mantle of development, is to confront the destruction wrought by progress. In other words, in the face of the corruption of 'Third World' leaders, trusteeship – though none dare speak its name – will have to be exercised by those who represent themselves as knowing and moral on behalf of those who are taken to be ignorant and corrupt.[87]

The assembled ghosts of the Saint-Simonians, Comte, Mill and Newman would be much amused. So, too, would be the spirit of Marx, seeing his own work consigned to the rubbish bin of history by those who scavenge through decidedly riper tips for choice intellectual morsels to be offered up as new fare.

DEVELOPMENT AND CAPITALISM

Marx might have had Goethe's Mephisto in mind when he wrote: 'Modern bourgeois society, a society that has conjured up such gigantic means of production and exchange, is like the sorcerer who is no longer able to control the powers of the underworld that he has called up by his spells.'[88] The doctrine of development first appeared in the nineteenth century at the onset of the conjuring-up of 'gigantic means of production and exchange'. This doctrine centred on the belief that the 'powers of the underworld' – the powers of productive force – could be controlled through an understanding of what had made the progress of productive force possible. If progress had been made possible by the creative knowledge of science with regard to things, then the generalised knowledge of people, upon the same basis of science, could give social purpose to the powers of productive force. It was only the way in which this force was asserted that made its powers extend beyond the compass of the belief of social control for progressive development.

Marx's abiding principle was that productive force was the expression of human subjectivity which inhered in the power to, rather than the force of, labour. It is the one distinctive assertion of productive force which finds an absolute purpose in the future potential of free development. In setting the power to labour free from the coercive force of production, free development would also diminish the abstraction of labour as a part of human activity. When the development of capitalism is known to be an objective and coercive process, it is because it is under the force of labour, as the object of things, that the power to labour is given value and thus the means to actively command power over things through exchange.

An objective process of development separated the power of technique in things from the power of labour in people. However much modernity contained both the consciousness of the idea of science and that of a population which possessed need, the two could not be brought together without the coercive force of production. The denizens of the 'underworld' were the unemployed, those who had the power to labour but who could not find the means of exchange because they were excluded from the force of labour in work. Mass unemployment and poverty were the evil powers which the idea of progressive development sought to exorcise from what had become an objective process of development. Development doctrine, stemming from the idea of progressive development, was set out as the positive means by which the productive force lost through the process of capitalist development could be regained. Thus development came to be intent and not process. While the intent of doctrine has since come to be conflated with that of the process itself, as when development is made identical to modernisation and Westernisation, the mid-nineteenth-century idea of development did not elide intent into imminence. Newman, for instance, reacted against the idea of progressive development and any doctrine which made constructivist claims for progress or which, in present-day parlance, accounted for development upon an instrumental basis. It was this tradition, with development as process but with no final end of progress, which reappeared in the twentieth century and of which Whitehead is an exemplar.

In the powerful final chapter of his 1925 *Science and the Modern World*, Whitehead set out a critique of how the belief in productive force was associated with a particular view of science. Science was socially creative, to repeat, when it was a method by which the relative attributes of the organism might be investigated and not simply the rational pursuit of the absolute and objective knowledge of matter. At the heart of Whitehead's critique of science lay political economy after Adam Smith. Political economy, Whitehead argued,

> did more harm than good. It destroyed many economic fallacies, and taught how to think about the economic revolution then in progress. But it riveted on men a certain set of abstractions which were disastrous in their influence on modern mentality. It dehumanised industry.

These were, Whitehead explained, 'the abstractions in terms of which commercial affairs are carried on. Thus all thought concerned itself with social organisation expressed itself in terms of material things and of capital. Ultimate values were excluded.'[89] Development, without the word, was put in the cusp of a recurrent historical cycle wherein the possibility of 'rapid scientific and technological advance' was followed by one of social decadence, characterised by 'the malignant use of material power', the loss of religious faith and 'the infertility of its best intellectual types'. In appraising the early twentieth century to be a period of social decadence, as a century earlier the positivists construed their age to be 'critical' or 'transitional', Whitehead implored the 'need for preservative action' as the positivist had offered constructive development.[90] All modern development theory has followed this theme.

Nineteenth-century ideas of constructivist economics and development, from the Saint-Simonians to List and to Chamberlain, were based on a reaction against the abstractions of political economy, including those of *laissez-faire*, as opposed to mercantalist postulates of Adam Smith. Schumpeter's theory of economic development was a reaction against the abstractions of economic doctrine which imposed routines of commercial criteria upon social life. Innis, after Veblen and Wallas, was concerned about how innovation in productive force could be used constructively, in the institutions of industry rather than business, to preserve cultural values. Keynes, in his attack upon the post-Smith economics of 'classical theory', took a benign view of what he called the 'economics of the underworld',[91] and Burkean-wise, he aspired to bring the habit of prehension back into the state. Marglin and the other malcontents, with an 'alternative development', are only a latter-day variation on the theme of modern development theory.

If it is the state which makes the doctrine of development possible, it is the development of capitalism which stands between the ultimate values of development theory and the application of development doctrine. It is striking, therefore, that the image of capitalism upon which modern development theory is projected is that of the logical origin of, or historical transition to, the development of capitalism. In other words, the theory projects its values onto objects of the past. Take Whitehead again:

A factory, with its machinery, its community of operatives, its social service to the general population, its dependence upon organising and designing genius [of society], its potentialities as a source of wealth to the holders of its stock is an organism exhibiting a variety of vivid values.[92]

Vivid values, meaning to represent the free development of the person, are here to be re-created through the medium of the past. For the factory, read the feudal estate. For the capitalist, in Marglin's case of the type who subjectively stands between producers and the market to denude workers of techne and so find a specious function of episteme, read the feudal landowner. It is the the separation of labour power from productive force, the logical origin of the development of capitalism, that modern development seeks to repair. It attempts to do so by reproducing the origin as if it were the process of capitalist development. It is in this way that development doctrine comes to be made consistent with the development of capitalism itself.

As Rosdolsky pointed out years ago, the origin of capitalism is perpetually reproduced.[93] The primary accumulation of capital – the amassing of money-capital as a potential source of productive investment that paralleled the separation of producers from their means of production – was not merely a set of historical episodes but a universal condition for the perpetuation of capitalism. Labour power must be perpetually separated from productive force. People must be moved rapidly from countryside to town and city and from one country to another. While in one place and time the labour force is expelled from the production process, in another, its potential is realised. The welfare of the relative surplus population became the aspiration of development.

Doctrine has come to inflect the values of theory. Affiliated to the 'community' of nation and region, of locale and area, and in becoming the goal of policy, development has been the means whereby the potential of labour was suspended in the interstices of vacant productive force. Whether through agrarian doctrine, or the Listian doctrine for a national economy, or that of 'learning' to make the factory efficient in the international economy, development is the bridge between imposing the order of community upon a population for the potential of productive force and thus what is inscribed in the immanent process of capitalism. Doctrine invokes ideals of stasis and fixity to counter what the idea, or principle, of development, denoted as movement and fluidity.

Community is the real abstraction of modern development theory, signalling the medium in which progress may be achieved. We have seen the variation in the characterisation of crisis or decadence which precipitates action for development. From the threat of emigration from Quebec or Victoria in the mid-nineteenth century to that of de-industrialisation in late-century Britain or rural unemployment in Kenya a century later, the focus is on the disordered movement of population. Doctrine appears out of a theory that contrives to marry progress with order, but one doctrine, with a different variant of community, comes to subvert the purpose of the other. As we have shown in the interplay of doctrine in India, the conservative application of policy brought the reaction of a radical-liberal doctrine whose positive source of development countermanded the negative dimension of the former. Whether community is represented by the state or the village, the confrontation is with the disordered potential of productive force, a potential that is immanent in the development of capitalism.

When we nowadays think of community, we are led to believe that it is the positive antithesis of development. Property speculators are said to disruptively 'develop' real estate which is the proper province of community, but the real estate in question is often as not the property nationalised by a past doctrine of development. Conservative governments radically put welfare services 'into the community' to denote their disestablishment of state development. Would-be Labour governments hanker after 'community development'. In either case, and through much confusion, the community comes to represent the possibility of true development on the back of another but corrupted doctrine of development. This is the nineteenth-century impression of development. It belongs to the past but it is perpetually re-created, in an utterly different world, as a truly corrupted vision of the future. If we are to take our visions from the future, then they should be visions of free development and not the doctrines of development. And so, let us try to answer the question which will inevitably be asked by those who cannot comprehend our renunciation of trusteeship.

> He who asks what is the goal of an emancipated society is given answers such as the fulfilment of human possibilities of the richness of life. Just as the inevitable question is illegitimate, so the repellent assurance of the answer is inevitable. . . . There is tenderness only in the coarsest demand: that none shall go hungry anymore.

And:

> Perhaps the true society will grow tired of development. . . . A mankind which no longer knows want will begin to have an inkling of the delusory, futile nature of all the arrangements hitherto made in order to escape want, which used wealth to reproduce want on a larger scale.[94]

Adorno's question as well as his speculative answer are about doctrines of development. The true alternative to these doctrines, that of development itself, awaits all of us.

ACKNOWLEDGEMENTS

There are many people who have helped us during the course of writing this book. They have shared, in some cases even without their knowing it, ideas with us. Some have pointed us towards essential references and suggested points of interest which we have been encouraged to pursue.

We are especially grateful to those who have taken the time to read drafts of chapters and care to have made comments which have done so much to assist us in preparing the manuscript. Many would disagee with our view of development but all have given us encouragement.

We owe thanks and gratitute to the following: John Aldridge, Jan-Otto Andersson, Jonathan Barker, Ernest Benz, Christine and Vince Brown, Leslie Budd, Peter Campbell, Elsie Chenier, Chris Dixon, Colin Duncan, Peter Dutkiewicz, Liz Francis, Judith Heyer, Tony Humphries, August Gaechter, Elsa Guzman-Flores, Paul Idahosa, Geoff Kay, Riitta Launonen, Catherine LeGrand, Scott Macwilliam, Chris Martin, Glenn McKnight, Suzanne Mueller, Sandra den Otter, David Parker, Bryan Palmer, Tristan Palmer, George Rawlyk, Carmen Schiffelite, Amartya Sen, Matthew Smith, Philip Steenkamp, John Solomos, Gerald Tulchinsky, Gavin Williams, David Wilson.

Special thanks are due to August Gaechter and Scott Macwilliam for providing references to Austrian sources on Austrian economic history and Australian colonial nationalism.

Much gratitude is due to our own editor, Naomi Frankel, to Sophie Richmond and Caroline Cautley, at Routledge, for their work on the manuscript.

We acknowledge a research grant from the Social Science and Humanities Council, Canada, and some research funding provided by Queens' University, Ontario and London Guildhall University.

Last but not least we thank the many libraries we have used while being engaged in the work for the book. In particular, we are especially grateful to the librarians at London Guildhall University.

NOTES

PREFACE

1. Adorno 1973 [1964].

2. For instance, Toye, 1987, p. 142, 1983, pp. 51–4 (quoting from Friedrich 1949, p. 124).

3. Arndt, 1981.

1 THE INVENTION OF DEVELOPMENT

1. Chambers, 1969 [1844], p. 360.

2. As quoted in the *Financial Times*, London, 10 January 1992.

3. Staudt, 1991, pp. 28–9.

4. 'Introduction' in H. Bernstein (ed.), 1973, p. 13.

5. Barnett, 1988, pp. 5, 6.

6. Bernstein, 1973, p. 13.

7. Aseniero, 1985, pp. 54–5.

8. Harris, 1989, pp. 4–11.

9. Thomas, 'Introduction', in Allen and Thomas (eds), 1992, p. 7, emphasis ours.

10. See Chapter 2.

11. Esteva, 1992, p. 7.

12. Quoted by Esteva, 1992, p. 7. Five of the twenty essays in *The Development Dictionary* start with Truman's inaugural address; Truman is quoted ten times in the dictionary.

13. Esteva, 1992, p. 7. Esteva quotes Benson, 1942.

14. For Bourdillon, Campbell Bannerman and Chamberlain, see Cowen and Shenton, 1991b, pp. 145–7, 164–5. See Chapter 5.

15. Kay, 1975, pp. 1–2.

16. Brunner, 1948, p. 55; Baillie, 1950, p. 95. Both are quoted in Wager, 1967, pp. 65–6.

17. Quoted in Chadwick, 1957, pp. ix–x.

18. See Cowen and Shenton, 1991a.

19. Musson, 1978, p. 149.

20. Hobsbawm, 1968, pp. 58–9.

21. Louis Blanc, *Organisation du Travail* (1841 edn), cited in Sewell, 1980, p. 233.

22. ibid., p. 249.

23. Hobsbawm, 1988, p. 13; see also Hobsbawm 1968, p. 55.

24. Meek, 1976, p. 255 (quoted in Skinner, 1982, p. 91); also see Toye, 1980, p. 17.

25. 'Of all the metaphors in Western thought on mankind and culture, the oldest, most powerful and encompassing is the metaphor of growth. When we say that a culture or institution or nation "grows" or "develops", we have reference to change in time, but change of a rather distinctive or special type. We are not referring to random and adventitious changes, to changes induced by some external deity or other being. We are referring to change that is intrinsic to the entity, to change that is held to be as much a part of the entity's nature as any purely structural element. Such change may require activation or nourishment from external agencies, just as does growth in a plant or organism. But what is fundamental and guiding is nonetheless drawn from within the institution or culture.' (Nisbet, 1969, p. 7.)

26. We take note of the dispute between Nisbet and Bury as to whether cyclicality was progressive.

The point is, however, that even if cyclicality can be taken to be progressive, it was not progressive in the same sense that latter avowed views of progress were. The ground surveyed in this chapter is obviously well trodden. For instance, see Bury, 1921; also see Pollard 1971 [1968] and Aseniero, 1985. It should be noted, however, that Pollard's work does not take development as its focus or problematic.

27. Ferguson, *An Essay on the History of Civil Society*, [1782] (quoted in Bock, 1979, p. 57).

28. Bock, 1979, p. 57.

29. Smith, 1937 [1776], p. lix.

30. ibid., p. 250.

31. Bock, 1979, p. 57.

32. Smith, 1937, p. 508. The sense of this is conveyed in paragraph 2.

33. See, for instance, Hirschman, 1977.

34. Smith, 1937, p. 14.

35. See, for instance, Choi, 1990; Martin, 1990; Johnson, 1990.

36. Smith, quoted in Martin, 1990, p. 279.

37. Thompson, 1965, pp. 225–7 (quoted in Martin, 1990, pp. 275–6).

38. Hont, 1983, p. 302.

39. Malthus, 1986a [1798], p. 146.

40. Wrigley and Souden, 1986, p. 107.

41. Malthus, 1986a, pp. 183–4.

42. ibid., p. 184.

43. ibid., pp. 183–5.

44. ibid., p. 186.

45. ibid., pp. 198–9.

46. Malthus, 1986b [1830] pp. 245–6.

47. Sraffa and Dobb, 1966, p. 303.

48. Kitching, 1989 [1982], p. 22–6.

49. George Iggers makes this point in the introduction to his translation of Saint-Simon, 1958, pp. xxii–xxv.

50. One interesting example of the way in which concerns with Saint-Simonian thought, British radicalism and African nationalism have come together is seen in the work *The Saint-Simonians, Mill, and Carlyle*. Richard Pankhurst, descendant of the feminist Pankhurst sisters, wrote the book and it was published by Lalibela Press, which was named after the famous stone churches of Ethiopia; this book was dedicated to the Kenyan nationalist Mbiyu Koinange.

51. Iggers, 1958 [1829], p. 28.

52. ibid. and p. 79.

53. ibid., p. 28.

54. ibid., p. 12.

55. ibid., p. 13.

56. ibid., pp. 14–15.

57. ibid., pp. 13–14.

58. ibid., p. 14.

59. See, for instance, Thompson, 1988; Claeys, 1987, 1989.

60. Thompson, 1988, p. 19.

61. Iggers, 1958, p. 15.

62. ibid., pp. 118–19, 132–6.

63. ibid., p. 89.

64. ibid., pp. 103–10.

65. ibid., p. 103.

66. ibid., p. 107.

67. ibid., pp. 103–5.

68. ibid., p. 111.

69. Lévy-Bruhl, 1903, pp. 4–5, 10–11; also see Charlton, 1959, pp. 25–7; Evans-Pritchard, 1970, p. 19; 1981, p. 57.

70. Lenzer, 1983, p. 233.

71. Comte, 1875, p. 83.

72. ibid., pp. 83–4.

73. ibid., p. 84.

74. Comte, 1876b, chs 1 and 2.

75. Ryan, 1974, pp. 228–9; Keat, 1981, p. 17.

76. Giddens, 1979, pp. 240–1.

77. Comte, 1876b, ch. 1, esp. pp. 24–30.

78. ibid., pp. 32–3.

79. ibid., p. 17.

80. Comte, 1876a, ch. 7, esp. pp. 376–7.

81. Comte, 1876b, ch. 1, esp. pp. 52–6.

82. ibid., p. 59.

83. Ginsburg, 1953, pp. 26–9.

84. Comte, 1876b, pp. 46–7.

85. ibid., p. 48.

86. Comte, 1876a, ch. 2, esp. p. 129.

87. ibid., p. 132.

88. ibid., p. 133.

89. ibid., p. 137.

90. ibid., pp. 140–4.

91. Comte, 1876b, pp. xxxiii, xxxv.

92. See, for instance, Giddens, 1979, pp. 239–40; Gouldner, 1971, p. 113; Hawthorn, 1987, pp. 68–9.

93. Simon, 1964, pp. 162, 167.

94. Comte, 1875, p. 12.

95. Lenzer, 1983, p. 373.

96. Comte, 1875, p. 257.

97. ibid., pp. 45–6.

98. Andreski, 1974, pp. 200–1.

99. The following paragraphs on Mill have benefited immeasurably from the work of Adelaide Weinberg, whose *The Influence of Auguste Comte on the Economics of John Stuart Mill*, privately published by her husband after her death, is one of the more remarkable instances of hidden knowledge. A copy of this work may be found in the Senate House Library, University of London. See also, Wright, 1986.

100. J.S. Mill, 1989 [1873], p. 131–2.
101. J.S. Mill, 1942, p. 35.
102. ibid., pp. 36–7.
103. ibid., pp. 6–7, emphasis ours.
104. ibid., pp. 9–10.
105. ibid., p. 13.
106. J.S. Mill, 1974, p. 833.
107. ibid., p. 845; ch. 3, *passim*.
108. ibid., p. 861; ch. 4, *passim*.
109. ibid., p. 869; ch. 5, *passim*.
110. ibid., p. 869.
111. ibid., p. 897. This reference was deleted in later editions of Mill's *Logic* along with the vast majority of other glowing references to Comte.
112. Indeed, in as much as 'political economy', rejected as 'metaphysics' by Comte, found a place within 'ethology', Mill can be justly pronounced the first development economist.
113. J.S. Mill, 1965 [1859], p. 307. It would be unfair to Mill not to reproduce the following footnote from his *Logic*, 1974, p. 837:

The pronoun *he* is the only one available to express all human beings, none having yet been invented to serve the purpose of designating them generally, without distinguishing them by a characteristic so little worthy of being made the main distinction as that of sex. This is more than a defect of language; tending greatly to prolong the almost universal habit, of thinking and speaking of one-half the human species as the whole.

114. ibid., p. 312.
115. ibid., p. 297.
116. See, for instance, Hicks, 1983.
117. J.S. Mill, 1985 [1848], pp. 115–17.
118. J.S. Mill, 1990, p. xx.
119. Mun, 1621; 1628.
120. Steuart, 1967 [1772], 1805; see also Barber, 1975.
121. Naoroji, 1901 [1871]; Dutt, 1970 [1904].
122. On Keynes, see Barber, 1975, p. 231; on the general point see also Bearce, 1961; Stokes, 1979 [1959]; Ambirajan, 1978; Ganguli, 1979.
123. James Mill, 1975 [1858], pp. 231–2.
124. Newman, 1927, p. 155.
125. O'Gorman, 1973, p. 53. Burke is here referring to England but the general point holds.
126. Newman, 1927, p. 189.
127. ibid., p. 172.
128. Wickwire and Wickwire, 1980, p. 71.
129. O'Gorman, 1973, p. 53.
130. Wickwire and Wickwire, 1980, pp. 71–2.
131. James Mill, 1975, p. 481.
132. Wickwire and Wickwire, 1980, pp. 71–2.
133. Stokes, 1979, p. 88.
134. Malthus, 1970 [1815], p. 179.
135. Sraffa and Dobb, 1966, pp. 197–8.
136. Barber, 1975, p. 154.
137. Ambirajan, 1978, p. 151.
138. ibid., pp. 151, 153.
139. ibid., p. 159.
140. Stokes, 1979, p. 88.
141. Hollander, 1931; Smith, 1872; McCulloch, 1843.
142. James Mill, 1975, p. 408.
143. ibid., pp. 485–6.
144. ibid., pp. 491–2.
145. ibid., p. 518.
146. ibid., p. 519.
147. ibid., p. 575.
148. ibid., pp. 493–4.
149. Barber, 1975, p. 175.
150. J.S. Mill, 1968 [1858], ch. 3.
151. ibid., p. 52.
152. ibid., p. 10.
153. J.S. Mill, 1990, pp. 215–16.
154. ibid., pp. 216–17.
155. ibid., p. 221.
156. ibid., p. 222.
157. ibid., p. 225.
158. ibid., p. 50.
159. ibid., p. 45.
160. ibid., p. 45.
161. ibid., p. 85.
162. Strachey and Strachey, 1882, pp. 401–2.
163. ibid., p. 406.
164. J.S. Mill, *Principles of Political Economy*, cited in Strachey and Strachey, 1882, pp. 408–9.
165. Strachey, 1985, p. 151.
166. Ganguli, 1979, p. 33.
167. ibid., p. 42.
168. On the establishment of 'Community Development' in India, see Hicks, 1961, chs 10 and 21.
169. K. Marx, 'The British rule in India', *New York Daily Tribune*, 25 June 1853, reprinted in Marx, 1979a, pp. 131, 132. Kenya readers may note that in the 1979 edition of the *Collected Works*, 'Hanuman' is corrupted as 'Kanuman'.
170. Marx, 'The future results of British rule in India', *New York Daily Tribune*, 8 August 1853, reprinted in Marx, 1979b, pp. 221, 222.

2 DEVELOPMENT WITHOUT PROGRESS: UNDERDEVELOPMENT AND J.H. NEWMAN

1. Baran, 1973 [1957], p. 402 (quoted in Little, 1982, p. 219).

2. See Frank, 1971 [1967]; the most succinct statement is to be found in Frank, 1983a, 1983b.

3. Frank, 1983b, p. 186.

4. ibid., p. 198.

5. Frank, 1972a, pp. xiii–xiv, 1972b, 1972c, pp. 35–6.

6. Prebisch, 1980, p. 25 (quoted in Chilcote, 1984, p. 26).

7. Frank, 1972c, p. 35.

8. Glade, 1969, p. 204.

9. Glade, 1969, p. 216. See also Burns, 1990, pp. 75–6.

10. Glade, 1969, p. 235.

11. ibid., pp. 242–5.

12. ibid., pp. 242–3, emphasis ours.

13. ibid., pp. 235–6, emphasis ours.

14. Davis, 1972, p. 9.

15. See Orgaz, 1934.

16. Zea, 1963, p. 69.

17. Esteban Echeverria, *Dogma Socialista de la Asociación de Mayo*, (Montevideo, 1846) as found in Crawford, 1963; and in Zea, 1963. The similarity in language to Ngugi wa'Thiongo's *De-colonizing the Mind* (1986), is uncanny.

18. Esteban Echeverria, *Dogma socialista de la Asociación de Mayo*, as noted in Zea, 1963, pp. 62–3.

19. Sarmiento, 1972 [1868], as reproduced in Burns, 1993, p. 77. See also Bunkley, 1948.

20. Burns, 1993, p. 78.

21. ibid., p. 79.

22. ibid., pp. 79–80.

23. ibid., p. 80.

24. Bilbao, *La America en Peligro*, quoted in Zea, 1963, pp. 74–5.

25. Alberdi, quoted in Zea, 1963, pp. 87–8.

26. ibid.

27. Lastarria, *Estudios literarios*, quoted in Zea, 1963, pp. 65–6.

28. ibid., pp. 65–6.

29. Alberdi, cited in Zea, 1963, pp. 87–8.

30. Lastarria, *Recuerdos literarios*, in Zea, 1963, p. 136.

31. Although he later claimed not to have gained a 'thorough knowledge' of Comte's writings before 1868, Lastarria noted his enthusiasm for Comte as early as 1833.

32. Barreda, cired in Zea, 1974, pp. 39–42.

33. Zea, 1974, pp. 51–2.

34. On this point see Hale, 1989, especially ch. 7. See also Hale, 1986.

35. This is Zea's view, with which we agree despite the recent criticisms by Hale.

36. Williamson, 1992, p. 299.

37. ibid., pp. 267–9.

38. ibid., p. 380.

39. ibid., pp. 399–400.

40. ibid., p. 399.

41. ibid., p. 400.

42. Zea, 1963, pp. 7, xiii, emphasis ours.

43. Zea, 1974, p. xxi.

44. Frank, 1969.

45. See, from among the many surveys, Palma, 1981; Brewer, 1980, part 3; Roxborough, 1979; Larrain, 1989.

46. Warren, 1980, p. 113, emphasis ours.

47. Banaji, 1983, pp. 108–9, 102.

48. Brenner, 1977, p. 27. Brenner, here, elides ideas of progress with development when he treats Smith's thesis of progress as if it were a theory of development.

49. ibid., p. 83.

50. ibid., p. 91.

51. ibid., p. 92.

52. Frank, 1967, p. 13 (quoted in Booth, 1975, p. 69).

53. Booth, 1975, p. 77, emphasis ours.

54. ibid., p. 77, emphasis ours.

55. Frank, 1972d, p. 92 (quoted in Booth, 1975, p. 77).

56. Frank, 1967 (quoted in Booth, 1975, p. 69). The hypothesis is repeated by Frank, 1972b, p. 9.

57. Mao Tse-tung, 'On Contradiction', 1955a (quoted in Frank, 1978, p. 3). See Arrighi, 1970, pp. 22–3 (quoted in Frank, 1978, p. 6).

58. Meisner, 1982, p. 55; also see Schram, 1969 [1963], pp. 210–14.

59. Mao, 'On Contradiction', 1955a, pp. 300, 302–3, 316. The most complete edition of 'On Contradiction', in translation, is to be found in N. Knight (ed.), *Mao Zedong on Dialectical Materialism*, 1990. For Mao on identity, therefore, also see Knight (ed.) 'On contradiction', pp. 158–79 and 190–1.

60. See, for instance, Wittfogel, 1963, p. 252; Schram, 1988, p. 62; Cohen, 1964, pp. 19–21, 27.

61. See Mao Tse-tung, 'On Practice', 1955b, pp. 282–97; also see Soo, 1981, pp. 40–2.

62. Mao Tse-tung, 'On the Mass Line', in 1965 (excerpts reprinted in Schram, 1969, pp. 316–17).

63. Yu-Ning, 1971; Bernal, 1976.

64. See Schram, 1969, pp. 21–32, 1988, p. 19; Snow, 1968 [1938], fn. 2, p. 422. His 12,000 words of notes on the *System of Ethics* indicated the extent, Mao told Edgar Snow, to which Paulsen had taught him the value of discipline, self-control and will power.

65. Paulsen, 1899 [1888], pp. x–xi, 191–2; Paulsen, 1938, pp. 228, 237, 248–50.

66. Paulsen, 1899, pp. 341, 343, 345.

67. ibid., pp. 370, 371, 368.

68. Schram, 1988, p. 20.

69. Mao, 1955a, pp. 298, 305 (quoting from Lenin, 'On the Question of Dialectics' 1961a [1915] p. 359).

70. Mao, 1955a, p. 335.

71. Lenin, 1961a, p. 360.

72. Lenin, 1961b, p. 483.

73. Lenin, 1961a, pp. 360, 363.

74. Mao, 1955a, p. 330.

75. ibid., pp. 298, 305–6, 307.

76. Quoted by Pannekoek, 1975 [1938], p. 8.

77. Bakhurst, 1991, p. 30.

78. Bakhurst, 1991, pp. 39–42.

79. Lenin, 1961b, p. 483.

80. ibid., pp. 477, 484.

81. Mao, 1955a, p. 300; Lenin, 1961b, p. 484.

82. Bakhurst, 1991, pp. 47–9, 52.

83. ibid., p. 35.

84. Jensen, 1978, p. 103 (quoting from Bogdanov, 1923 [1913], p. 242).

85. Bakhurst, 1991, pp. 35–6.

86. ibid., pp. 33–5.

87. ibid., p. 55.

88. Knight (ed.) 'On Contradiction', 1990, p. 159.

89. Lenin, 1961a, p. 363.

90. Quoted in Chadwick, 1957, pp. ix–x.

91. Fairbairn, 1893, p. 35.

92. ibid., p. 35.

93. ibid., p. 34.

94. ibid., pp. 40, 41.

95. ibid., pp. 42–5.

96. ibid., p. 34.

97. For instance, see Ker, 1990, p. 20.

98. See Ker, 1990, ch. 6, esp. pp. 285–7, 303–6, 311–12 and Ker's 'Introduction', pp. 18–19, 30; Gilley, 1990, p. 230; Lash, 1975, pp. 73, 106–8.

99. Willey 1973 [1949], p. 90; Chadwick, 1957, pp. 153–5.

100. Stephen, 1877, pp. 690, 693.

101. Newman, 1918 [1850] (quoted in Brinton, 1933, p. 155).

102. Chadwick, 1957, pp. 136–7; Lash, 1975.

103. Beer, 1983, p. 3.

104. Chadwick, 1957, p. 97; Lash, 1975, p. 62.

105. Allen, 1975, p. 16 (quoted in Coulson, 1990.

106. Grave, 1989, p. 2.

107. Newman, 1845, p. 44.

108. Lash, 1975, p. 72.

109. Barry, 1904, p. 278. Barry wrote that we cannot but see 'on every page of the *Development* Darwin's advancing shadow'.

110. Stephen, 1887, p. 809.

111. ibid., p. 690.

112. Newman, 1845, p. 55.

113. ibid., p. 37.

114. ibid., p. 44.

115. ibid., p. 63.

116. ibid., pp. 63–4, emphasis ours.

117. Furtado, 1963 [1959]; Frank, 1971, pp. 174, 229–30.

118. Furtado, 1964 [1961], p. 90.

119. ibid., p. 142.

120. ibid., p. 143.

121. ibid, pp. 52, 171.

122. Theobald, 1990, p. 160. Also see, for instance, Alatas, 1990; Ward, 1989.

123. See Hirschman, 1977, p. 40, fn. q: '"Corruption"', noted Hirschman, 'has had a similar semantic trajectory' to that of the 'interests'.

124. Viscount Bolingbroke, quoted by Spadafora, 1990, p. 14.

125. Temple 1972 [1673] (quoted by Hont, 1983, p. 279).

126. Hont, 1983, p. 275.

127. See, for instance, Pocock, 1985, 1992. Various interpretations of the tradition of 'civic humanism' are set out by Cotton, 1991, ch. 1.

128. See Dickinson, 1977, pp. 102–18, 169–88; also see Brewer, 1989.

129. Newman, 1845, p. 45.

130. See, for instance, Kenny, 1957, pp. 84, 92.

131. Quoted in Soltau, 1959 [1931], p. 46.

132. Soltau, 1959, pp. 48, 49.

133. Newman, 1845, p. 451, quoted in Dessain, 1980 [1966], p. 80. For the development of corruption, also see Brinton, 1933, p. 162 and Willey, 1973 [1949], p. 94.

134. Mitchell, 1990, p. 244.

135. Newman, 1845, pp. 64–5.

136. ibid., pp. 57–93; the tests are also laid out in Gilley, 1990, pp. 232–3.

137. Yearley, 1978, p. 101.

138. Newman, *Essay on Development*, 2nd edn, 1878, p. 111, 1967, p. 98. Newman had read the third edition of Mill's *System of Logic* in 1857 and it was as a reaction against Mill that Newman began to develop the basic concepts for the *Grammar of Assent*. Sillem has written that, like Locke earlier, 'Mill was useful to Newman as a source from which he obtained an insight into the way of thinking of his Liberal philosophical opponents.' See Newman, 1969 (ed. Sillem), p. 226.

139. See Brinton, 1933, p. 163.

140. For Pragmatism, see Brinton, 1933, p. 164 and J. Newman, 1986, p. 28; for Wittgenstein, who read Newman's writings carefully but which he disliked, Gilley, 1990, pp. 361–2 and Malcolm, 1984, p. 59; for Whitehead, Atkins, 1966, pp. 539–40.

141. Newman, 1870, p. 70 (quoted in Brinton, 1933, p. 163).

142. For succinct interpretations of the *Grammar* see, for instance, Brinton, 1933; Cameron, 1960, 1967; Chadwick, 1957; Lash, 1975; Gilley, 1990; Grave, 1989; Kenny, 1990; J. Newman, 1986; Yearley, 1978.

143. Pattison, 1991, p. 145.

144. Newman, 1870, p. 92.

145. Pattison, 1991, p. 194.

146. Selby, 1975, p. 115. Newman, 1870, p. 337.

147. Newman, 1870, p. 93.

148. Gilley, 1990, p. 360.

149. Selby, 1975, pp. 22, 45–8, 53–66; also Pattison, 1991, pp. 159–60. Both Selby and Pattison show how Newman was intrigued by the analogy of differential calculus in working out the method of economy as approximation to truth.

150. Selby, 1975, p. 83.

151. Lash, 1975, p. 73.

152. Yearley, 1978, p. 10; also see Miller, 1992.

153. For Newman and Paley, see Cameron, 1960, pp. 109–10, 1967, p. 79; for Paley, see, for instance, Young, 1985, ch. 2.

154. See, from different vantage points, Norman, 1990a, pp. 157–61, 1990b, pp. 460-1; also, Misner, 1985; Pattison, 1991, ch. 5.

155. Yearley, 1978, pp. 93–117.

156. See, for instance, Stephen, 1877, p. 684; Lash, 1975, pp. 92–4; Chadwick, 1957, p. 35.

157. See Chadwick, 1957, p. 131; Yearley, 1978, p. 107.

158. See, for instance, Cameron, 1960, pp. 114–15; 1967, p. 93; Chadwick, 1957, pp. 157–60, 195; Lash, 1975, p. 100; Kenny, 1990, pp. 117–19; Ker, 1990, p. 37; Grave, 1989, pp. 180–4.

159. Stephen, 1877, p. 680.

160. Fairbairn, 1893, pp. 44, 45.

161. ibid., p. 44.

162. Annan, 1984, pp. 279–88; also see Pattison, 1991, pp. 189–91.

163. Barry, 1904, p. 278.

164. Annan, 1984, p. 289.

165. Stephen, 1877, p. 810.

166. Figgis, 1914, p. 214.

167. Hirst, 1989, p. 10.

168. Figgis, 1914, pp. 214, 259–60.

169. There is no mention of Newman in Hirst, 1989. Newman is given some recognition in Nicholls, 1975, pp. 22, 46, 68, 114. Hirst leans heavily on Nicholls for his account of English pluralism. Yearley, 1978, pp. 136–41, discusses the extent to which Newman presaged twentieth-century pluralism.

170. Figgis, 1914, p. 221. Figgis quoted Newman through W.G. Ward, *Life of Newman*, 1912, p. 367.

171. Newman, 1870, p. 349 (quoted in Yearley, 1978, p. 14).

172. Figgis, 1913, pp. 88–90.

173. Figgis, 1914, p. xii. Also, Figgis, 1917, pp. 2–3. Here, Figgis suggests that Comte's altruism is equivalent to Smith's sympathy in that both are derived from Christian values.

174. Figgis, 1914, pp. 141–4. Also, see Figgis, 1917.

175. Figgis, 1914, p. 136; pp. 124–39.

176. Figgis, 1896, p. 249.

177. ibid., p. 250.

178. Newman, 1993, p. 40.

179. Laski, 1917, p. 202 (quoted in Nicholls, 1975, p. 22).

180. Hirst, 1989, pp. 18, 29.

181. The case is discussed in Figgis, 1913, pp. 18–22; also, Nicholls, 1975, pp. 66–8; Hirst, 1989, p. 18.

182. Haldane, 1883, p. 58. For Haldane and his idea of development, as inflected from the influence of T.H. Green, see Cowen and Shenton, 1994.

183. Laski, 1917, p. 205; Hirst, 1989, pp. 17–38.

184. *Correspondence of John Henry Newman with John Keble and Others, 1839–45*, 1917, pp. 375–7 (quoted in Kenny, 1957, p. 125).

185. See Zylstra, 1970 [1968], pp. 42–3.

186. ibid., p. 94; Nicholls, 1975, p. 46.

187. Figgis, 1913, pp. 84–5.

188. Laski, 1921, p. 71 (referred to in Hirst, 1989, p. 36). Laski referred to Canada and Australia

as being 'in juristic fact' the simple 'immense instances of decentralisation'.

189. Fawkes, 1913, pp. 30–1, 277.

190. Lash, 1990, p. 453.

191. Coulson and Allchin, 1967, p. xx (quoted in Lash, 1990, p. 449).

192. Misner, 1973, p. 691 (quoted in Lash, 1990, p. 453).

193. Lash, 1990, p. 454.

3 THE DEVELOPMENT OF PRODUCTIVE FORCE: MARX AND LIST

1. Cohen, 1978, p. 134.
2. Marx, 1975a, p. 12.
3. Marx, 1975b, p. 364.
4. Hegel, 1975 [1837], p. 126.
5. ibid., pp. 128, 126–7.
6. ibid., p. 126.
7. Meikle, 1985, p. 36.
8. Hegel, *Science of Logic*, (quoted in Shamsavari, 1991, p. 121).
9. Hegel, 1975, pp. 129, 134, 138.
10. See, for instance, Shamsavari, 1991, pp. 119–21.
11. Hegel, 1975, pp. 134, 19, emphasis ours.
12. See Arthur, 1986.
13. ibid., p. 127.
14. ibid., p. 125.
15. Hegel, 1977 [1807], p. 8.
16. Hegel, 1975, pp. 126, 124, 126.
17. ibid., p. 129.
18. ibid., p. 125.
19. ibid., pp. 126, 127.
20. See, for instance, Hegel, 1971, pp. 8–20, 1977, pp. 1–44, 1986 [1840], p. 125, emphasis ours.
21. Hegel, 1971, pp. 20, 183.
22. Hegel, 1986, p. 125.
23. ibid., p. 134.
24. ibid., pp. 126, 127.
25. ibid., p. 134.
26. ibid., p. 146.
27. ibid., p. 127.
28. ibid., p. 130.
29. ibid., p. 131.
30. ibid., p. 154.
31. ibid., p. 155.
32. ibid., p. 131.
33. Hegel, 1977, p. 10, emphasis ours.
34. Cohen, 1978, p. 2.
35. Lowith, 1965 [1941], p. 265 (quoted in Arthur, 1986, p. 85).
36. Arthur, 1986, p. 87.
37. Hegel, 1971, pp. 183, 184.
38. Hegel, 1975, p. 127.
39. ibid., p. 131.
40. ibid.
41. Kain, 1982, pp. 46, 45.
42. Marx, 1967, p. 61 (quoted by Murray, 1988, p. 97).
43. Marx and Engels, 1975, p. 7, emphasis ours.
44. ibid., p. 139, emphasis ours.
45. ibid., pp. 84–5.
46. ibid., pp. 83–4, emphasis ours.
47. ibid., p. 84, emphasis ours.
48. Hegel, 1977, pp. 10, 12.
49. Marx and Engels, 1975, p. 82, emphasis ours.
50. ibid., p. 86, emphasis ours.
51. ibid., emphasis ours.
52. Marx and Engels, 1976, p. 474.
53. ibid., p. 474, emphasis ours.
54. ibid.
55. ibid., p. 520.
56. Hegel, 1977, p. 255.
57. Marx and Engels, 1975, p. 23, emphasis ours.
58. Marx and Engels, 1976, pp. 457–63.
59. ibid., p. 475, emphasis ours.
60. ibid., p. 476, emphasis ours.
61. Nicolaus, 1973 [1939], p. 43.
62. Marx, 1973, p. 101.
63. Marx and Engels, 1976, p. 473, emphasis ours.
64. Marx, 1973, p. 687, 1981 [1894], p. 368.
65. Marx, 1981, p. 368, 1976 [1867], p. 284.
66. Marx and Engels, 1976, pp. 516, 517.
67. Marx, 1976 [1867], p. 284.
68. ibid., pp. 283–4.
69. Marx and Engels, 1976, pp. 515, 517.
70. Marx, 1976, p. 284.
71. For a recent, original treatment of social labour in Marx, see Wilson, 1986, 1989, 1991.
72. Marx and Engels, 1976, p. 519.
73. Marx, 1976, p. 285. But see, for instance, the first English edition of 1887, edited by Engels and reproduced as the 'Moscow' edition where, significantly, as elsewhere 'presupposes' is translated as 'implies' in the first sentence and development of the 'labour process' by 'labour' in the second sentence (Marx, 1965, p. 179).

74. Marx, 1973, pp. 277, 278, emphasis ours.
75. Cohen, 1978, p. 134, 1988, pp. 83–4.
76. Cohen, 1978, p. 96.
77. Marx and Engels, 1976, pp. 31, 32.
78. Marx, 1973, p. 158.
79. Cohen, 1978, p. 98.
80. Cohen, 1988, pp. 149, 140, 142, 1978, p. 131.
81. Cohen, 1988, p. 149.
82. Marx, 1976, pp. 648–9.
83. ibid., p. 651.
84. Uchida, 1988, p. 14; also, Murray, 1988, pp. 97, 98.
85. Marx, 1973, pp. 332–3.
86. Uchida, 1988, p. 6.
87. Marx, 1973, p. 278.
88. Uchida, 1988, pp. 4, 11–12, emphasis ours.
89. Marx, 1981 [1894], pp. 957–9.
90. Cohen, 1978, pp. 85–6, 152ff.
91. ibid., pp. 100–1.
92. Marx, 1973, p. 109.
93. Shamsavari, 1991, pp. 2–9, 276–80.
94. Marx, 1973, p. 604.
95. Cohen, 1988, p. 148.
96. Elster, 1985, pp. 261–2 (quoted by Cohen, 1988, p. 154).
97. Marx, 1976, pp. 1072, 1071.
98. Marx, 1973, pp. 278–9, emphasis ours.
99. ibid., p. 278.
100. ibid., p. 605; Marx, 1976, pp. 781–802, emphasis ours.
101. Marx, 1973, p. 606.
102. ibid., pp. 604, 607.
103. ibid., p. 609, emphasis ours.
104. ibid., pp. 279, 609–10.
105. List, 1991 [1885]. The 1856 translation by G.A. Matile, published by J.B. Lippincott & Co. of Philadelphia, is in some respects considerably different and will be cited below where appropriate.
106. Marx, 1975c [1845].
107. Hamilton, 1966.
108. Tribe, 1988a, 1988b; Samuels, 1990.
109. Miller, 1959, p. 225.
110. ibid., ch. 16.
111. ibid., chs 17, 18.
112. Hamilton, 1966, p. 289.
113. ibid., pp. 236–7.
114. ibid., p. 239.
115. ibid., p. 249.
116. ibid., pp. 262–3.
117. ibid., p. 253. The editors write, in an explanatory footnote accompanying this passage that: 'Advocacy of child labour and the industrial employment of women was, of course, common in the eighteenth century in Europe and America.' No further comment is necessary.
118. ibid., p. 253.
119. ibid., pp. 287–305, 307.
120. Miller, 1959, ch. 20.
121. Kitching, 1989 [1982], p. 143.
122. Tribe, 1988a, pp. 1–16. Also see, for List's life, Henderson, 1983. See also the American influence evident in List's *National System of Political Economy*, 1991 and in his 'Outlines of American political economy in a series of letters to Charles J. Ingersoll, Esq.', 1827, folio 3960/67822, New Hampshire State Library.
123. List, 1991, p. xxix.
124. Tribe, 1988a, pp. 1–16.
125. See, for instance, Szporluk, 1988, pp. 99–100.
126. List, 1856, pp. 209, 229–31, 74.
127. List, 1991, pp. 149–50.
128. List, 1856, p. 233–4.
129. ibid., p. 76.
130. List, 1991, pp. 174–5.
131. List, 1856, pp. 63–4.
132. List, 1991, pp. 197–98.
133. ibid., p. 197.
134. List, 1856, pp. 241–2.
135. ibid., pp. 262–3, 147. List, 1991, pp. 174, 347–51.
136. List, 1856, p. 72
137. List, 1991, p. 360.
138. ibid.; List, 1856, pp. 263, 191, 77.
139. List, 1991, p. 131.
140. List, 1856, pp. 74, 309.
141. List, 1991, pp. 171–2.
142. List, 1856, p. 227, 307.
143. ibid., p. 205.
144. ibid., pp. 75–6, 278–9.
145. ibid., pp. 77–8, 199.
146. Marx, 1975c, p. 278.
147. ibid., pp. 277–9.
148. ibid., p. 280.
149. ibid., pp. 282–3.
150. ibid., p. 281.
151. ibid.
152. ibid., p. 285.
153. Engels, 1975 [1845], pp. 257–64.
154. ibid., pp. 258–62.
155. ibid., p. 263.

4 IMMANENT AND INTENTIONAL DEVELOPMENT: THE ORIGINS OF DEVELOPMENT IN AUSTRALIA AND CANADA

1. Arndt, 1981, p. 460.
2. ibid., p. 462.
3. Paradoxically, Arndt himself once commented that he made the 'first estimate of agricultural surplus population in under-developed regions'. For the genesis of Arndt, a well-known Fabian economist, see Groenewegen and McFarlane, 1990, pp. 180, 180–4.
4. Goodwin, 1966, pp. 12–13.
5. Serle, 1963, pp. 382, 385, 388–9; Cotter, 1967; Sinclair, 1976, p. 81; Butlin, 1986, p. 113.
6. *Argus*, 11 May 1860, quoted in Serle, 1963, p. 241.
7. Mayes, 1859, p. 3; Serle, 1963, p. 240.
8. Serle, 1963, p. 234.
9. Mayes, 1861, p. 249.
10. ibid., pp. 253, 255.
11. ibid., p. 254, emphasis ours.
12. ibid., p. 256.
13. ibid., pp. 251, 252, 253.
14. Marx, 1976, p. 1072.
15. Mayes, 1859, p. 3.
16. Mayes, 1861, p. 247.
17. Schedvin, 1990, p. 534.
18. Sinclair, 1976, pp. 83, 91, 109.
19. Thompson, 1970, pp. 86, 90.
20. Butlin and Sinclair, 1986, p. 132; Sinclair, 1976, pp. 91ff.
21. Index of average money wages in industry, Victoria, 1861–99 (1890 = 100)

1861–4	*1865–8*	*1869–72*	*1873–6*	*1877–80*	*1881–4*	*1885–8*	*1889–92*	*1893–5*	*1896–9*
88.3	82.2	86.0	93.4	93.3	92.3	94.3	98.0	90.1	85.8

Source: Mitchell, 1983, Table C4, p. 184.

22. Sinclair, 1971, p. 78, 1976, p. 98.
23. Serle, 1963, p. 247.
24. Sinclair, 1976, p. 98.
25. Fogarty, 1966, p. 41. (Fogarty quotes from Gollan, 1960, p. 37.)
26. Cowen and Shenton, 1992, pp. 13–18.
27. Irving, 1974, p. 156.
28. Cotter, 1967, p. 131.
29. Serle, 1963, pp. 247, 269 (Serle quotes from 'Old Colonist', *Land and Labour in Victoria*, 1856, p. 14).
30. Irving, 1974, p. 155; McMichael, 1984, pp. 89–90.
31. ibid., p. 220, emphasis ours.
32. Serle, 1963, pp. 296, 299 (Serle quotes Brodribb, *Recollections of an Australian Squatter*, p. 129).
33. McMichael, 1984, pp. 221–8; Sinclair, 1976, p. 92.
34. Fogarty, 1966, p. 39.
35. Sinclair, 1976, p. 103.
36. Serle, 1963, p. 375 (Serle quotes from Fitzpatrick, 1946, p. 56).
37. Pomfret, 1981a, p. 141.
38. Phillips, 1967, p. 55; Ormsby, 1958, p. 159.
39. Phillips, 1967, pp. 46, 59.
40. See, for instance, Woodcock, 1975, p. 64, 1990, p. 105; Phillips, 1967, fn. 45, p. 224; Kendrick, 1967.
41. Jebb, 1905, pp. 82–3 (quoted in Schreuder, 1988, pp. 74–5); also see Miller, 1956; Trainor, 1994.
42. Anderson, 1983, pp. 88–9.
43. Cole, 1971, pp. 177–8.
44. Davidson, 1991, pp. xii, 190.
45. Cole, 1971, pp. 164, 166.
46. Hobsbawm, 1990, p. 78.
47. Cole, 1971, p. 167.
48. See, for example, Ward, 1976.
49. Blackton, 1955, p. 122.
50. Davidson, 1991, p. xiv.
51. Blackton, 1955, p. 123; also, for example, Gollan, 1955, pp. 43–6; Quaife, 1967, pp. 223–4, 229; Cole, 1971, p. 167; Irving, 1974, pp. 137, 144–5.
52. Nadel, 1957, pp. 111–13, 118, 121, 175–8, 267–9.
53. For instance, see Kociumbas, 1992, ch. 8, especially pp. 216–17.

54. Nadel, 1957, p. 268.

55. Blackton, 1955, p. 138; also, La Nauze, 1967, p. 94.

56. Kociumbas, 1992, p. 216; Irving, 1974, p. 149.

57. Dilke, 1868, p. 52 (quoted in Goodwin, 1974, pp. 64–5).

58. 'Democratic government in Victoria', *Westminster Review* 33 (quoted by Davidson, 1991, pp. 189–90).

59. ibid., p. 190.

60. Kay and Mott, 1982, p. 128.

61. Davidson, 1991, p. 190.

62. Woodcock, 1990, p. 105; also see Barman, 1991, ch. 5; Hendrickson, 1981.

63. Patterson, 1990, pp. 13, 141, 145, 152–6; Woodcock, 1990, pp. 106, 121 and 1975, pp. 64, 112.

64. Patterson, 1990, p. 159.

65. See Greer and Radforth, 1992.

66. J.-M. Fecteau, 'Etat et associationnisme au XIXe siècle québecois: éléments pour une problématique des rapports Etat/société dans la transition au capitalisme', in Greer and Radforth, 1992, p. 144. Our translation.

67. Zeller, 1987, pp. 8–9; for Australia, albeit 'the science of man', see Kociumbas, 1992, ch. 10.

68. Province of Canada, 'Report of the Select Committee appointed to inquire into the causes and importance of the emigration which takes place annually from Lower Canada to the United States; etc.', Appendix (A.A.A.A.) to Volume 8, *Journals of the Legislative Assembly*, Montreal, 1849, p. 1. Henceforth, Chauveau Report.

69. Through successive constitutional changes, a process that itself was part of the 'development' of Canada as a national entity, the present province of Quebec has been successively known as New France until the British conquest of 1760, Lower Canada until the union with Upper Canada in 1841, Canada East as part of the United Province of Canada, with self-government in 1848, and then Quebec Province when separated from Ontario Province as part of the confederation of British North America that created the present state of Canada in 1867.

We refer to the province as either Lower Canada or Quebec. We have also adopted the contemporary use of 'anglo' or 'anglophone' and 'francophone' when referring to English-speaking and French-speaking Canadians.

70. See, for instance, Burroughs, 1984.

71. Grey, *The Colonial Policy of Lord John Russell's Administration*, 1853, I, p. 33 (quoted by Burroughs, 1990, p. 45).

72. Buckner, 1985, p. 6; see more generally, for instance, Burroughs, 1971, 1984; Ward, 1976; Cell, 1970; Morrell, 1966 [1930].

73. Careless, 1971.

74. Halpenny and Hamelin, 1982, p. 196. Our translation.

75. Monière, 1981 [1977], p. 141.

76. Little, 1989, p. 5.

77. Morisonneau, 19, p. 124 (quoted in Little, 1989, p. 5).

78. Little, 1989, pp. xii, 32.

79. ibid., p. 205.

80. Young, 1978, p. 24.

81. Little, 1985, p. 526. Little does not refer to Young but he does refer to Pentland and Palmer. See Pentland, 1981, pp. 110–12 (part of which is reprinted as 'The Transformation of Canada's Economic Structure' in Laxler, 1991; Palmer, 1983, p. 12.

82. Little, 1989, p. 127.

83. A point made widely in the literature and repeated by Little, 1989, p. xii.

84. Ramirez, 1991, p. 82.

85. Little, 1989, p. 89. Little remarks that Chauveau was then chairing the select committee on emigration, but the Legislative Assembly resolved to appoint the committee on 1 February 1849, more than six months after the July 1848 meeting of the Quebec Association.

86. Chauveau Report, p. 1.

87. ibid.

88. See, for instance, Semmel, 1961; Morrell, 1966, ch. 1.

89. Marx, 1976, p. 932, our emphasis.

90. Wakefield, 1833, vol. 1, p. 247 (quoted in Marx, 1976, p. 935).

91. Norman, 1963, p. 7. In 1831, Wakefield published a work called *Swing Unmasked, or The Causes of Rural Incendiarism*, a condemnation of rural poverty.

92. See Macdonald, 1939, pp. 3–4; for Mill and Wakefield, see Wood, 1983, pp. 9–13.

93. Wakefield, 1833, vol. 2, pp. 132, 149.

94. Marx, 1976, p. 936.

95. Quoted from *The Edinburgh Review*, 1840, 71, pp. 541–3 by Norman, 1963, p. 9.

96. Morrell, 1966, p. 6.

97. Wakefield, 1833, vol. 2, p. 183.

98. Norman, 1963, p. 2.

99. Lucas 1912, vol. 2, p. 16 (quoted in Norman, 1963, p. 11).

100. Bloomfield, 1961, chs 11–12, 15; MacDonell, 1924, pp. 1, 23, 26, 35–7; Norman, pp. 11–13; Monet, 1966, p. 264; 1969, p. 275.

101. Morrell, 1966, p. 6.

102. MacDonell, 1924, pp. 2–20; Norman, 1963, p. 12.

103. Wakefield, 1833, vol. 2, pp. 179–80.

104. ibid., p. 132.

105. See Riddell, 1937, 1939.

106. Macdonald, 1939, p. 512.

107. ibid., p. 21.

108. Teeple, 1972, p. 56; Fingard, 1986, p. 263; Strong, 1930, p. 36. Fingard refers to a seasonal house of industry that was established in Montreal in the winter of 1836–7, but Strong reports that a house of industry was first incorporated in Montreal in 1818.

109. Acheson, 1990, p. 17; Careless, 1971, p. 248.

110. Young and Dickinson, 1988, p. 163; Young, 1981, p. 89.

111. See, for instance, Tucker, 1964 [1936], pp. 129–34.

112. Faucher, 1973, p. 196.

113. Young and Dickinson, 1988, p. 157; Easterbrook and Aitken, 1956, p. 367; Monet, 1971, p. 221.

114. Piva, 1985, p. 203, 1992a, p. 66.

115. Seccareccia, 1992, p. 11.

116. Strong, 1930, p. 34.

117. Acheson, 1990, p. 28; Ouellet, 1991, p. 212.

118. Innis, 1937, p. 380 (republished in Innis, 1956, p. 206).

119. Palmer, 1983, pp. 57–8, 1985, pp. 53–4; also Langdon, 1973 (reprinted in Bumstead, 1986).

120. Palmer, 1983, p. 60.

121. Young, 1992, pp. 58, 59; also see Johnson, 1987.

122. See, for instance, Bernard, 1971, pp. 61–73; Monet, 1966.

123. Quoted in Tucker, 1964, p. 135.

124. Dessaulles, 1851 (quoted by Bernard, 1971, p. 69). Our translation.

125. Elgin to Grey, 23 April 1849, in Doughty, 1937, p. 349. For Elgin, see Francis, 1992, ch. 11.

126. See, for instance, Easterbrook and Aitken, 1956, pp. 361–5; Tucker, 1964, ch. 6; Stuart, 1988, ch. 9.

127. Tucker, 1964, p. 102.

128. See, for instance, Stokes, 1959, pp. 248–51.

129. Dalhousie to Bathurst, 19 June 1826, Public Archives of Canada, Quebec, vol. 176, pp. 499–505 (quoted in Macdonald, 1939, p. 288).

130. Hincks 'To the electors of the county of Oxford', Toronto, 20 November 1839, in Hincks, 1884, p. 45.

131. ibid., p. 49.

132. See Gallagher and Robinson, 1953; Ouellet, 1991, p. 228.

133. For instance, Piva, 1985, 1992a, chs. 3–4; 1992b.

134. Hincks, 'Memorandum on Immigration and on Public Works as connected therewith', 20 December 1848, Public Archives of Canada, Executive Council Office, state book 1, vol. 71, p. 400 (quoted in Piva, 1985, p. 187).

135. Piva, 1985, p. 210.

136. Pomfret, 1981b, p. 99.

137. Tulchinsky, 1977, chs. 7–8.

138. Young, 1978, pp. xii, 36.

139. Sweeny, 1976, p. 70.

140. Young, 1978, p. 52.

141. Tulchinsky, 1977, p. 125.

142. Young, 1978, p. 54.

143. Armstrong, 1984, p. 133; Pomfret, 1981b, p. 100; McCallum, 1980, p. 79; Piva, 1992a, pp. 76–81.

144. Roman, 1991, p. 12; for an account of financing the Grand Trunk railway, see Platt and Adelman, 1990; Piva, 1992a, ch. 4.

145. Hincks, 1849, p. 10.

146. ibid., pp. 13–16, 20.

147. Morrell, 1966, p. 438.

148. Baring Brothers to Hincks, 27 December 1850, Baring Papers (quoted in Tucker, 1964, pp. 60–1).

149. Tucker, 1964, p. 62.

150. Sweeny, 1976, p. 89; see Myers, 1972 [1914], chs 10–15.

151. Easterbrook and Aitken, 1956, p. 367.

152. For instance, Macintosh, 1923 (reprinted in Laxer, 1991).

153. See, for instance, Richards, 1985; Neill, 1991, ch. 8; Findlay, 1985; Ostrander, 1983; Findlay and Lundahl, 1992.

154. Smith, 1937 [1776], p. 392 (quoted in Innis, 1937, p. 375).

155. Innis, 1937, p. 376.

156. Hopkins, 1973, p. 125.

157. For instance, see Gray, 1981, p. 102 (quoting from Innis, 'The Work of Thorstein Veblen' [1929] in Innis, 1956, p. 24).

158. Patterson, 1990, pp. 4–5, 6, 38, 70, 79, 135.

159. McLuhan, 1953, p. 385 (quoted by Patterson, 1990, p. 12).

160. Patterson, 1990, pp. 80–8.

161. Innis, 'The Problem of Space' [1951] in *The Bias of Communication*, 1964, p. 92 (quoting from

Cornford, 'The Invention of Space' in *Essays in Honour of Gilbert Murray*, and quoted in Patterson, 1990, pp. 5, 72). As Patterson shows, Innis had read widely in accounts of post-Einstein quantum physics and seems to have been especially interested in interpretations which took the new theory back to the ancient world.

162. Smith, 1937, p. 392 (Innis, 1937, p. 375).

163. Innis, 1937, p. 382.

164. Aitken, 1959, pp. 3, 4.

165. Aitken, 1967, pp. 183–4. Durham, 1912 [1839].

166. Creighton, 1967, p. 4.

167. Aitken, 1967, p. 221.

168. Aitken, 1959, p. 12.

169. Ouellet, 1980a [1966], ch. 12.

170. Ouellet, 1991, p. 241.

171. Ouellet, 1980a, p. 441 (quoted in Gagnon, 1985, p. 93); Ouellet, 1991, p. 245.

172. Watkins, 1963, pp. 144,147,146 (reprinted in Laxer, 1991, p. 85).

173. Watkins, 1977, p. 90.

174. Innis, 'The Strategy of Culture', in *Changing Concepts of Time*, 1952, pp. 19, 18, 20.

175. ibid., p. 2.

176. Panitch, 1981 (reprinted in Laxer, 1991, pp. 270–1). Panitch quoted from Levitt, 1970, pp. 27–8. The same point follows from the contention that Schumpeter emphasised indigenous, local entrepreneurship as the basis of development, since it confuses local and foreign with what the Austrian, after Marx, meant by capitalism as an internal process of development. See Chapter 7.

177. See, for instance, Dente, 1977, pp. 7–17; Roll, 1961 [1938], pp. 439–54.

178. Veblen, 1919, p. 141 (quoted in Roll, 1961, p. 444).

179. Watkins, 1977, p. 90.

180. See Neill, 1972, p. 47.

181. Naylor, 1972, pp. 3, 6. Also see Naylor, 1975, I, pp. 1–4.

182. Naylor, 1972, p. 3.

183. See Neill, 1972, p. 45.

184. Innis, 1962 [1930] pp. 402, 401 (reprinted in Laxer, 1991, p. 65).

185. ibid., p. 401.

186. Parker, 1981, p. 141, 1985, pp. 76–84; Drache, 1982 (reprinted in Laxer, 1991).

187. Innis, 1956, p. 260 (quoted in Parker, 1985, p. 85.).

188. Clark, 1962, p. 15 (quoted in Parker, 1985, p. 85).

189. Neill, 1991, p. 136.

190. Innis, 1962, p. 401.

191. See Neill, 1972, pp. 64, 110, 123; 1991, pp. 134–9; Roll, 1961, pp. 444–5.

192. Roll, 1961, p. 445.

193. Innis, 1929 (quoted by Neill, 1972, pp. 36–7).

194. Innis, 'Snarkov Island', in Neill, 1972, p. 5 (and quoted by Neill, ibid., p. 64).

195. ibid., pp. 74–6.

196. Innis, *The Canadian Economy and its Problems*, 1934, p. 24 (quoted in Neill, 1972, p. 67).

197. Innis, 'Snarkov Island', in Neill, 1972, p. 9.

198. See the scattered references to the oral tradition and time in Innis, 1980. See also Neill, 1972, p. 99; Crowley, 1981, p. 236; Heyer, 1981, pp. 252–5; Patterson, 1990, p. 20.

199. ibid., p. 37.

200. See, for instance, Eff, 1989; Edgell and Tilman, 1989.

201. See, for instance, Bell, 1963, 32(4), p. 616; Edgell and Tilman, 1989, pp. 1010–14; Lipow, 1982, pp. 76, 89.

In *The 'Idea File'*, 1980, p. 113, Innis made a perfunctory note of Comte and also commented: 'Comte an illustration of French interest in system and unity – i.e. of work of scholastics in making language flexible.'

202. Neill, 1972, p. 44.

203. See, for instance, Fowke, 1947, pp. 18–21; Brunet, 1958, pp. 146–7 (abridged and translated in Miquelon, 1977, p. 167).

204. Ouellet, 1991, pp. 180–4.

205. It is clear that Innis was fascinated by Mark Pattison, the devotee of Newman whom we met in Chapter 2. Innis closely read both volumes of Pattison's *Essays* (1899), his *Memoirs* (1885) and the 1932 edition of his *Milton*. Indeed, there are more references to Pattison in a list of Innis' readings (not in *The 'Ideas File'*) than to any other nineteenth-century author (University of Toronto Archives, Innis Papers: B72-0003/012 (33)).

Innis made use of Pattison during the course of a 1947 lecture on 'The Church in Canada' (in Innis, 1956, pp. 385–93) when he claimed that the bureaucratic organisation of the state failed to appreciate the importance of 'training of character'. The Church, now bereft of an interest in ideas and concerned for social action – 'vast amount of planning for others and pushing others around' – was partly responsible for the same neglect in social sciences, the handmaiden to the system by which 'so-called brain workers' exploit those who work with their hands. Development of character, Innis insisted, was 'essential to an appreciation of the danger in interfering in other people's lives' and 'no one can undertake the task of pushing people around without adequate discipline and training'. Innis concluded: 'We have developed an amazing aptitude for knowing what the other fellow ought to be doing.'

206. Berger, 1976, p. 105.

207. Neill, 1972, p. 102. For a recent treatment which invokes Innis with reference to the values of development see Wright, 1993, esp. ch. 4. Wright's analysis dovetails, in its stress on the need to liberate subjugated 'local knowledge' with the works discussed in our conclusion.

208. ibid., p. 114.

209. Innis remarked: 'I am aware that I am only presenting a footnote on the work of Graham Wallas, but it should be said that the subject of his work was inherently neglected.' Innis had met Wallas and was 'invited for talks with Sidney and Beatrice Webb' during a visit to Europe in 1922. Innis, 'The press, a neglected factor in the economic history of the twentieth century', in Innis, 1952, p. 78; Berger, 1976, p. 90.

210. See Parker, 1985, fn. 37, p. 90; Mackenzie and MacKenzie, 1979 [1977], pp. 401, 409–10; Weiner, 1971, p. 143; Qualter, 1980, pp. 144–53. Weiner argued that, contra Webb and for Wallas, social engineering, or social 'efficiency', was a means to the end of an ideal of democracy; for Webb it was the other way around.

211. Innis, 1964, p. 85 (quoted in Parker, 1985, p. 92).

212. Innis, 1962, p. 383 and in Laxler, 1991, p. 50.

213. Neill, 1972, p. 101.

214. Innis, 'The Economic Aspect', p. 15 (quoted in Neill, 1972, p. 75).

215. Neill, 1972, p. 100.

216. University of Toronto, Innis Papers: B72-0003/013 (06); Innis, 1980, p. 28.

217. For cyclonics, see Innis, 'The Economic Destiny of Canada' (unpublished ms, 1927–34) in Neill, 1972, pp. 62–5.

218. Parker, 1985, p. 82.

219. Innis, 1962, p. 383.

220. ibid., p. 385.

221. ibid., p. 385.

222. Macdonald, 1975; also, Tulchinsky, 1977, pp. 216–17.

223. Ouellet, 1991, p. 135.

224. ibid., p. 144; Greer, 1985.

225. See Hirschman, 1958; Schedvin, 1990, p. 534.

226. Analogues, 1977, p. 674; McCallum, 1980, pp. 104–14; Seccareccia, 1992, p. 12.

227. Platt and Adelman, 1990, pp. 209, 208.

228. McInnis, 1982, p. 27; the same point is made by Paquet and Smith, 1983, p. 433.

229. Tulchinsky, 1977, pp. 209–28, 226.

230. Craven and Traves, 1983, reprinted in McCalla, 1987, pp. 139, 118.

231. Tulchinsky, 1977, pp. 11–14. For Upper Canada, see Baskerville, 1981.

232. For a summary of the debate, see Gagnon, 1985, ch. 3; Clement and Drache, 1978, pp. 35–40, 1985, ch. 6.

233. Tulchinsky, 1977, pp. 14–66, 67; Young, 1986, also 1978, 1981.

234. Ouellet, 1991, pp. 242–3.

235. Craven and Traves, 1983, p. 140; Langdon, 1973, p. 346.

236. McCallum, 1980, p. 93.

237. Pentland, 1981, pp. 146, 148, 159–60, 168.

238. Tulchinsky, 1977, p. 104.

239. Falardeau, 1975, p. 30.

240. Neill, 1991, pp. 74–90.

241. See Forster, 1986.

242. ibid., p. 49.

243. Buchanan, *Relations of the Industry of Canada*, (quoted in Goodwin, 1961, p. 51). Emphasis ours.

244. Marx, 1976, vol. 1, p. 932.

245. Ramirez, 1991, p. 147.

246. Brunet, abridged and trans. in Miquelon, 1977, p. 165.

247. Ouellet, 1980a, p. 607; also, for instance, 1980b, ch. 13.

248. See, for instance, Dumont, 1973.

249. Ramirez, 1991, pp. 84–5 (Ramirez also quoted from Bouchard, 1983, p. 21).

250. Quoted in Dumont, 1973, p. 76. Our translation. Also quoted in Ramirez, 1991, p. 84.

251. Neill, 1991, p. 51.

252. Monière, 1981 [1977], p. 101.

253. Chauveau Report, p. 9.

254. Brunet, abridged and trans. in Miquelon, 1977, p. 164.

255. Angers, 1961, pp. 206–7, 217; also Angers, 1960.

256. Parent, 1969, p. 90.

257. See, for instance, Parizeau, 1975, pp. 415–62; Falardeau, 1975, pp. 8–26.

258. Parent, 1969, p. 89.

259. Neill, 1991, p. 44.

260. Ouellet, 1980a, p. 608.

261. Dever, 1976, p. 63.

262. Young, 1981, p. 62.

263. Forster, 1986, p. 27.

264. Cooper, 1937, p. 530.

265. Den Otter, 1982, pp. 161, 169; Barnett, 1976, p. 403; Cooper, 1937, p. 530; Forster, 1986, pp. 41–51.

266. Neill, 1991, p. 43.

267. Chauveau Report, p. 10; Pomfret, 1981b, p. 69; Forster, pp. 27–31.

268. Ouellet, 1991, p. 246.

269. Young, 1986; Strong, 1930, p. 35; also see Fecteau, 1992.

270. Chauveau Report, p. 3; the 1857 inquiry on emigration reported that the incidence of emigration from the Montreal region and the Eastern Townships was significantly higher than from other regions. See Lavoie, 1981, Table 2, p. 22.

271. Little, 1989, pp. xi, 4, 80, 88; more generally, see Hamelin and Roby, 1971, pp. 163–84.

272. Little, 1989, pp. 17, 6, also 1978, p. 93; Ramirez, 1991, pp. 77, 32; Drapeau, 1863, p. 550.

273. Fowke, 1947, p. 137.

274. Ramirez, 1991, p. 79.

275. Little, 1989, pp. 8–12.

276. For the debate over the agricultural crisis, see, for instance, Le Goff, 1974; Paquet and Wallot, 1972; Courville and Seguin, 1989; Altman, 1981; Gagnon, 1985; McCallum, 1980; Ouellet, 1980a, 1980b; McInnis, 1982; Lewis and McInnis, 1980.

277. Macdonald, 1939, p. 12 (quoted by Little, 1989, p. 99).

278. Little, 1989, p. 99.

279. Altman, 1981, pp. 124–5.

280. For instance, see Gagnon, 1985, ch. 5; McInnis, 1982; Lewis and McInnis, 1980.

281. Lewis and McInnis, 1980, p. 502: 'We do not deal directly with the question of whether French farmers may have been *poorer* than the English. It is quite possible that they were, if they had fewer resources to work with.'

282. Drapeau, 1863, p. 550; Hamelin and Roby, 1971, p. 163.

283. Ramirez, 1991, p. 27.

284. ibid., pp. 114–17.

285. A useful account of different theses to explain Canadian emigration to the United States can be found in Marr and Patterson, 1980, pp. 178–80.

286. Average Price Indices, Canada and USA, 1848–1901

	Canada Wholesale (1900 = 100)	*USA Retail (1910–14 = 100)*
1848–50	86	83
1851–53	95	89
1854–56	146	108
1857–59	129	100
1860–62	101	95
1863–65	111	170
1866–68	124	165
1869–71	129	139
1872–74	143	132
1875–77	125	111
1878–80	110	94
1881–83	115	104
1884–86	103	87
1887–89	105	86
1890–92	105	80
1893–95	96	73
1896–98	92	69
1899–1901	100	80

Source: Mitchell, 1983, Table H1, p. 690.

Since base years are different and because wholesale prices are the only available consistent price series for Canada, the indices should be treated with caution. However, the inflationary effect of the American Civil War is captured to the extent that any snapshot evidence for the period 1862–74 gives a mistaken impression of the general tendency for average costs of living to have been lower in the United States than in Canada. To take one example, in 1874, Ontario consumer prices for food and rent were reported as being significantly lower than in New York State (Panitch, 1981, Table 2, p. 278). While early twentieth-century evidence found 'food, fuel and rent costs for Boston to be higher in more than average degree than in Montreal', the lower prices of manufactured consumer goods in the United States led to the conclusion that there was 'striking superiority in the economic position of the American worker'. The superiority of the average real wage, one-quarter to one-third, over the Canadian real wage corresponded to the ratio of wage costs to value added in manufacture and was explained, therefore, by higher labour productivity in manufacturing in the United States (Logan, 1937, pp. 201,179, 203).

287. Ramirez, 1991, pp. 36, 45–6.

288. Little, 1978, p. 98.

289. Marx, 1976, vol. 1, p. 940.

5 DEVELOPMENT DOCTRINE IN 'UNDERDEVELOPED' BRITAIN

1. See the comment by Saul, 1979, p. 55. He states: 'this at least is clear: the sooner the "Great Depression" is banished from the literature, the better'.

2. The entry for 'Agriculture' in the famed 11th edition of the *Encyclopaedia Britannica*, published in 1910, states on p. 396 that 'The last quarter of the nineteenth century proved . . . a fateful period for British agriculture', while the sub-entry on 'Agricultural Population and Wages' (p. 413) takes note of the 'remarkable diminution of the British rural population' in the last half of the century. Although the wages of those who remained on the land increased, the number of 'Persons Engaged in Agriculture' is noted as having fallen from 3,453,500 in 1851 to 2,262,600 in 1901. The fall in 'agricultural labourers and shepherds', noted as a 'more precise index', was from 1,110,311 to 609,105 over the same period. Even those who agree with Saul stress this point. See Jones, 1968, pp. 38, 55.

3. See Seaman, 1973, pp. 264–6. On the debate over the impact of the Corn Laws see Fletcher, 1960–61, pp. 417–32 and Fairlie, pp. 562–75 as well as 1965, pp. 562–75.

4. Garvin, 1932, p. 220.

5. Heyck, 1974, pp. 5–6.

6. ibid., p. 6.

7. ibid., p. 245.

8. Garvin, 1932, p. 174.

9. Excerpt is from the authorised version of J. Chamberlain, 'Speeches, 1885' as reported in Maccoby, 1938, pp. 298–302.

10. ibid.

11. ibid.

12. For what follows, see Cowen and Shenton, 1992.

13. See Chase, 1988, pp. 173–7; Barry, 1965, pp. 29–40.

14. Gammage, 1969 [1854], p. 456.

15. Quoted in Barry, 1965, p. 31.

16. Chase, 1988, pp. 46–56.

17. Quoted in Chase, 1988, p. 1.

18. Buxton, 1910, 1914.

19. Clements, 1983, p. 85.

20. Anon., 1892.

21. Lipow, 1982, p. 8.

22. George, 1981 [1879]; 1947 [1891]. For George's influence see, among many others, Saville, 1960, p. 321.

23. Wallace, 1897, p. 11.

24. Wallace, 1892, p. 4.

25. Wallace, 1897, p. 12.

26. See Harris, 1972; Collini, 1979; Marsh, 1982.

27. Collini, 1976, p. 110; 'Introduction' in Rodman, 1964.

28. Hylton, 1984, p. 383.

29. MacIntyre, 1972, p. 28 (quoted in Vincent and Plant, 1984, p. v).

30. See Vincent and Plant, 1984, p. 68; other useful sources on Green include Richter, 1964; Rodman, 1964; Vincent, 1986; Greengarten, 1981; Thomas, 1987; Quinton, 1972.

31. For a clear account of Green's theory of property, see Greengarten, 1981, ch. 5. Green, it is suggested, had translated part of Hegel's *Propaedeutic* during the 1850s and early 1860s. Hegel wrote the *Propaedeutic* between 1808 and 1811 as an introduction to philosophy for high school students and regarded it as the summation of his theory. See George and Vincent, 'Preface' and 'Introduction' to Hegel, 1986, pp. vii, xi.

32. Green, 1929 [1907], p. 41. The lectures this work comprises were first delivered between 1877 and Green's death in 1882.

33. ibid., p. 75.

34. ibid., p. 77.

35. ibid., p. 94.

36. ibid., p. 95.

37. ibid., p. 93, emphasis in original.

38. ibid., pp. 126–7.

39. ibid., p. 129.

40. ibid., pp. 201–2.

41. ibid., p. 203.

42. ibid., pp. 216–19.

43. ibid., pp. 231–2.

44. ibid., p. 225.

45. Green, 1941, p. 207.

46. ibid., pp. 208–9, emphasis ours.

47. ibid., pp. 209–10.

48. ibid., p. 219.

49. ibid., emphasis ours.

50. ibid., pp. 219–20.

51. ibid., p. 227.

52. ibid., pp. 226, 228.

53. ibid., p. 228.

54. ibid., p. 225.

55. ibid., p. 229.

56. Rodman, 1964, p. 13.

57. Greengarten, 1981, ch. 6.

58. Hegel, 1952, para. 51, 45, 1986, para. 12, 26. Together with *Philosophy of Mind*, 1971, these are Hegel's texts which are used in what follows. We have also used the following interpretations: Avineri, 1972; Dickey, 1987; Riedel, 1984; Pelczyinski, 1984a.

59. Hegel, 1986, p. 74.

60. Riedel, 1984, 52–53 from Hegel, 1952, paras 235, 243–4, 147, 149–50.

61. Emphasised by Avineri, 1972, pp. 109, 147–8, 153, 154; also see Plant, 1977, where he concluded that 'Hegel's self-acknowledged failure to explain how the problem of poverty' could be resolved 'demonstrated very clearly the limitations, even on its own terms', of the Hegelian enterprise (104:113).

62. See Ilting, 1984, p. 211; Hegel, 1986, paras 8, 24, paras 18–54, 68–73, paras 11–55, 107–17.

63. Pelczynski, 1984b, p. 55.

64. Collini, 1976, pp. 109, 106, fn. 68, 103.

65. See, for instance, Durbin, 1984; Ricci, 1969; Stigler, 1965 [1959]; De Vivo, 1987.

66. Amery, 1953, vol. 1, pp. 247–51. See also Amery, 1908.

67. Garvin, 1905, p. 24.

68. Elbaum, 1986, p. 57. See also Dintenfass, 1992.

69. Garvin, 1905, p. 28.

70. ibid., pp. 6, 26.

71. ibid., p. 2.

72. Maccoby, 1938, vol. 4, pp. 269–70.

73. Garvin, 1932, pp. 366–7, n. 1.

74. ibid., p. 366.

75. Heyck, 1974, p. 151.

76. Hooper, 1986, p. 7.

77. Amery, 1953, p. 261.

78. Kubicek, 1969, p. 68.

79. Boyd, 1914, vol. 2, p. 248.

80. Judd, 1977, p. 190.

81. The source for this discussion of Chamberlain's activities regarding colonial development is Kubicek, 1969, pp. 72–3 and esp. ch. 4.

82. Koss, 1973, p. xxv.

83. ibid., p. xxix.

84. Cain and Hopkins, 1993, pp. 15–16, 200 (quoting from Hobson, 1910, p. 113).

85. Belloc, 1904.

86. Koss, 1973, pp. 26–7.

87. Cain and Hopkins, 1993, p. 214.

88. Cain and Hopkins, 1993, p. 216.

89. Koss, 1973, pp. 43, 50, 55.

90. ibid., p. 94.

91. ibid., pp. 60–1.

92. See Louis and Stengers, 1968.

93. Amery, 1953, p. 254.

94. Speech quoted in Harris, 1972, p. 227; also in Winfrey, 1907, p. 8.

95. McBriar, 1987, p. 35.

96. ibid., pp. 36–7.

97. ibid., p. 48.

98. ibid., p. 49. McBriar is generally reliable but we would quarrel with him on the philosophical origins of Fabians and explicit connection of Bosanquet to Green.

99. ibid., ch. 3.

100. ibid., p. 83.

101. The source which makes the Fabian debt to Comtean positivism most clear is Wolfe, 1975.

102. McBriar, 1987, pp. 95–6.

103. Harris, 1972, p. 273.

104. ibid., p. 343.

105. ibid., pp. 344–5.

106. ibid., p. 340.

107. ibid., p. 338.

108. Emy, 1973, pp. 136–7.

109. Hobhouse, 1927 [1913], pp. xxi–xxii. Hobhouse saw this work as both summarising and extending his earlier *Mind in Evolution*, 1901, and *Morals in Evolution*, 1929 [1906].

110. Hobhouse, 1927, pp. xxi–xxiii.

111. ibid., p. 181.

112. ibid., pp. 11–12.

113. ibid., p. 17.

114. ibid., p. 64.

115. ibid., pp. 92–3.

116. ibid., p. 147.

117. ibid., p. 153.

118. ibid., p. 161.

119. ibid., p. 179.

120. For Hobhouse's views on the Poor Law, see McBriar, 1987, pp. 362–3. On his criticism of the Webbs see Clarke, 1978, p. 65.

121. See Cowen and Shenton, 1991a; 1991b; also see Shenton, 1986.

122. For radical liberals in England, see Maccoby, 1961; Emy, 1973; Rowland, 1968; Morris, 1974; Bernstein, 1986; Clarke, 1978.

For radicals in the African setting, see Porter, 1968; Phillips, 1989; Omosini, 1968–9; Nworah, 1966.

For Lugard, see Shenton, 1986; for Morel, see Porter, 1968; Phillips, 1989; Omosini, 1968–9; Nworah, 1966, but also Cookey, 1968; Cline, 1974; Adams, 1957, ch. 7; Morel's own works, esp. *Affairs of West Africa*, 1902. For Olivier, see Lee, 1988.

123. United Kingdom Parliamentary Papers, House of Lords number 158 of 1907 (quoted in McGregor Ross, 1927, p. 70); PRO:CO 879/95, Colonel J.A. Montgomery, Commissioner of Lands, to HM Commissioner, Mombasa, 21 August 1906.

124. The renowned critique of state development, Hayek, *The Road to Serfdom*, 1944, is one source of inspiration for the current period of liberalisation. For the general critique of state development in Africa, see Fieldhouse, 1986; for a particular case, see Collier and Lal, 1986.

6 DEVELOPMENT DOCTRINE IN AFRICA: THE CASE OF KENYA

1. PRO:CO 583/243/30415, p. 8, Bourdillon to Secretary of State for the Colonies, 5 April 1939.

2. Bourdillon, 1937, p. 75.

3. See, for instance, Shenton, 1985, ch. 6; Kitching, 1980; Sender and Smith, 1986, chs 2–3.

4. Buxton, 1937, p. 19.

5. Bourdillon, 1937, pp. 75–6. On this point see also Auchinleck, 1927a and 1927b.

6. See, for instance, Clayton and Savage, 1974, p. 348.

7. Kenya National Archives (KNA): LAB/27/1/D/267 and 207, Special Commissioner, Central Province to Provincial Commissioner, Rift Valley Province, 11 August 1956, and to Senior Labour Officer, Nyeri, 8 May 1956.

8. KNA, *Annual Report*, Nyeri District, 1958.

9. KNA, *Annual Report*, Nyeri District, 1959.

10. Kanogo, 1987, p. 152; KNA:LAB/27/1/D/310, Secretary of Agriculture, Nairobi, to Special Commissioner, Central Province, 24 September 1956.

11. ibid.; KNA: LAB/27/1/D/23G, District Commissioner, Kiambu District, to Provincial Commissioner, Central Province, 20 June 1956.

12. KNA, *Annual Report*, Nyeri District, 1961.

13. Kenya Colony, 1960, pp. 1, 2–3. Henceforth Dalgleish Report.

14. For an earlier view, Rao, 1952; for more recent views, Seers, 1983; Lipton, 1993. For a critical overview, see Toye, 1987, ch. 2.

15. Collier and Lal, 1986, pp. 50–1. Collier and Lal quote extensively, including the quotation here, from Dalgleish.

16. Chandavarkar, 1993, pp. 140–1 (quoting from Keynes, 1983, pp. 28–9); also, Chandavarkar, 1989, pp. 40–1.

17. Toye, 1993, 'Discussion', p. 174 and 'Keynes, Russia and the State in Developing Countries', p. 247 (quoting Keynes, 1972a, pp. 263–4).

18. ILO, 1950.

19. Dalgleish Report, p. 3.

20. ibid., p. 9.

21. *The Times* (London), 21 May 1962 and 14 July 1964.

22. See, for instance, Clayton and Savage, 1974, pp. 437–45; Amsden, 1971, pp. 100–1.

23. Heyer, Ireri and Moris, 1971, p. 7.

24. Kenya Republic, 1970, pp. 2, 8–9. Henceforth Mwiciji Report.

25. ILO, 1972.

26. Kenya Republic, 1973, p. 15.

27. ibid., p. 17.

28. Livingstone, 1986, Table 8.1, p. 146.

29. Kenya Republic, 1991a, Table 2.2, p. 21. Henceforth Ndegwa Report.

30. ibid., Table 3.1, p. 36. Also see MacWilliam, Desaubin and Timms, 1995, p. 41.

31. Hunt, 1984, p. 4.

32. Livingstone, 1986, p. 2.

33. ibid., p. 1.

34. ibid., p. xix.

35. Mwicigi Report, p. 11; ILO, 1972, pp. 167, 170–1, 435–6; Hunt, 1984, p. 4; Livingstone, 1986, p. 248.

36. For instance, Kenya Republic, 1978. This report, presumably taking advantage of the 1978 succession of Moi to the presidency when the new president engaged in a bout of rhetorical populism to disestablish the Kenyatta regime, referred to 'the growing inequality of landholdings and numbers of landless' and propensities for holding land for speculative purposes following the issue of freehold land titles (quoted in Livingstone, 1986, p. 249).

37. Kenya Republic, 1973, p. 34.

38. Kenya Republic, 1983, p. 5, henceforth Wanjigi Report).

39. ibid., pp. 12, 19.

40. Ndegwa Report, pp. 2, 38, 63.

41. ibid., pp. 88, 92, 94, 139.

42. Kenya Republic, 1992, p. 5; quoted by Gatama, 1986, p. 66 (reproduced in King and Abuodha, 1991, p. 5).

43. Kenya Republic, 1992, p. 1.

44. King and Abuodha, 1991, pp. 11–12.

45. Kenya Republic, 1992, p. 2.

46. IPS, 'Kenya-Economy', 17 June, 1994; Xinhua Press Agency, 'Kenya Minister on Development', 13 July 1994.

47. Ndegwa Report, pp. 88, 92, 94, 139.

48. IPS, 'Kenya-Economy'.

49. Dalgleish Report, p. 3.

50. Advisory Committee on Native Education in the British Tropical African Dependencies, 'Educational policy in British Tropical Africa', Cmnd 2374, xxi, 1925.

51. United Nations, 1955.

52. 'Mass Education in African Society', Colonial no. 186, HMSO, 1944.

53. Quoted in Hill, 1991, p. 24.

54. For *mwethya*, see Hill, 1991, pp. 36–8; for the similar experience of *ngwatio*, KNA, Nyeri District, Annual Report, 1959,1960: In December 1959, 230 *ngwatio* or 'self-help' neighbourhood groups, each of 30–40 household members, were reported to have done 'excellent work' on agricultural improvement in the Othaya and North Tetu divisions of Nyeri District. Yet a year later and 'for political reasons', the number of groups had fallen to 30 and they were confined to building one-third of the new houses in North Tetu during 1960.

55. Ngau, 1987, Tables 1 and 2, pp. 526–7; Orora and Spiegel, 1979, p. 249.

56. Hill, 1991, pp. 39–40.

57. Mbithi and Rasmusson, 1977, pp. 13, 16.

58. For instance, Thomas, 1985, 1987.

59. Mbithi and Rasmusson, 1977, Table 1, p. 15; Hill, 1991, p. 41; Chehu, 1987, p. 129; Abreu, 1982, pp. 217–18; Gould, 1989.

60. See, for instance, Gyllstrom, 1991; Ouma, 1980; Bager, 1980; Holmquist, 1975; Karanja, 1974.

61. Thomas, 1985, p. 10; Mbithi and Rasmusson, 1977, pp. 56–7.

62. Ngau, 1987, p. 529.

63. See Taylor, 1987, pp. 24–5; Bayart, 1993 [1989], ch. 8.

64. Wilson, 1992, pp. 11–15.

65. Quoted in Porter, Allen and Thompson, 1991, p. 154.

66. For instance, Maas, 1986; Davison, 1989; Pala, 1978; Mutiso, 1975, ch. 13; Udvardy, 1988.

67. Widner, 1992, pp. 34–5, 60–6, 130, 142, 199–200.

68. Holmquist, 1984, pp. 86, 87.

69. B.P. Thomas, 1987, pp. 468–9.

70. See, for instance, Reynolds and Wallis, 1976; Delp, 1981; Evans, 1989.

71. For example, Hill, 1991; Mbithi and Rasmusson, 1977; Ngau, 1987; Holmquist, 1984; Orora and Spiegel, 1977.

72. Quoted in Orora and Spiegel, 1979, p. 245.

73. See, for instance, Godfrey and Mutiso, 1979.

74. Goldsworthy, 1982, pp. 234–7.

75. Atieno-Odhiambo, 1987, pp. 195, 196.

76. Murray-Brown, 1972, p. 220.

77. Nelkin, 1964, p. 70; Zolberg, 1964, p. 111.

78. Roberts, 1964, p. 83.

79. See Lewis, 1993, pp. 130–55. Gustav von Schmoller, as Du Bois' doctoral supervisor at the University of Berlin, suggested that he write a thesis on 'The large and small-scale system of agriculture in the southern United States, 1840–1890'.

80. Gilroy, 1993, p. 137. Gilroy takes this to be a one-sided view of Du Bois, a view to be found in West, 1989.

81. Gilroy, 1993, p. 35.

82. See Hutchful, 1991.

83. Quoted by Padmore, 1956, p. 130.

84. ibid., p. 131.

85. ibid., pp. 130, 139; Nelkin, 1964, p. 64.

86. Hooker, 1967, pp. 70, 105; also see Makonnen, 1973, pp. 178–81.

87. Hooker, 1967, p. 22.

88. Padmore, 1956, pp. 139, 148, 128.

89. For instance, Padmore followed Du Bois in numbering the congresses from 1919; Legum, Nelkin and Logan number from 1900 and we follow this sequence. See Legum, 1965 [1962], ch. 2; Nelkin, 1964; Logan, 1962.

90. Logan, 1962, p. 38. Emphasis ours.

91. Murray-Brown, 1972, p. 164 and ch. 17; Goldsworthy, 1982, p. 106.

92. Goldsworthy, 1982, pp. 260–1; Nyong'o, 1989, p. 241.

Goldsworthy quotes from Colin Legum who wrote of Mboya as a 'democratic socialist' and John Hatch that if he had lived longer, Mboya could have formed a 'new radical socialist group, broadly following Nyerere's ideals'.

93. Nyong'o, 1989, p. 237.

94. Gadzey, 1992, pp. 480, 456.

95. Leonard, 1991, p. 282.

96. ibid.

97. Lofchie, 1989, p. 189 (quoted and criticised by MacWilliam, Desaubin and Timms, 1995).

98. Njonjo, 1981.

99. Leonard, 1991, pp. 56–7.

100. ibid., p. 225.

101. ibid., pp. 232–3.

102. 'Corruption probe jeopardises Kenya aid', *Financial Times*, 2 November 1993, p. 4.

103. See, for instance, Swainson, 1980, part 3; Leys, 1978.

104. Africa Watch, 1993, p. 71.

105. *The Times*, 29 January 1962.

106. Basic sources are Wasserman, 1976; Njonjo, 1977; Leo, 1984.

107. Africa Watch, 1993, p. 77.

108. Bayart, 1993, pp. 252–9.

109. See, for instance, Widner, 1992, pp. 183–6; MacWilliam, Desaubin and Timms, 1995.

110. Mboya, 1963a, reprinted in Friedland and Rosberg, 1964, pp. 251, 252, 253 (partly quoted as Burkean by Burke, 'Tanganyika' in Friedland and Rosberg, 1964, p. 206).

111. Mboya, in Friedland and Rosberg, 1964, p. 257; Goldsworthy, 1982, p. 255.

112. Kenya Republic, 1965, p. 3.

The same argument, including the principles of political democracy and mutual social responsibility, was set out by Mboya in *Freedom and After*, 1963b, pp. 167–70 (quoted in Goldsworthy, 1982, pp. 252–3).

113. For one example from many, see Ambler, 1988.

114. Mboya, in Friedland and Rosberg, 1964, pp. 4, 5.

115. ibid., p. 257.

116. ibid., pp. 11, 12.

117. ibid., pp. 6, 8, 36, 37, 49, emphasis ours.

118. Mohiddin, 1973, pp. 196–223; Leys, 1975, p. 221 (both quoted in Goldsworthy, 1982, pp. 254, 235).

119. Roberts, 1964, p. 85.

120. Mwicigi Report, p. 13.

121. Wanjigi Report, pp. 4, 17; Ndegwa Report, p. 2.

122. Leonard, 1991, pp. 71, 86, 202, 184–6, 183.

123. Elkan, 1970, p. 517.

124. See, for instance, Arrighi and Saul, 1973.

125. Elkan, 1970, pp. 523, 525, 526–7.

126. ibid., p. 527. Emphasis ours.

127. Rempel and House, 1978, pp. 161–82.

128. Dalgleish Report, p. 9.

129. Mwicigi Report, p. 10; Wanjigi Report, p. 123; Ndegwa Report, ch. 8.

130. Wanjigi Report, p. 84 (quoting from a 1982 World Bank Report).

131. See, for instance, Wrigley, 1965, pp. 213–21.

132. These arguments can be found, for instance, in Brett, 1973, ch. 6; Woolf, 1974; van Zwanenberg, 1975; Berman, 1990, ch. 4.

133. Smith, 1976, p. 123.

134. Clayton and Savage, 1974, p. 337; Hinga and Heyer, 1976, Appendix 1, p. 253.

135. Quoted in van Zwanenberg with King, 1975, p. 268.

136. Quoted, but neither referenced nor dated, in van Zwanenberg with King, 1975, p. 213 (partly reproduced in Berman, 1990, p. 236).

137. For an overview of the Carter Commission, see Breen, 1976.

138. United Kingdom Cmd 4556, *The Kenya Land Commission: Report* (Carter Commission Report), Nairobi, Government Printer, 1934, pp. 25–6, 142.

139. ibid., p. 42.

140. See, for instance, Okoth-Ogendo, 1991; Sorrenson, 1967, part 1.

141. ibid., pp. 438–9, 440, 444.

142. Presbyterian Church of East Africa archive: G/1. G.A.S. Northcote, acting PC, Nairobi to Kinyanjui wa Githirimu, 28 October 1919.

143. See Kitching, 1980; Cowen and Kinyanjui, 1977.

144. Carter Report, 1934, p. 146.

145. ibid., pp. 146, 147.

146. ibid., p. 148.

147. Muriuki, 1974, p. 35.

148. Cowen, 1981.

149. Cowen, 1979, pp. 71–5.

150. KNA: Ministry of Agriculture, 4/113, Central Province, Half-yearly Report, January–June, 1948.

151. Oxford Development Records Project (ODRP), 183: G. Yates, f. 68 (quoted in Thurston, 1987, pp. 112–13).

152. ODRP, 133A: R.E. Robinson, Africa tour journal, ff. 43–4 (quoted in Thurston, 1987, p. 108).

153. Cowen, 1983.

154. For two detailed studies of Kutus in Kirinyaga District, see Lewis and Thorbecke, 1992; Evans and Ngau, 1991.

155. King, 1977, p. 103 (quoted by Kitching, 1980, pp. 400–1).

156. Rempel and House, 1978, p. 168; House, 1981, p. 358.

157. Mcwilliam, 1976, pp. 273, 274. See also Tignor, 1993, and especially in view of the Himbara Thesis, pp. 55, 64: 'While the colonial government of Kenya was often willing to endorse British business firms and to purchase their products, similar endorsements of Asian enterprises did not occur.' Furthermore, Tignor concluded: 'While it is certainly true that Asians were not able to constitute themselves as an indigenous and powerful bourgeoisie (to agree with Leys), they did persevere, overcoming obstacles

to their entry into the industrial sector, including lack of government enthusiasm.'

158. See, for instance, Sharpley and Lewis, 1988; Coughlin and Ikiara, 1988.

159. Sharpley and Lewis, 1988, pp. 17, 19–20, 4.

160. ibid., pp. 45, 48, 56, 71.

161. Himbara, 1994, pp. 103–7; 1992, p. 386.

162. Himbara, 1994, p. 70.

163. KNA: Vasey Papers. E.A. Vasey, 'Development: Economic and Political Planning in Kenya', paper read to Economics Club of Kenya, 21 July 1955 (quoted in Himbara, 1994, p. 104).

164. Mcwilliam, 1976, p. 274.

165. ibid., pp. 274–5.

166. Himbara, 1994, p. 8; 1992, pp. 16, 95.

167. Ndele, 1991, p. 5.

168. Himbara, 1994, pp. 45-51; 1992, pp. 19, 120, 375, 150.

169. KNA: MCI/6/782. 'The African in business: memorandum for meeting of Provincial Commissioners', February 1950 (quoted by Himbara, 1994, p. 81).

170. See, for instance, Ikiara, Jama and Amadi, 1993, Table 5, pp. 90, 91; Sharpley, 1986, Table 7.3, pp. 92, 93.

171. Swainson, 1980, pp. 190, 224–6.

172. World Bank, 1992, pp. 69–72.

173. Himbara, 1994, p. 76; World Bank, 1989, p. 38.

174. Himbara, 1994, pp. 90–1; 1992, pp. 247, 250, 118.

175. Himbara, 1994, p. 70; 1992, p. 207.

176. For instance, Siggel, 1992, pp. 365–7, 372–3, 375; also, Lall, Khanna and Alikhani, 1987.

177. Sharpley and Lewis, 1988, p. 2.

178. The overwhelming body of econometric studies in Kenya are based on neo-classical production functions, broadly indicating that productivity change is to be explained by capital-good bias in production. This conventional view, that there is little historic substitution of labour for capital equipment inputs in production and that 'capital–labour ratio' rises with firm size in manufacturing is best represented by Maitha and Manundu, 1981. Howard Pack's alternative view is that gains in task-level productivity, representing labour productivity gains without capital deepening and due to labour effort and discipline in production, has outstripped firm-level technological mastery in large-scale manufacturing. See Pack, 1976, 1977, 1987. For a summary of these views, see Siggel, 1992, pp. 366–7.

179. For instance, Gershenberg, 1986; Grosh, 1988; Ruotsi, 1992; Matthews, 1991.

180. King and Abuodha, 1991, p. 9; Himbara, 1994, pp. 92, 155, 162.

181. Livingstone, 1991, p. 663. Livingstone has used unpublished data to estimate that very small enterprise grew at an annual average rate of 11 per cent over the 1985–8 period.

182. King and Abuodha, 1991, p. 28.

183. Norcliffe, 1983.

184. Livingstone, 1986, pp. 59, 60; 1991, p. 667; King and Abuodha, 1991, p. 19.

185. Kenya Republic, 1992, pp. 1, 4.

186. King and Abuodha, 1991, p. 47.

187. Aboagye, 1986 (quoted by Livingstone, 1991, pp. 661–2).

188. Kenya Republic, 1989, Table 1; Kenya Republic, *Statistical Abstract*, 1987, Tables 213, 219.

189. Billetoft, 1989 (quoted by Livingstone, 1991, p. 662).

190. King and Abuodha, 1991, p. 45.

191. World Bank, 1983, pp. 454, 456, 457; IPS, 'Kenya-Economy: Small Businesses Struggle to Adjust', 21 September 1994.

192. World Bank, 1983, p. 463.

193. IPS, 'Kenya-Economy'.

194. World Bank, 1983, p. 459.

195. Kenya Republic, 1989, Table 14.1.

196. Kenya Republic, 1982, Tables 7.33, 7.34; 1991b, 1989.

197. Hebnick, 1990, Table 6.8, p. 185; Lavrijsen, 1984, Table 27, p. 87.

198. Francis and Hoddinott, 1993, Table 3, p. 131; Francis, 1991, Table 15, p. 132; Hoddinott, 1989, Table II.2, p. 27, Appendix 5.1, p. 187; Orvis, 1989, Table 6.1, p. 252; Paterson, 1984, Table 1.8, p. 25.

199. Hebinck, 1990, Table 6.8, p. 185; Lavrijsen, 1984, Tables 23, 24, pp. 84–5.

200. Francis and Hoddinott, 1993, pp. 134–42; Francis, 1991, pp. 130–2, 143, 95–6, 127, 88; Hoddinott, 1989, pp. 182–4.

201. See, for instance, Paterson, 1984, pp. 92, 107.

202. Francis, 1991, p. 134. All the foregoing western Kenya studies emphasise this cycle of reproduction; also see Hoddinott, 1994; for the same results from the Kutus, Central Province case, see Evans and Ngau, pp. 523–9.

203. King and Abuodha, 1991, p. 40.

204. Kenya Republic, 1992, pp. 18–19.

205. Livingstone, 1991, p. 665.

206. Ndegwa Report, p. 125.

207. King and Abuodha, 1991, pp. 96–8.

208. Livingstone, 1991, p. 666.

209. Kenya Republic, 1992, p. 2.

210. King and Abuodha, 1991, p. 10.

211. Himbara, 1994, pp. 161, 163.

212. Lonsdale, 1992, p. 316.

213. Anderson, 1983, ch. 7.

214. Lonsdale, 1992, p. 316.

215. ibid., p. 317 (quoting from Mboya, 1963b, pp. 61–5).

216. *African Mail*, 11 February 1910 (quoted in Nworah, 1966, p. 320).

217. Morel Papers: F.9. E.D. Morel to Sir W. Lever, 19 April 1911; Morel to Lord Crewe, Colonial Secretary, April 1910 (quoted in Nworah, 1966, pp. 317, 331).

218. *African Mail*, 18 February 1910 (quoted in Nworah, 1966, p. 335).

219. *African Mail*, 11 February 1910 (quoted in Nworah, 1966, p. 320).

220. PRO: CO 96/504. Morel to Crewe, 27 June 1910; *African Mail*, 29 July 1910 (quoted in Nworah, 1966, pp. 320, 342, 348).

221. Nworah, 1966, p. 20.

222. John Holt to Morel, 28 October, 25 October 1906 (quoted in Nworah, 1966, p. 497); Nworah, 1966, p. 22.

223. Morel, open letter, 'The future of the peoples of West Africa in relation to the forces of European capital and industry, with particular reference to the occupancy and enjoyment of the land', London, August 1912' (quoted in Nworah, 1966, pp. 412–3).

224. For instance, see Weinroth, 1974, p. 220.

225. John Holt Papers: 16/3. Mary Kingsley to Holt, 29 January 1899 (quoted in Nworah, 1966, p. 37).

226. See Cowen and Shenton, 1991a; 1994.

227. PRO: CO 554/10. C. Strachey, minute, 9 May 1909; Morel Papers: F8/4. Holt to Morel, 20 January 1910 and Morel to G.A. Moore, 29 May 1913; *Gold Coast Leader* 6 December 1913 (quoted in Nworah, 1966, pp. 259, 337, 368, 371).

228. Lonsdale, 1992, pp. 317, 467.

229. Himbara, 1992, p. 87; 1994, pp. 161–2.

230. Himbara, 1994, pp. 135, 147–52.

231. Absolute totals of unauthorised expenditure for selected years in Himbara, 1992, p. 311.

232. Unfunded external debt as per cent of GDP

1963–6	*1967–70*	*1971–4*	*1975–8*	*1979–82*	*1983–6*	*1987–90*
11.5	11.5	11.8	11.8	20.5	32.6	40.1

Sources: Kenya Republic, *Statistical Abstract*, 1970 (Tables 44, 178), 1974 (Tables 46, 209), 1978 (Table 44), 1980 (228), 1982 (40), 1986 (37, 201), 1991 (40, 200).

233. Himbara, 1994, pp. 121–2.

234. Sharpley, 1986, Table 7.1, p. 86.

235. See Cowen, 1986.

236. King, 1979, p. 46.

237. Kenya Republic, *Statistical Abstract*, 1974, 1991 (Tables 209, 201).

7 JOSEPH SCHUMPETER AND 'FAUSTIAN' DEVELOPMENT

1. Schumpeter was self-conscious about the generality of his general theory of development. Bitter that Keynes had stolen the twentieth-century thunder of innovation in theory, Schumpeter chastised the English economist for being fixated by his immediate 'realities' and policy. From his obituarial 1946 reading of Keynes' 1923 *Tract on Monetary Reform*, Schumpeter generalised thus:

> It cannot be emphasised too strongly that Keynes' advice was in the first instance English advice, born of English problems even where addressed to other nations. Barring some of his artistic tastes, he was surprisingly insular, even in philosophy, but nowhere so much as in economics. And he was fervently patriotic – of a patriotism which was indeed quite untinged by vulgarity but was so genuine to be subconscious and therefore all the more powerful to impart a bias to his thought and to exclude full understanding of foreign (also American) viewpoints, conditions, interests, and especially creeds.

And, Schumpeter reflected, as List had done of Adam Smith, that: 'Like the old free-traders, he always exalted what was at any moment truth and wisdom for England into truth and wisdom for all times and places.' See Schumpeter, 1948, p. 85.

However much he might have wished to implicitly confirm that his own theory was genuinely general whereas that of Keynes was not, Schumpeter's passages gave hostage to fortune. We should not be surprised to find the same kind of question being asked of Schumpeter's endeavour to create a modern theory of development and, in particular, the Austrian background to the theory.

2. Nurkse, 1953, p. 11 (quoted in Pal, 1955, and reprinted in Wood, 1991, pp. 85–6 and in Rimmer, 1961).

3. Berman, 1983, p. 75.

4. ibid, pp. 299, 290–312. Also, see Caro, 1974.

5. Berman, 1983, p. 305.

6. ibid., p. 308.

7. ibid., p. 40.

8. Anderson, 1984, p. 98.

9. Berman, 1983, p. 309.

10. ibid., p. 61.

11. ibid., pp. 72–4.

12. ibid., p. 72.

13. ibid., pp. 49–50, 74.

14. ibid., p. 68.

15. See, for instance, Johnston, 1972; Schorske, 1980; Laszlo and Pynsent, 1988.

16. Menger, 1963 [1883], p. 119 (quoted by Hutchinson, 1973, p. 34). Also, Bostaph, 1978.

17. Cubeddu, 1993, p. xi. Also, for instance, see Johnston, 1972, pp. 78–81, and essays in Hicks and Weber, 1973; Grassl and Smith, 1986; Littlechild, 1990; Special Issue, *Atlantic Economic Journal* 1978, 6(3).

18. See, for instance, Bottomore, 'Introduction' to Bottomore and Goode, 1978; Bottomore, 'Introduction' to Hilferding, 1985 [1910], p. 4; Bottomore, 1992, ch. 1.

19. Bottomore, 1978, p. 20.

20. Adler, 1978a, p. 78, 1978b [1914], p. 61.

21. Adler, 1978c [1927], p. 72; Bottomore, 1978, p. 11.

Oldroyd has suggested that in the period between Comte and the logical positivists 'we have a number of moderately distinct schools or "isms" such as pragmatism, conventionalism and instrumentalism, which may nonetheless be classified more or less satisfactorily as different manifestations of positivism'. He could have added associationism which we meet below. See Oldroyd, 1986, p. 187 (quoted in Shionoya, 1990b, fn. 1, p. 187).

22. For an effective account, see Shionoya, 1990b, pp. 192–6.

23. T. Bottomore, 1992, p. 21.

24. Shionoya, 1990b, p. 210 (quoting in translation from Schumpeter, 1908, p. 42).

25. O.Bauer, 1978 [1924], p. 215.

26. Schumpeter, 1908, p. xvi (quoted in Shionoya, 1990b, p. 197).

27. See, for instance, Robbins, 1966.

28. Newman, 1851, pp. 1–2; 14, 44.

29. Marx, 1981, pp. 729–30, 735, 738.

30. Kindleberger, for instance, has reported that recent financial history for the nineteenth century shows that 'France lagged about a hundred years behind Britain in the development of modern institutions – a central bank, reform of national finances, use of bank notes and deposits, insurance and so on'. See Kindleberger, 1984, p. 4.

31. ibid., p. 741 (quoting *Doctrine de Saint-Simon: Exposition, Première année, 1828/29*).

32. Marx, 1981, p. 572.

33. ibid., p. 742.

34. ibid., pp. 572–3.

35. ibid., p. 739. For the most basic general surveys, see Kindleberger, 1984, chs 5, 6; Cameron, 'England, 1750–1844' and 'France, 1800–1870' in Cameron, 1967.

36. Marx, 1981, pp. 741, 742.

37. ibid., p. 744 (quoting Pecqueur, *Théorie nouvelle d'économie sociale et politique*, pp. 433, 434).

38. Anderson, 1984, p. 104.

39. ibid., pp. 104–5.

40. Eddie, 1989, p. 886; Rudolph, 1976, Tables A8–17, pp. 217–30, Table 1.7, p. 34; Marz, 1984 [1981], p. 9.

41. John and Lichtblau, 1993, pp. 13–15; Bolognese-Leuchtenmuller, 1978, pp. 40–1; Helczmanovski, 1973, Table 3, p. 122; Marz, 1984, pp. 20, 16, 3.

42. Rothschild, 1961, p. 64; Koren, 1961, p. 265 (both quoted in Marz, 1984, pp. 16, 22).

43. Marz, 1991, p. 118; Rudolph, 1976, p. 195. Also, Gross, 1983.

44. Marz, 1984, p. 2, ch. 2; 1991, p. 39; Rudolph, 1976, p. 17; Eddie, 1989, pp. 865–75.

45. Marz, 1984, p. 5.

46. Cisleithania yields of wheat and rye, in hundred kilograms per hectare, rose from an average 10.3 and 10, over 1870–95, to 12.3 and 11.4 over 1896–1908. Barley and oats yields rose in the same proportion. Average yield per hectare, for wheat in 1907, was 60 per cent of the average German yield; corresponding yields for rye, barley, oats and potatoes were 73, 72, 61 and 84 per cent of the German yield in the same year. Yields in Austria may have been higher since the average was lowered by very low cereal yields in Galicia, Bukowina, and in the mountainous parts of Croatia and Bosnia. See Dinklage, 1973, pp. 450, 451.

47. Eddie, 1989, p. 863, Rudolph, 1976, pp. 198–9.

48. ibid., pp. 76, 120.

49. ibid., p. 120; Marz, 1984, ch. 6.

50. Marz, 1983, pp. 118–19; Rudolph, 1976, p. 69.

51. Berend and Ranki, 1974, p. 62 (quoted in Rudolph, 1976, p. 3).

52. Marz, 1984, p. 45, also, 1983, p. 122.

53. Hilferding, 1985 [1910], p. 213.

54. Marz, 1983, p. 122; for Hilferding on commercial and industrial profit, see his 1985, ch. 11, esp. pp. 193ff.

55. Rudolph, 1976, p. 87.

56. Riesser, 1911, p. 259 (quoted in Rudolph, 1976, p. 87); for general surveys of the 'great German banks', see Kindleberger, 1984, ch. 7; Tilley, 1967 and 1986 (reprinted in Cameron, 1992).

57. Rudolph, 1976, pp. 80, 163–4.

58. ibid., pp. 191, 192; also see Rudolph, 1972, pp. 26–9.

59. Marz, 1984, fn. 1, p. 47 and fn. 12, p. 47; 1983, p. 119.

60. Gerschenkron, 1965 [1962], p. 14.

61. Quoted in Soltau, 1959 [1931], p. 140.

62. Gerschenkron, 1965, pp. 23–4.

63. ibid., p. 25.

64. Gerschenkron, 1977, pp. 31, 155.

65. Gerschenkron, 1965, p. 20.

66. Gerschenkron, 1977, lecture 1.

67. Rudolph, 1976, p. 192.

68. Gerschenkron, 1977, p. 52.

69. ibid., p. 77.

70. Pasvolsky, 1928, pp. 11–13 (reproduced in Marz, 1984, Table 8, p. 32 and Rudolph, 1976, Table 6.9, p. 177); for trade and payments accounts, see Rudolph, 1976, p. 198 and Tables B.2, B.3, pp. 230–1.

71. quoted in Gerschenkron, 1977, p. 24.

72. ibid., p. 30.

73. Gerschenkron, 1977, p. 71.

74. ibid., pp. 56–9.

75. ibid., pp. 103–6, 113.

76. Marz, 1984, p. 30.

77. Gerschenkron, 1977, pp. 29, 65–8.

78. See, for instance, Madarasz, 1980, p. 340.

79. Helburn, 1986, p. 156.

80. Heilbroner, 1988, p. 175.

81. ibid., pp. 176, 175.

82. Swedberg, 1991a, pp. 52, 53 (quoting in translation from Schumpeter, 1985, pp. 271, 302).

83. Translation from the original French record by Prime and Henderson, 1975, pp. 296–8 (and quoted by Raines and Leathers, 1992, pp. 53–4). Also, see Cramer and Leathers, 1981; Walters, 1961 (reprinted in Wood, 1991, vol. 2).

84. Morishima, 1992, p. 69.

85. Streissler, 1981, pp. 79–80.

86. ibid., p. 80.

87. ibid.

88. Schumpeter, 1991a [1918], pp. 129, 130.

89. For instance, Catephores, 1994, p. 205.

90. Catephores, 1994, pp. 29–30.

91. Schumpeter, 1991b [1918–19], pp. 180, 183, Schumpeter, 1991c [1927], p. 274.

92. Schumpeter, 1991a, p. 130, 1991b, pp. 209–10.

93. Schumpeter, 1991c, p. 269.

94. Schumpeter, 1991b, p. 209.

95. ibid., pp. 209, 143, 212.

96. ibid., p. 149: 'Chamberlain was unquestionably serious. A man of great talent, he rallied every ounce of personal and political power, marshalled tremendous resources, organised all the interests that stood to gain, employed a consummate propaganda technique – all this to the limits of the possible. Yet England rejected him, turning over the reins to the opposition by an overwhelming majority.'

Schumpeter regarded his year of 1906–7 in England as the 'happiest year of his life'. He coupled socialising in the company of high society with intense study of economics texts at the British Museum. See Swedberg, 'Introduction: the man and his work', in Swedberg, 1991a, pp. 8–9; Swedberg, 1991b, p. 52; Allen, 1991, pp. 59–60.

97. Schumpeter, 1991b, p. 212.

98. Schumpeter, 1991a, pp. 130, 131.

99. ibid., p. 131.

100. ibid., p. 129; Schumpeter, 1956 [1917–18] p. 210.

101. Schumpeter, 1964 [1939], pp. 85, 86.

102. ibid., p. 87.

103. Janeway, 1986, p. 437; Minsky, 1990, p. 59.

104. Heilbroner, 1988, p. 173.

105. For what has become a classic comment on Schumpeter and the banker as entrepreneur, see Cameron, 1963.

106. Schumpeter, 1964, pp. 90, 91, 92.

107. Schumpeter, 1956, pp. 205–6.

108. ibid., p. 159.

109. ibid., p. 206.

110. Schumpeter, 1951, pp. 48, 49.

111. See, for instance, Oakley, 1990, pp. 13–14; Dyer, 1988, p. 28.

112. Schumpeter, 1951, pp. 48, 49.

113. Schumpeter, 1991c, pp. 258–69; Dyer, 1988, p. 32.

114. ibid., pp. 33, 40.

115. Brouwer, 1991, p. 2.

116. Schumpeter, 1934 [1912], pp. 3–9.

117. See, for instance, Shionoya, 1990b, pp. 209ff; also, Goodwin, 1990.

118. See, for instance, Haines, 1987, p. 20.

119. Schumpeter, 1954, pp. 446–7.

120. Jones, 1980, p. 64.

121. ibid., p. 7; also, see Ryan, 1974, pp. 26, 33, 117–18.

122. For instance, see Marshall, 1961 [1890] p. 207: 'Man's prerogative extends to a limited but effective control over natural development by forecasting the future and preparing the way for the next step.

Thus progress may be hastened by thought and work; by the application of the principles of Eugenics to the replenishment of the race from its higher rather than its lower strains, and by the appropriate education of the faculties of either sex: but however hastened it must be gradual and relatively slow. . . . Thus progress itself increases the urgency of the warning that in the economic world, *Natura non facit saltum*.'

123. Schumpeter, 1934, pp. 10, 42.

124. Schumpeter, 1991b, pp. 200, 189, 190.

125. Schumpeter, 1934, pp. 19–50.

126. ibid., pp. 30–1.

127. ibid., p. 8.

128. ibid., pp. 32–4.

129. It is to be noted that while the more recent distinctions which have been made with regard to frictional and structural unemployment derive from this tradition of analysis it was more in keeping with Schumpeter's purpose of development that the 'structural' failing of the economic system derived not from the economic system as such but from the social order of capitalism. See, for instance, Schumpeter, 1991d [1941], pp. 355–63.

130. Schumpeter, 1951, fn. 1, p. 60.

131. Shionoya, 1986, pp. 740, 743, 1990a, p. 321 (both sources quoting from Schumpeter, 1912, p. 489).

132. Schumpeter, 1951, pp. 61–2.

133. ibid., fn. 2, pp. 62–3.

134. Schumpeter, 1934, p. 95.

135. Schumpeter, 1951, pp. 63–4.

136. Schumpeter, 1934, p. 67.

137. Schumpeter, 1956, pp. 205–6.

138. Schumpeter, 1954, p. 41.

139. ibid., p. 573 (partly quoted in Arndt, 1981, p. 459).

140. Schumpeter, 1946, pp. 205–7.

141. Heilbroner, 1988, p. 178.

142. Schumpeter, 1934, pp. 122, 126.

143. ibid., pp. 117, 116.

144. Heilbroner, 1988, p. 167 (quoted in Oakley, 1991, p. 81); also Brouwer, 1991, pp. 49–53.

145. Oakley, 1990, pp. 83–4.

146. Schumpeter, 1934, ch. 1, 1964, chs. 2–3; Schumpeter's scheme is also laid out, for instance, in Oakley, 1990, chs. 4–7 and Brouwer, 1991, ch. 1.

147. Schumpeter, 1946, p. 210.

148. Brouwer, 1991, p. 58.

149. Schumpeter, 1934, p. 114.

150. ibid., pp. 210–11; also see Bellofiore, 1985a, p. 28, 1985b, p. 69; Oakley, 1990, p. 109.

151. Schumpeter, 1934, pp. 85–6, 107, 1964, p. 102.

152. Schumpeter, 1934, p. 51.

153. Morishima, 1992, pp. 3–9; Bellofiore, 1985a, pp. 29–30, and 1985b, p. 69.

154. Schumpeter, 1934, p. 102.

155. ibid., pp. 107, 110.

156. ibid., p. 74.

157. Schumpeter, 1951, p. 70 [384].

158. From a wealth of literature, see Scherer, 1984; Brouwer, 1991, chs. 2–4; Mayhew, 1980; Nelson and Winter, 1982; Freeman, 1990.

159. Schumpeter, 1949 (reprinted in Schumpeter, 1951, pp. 255–6 and quoted in Swedberg, 1991b, p. 173).

160. Schumpeter, 1951, p. 71.

161. ibid., pp. 70–1.

162. Schumpeter, 1934, p. 67.

163. Swedberg, 1991b, pp. 148, 141.

164. Schumpeter, 1964, pp. 86, 87.

165. Schumpeter, 1946, p. 208.

166. ibid., pp. 209, 208.

167. H. Wallich, 1958 [1952]; Singer, 1953 (Wallich and Singer quoted in Rimmer, 1961, p. 361); Dass, 1962 and in Wood, 1991, vol. 2; Pal, 1955, p. 86.

168. Johnson, 1982, p. 307 (quoted in Smith, 1986 [1984], p. 259).

169. See, for example, Smith, 1986, p. 233. In the conclusion to his ch. 12, entitled 'Behind The Japanese Miracle', Smith wrote: 'The evidence seems very strong: the heavily interventionist industrial policy which has been described has worked spectacularly well in the face of economic problems not unlike those which face Britain today; we should draw the appropriate conclusions.' In an afterword of 1986, Smith supported his view against critics who argued for cultural relativism and referred (p. 259) to Chalmers Johnson's developmental state. Also see Freeman, Clark and Soete, 1982. They end their texts with a question, 'Lessons from Japan?'.

170. Schumpeter's influence on Japan is twofold. First, it was alleged by his Harvard colleagues such as

Gottfried Haberler and Arthur Smithies that Schumpeter had 'many devoted disciples' in Japan, 'the country that has probably extended [him] more esteem and admiration than any other'. This view is regarded, by McCraw, as being 'overstated' and 'certainly any direct influence of ideas on policy is notoriously hard to trace'. The second source of influence is that, as in Austria, Schumpeter's theory can be powerfully used to explain the development of post-war Japan. Such is the view of, for instance, Ozawa, Prime and Henderson, McCraw himself, and those such as Freeman, who employ Japan as the exemplar of industrial innovation. An obvious question to ask of both sources of influence is 'Which Schumpeter?' – that of the paradigm of innovation and technology, the Schumpeter of the restricted domain of endogenous economic growth or the Schumpeter of Faustian development within the general domain? See, for instance, McCraw, 1991, pp. 389–92; Ozawa, 1974, pp. 2–5; Prime and Henderson, 1975. A succinct, critical review of the above is to be found in Raines and Leathers, 1992.

171. Scherer, 1992, p. 1426. For MITI, see Johnson, 1982.

172. Allen, 1981, p. 73; also, Raines and Leathers, 1992, p. 57. Ironically enough, when Raines and Leathers contest the view that post-war Japan fits into Schumpeter's alleged endorsement of 1945 Catholic corporatism, they make the same argument: MITI, the Ministry of Finance, the Export–Import Bank and the Japan Development Bank in providing 'long-term, low interest loans to basic industries and export industries have been directly responsible for technological developments in Japanese industry'.

173. ibid., pp. 74–5, 85; also see, for instance, Lockwood, 1965, and Morishima, 1982, ch. 5.

174. Amsden, 1989, p. 145.

175. ibid., p. 146.

176. ibid., p. 153. Amsden refers to 'the implied perversities' of relationships, such as those between government intervention and fast growth and comments that 'perversity has a long tradition in economic theory'.

177. Amsden, 1990, p. 25.

178. ibid., pp. 14–16, and 1989, ch. 1.

179. Amsden, 1989, p. 16.

180. ibid., p. 144 and 1990, p. 16.

181. Amsden, 1989, pp. 15, 16.

182. Amsden, 1990, p. 24, and 1989, pp. 9, 38–41, 246.

183. Quoted in Smith, 1986, p. 233.

184. Amsden, 1989, p. 141.

185. ibid., pp. 22, 23.

186. For instance, see Schumpeter, 1934, fn., p. 77.

187. Lippman, 1961 [1914], p. 98 (quoted in Lasch, 1965, p. 163).

188. Lipton, 1986, p. 726.

189. Chambers, 1989, p. 7.

190. ibid., pp. 14, 16.

191. ibid., p. 20.

192. See, for instance, Green, 1989, p. 45. Green concluded: 'One "resource" of which less is needed is pre-emptive or coercive advice. Imposed programmes rarely survive the crisis stick or the resource carrot – even if they are inherently sound. Only Africans are primarily concerned about and able to achieve the development of Africa. The record of foreign (or citizen) experts' export model solutions to Africans' problems is – to say the least – not very good, partly at least because these models almost always lack contextual, temporal and technical knowledge Africans (often poor Africans) possess and their expert designers did not.'

193. J. Harris to E.D. Morel, July 1912 (quoted in Nworah, 1966, p. 398).

194. Goethe, quoted in Berman, 1983, p. 47.

8 CONCLUSION: THE JARGON OF DEVELOPMENT

1. Adorno, 1973 [1964], p. 12.
2. Coleman and Nixson, 1986 [1978], p. 2.
3. Todaro, 1992, p. 7.
4. Coleman and Nixson, 1986, p. 7.
5. Todaro, 1992, p. 102.
6. Coleman and Nixson, 1986, p. 7.
7. Hobhouse, 1927 [1913], pp. 244, 472.
8. Garvin, 1905, p. 2.
9. Thirlwall, 1989 [1972], p. 8.
10. Coleman and Nixson, 1986, p. 2.
11. Todaro, 1992, p. 10.
12. Coleman and Nixson, 1986, p. 6.
13. Todaro, 1989, pp. 12, 23.
14. Todaro, 1992, p. 101.
15. Goulet, 1977 [1971], pp. 123–5 (referred to in Todaro, 1992, p. 101).
16. Fukuyama, 1992, part 3; Lewis, 1963, p. 420 (referred to by Todaro, 1992, p. 102).
17. Goulet, 1977, pp. 125–6.

18. Todaro, 1992, p. 102.

19. Seers, 1979, p. 10.

20. ibid.; also Seers, 1983.

21. See, for instance, Sen, 1983a, 1983b, 1984a, 1988a, 1988b. Also, see Sen, 1985, 1987.

There is no reference to Amartya Sen in Todaro, 1989; Thirlwall, 1989; Coleman and Nixson, 1986. An exception is Ingham, 1995, pp. 49–51.

22. Sen, 1984b, p. 497.

23. Marx and Engels, 1947 [1845–6], p. 22 (quoted in Sen, 1988b, p. 271).

24. See, for instance, Cohen, 1988, pp. 141–4; Berki, 1983; MacGregor, 1984.

25. See, for instance, Gouldner, 1980; Marcuse, 1986 [1941], part 2.

26. Sen, 1988a, p. 21; 1988b, pp. 278–92.

27. Sen, 1984b, 551; also see pp. 326 and 325 where Sen defines '*absolute* deprivation in terms of a person's capabilities' and absolute deprivation 'relates to *relative* deprivation in terms of commodities, incomes and resources'. Or, poverty is 'an absolute notion in the space of capabilities but very often it will take a relative form in its space of commodities or characteristics'.

28. ibid., p. 497.

29. ibid., pp. 514, 512.

30. ibid., p. 498.

31. Blackburn, 1990.

32. Banuri, 1990a, p. 96 (Deleuze is quoted from Sheridan, 1980, p. 114).

33. Marglin, 1990a, p. 11.

34. Nandy and Visvanathan, 1990, p. 175.

35. Marglin, 1990a, pp. 24–5; Marglin, 1990b, pp. 232–5.

36. Banuri, 1990a, pp. 74–82.

37. Marglin, 1990b, p. 241.

38. ibid., pp. 218–56.

39. Pippin, 1991, p. 5.

40. See Pippin, 1991, ch. 2; also, Berman, 1983.

41. Banuri, 1990a, p. 97.

42. ibid., p. 89, 1990b, p. 66.

43. Marglin, 1990a, p. 15.

44. The literature is extensive. See, for example, Keynes, 1972b [1933]; Shionoya, 1991; Skidelsky, 1983, ch. 6; Harrod, 1972 [1951], chs. 2–3; Moggeridge, 1992, ch. 5; Carabelli, 1988, ch. 3; Fitzgibbons, 1988, ch. 2; O'Donnell, 1989, ch. 16.

45. For Keynes' scathing remarks on Whitehead, the referee for his 1907 draft of the dissertation on probability, Whitehead's 'conversion' and later reverse influence on Keynes, see Harrod, 1972, pp. 148, 160; Skidelsky, 1983, pp. 182, 198; Moggeridge, 1992, pp. 185, 364; Carabelli, 1988, pp. 10, 136, 149. For Keynes' acknowledgement to Whitehead, see Keynes, 1973a [1921], fn., p. 110.

For Whitehead's involvement with Moore and Keynes' Bloomsbury Circle, see Lowe, 1985, pp. 132–4. For the view that Keynes came to adopt Whitehead's scheme of organic interdependence, see Winslow, 1986, 1989a.

For a critique of Winslow and the contrary view, that Keynes was no organicist but continued to employ Moore's organic principle to defend individual autonomy and that interdependence is external to active, individual self-interest, see Bateman, 1989 (and rejoinder by Winslow, 1989b) and Bateman, 1991a, 1991b; Davis, 1989.

46. Lucas, 1989, pp. 81, 86, 147 (quoting from Whitehead, 1929, p. 239).

47. Whitehead, 1929, p. 30. Whitehead's influence on Keynes becomes clear when he writes that 'every prehension consists of three factors: (a) the "subject" which is prehending, namely the actual entity in which the prehension is a concrete element; (b) the "datum" which is prehended; (c) the "subjective form" which is *how* that subject prehends that datum' (p. 31).

48. Bateman, 1991a, p. 32. This view of an 'epistemological break' in Keynes is criticised as part of 'the continuity issue' in O'Donnell, 1991a; also see O'Donnell, 1989 and Gerrard, 1992.

49. Davis, 1989, pp. 1162–3.

50. For instance, Coddington, 1983; for a generalised critique of Coddington, see Lawson, 1985, 1987; Runde, 1991.

51. Lawson, 1985, p. 916.

52. Keynes, 1973b, p. 114 (quoted in Runde, 1991, p. 138).

53. The difference between extreme uncertainty, where knowledge is very incomplete, is different, O'Donnell insists, from 'irreducible complete uncertainty' of the future but the two concepts can logically 'overlap'. O'Donnell, 1991, p. 82.

54. Keynes, 1973c [1936], pp. 161–3.

55. ibid., p. 114 (p. 138).

56. Lawson, 1985, p. 917; for the difference between true or absolute and relative belief, see Lawson, 1987, pp. 968–9; also, Lawson, 1991 and Hollis, 1991; see also, Lawson, 1993.

57. Keynes, 1973c, p. 153.

58. ibid., p. 155: 'Of the maxims of orthodox finance none, surely, is more anti-social than the fetish of liquidity, the doctrine that it is a positive virtue on the part of investment institutions to concentrate their resources upon the holding of "liquid" securities. It forgets that there is no such thing as liquidity for

investment for the community as a whole. The social object of skilled investment should be to defeat the dark forces of time and ignorance which envelop our future.'

59. Carabelli, 1988, pp. 163–7; also, Carabelli, 1985, pp. 173–4, 1991.

60. Winslow, 1989a, p. 1175; Carabelli, 1988, p. 170.

61. Carabelli, 1985, p. 173 (quoting from Keynes, 1973a, pp. 271–4).

62. Quoted in Moggeridge, 1976, p. 47.

63. Keynes, 1973c, p. 146.

64. Fitzgibbons, 1988, ch. 4; also, Skidelsky, 1983, pp. 154–7; Helburn, 1991.

65. Burke, 1968 [1790], p. 281 (quoted in Freeman, 1980, p. 30).

66. O'Gorman, 1973, p. 54.

67. Fitzgibbons, 1988, p. 54.

68. Marglin, 1990b, p. 231; 1990a, pp. 11, 23.

69. Banuri, 1990a, pp. 87, 74.

70. For instance, Kuntz, 1984, pp. 83, 87; Ford, 1984.

71. Whitehead, 1953 [1925], p. 201.

72. Kuntz, 1984, pp. 91, 87 (quoting from Whitehead, *Adventures in History* pp. 5, 24); Leclerc, 1984, pp. 133–4.

73. Whitehead, 1953, p. 205.

74. ibid., p. 207.

75. Marglin, 1990a, pp. 4–5.

76. Whitehead, 1953, pp. 206–7.

77. Marglin, 1990a, p. 16.

78. Marglin, 1990b, p. 265, generally, pp. 256–77.

79. Marglin, 1990a, p. 10; 1990b, p. 278.

80. Bearce, 1961, pp. 131–2.

81. Hettne, 1990, chs 1–2.

82. ibid., ch. 3.

83. ibid., p. 195.

84. Quoted in ibid., p. 195.

85. ibid., pp. 195–6.

86. ibid., p. 155.

87. ibid.

88. Marx and Engels, 1968 [1848], pp. 66–7.

89. Whitehead, 1953, pp. 200, 202–3.

90. ibid., pp. 200, 205.

91. Keynes, 1973c, pp. 353–71. Keynes was referring to the non-academic body of writing from Mandeville and Malthus to Hobson which rested upon the idea of underconsumptionism. He also included Major Douglas' concept and political movement of Social Credit and was particularly taken by one successor to Henry George, namely Silvio Gesell's 'semi-religious' radical movement which was founded upon his *Natural Economic Order.* This, Keynes suggested, provided the 'answer to Marxism'.

92. Whitehead, 1953, p. 200.

93. Rosdolsky, 1977 [1968].

94. Adorno, 1993 [1951], pp. 156–7.

BIBLIOGRAPHY

Aboagye, A. (1986) *Informal Sector Employment in Kenya*, Addis Ababa: ILO/JASPA.

Abreu, E. (1982) *The Role of Self-Help in the Development of Education in Kenya 1900–1973*, Nairobi, Kenya Literature Bureau.

Acheson, T.W. (1990) 'The Great Merchant and Economic Development in Saint John 1820–1850', in D. McCalla (ed.) *The Development of Canadian Capitalism: Essays in business history*, Toronto: Copp Clark Pitman.

Adams, W.S. (1957) *Edwardian Portraits*, London: Secker and Warburg.

Adler, M. (1978a) [1904] 'Causality and Teleology', in T. Bottomore and P. Goode (eds) *Austro-Marxism: texts*, Oxford: Clarendon Press.

——— (1978b) [1914] 'The Sociological Meaning of Karl Marx's Thought', in T. Bottomore and P. Goode (eds) *Austro-Marxism: texts*, Oxford: Clarendon Press.

——— (1978c) [1927] 'A Critique of Othmar Spann's Sociology', in T. Bottomore and P. Goode (eds) *Austro-Marxism: texts*, Oxford: Clarendon Press.

Adorno, T.W. (1973) [1964] *The Jargon of Authenticity*, Evanston: Northwestern Press.

——— (1993) [1951] *Minima Moralia*, London: Verso.

Advisory Committee on Native Education in British Tropical African Dependencies (1923) *Educational Policy in British Tropical Africa*, Cmd 2374, XXI 1925, London: HMSO.

Africa Watch (1993) *Divide and Rule: state sponsored ethnic violence in Kenya*, New York: Human Rights Watch.

Aitken, H.G.J. (1959) 'The Changing Structure of the Canadian Economy: with reference to the influence of the United States', in H.G.J. Aitken (ed.) *The American Impact on Canada*, Durham, NC: Duke University Press.

——— (1967) 'Defensive Expansion: the state and economic growth in Canada', in W.T. Easterbrook and M.H. Watkins (eds) *Approaches to Canadian Economic History*, Toronto: McClelland and Stewart.

Alatas, S.H. (1990) *Corruption: its nature, causes and functions*, Aldershot: Avebury.

Allen, G.C. (1981) 'Industrial Policy and Innovation in Japan', in C. Carter (ed.) *Industrial Policy and Innovation*, London: Heinemann.

Allen, L. (ed.) (1975) *John Henry Newman and the Abbé Jager*, Oxford: Oxford University Press.

Allen, R.L. (1991) *Opening Doors: the life and work of Joseph Schumpeter. Volume One: Europe*, New Brunswick, NJ: Transaction.

Allen, T. and Thomas, A. (eds) (1992) *Poverty and Development in the 1990s*, Oxford:

Oxford University Press.

Altman, M. (1981) 'Economic Aspects of Agricultural Productivity and the Seigniorial Land-holding System in Quebec, 1780–1850: the testing of causal hypotheses', unpublished MA thesis, McGill University.

Ambirajan, S. (1978) *Classical Political Economy and British Policy in India*, New York: Cambridge University Press.

Ambler, C.H. (1988) *Kenyan Communities in the Age of Imperialism: the central region in the late nineteenth century*, New Haven: Yale University Press.

Amery, L.S., (1908) *Fundamental Fallacies of Free Trade*, London: Love.

——— (1953) *My Political Life*, vol. 1, London: Hutchinson and Co.

Amsden, A.H. (1971) *International Firms and Labour in Kenya, 1945–1970*, London: Cass.

——— (1989) *Asia's Next Giant: South Korea and late industrialization*, New York: Oxford University Press.

——— (1990) 'Third World Industrialisation: "global fordism" or a new model', *New Left Review* 182.

Analogues, J. (1977) 'Agriculture, Balanced Growth and Social Change in Central Canada since 1850', *Economic Development and Cultural Change* 25(4).

Anderson, B. (1983) *Imagined Communities: reflections on the origin and spread of nationalism*, London: Verso.

Anderson, P. (1984) 'Modernity and Revolution', *New Left Review* 144.

Andreski, S. (ed.) (1974) *The Essential Comte: selected from the Cours de Philosophie Positive*, London: Croom Helm.

Angers, F.-A. (1960) 'Nationalisme et vie économique', *Revue d'Histoire de l'Amérique Française* 14(4).

——— (1961) 'Naissance de la pensée économique au Canada Français', *Revue d'Histoire de l'Amérique Française* 15(2).

Annan, N. (1984) *Leslie Stephen: the godless Victorian*, New York: Random House.

Anon. (1892) 'The Land for the People', London: Land Nationalisation Society (LNS Tract 2).

Armstrong, R. (1984) *Structure and Change: an economic history of Quebec*, Toronto: Gage.

Arndt, H.W. (1981) 'Economic Development: a semantic history', *Economic Development and Cultural Change* 29(3).

Arrighi, G. (1970) 'Struttura di classe e struttura coloniale nell' analisi del sottoviluppo', *Problemi del Socialismo* 14(10).

Arrighi, G. and Saul, J. (1973) *Essays in the Political Economy of Tropical Africa*, New York: Monthly Review Press.

Arthur, C.J. (1986) *Dialectics of Labour: Marx and his relation to Hegel*, Oxford: Blackwell.

Aseniero, G. (1985) 'A Reflection on Developmentalism: from development to transformation', in H. Addo (ed.) *Development as Social Transformation: reflections on the global problématique*, London: Hodder and Stoughton.

Atieno-Odhiambo, E.S. (1987) 'Democracy and the Ideology of Order in Kenya', in M.G. Schatzberg (ed.), *The Political Economy of Kenya*, New York: Praeger.

Atkins, A. (1966) 'Religious Assertions and Doctrinal Development', *Theological Studies* 27(4).

Auchinleck, G. (1927a) 'The Oil Palm Plantation Industry in Sumatra and Malaya', *Proceedings of the West African Agricultural Conference of 1927*, London, Foreign and Commonwealth Office Library.

——— (1927b) 'General Discussion of the Oil Palm Papers and the Eastern Menace to the West African Trade', *Proceedings of the West African Agricultural Conference of 1927*, London, Foreign and Commonwealth Office Library.

Avineri, S. (1972) *Hegel's Theory of the Modern State*, London: Cambridge University Press.

Bager, T. (1980) *Marketing Cooperatives and Peasants in Kenya*, Uppsala: Scandinavian Institute of African Studies.

Baillie, J. (1950) *The Belief in Progress*, London: Oxford University Press.

Bakhurst, D. (1991) *Consciousness and Revolution in Soviet Philosophy: from the Bolsheviks to Evald Ilyenkov*, Cambridge: Cambridge University Press.

Banaji, J. (1983) 'Gunder Frank in Retreat?', in P. Limqueco and B. McFarlane (eds) *Neo-Marxist Theories of Development*, London: Croom Helm.

Banuri, T. (1990a) 'Modernisation and its Discontents: a cultural perspective on the theories of development', in F.A. Marglin and S.A. Marglin (eds) *Dominating Knowledge*, Oxford: Clarendon Press.

——— (1990b), 'Development', in F.A. Marglin and S.A. Marglin (eds) *Dominating Knowledge*, Oxford: Clarendon Press.

Baran, P.A. (1973) [1957] *The Political Economy of Growth*, Harmondsworth: Penguin.

Barber, W.J. (1975) *British Economic Thought and India: 1600–1858*, Oxford: Oxford University Press.

Barlow, A.R. (1932) 'Kikuyu Land Tenure and Inheritance', *Journal of the East Africa and Uganda Natural History Society* 11(2).

Barman, J. (1991) *The West Beyond the West: a history of British Columbia*, Toronto: University of Toronto Press.

Barnett, D.F. (1976) 'The Galt Tariff: incidental or effective protection?' *Canadian Journal of Economics* 9(3).

Barnett, T. (1988) *Sociology and Development*, London: Hutchinson.

Barry, E.E. (1965) *Nationalization in British Politics: the historical background*, London: Cape.

Barry, W. (1904) *Newman*, London: Hodder and Stoughton.

Baskerville, P. (1981) 'Americans in Britain's Backyard: the railway era in Upper Canada, 1850–1880', *Business History Review* 55(3).

Bateman, B.W. (1989) '"Human Logic" and Keynes's Economics: a comment', *Eastern Economic Journal* 15(1).

——— (1991a) 'Hutchinson, Keynes and Empiricism', *Review of Social Economy* 49(1).

——— (1991b) 'The Rules of the Road: Keynes's theoretical rationale for public policy', in B.W. Bateman and J.B. Davis (eds) *Keynes and Philosophy: essays on the origins of Keynes's thought*, Aldershot: Elgar.

Bauer, O. (1978) [1924] 'The World View of Organised Capitalism', in T. Bottomore and P. Goode (eds) *Austro-Marxism: texts*, Oxford: Clarendon Press.

Bayart, J.-F. (1993) [1989] *The State in Africa: the politics of the belly*, London: Longman.

Bearce, G.D. (1961) *British Attitudes Towards India: 1784–1858*, London: Oxford.

Beer, G. (1983) *Darwin's Plots: evolutionary narrative in Darwin, George Eliot and nineteenth-century fiction*, London: Routledge and Kegan Paul.

Bell, D. (1963) 'Veblen and the New Class', *The American Scholar* 32(4).

Belloc, H. (1904) *Emmanuel Burden: merchant*, London: Methuen.

Bellofiore, R. (1985a) 'Money and Development in Schumpeter', *Review of Radical Political Economics* 17(1/2).

——— (1985b) 'Marx after Schumpeter', *Capital and Class* 24.

Benson, W.B. (1942) 'The Economic Advancement of Underdeveloped Areas', in *The Economic Basis of Peace*, London: National Peace Council.

Berend, I.T. and Ranki, G. (1974) *Economic Development in East-Central Europe in the 19th and 20th Centuries*, New York: Columbia University Press.

Berger, C. (1976) *The Writing of Canadian History: aspects of English Canadian historical writing since 1900*, Toronto: Oxford University Press.

Berki, R.N. (1983) *Insight and Illusion: the problem of communism in Marx's thought*, London: Dent.

Berman, B. (1990) *Control and Crisis in Colonial Kenya: the dialectic of domination*, London: James Currey.

Berman, M. (1983) *All that is Solid Melts into Air: the experience of modernity*, London: Verso.

Bernal, M. (1976) *Chinese Socialism to 1907*, Ithaca: Cornell University Press.

Bernard, J.P. (1971) *Les Rouges: liberalisme, nationalisme et anticlericalisme au milieu du XIXè siècle*, Montreal: Les Presses de l'université du Québeç.

Bernstein, G.L. (1986) *Liberalism and Liberal Politics in Edwardian England*, London: Allen and Unwin.

Bernstein, H. (ed.) (1973) *Underdevelopment and Development: the Third World today*, Harmondsworth: Penguin.

Billetoft, J. (1989) *Rural Nonfarm Enterprises in Western Kenya: spatial structure and development*, CDR Project Paper 89.3, Copenhagen: Centre for Development Research.

Blackburn, R.J. (1990) *The Vampire of Reason: an essay in the philosophy of history*, London: Verso.

Blackton, C.S. (1955) 'The Dawn of Australian National Feeling 1850–1856', *Pacific Historical Review* 24(2).

Block, E. (ed.) (1992) *Critical Essays on John Henry Newman*, English Critical Studies, British Columbia: University of Victoria.

Bloomfield, P. (1961) *Edward Gibbon Wakefield*, London: Longman.

Bock, K. (1979) 'Theories of Progress, Development, Evolution', in T. Bottomore and R. Nisbet (eds) *A History of Sociological Analysis*, London: Heinemann.

Bogdanov, A.A. (1923) [1913] *Filosophiia zhivogo optya* [The Philosophy of Living Experience], Petrograd and Moscow.

Bolognese-Leuchtenmuller, B. (1978) *Bevolkerungsentwicklung und Beufsstruktur: Gesundheits- und Fürsorgewesen in Österreich 1750–1918*, Vienna: Verlag für Geschichte und Politik.

Booth, D. (1975) 'Andre Gunder Frank: an introduction and an appreciation', in I. Oxaal, T. Barnett and D. Booth (eds) *Beyond the Sociology of Development: Economy and Society in Latin America and Africa*, London: Routledge and Kegan Paul.

Bostaph, S. (1978) 'The Methodological Debate Between Carl Menger and the German Historicists', *Atlantic Economic Journal* 6(3).

Bottomore, T. (1978) 'Introduction', in T. Bottomore and P. Goode (eds) *Austro-Marxism: texts*, Oxford: Clarendon Press.

——— (1981) [1910] 'Introduction', in R. Hilferding, *Finance Capital: a study of the latest phase of capitalist development*, London: Routledge and Kegan Paul.

——— (1992) *Between Marginalism and Marxism: the economic sociology of J.A. Schumpeter*, New York: St Martin's Press.

——— and Goode, P. (eds) (1978) *Austro-Marxism: texts*, Oxford: Clarendon Press.

Bouchard, G. (1983) 'Anciens et nouveaux Québecois? Mutations de la société rurale et problèmes collective au XXe siècle', *Questions de Culture* 5.

Bourdillon, B. (1937) 'The African Producer in Nigeria', *West Africa*, 30 January.

Boyd, C.W. (1914) *Mr. Chamberlain's Speeches*, vol. 2, London: Constable.

Bradley, I. and Howard, M. (eds) (1982) *Classical and Marxian Political Economy: essays in honour of R.L. Meek*, London: Macmillan.

Breen, R.M. (1976) 'The Politics of Land: the Kenya Land Commission (1932–33) and its effects on land policy in Kenya', unpublished PhD thesis, Michigan State University.

Brenner, R. (1977) 'The Origins of Capitalist Development: a critique of Neo-Smithian Marxism', *New Left Review* 104.

Brett, E.A. (1973) *Colonialism and Underdevelopment in East Africa: the politics of change 1919–39*, London: Heinemann.

Brewer, A. (1980) *Marxist Theories of Imperialism: a critical survey*, London: Routledge and Kegan Paul.

Brewer, J. (1989) *The Sinews of Power: war, money and the English state, 1688–1793*, London: Unwin Hyman.

Brinton, C.C. (1933) *English Political Thought in the Nineteenth Century*, London: Ernest Benn.

Brodribb, W.A. (1883) *Recollections from an Australian Squatter, or Leaves from my Journal since 1835*, Sydney: John Woods.

Brouwer, M. (1991) *Schumpeterian Puzzles: technological competition and economic evolution*, Hemel Hempstead: Harvester Wheatsheaf.

Brunet, M. (1958) 'Trois dominantées de la pensée Canadienne-française: l'agriculturalisme, l'anti-étatisme et la messianisme', in *La Présence anglaise et les Canadiens*, Montreal: Beauchemin.

Brunner, E. (1948) *Christianity and Civilisation: Part One*, London: Nisbet.

Buckner, P.A. (1985) *The Transition to Responsible Government: British policy in British North America, 1815–1850*, Westport, CT: Greenwood Press.

Bumstead, J.M. (ed.) (1986) *Interpreting Canada's Past: Volume 1, Before confederation*, Toronto: Oxford University Press.

Bunkley, A.W. (ed.) (1948) A *Sarmiento Anthology*, Princeton: Princeton University Press.

Burke, E. (1968) [1790] *Reflections on the Revolution in France*, Harmondsworth: Penguin.

Burns, E. Bradford (1990) *The Poverty of Progress: Latin America in the nineteenth century*, Berkeley: University of California Press.

——— (1993) *Latin America: Conflict and Creation: a historical reader*, Englewood Cliffs, NJ: Prentice-Hall.

Burroughs, P. (1971) *British Attitudes towards Canada 1822–1849*, Scarborough, Ontario: Prentice Hall.

——— (1984) 'Colonial Self-government', in C.C. Eldridge (ed.) *British Imperialism in the Nineteenth Century*, Basingstoke: Macmillan.

——— (1990) 'Liberal, Paternalist or Cassandra? Earl Grey as a critic of colonial self-government', *Journal of Imperial and Commonwealth History* 18(1).

Bury, J.B. (1921) *The Idea of Progress: an inquiry into its origin and growth*, London: Macmillan.

Butlin, N.G. (1986) 'Contours of the Australian Economy 1788–1860', *Australian Economic History Review* 26(2).

——— and Sinclair, W.A. (1986) 'Australian Gross Domestic Product 1788–1860: estimates, sources and methods', Appendix to N.G. Butlin, 'Contours of the Australian Economy 1788–1860', *Australian Economic History Review* 26(2).

Buxton, C.R. (1910) *Minimum Wages for Agricultural Labourers*, Pamphlet 8, London: The National Land and Home League.

——— (1914) *Sixty Rural Points on Liberal Land and Housing Policy*, London: Central Land and Housing Council.

——— (1937) *The Race Problem in Africa*, London: Hogarth Press.

Cain, P.J. and Hopkins, A.G. (1993) *British Imperialism: innovation and expansion 1688–1914*, London: Longman.

Cameron, D. (ed.) (1985) *Explorations in Canadian Economic History: essays in honour of Irene M. Spry*, Ottawa: University of Ottawa Press.

Cameron, J.M. (1960) 'The Night Battle: Newman and empiricism', *Victorian Studies* 4(2).

——— (1967) 'Newman and the Empiricist Tradition', in J. Coulson and A.M. Allchin (eds) *The Rediscovery of Newman: an Oxford Symposium*, London: SPCK.

Cameron, R. (1963) 'The Banker as Entrepreneur', *Explorations in Entrepreneurial History* 1(1).

——— (1967) 'England, 1750–1844' and 'France, 1800–1870', in R. Cameron (ed.) *Banking in the Early Stages of Industrialisation*, New York: Oxford University Press.

——— (ed.) (1992) *Financing Industrialisation*, vol. 1, Aldershot: Edward Elgar.

Carabelli, A. (1985) 'Keynes on Cause, Chance and Possibility', in T. Lawson and H. Pesaran (eds) *Keynes' Economics: methodological issues*, London: Croom Helm.

——— (1988) *On Keynes' Method*, Basingstoke: Macmillan.

——— (1991) 'The Methodology of the Critique of Classical Theory: Keynes on organic interdependence', in B.W. Bateman and J.B. Davis (eds) *Keynes and Philosophy: essays on the origins of Keynes's thought*, Aldershot: Elgar.

Careless, J.M.S. (1971) 'The 1850s', in J.M.S. Careless (ed.) *Colonists and Canadiens 1760–1867*, Toronto: Macmillan.

Caro, R. (1974) *Robert Moses and the Fall of New York*, New York: Knopf.

Carter Report, see United Kingdom, 1934.

Catephores, G. (1994) 'The Imperious Austrian: Schumpeter as bourgeois marxist', *New Left Review* 205.

Cell, J.W. (1970) *British Colonial Administration in the Mid-Nineteenth Century: the policy making process*, New Haven: Yale.

Chadwick, Owen (1957) *From Bossuet to Newman: the idea of doctrinal development*, Cambridge: Cambridge University Press.

Chambers, R. (1969) [1844] *Vestiges of the Natural History of Creation*, New York: Humanities Press.

Chambers, R. (1989) 'The State and Rural Development: ideologies and an agenda for the 1990s', Institute of Development Studies, University of Sussex, Discussion Paper 269.

Chandavarkar, A. (1989) *Keynes and India: A study in economics and biography*, Basingstoke: Macmillan.

——— (1993) 'Keynes and the Role of the State in Developing Countries', in D. Crabtree and A.P. Thirlwall (eds) *Keynes and the role of the State: the tenth Keynes seminar*, Basingstoke: Macmillan.

Chase, M. (1988) *'The People's Farm': English radical agrarianism 1775–1840*, Oxford: Clarendon Press.

Charlton, D.G. (1959) *Positivist Thought in France during the Second Empire 1852–1870*, Oxford: Clarendon Press.

Chauveau Report, see Province of Canada.

Chehu, F. (1987) *Dependence, Underdevelopment and Unemployment in Kenya: school leavers in a peripheral capitalist economy*, Lanham, MD: University Press of America.

Chilcote, R.H. (1984) *Theories of Development and Underdevelopment*, Boulder: Westview Press.

Choi, Y.B. (1990) 'Smith's View on Human Nature: a problem in the interpretation of *The Wealth of Nations* and *The Theory of Moral Sentiments*', *Review of Social Economy* 48(3).

Claeys, G. (1987) *Machinery, Money and Millennium: from moral economy to socialism, 1815–60,* Cambridge: Cambridge University Press.

——— (1989) *Citizens and Saints: politics and anti-politics in early British socialism*, Cambridge: Cambridge University Press.

Clark, J.M. (1962) *Studies in the Economics of Overhead Costs*, Chicago: University of Chicago Press.

Clarke, P. (1978) *Liberals and Social Democrats*, Cambridge: Cambridge University Press.

Clayton, A. and Savage, D.C. (1974) *Government and Labour in Kenya 1895–1963*, London: Cass.

Clement, W. and Drache, D. (1978) *A Practical Guide to Canadian Political Economy*, Toronto: James Lorimer.

——— and ——— (1985) *The New Practical Guide to Canadian Political Economy*, Toronto: James Lorimer.

Clements, H. (1983) *Alfred Russel Wallace: biologist and social reformer*, London: Hutchinson.

Cline, C.A. (1974) 'E.D. Morel: from the Congo to the Rhine', in A.J.A. Morris (ed.) *Edwardian Radicalism*, London: Routledge and Kegan Paul.

Coddington, A. (1983) *Keynesian Economics: the search for first principles*, London: Allen and Unwin.

Cohen, A.A. (1964) *The Communism of Mao Tse-tung*, Chicago: Chicago University Press.

Cohen, G.A. (1978) *Karl Marx's Theory of History: a defence*, Oxford: Clarendon Press.

——— (1988) *History, Labour and Freedom: themes from Marx*, Oxford: Clarendon Press.

Cole, D. (1971) 'The Problem of "Nationalism" and "Imperialism" in British Settlement', *Journal of British Studies* 10(1).

Coleman, D. and Nixson, F. (1986) [1978] *Economics of Change in Less Developed Countries*, Oxford: Philip Allan.

Collier, P. and Lal, D. (1986) *Labour and Poverty in Kenya: 1900–1980*, Oxford: Clarendon Press.

Collini, S. (1975) 'Idealism and "Cambridge Idealism"', *Historical Journal* 18(1).

——— (1976) 'Hobhouse, Bosanquet and the State: philosophical idealism and political argument in England 1880–1918', *Past and Present* 72(110).

——— (1979) *Liberalism and Sociology: L.T. Hobhouse and political argument in England 1880–1914*, Cambridge: Cambridge University Press.

Comte, A. (1875) *System of Positive Polity*, vol. 1, London: Longmans, Green.

——— (1876a) *System of Positive Polity*, vol. 2, London: Longmans, Green.

——— (1876b) *System of Positive Polity*, vol. 3, London: Longmans, Green.

Cookey, S.J.S. (1968) *Britain and the Congo Question: 1885–1913*, New York: Humanities Press.

Cooper, J.I. (1937) 'Some Early French-Canadian Advocacy of Protection: 1871–1873', *Canadian Journal of Economics and Political Science* 3(4).

Cotter, R. (1967) 'The Golden Decade', in J. Griffin (ed.) *Essays in Economic History of Australia*, Brisbane: Jacaranda Press.

Cotton, J. (1991) *James Harrington's Political Thought and its Context*, New York: Garland.

Coughlin, P. and Ikiara, G.K. (eds) (1988) *Industrialization in Kenya: in search of a strategy*, Nairobi: Heinemann.

Coulson, J. (1990) 'Was Newman a modernist?' in A.J. Jenkins (ed.) *John Henry Newman and Modernism*, Sigmaringendorf: Verlag Gluck und Lutz.

——— and Allchin, A.M. (eds) (1967) *The Rediscovery of Newman: an Oxford Symposium*, London: SPCK.

Courville, S. and Seguin, N. (1989) *Rural Life in Nineteenth-century Quebec*, Ottawa: Canadian Historical Association Booklet 47.

Cowen, M.P. (1979) 'Capital and Household Production: the case of wattle in Kenya's Central Province 1903–64', unpublished PhD thesis, University of Cambridge.

——— (1981) 'The Agrarian Problem: notes on the Nairobi Discussion', *Review of African Political Economy* 20.

——— (1983) 'The Commercialisation of Food Production in Kenya after 1945', in R. Rotberg (ed.) *Imperialism, Colonialism and Hunger: east and central Africa*, Lexington: D.C. Heath.

——— (1986) 'Change in State Power, International Conditions and Peasant Producers: the case of Kenya', *Journal of Development Studies* 22(2).

——— and Newman, J. (1975) 'Real Wages in Central Kenya, 1924–71', mimeo.

——— and Kinyanjui, K. (1977) *Some Problems of Capital and Class in Kenya*, University of Nairobi: Institute of Development Studies, Occasional Paper 26.

——— and Shenton, R.W. (1991a) 'The Origin and Course of Fabian Colonialism in Africa', *Journal of Historical Sociology* 4(2).

——— and ——— (1991b) 'Bankers, Peasants and Land in British West Africa', *Journal of Peasant Studies* 19(1).

——— and ——— (1992) 'Development and

Agrarian Bias Part 3: land nationalisation', Working Paper No. 21, Department of Economics, London Guildhall University.

——— and ——— (1994) 'British Neo-Hegelian Idealism and Official Colonial Practice in Africa: the Oluwa land case of 1921', *Journal of Imperial and Commonwealth History* 22(2).

——— and ——— (1995) 'The Invention of Development', in J. Crush (ed.) *The Power of Development*, London: Routledge.

Cramer, D.L. and Leathers, C.G. (1981) 'Schumpeter's Corporatist Views: links among his social theory, Quadragesimo Anno, and moral reform', *History of Political Economy* 13(4).

Craven, P. and Traves, T. (1987) 'Canadian Railways as Manufacturers, 1850–1880', in D. McCalla (ed.) *Perspectives on Canadian Economic History*, Toronto: Copp Clark Pitman.

Crawford, W.R. (1963) *A Century of Latin American Thought*, Cambridge, MA: Harvard University Press.

Creighton, D.G. (1967) 'Economic Nationalism and Confederation', in R. Cook, C. Brown and C. Berger (eds) *Confederation*, Toronto: University of Toronto Press.

Crowley, D. (1981) 'Harold Innis and the Modern Perspective of Communications', in W.H. Melody, L. Salter and P. Heyer (eds) *Culture, Communication, and Dependency: the tradition of H.A. Innis*, Norwood, NJ: Ablex Publishing.

Crowley, F.K. (ed.) (1974) *A New History of Australia*, Melbourne: Heinemann.

Cubeddu, R. (1993) *The Philosophy of the Austrian School*, London: Routledge.

Dalgleish Report, see Kenya Colony (1960).

Dass, S. (1962) 'Professor J.A. Schumpeter on Economic Development', *Indian Journal of Economics* 42(3).

Davidson, A. (1991) *The Invisible State: the formation of the Australian state 1788–1901*, Cambridge: Cambridge University Press.

Davis, H. (1972) *Latin American Thought: a historical interpretation*, Baton Rouge: Louisiana State University Press.

Davis, J.B. (1989) 'Keynes on Atomism and Organicism', *Economic Journal* 99(4).

Davison, J. (1989) *Voices from Mutira: lives of rural Gikuyu women*, Boulder: Lynne Rienner.

Delp, P. (1981) 'District Planning in Kenya', in T. Killick (ed.) *Papers on the Kenyan Economy: performance, problems and policies*, Nairobi: Heinemann.

Den Otter, A.A. (1982) 'Alexander Galt, the 1849 Tariff, and Canadian Economic Nationalism', *Canadian Historical Review* 63(2).

Dente, L.A. (1977) *Veblen's Theory of Social Change*, New York: Arno Press.

Dessain, C.S. (1980) [1966] *John Henry Newman*, Oxford: Oxford University Press.

Dessaulles, L.-A. (1851) *Six lectures sur l'annexion du Canada aux Etats-Unis*, Montreal: Gendron.

Dever, A.R. (1976) 'Economic Development and the Lower Canadian Assembly, 1828–1840', unpublished MA thesis, McGill University.

De Vivo, G. (1987) 'Marx, Jevons and Early Fabian Socialism', *Political Economy, Studies in the Surplus Approach* 3(1).

Dickey, L. (1987) *Hegel: religion, economics and the politics of the Spirit 1770–1807*, Cambridge: Cambridge University Press.

Dickinson, H.T. (1977) *Liberty and Property: political ideology in eighteenth century Britain*, New York: Holmes and Meier.

Dilke, C. (1868) *Greater Britain*, vol. 2, London: Macmillan.

Dinklage, K. (1973) 'Die landwirtschaftliche Entwicklung', in A. Brusatti, *Die Habsburgermonarchie 1848–1918 Band 1: Die wirtschaftliche Entwicklung*, Vienna: Verlag der Österreichischen Akademie der Wissenschaften.

Dintenfass, M. (1992) *The Decline of Industrial*

Britain, London: Routledge.

Doughty, A.G. (1937) *The Elgin–Grey Papers 1846–1852*, vol. 1, Ottawa: Government Printer.

Drache, D. (1982) 'Harold Innis and Canadian Economic Development', *Canadian Journal of Political and Social Theory* 6(1/2).

Drapeau, S. (1863) *Etudes sur les développements de la colonisation du Bas-Canada depuis dix ans: 1851–1861*, Quebec: Leger Brousseau.

Dumont, F. (1973) 'Idéologie et conscience historique dans la société Canadienne-Française du XIX siècle', in J.-P. Bernard (ed.) *Les Idéologies Québecoises au 19e Siècle*, Montreal: Boreal Express.

Durbin, E. (1984) 'Fabian Socialism and Economic Science', in B. Pimlott (ed.) *Fabian Essays in Socialist Thought*, London: Heinemann.

Durham, J.G.L., Earl of, (1912) [1839] *Lord Durham's Report of the Affairs of British North America*, Volume 2, edited by C.P. Lucas, Oxford: Clarendon.

Dutt, R. (1970) [1904] *The Economic History of India*, New York: Burt Franklin.

Dyer, A.W. (1988) 'Schumpeter as an Economic Radical: an economic sociology reassessed', *History of Political Economy* 20(1).

Easterbrook, W.T. and Aitken, H.G.J. (1956) *Canadian Economic History*, Toronto: Macmillan.

Eddie, S.M. (1989) 'Economic Policy and Economic Development in Austria-Hungary, 1867–1913', in P. Mathias and S. Pollard (eds) *Cambridge Economic History of Europe Volume 8: The industrial economies: the development of economic and social policies*, Cambridge: Cambridge University Press.

Eddy, J. and Schreuder, D. (1988) *The Rise of Colonial Nationalism: Australia, New Zealand, Canada and South Africa first assert their nationalities, 1880–1914*, Sydney: Allen and Unwin.

Edgell, S. and Tilman, R. (1989) 'The Intellectual Antecedents of Thorstein Veblen: a reappraisal', *Journal of Economic Issues* 23(4).

Eff, E.A. (1989) 'History of Thought as Ceremonial Genealogy: the neglected influence of Herbert Spencer on Thorstein Veblen', *Journal of Economic Issues* 23(3).

Elbaum, B. (1986) 'The Steel Industry before World War I', in B. Elbaum and W. Lazonick, (eds) *The Decline of the British Economy*, Oxford: Clarendon Press.

Elkan, W. (1970) 'Urban Unemployment in East Africa', *International Affairs* 46(3).

Elster, J. (1985) *Making Sense of Marx*, Cambridge: Cambridge University Press.

Emy, H.V. (1973) *Liberals, Radicals and Social Politics 1892–1914*, Cambridge: Cambridge University Press.

Encyclopaedia Britannica (1910) (11th edn), Cambridge: Cambridge University Press.

Engels, F. 1975 [1845] 'Speeches in Elberfield', in K. Marx and F. Engels, *Collected Works*, vol. 4, Moscow: Progress Publishers.

Esteva, G. (1992) 'Development', in W. Sachs (ed.) *The Development Dictionary: a guide to knowledge as power*, London: Zed.

Evans, H.E. (1989) 'National Development and Rural–Urban Policy: past experience and new directions in Kenya', *Urban Studies* 26(2).

——— and Ngau, P. (1991) 'Rural–Urban, Household Income Diversification and Agricultural Productivity', *Development and Change* 22(3).

Evans-Pritchard, E.E. (1970) *The Sociology of Comte: an appreciation*, Manchester: Manchester University Press.

——— (1981) *A History of Anthropological Thought*, London: Faber and Faber.

Fairbairn, A.M. (1893) *The Place of Christ in Modern Theology*, London: Hodder and Stoughton.

Fairlie, S. (1965) 'The Nineteenth-century Corn Law Reconsidered', *Economic History Review* 18(3).

——— (1969) 'The Corn Laws and British Wheat Production, 1829–76', *Economic History Review* 22(1).

Falardeau, J.-C. (ed.) (1975) *Etienne Parent*, Montreal: La Presse Ltee.

Faucher, A. (1973) *Québec en Amérique au XIXe Siècle: Essai sur les caractères économiques de la Laurante*, Montreal: Fides.

Fawkes, A. (1913) *Studies in Modernism*, London: Smith, Elder.

Fecteau, J.-M. (1992) 'Etat et associationnisme au XIXe siècle québecois: éléments pour une problématique des rapports état/sociéte dans la transition au capitalisme', in A. Greer and I. Radforth, *Colonial Leviathan: state formation in mid-nineteenth-century Canada*, Toronto: University of Toronto Press.

Fieldhouse, D.K. (1986) *Black Africa 1945–80: economic decolonization and arrested development*, London: Allen and Unwin.

Figgis, J.N. (1896) *The Theory of the Divine Right of Kings*, Cambridge: Cambridge University Press.

——— (1913) *Churches in the Modern State*, London: Longmans, Green.

——— (1914) 'John Henry Newman', Appendix A, *The Fellowship of the Mystery*, London: Longmans, Green.

——— (1917) *The Will to Freedom: or the Gospel of Nietzsche and the Gospel of Christ*, London: Longmans.

Findlay, R. (1985) 'Primary Exports, Manufacturing and Development', in M. Lundhal (ed.) *The Primary Sector in Economic Development*, London: Croom Helm.

——— and Lundahl, M. (1992) 'Natural Resources, "Vent for Surplus" and the Staple Theory', Columbia University Department of Economics Discussion Paper Series 585.

Fingard, J. (1986) 'The Winter's Tale: the seasonal contours of pre-industrial poverty in British North America, 1815–1860', in J.M. Bumstead (ed.) *Interpreting Canada's Past: Volume 1, Before confederation*, Toronto: Oxford University Press.

Fitzgibbons, A. (1988) *Keynes's Vision: a new political economy*, Oxford: Clarendon Press.

Fitzpatrick, B. (1946) *The Australian People, 1788–1945*, Carlton, Victoria: Melbourne University Press.

Fletcher, T.W. (1960–61) 'The Great Depression of English Agriculture, 1873–1896', *Economic History Review* 13(3).

Fogarty, J.P. (1966) 'The Staple Approach and the Role of Government in Australian Economic Development: the wheat industry', *Business Archives and History* 6(1).

Ford, L.S. (1984) 'The Concept of "Process": from "transition" to "concrescence"', in H. Holz and E. Wolf-Gazo (eds) *Whitehead and the idea of Process*, Freiburg: Verlag Karl Alber.

Forster, B. (1986) *A Conjunction of Interests: business, politics, and tariffs 1825–1879*, Toronto: University of Toronto Press.

Fowke, V.C. (1947) *Canadian Agricultural Policy: the historical pattern*, Toronto: University of Toronto Press.

Francis, E.M. (1991) 'Migration and Social Change in Koguta, Western Kenya', unpublished PhD thesis, University of Oxford.

——— and Hoddinott, J. (1993) 'Migration and Differentiation in Western Kenya: a tale of two sub-locations', *Journal of Development Studies* 30(1).

Francis, M. (1992) *Governors and Settlers: images of authority in the British colonies, 1820–60*, Basingstoke: Macmillan.

Frank, A.G. (1967) *Capitalism and Underdevelopment in Latin America: historical studies of Chile and Brazil*, New York: Monthly Review Press.

——— (1969) 'Sociology of Development and Underdevelopment of Sociology', in *Latin America: Underdevelopment or Revolution?*, New York: Monthly Review Press.

——— (1971) [1967] *Capitalism and Under-*

development in Latin America, Harmondsworth: Penguin.

——— (1972a) 'Introduction', in J.D. Cockcroft, A.G. Frank and D.L. Johnson (eds) *Dependence and Underdevelopment: Latin America's Political Economy*, New York: Doubleday.

——— (1972b) 'The Development of Underdevelopment', in J.D. Cockcroft, A.G. Frank and D.L. Johnson (eds) *Dependence and Underdevelopment: Latin America's Political Economy*, New York: Doubleday.

——— (1972c) 'Economic Dependence, Class Structure and Underdevelopment Policy', in J.D. Cockcroft, A.G. Frank and D.L. Johnson (eds) *Dependence and Underdevelopment: Latin America's Political Economy*, New York: Doubleday.

——— (1972d) *Lumpenbourgeoisie: lumpendevelopment – dependence, class and politics in Latin America*, New York: Monthly Review Press.

——— (1978) *Dependent Accumulation and Underdevelopment*, London: Macmillan.

——— (1983a) 'Introduction', in R.H. Chilcote and D.L Johnson (eds) *Theories of Development: mode of production or dependency*, Beverly Hills: Sage.

——— (1983b) 'Crisis and Transformation of Dependency in the World System', in R.H. Chilcote and D.L Johnson (eds) *Theories of Development: mode of production or dependency*, Beverly Hills: Sage.

Freeman, C. (1990) 'Schumpeter's Business Cycles Revisited', in A. Heertje and M. Perlman (eds) *Evolving Technology and Market Structure: studies in Schumpeterian economics*, Ann Arbor: University of Michigan Press.

———, Clark, J. and Soete, L. (1982) *Unemployment and Technical Innovation: a study in long waves and economic development*, London: Francis Pinter.

Freeman, M. (1980) *Edmund Burke and the Critique of Political Radicalism*, Oxford: Blackwell.

Friedland, W.H. and Rosberg, C.G. (1964) *African Socialism*, Stanford: Stanford University Press.

Friedrich, C.J. (ed.) (1949) *The Philosophy of Kant: Kant's moral and political writings*, New York: Modern Library.

Fukuyama, F. (1992) *The End of History and the Last Man*, New York: Free Press.

Furtado, C. (1963) [1949] *The Economic Growth of Brazil*, Berkeley: University of California Press.

——— (1964) [1961] *Development and Underdevelopment*, Berkeley: University of California Press.

Gadzey, Tuo Kofi A. (1992) 'The State and Capitalist Transformation in Sub-Saharan Africa: a development model', *Comparative Political Studies* 24(2).

Gagnon, S. (1985) *Quebec and its Historians: the twentieth century*, Montreal: Harvest House.

Gallagher, J. and Robinson, R. (1953) 'The Imperialism of Free Trade', *Economic History Review* 6(1).

Gammage, R.G. (1969) [1854] *History of the Chartist Movement 1837–1854*, London: Cass.

Ganguli, B.N. (1979) *Some Aspects of Classical Political Economy in the Nineteenth Century Indian Perspective*, Calcutta: Orient Longman.

Garvin, J.L. (1905) 'The Principles of Constructive Economics as applied to the maintenance of the Empire', *Compatriots' Club Lectures*, 1st series, London: Macmillan.

——— (1932) *The Life of Joseph Chamberlain, vol. 1: 1836–1885*, London: Macmillan and Co.

Gatama, W.M. (1986) *Business Education for Primary Schools: standard six: pupils book*, Nairobi: Transafrica.

——— (1947) [1891] *The Condition of Labour*, London: Land and Liberty Press.

George, H. (1981) [1879] *Progress and Poverty*, London: Land and Freedom Press.

Gerrard, B. (1992) 'From *A Treatise on Probability*

to *The General Theory*: continuity or change in Keynes's thought?', in B. Gerrard and J.Hillard (eds) *The Philosophy and Economics of J.M. Keynes*, Aldershot: Edward Elgar.

Gerschenkron, A. (1965) [1962] *Economic Backwardness in Historical Perspective: a book of essays*, New York: Praeger.

——— (1977) *An Economic Spurt That Failed: four lectures in Austrian history*, Princeton: Princeton University Press.

Gershenberg, I. (1986) 'Labour, Capital and Management Slack in Multinational and Local Firms in Kenyan Manufacturing', *Economic Development and Cultural Change* 35(1).

Giddens, A. (1979) 'Positivism and its Critics', in T. Bottomore and R. Nisbet (eds) *A History of Sociological Analysis*, London: Heinemann.

Gilley, S. (1990) *Newman and his Age*, London: Darton, Longman and Todd.

Gilroy, P. (1993) *The Black Atlantic: modernity and double consciousness*, London: Verso.

Ginsburg, M. (1953) *The Idea of Progress: a revaluation*, London: Methuen.

Glade, W. (1969) *The Latin American Economies*, New York: American Book.

Godfrey, E.M. and Mutiso, G.C.M. (1979) *Politics, Economics and Technical Training: a Kenyan case study*, Nairobi: KLB.

Goldsworthy, D. (1982) *Tom Mboya: the man Kenya wanted to forget*, Nairobi: Heinemann.

Gollan, R. (1955) 'Nationalism and Politics in Australia before 1855', *Australian Journal of Politics and History* 1(1).

——— (1960) *Radical and Working Class Politics*, Melbourne: Melbourne University Press.

Goodwin, C.D.W. (1961) *Canadian Economic Thought: the political economy of a developing nation 1814–1914*, London: Cambridge University Press.

——— (1966) *Economic Enquiry in Australia*, Durham, NC: Duke University Press.

——— (1974) *The Image of Australia: British perception of the Australian economy from the eighteenth to the twentieth century*, Durham, NC: Duke University Press.

Goodwin, R.M. (1990) 'Walras and Schumpeter: the vision reaffirmed', in A. Heertje and M. Perlman, *Evolving Technology and Market Structure: studies in Schumpeterian economics*, Ann Arbor: University of Michigan Press.

Gould, W.T.S. (1989) 'Technical education and migration in Tiriki, Western Kenya, 1902–1987', *African Affairs* 88(351).

Gouldner, A.W. (1971) *The Coming Crisis of Western Sociology*, London: Heinemann.

——— (1980) *The Two Marxisms: contradictions and anomalies in the development of theory*, London: Macmillan.

Goulet, D. (1977) [1971] *The Cruel Choice: a new concept in the theory of development*, New York: Athenaeum.

Grassl, W. and Smith, B. (eds) (1986) *Austrian Economics: historical and philosophical background*, London: Croom Helm.

Grave, S.A. (1989) *Conscience in Newman's Thought*, Oxford: Clarendon Press.

Gray, H.M. (1981) 'Reflections on Innis and Institutional Economics', in W.H. Melody, L. Salter and P. Heyer (eds) *Culture, Communication, and Dependency: the tradition of H.A. Innis*, Norwood, NJ: Ablex Publishing.

Green, R.H. (1989) 'Degradation of Rural Development: development of rural degradation – change and peasants in sub-Saharan Africa', Institute of Development Studies, University of Sussex, Discussion Paper 265.

Green, T.H. (1929) [1907] *Prolegomena to Ethics*, Oxford: Oxford University Press.

——— (1941) *Lectures on the Principles of Political Obligation*, London: Longmans, Green and Co.

Greengarten, I.M. (1981) *Thomas Hill Green and the Development of Liberal-Democratic Thought*, Toronto: University of Toronto Press.

Greer, A. (1985) *Peasant, Lord and Merchant: rural society in three Quebec parishes 1740–1840*,

Toronto: University of Toronto Press.

——— and Radforth, I. (1992) *Colonial Leviathan: state formation in mid-nineteenth-century Canada*, Toronto: University of Toronto Press.

Groenewegen, P. and McFarlane, B. (1990) *A History of Australian Economic Thought*, London: Routledge.

Grosh, B. (1988) 'Comparing Parastatal and Private Manufacturing Firms; would privatization improve performance?', in P. Coughlin and G.K. Ikiara (eds) *Industrialization in Kenya: in search of a strategy*, Nairobi: Heinemann.

Gross, N. (1983) 'Austria-Hungary in the world economy', in J. Komlos (ed.) *Economic Development in the Habsburg Monarchy in the Nineteenth Century: essays*, Boulder: East European Monographs.

Gyllstrom, B. (1991) *State-administered Rural Change: agricultural cooperatives in Kenya*, London: Routledge.

Haines, V.A. (1987) 'Biology and Social Theory: Parson's evolutionary theme', *Sociology* 2(1).

Haldane, R.B. (1883) 'The Relation of Philosophy to Science', in A. Seth and R.B. Haldane (eds) *Essays in Philosophical Criticism*, London: Longmans, Green.

Hale, C. (1986) 'Political and Social Ideas in Latin America, 1870–1930', in *The Cambridge History of Latin America*, Volume 8, edited by L. Bethell, Cambridge: Cambridge University Press.

——— (1989) *The Transformation of Liberalism in Late Nineteenth-century Mexico*, Princeton: Princeton Unversity Press.

Halpenny, F.G. and Hamelin, J. (eds) (1982) *Dictionnaire Bibliographique du Canada, Volume XI: 1881–1890*, Quebec: Les Presses de l'Université Laval.

Hamelin, J. and Roby, Y. (1971) *Histoire Economique du Québec 1851–1896*, Montreal: Fides.

Hamilton, A. (1966) 'Final Version of the Report on the Subject of Manufactures', in H.C. Syrett and J.E. Cooke (eds) *The Papers of Alexander Hamilton: Volume 1*, New York: Columbia University Press.

Harrington, J., (1992) [1656] *The Commonwealth of Oceana and a System of Politics*, Cambridge: Cambridge University Press.

Harris, G. (1989) *The Sociology of Development*, London: Longman.

Harris, J. (1972) *Unemployment and Politics: a study in English social policy 1886–1914*, Oxford: Clarendon Press.

Harrod, R.F. (1972) [1951] *The Life of John Maynard Keynes*, Harmondsworth: Pelican.

Hawthorn, G. (1987) *Enlightenment and Despair: a history of social theory*, 2nd edn, Cambridge: Cambridge University Press.

Hayek, F.A. (1944) *The Road to Serfdom*, Chicago: University of Chicago Press.

Hebnick, P.G.M. (1990) *Agrarian Structure in Kenya: state, farmers and commodity relations*, Nijmegen Studies in Development and Cultural Change, 5, Saarbrücken: Verlag.

Hegel, G.W.F. (1952) *Philosophy of Right*, ed. and trans. by T.M. Knox, Oxford: Clarendon Press.

——— (1971) *Philosophy of Mind*, ed. and trans. by A.V. Miller, Oxford: Clarendon Press.

——— (1975) [1837] *Lectures on the Philosophy of World History: Introduction: reason in history*, Cambridge: Cambridge University Press.

——— (1977) [1807] *Phenomenology of Spirit*, London: Oxford University Press.

——— (1986) [1840] *The Philosophical Propaedeutic*, with a Preface and Introduction by M. George and A. Vincent, Oxford: Blackwell.

Heilbroner, R.L. (1988) *Behind the Veil of Economics: essays in the worldly philosophy*, New York: Norton.

Helburn, S. (1986) 'Schumpeter's Research Programme', in S. Helburn and D.F. Bramhall (eds) *Marx, Keynes, Schumpeter: a centenary celebration of dissent*, Armonk, NY: M.E. Sharpe.

——— (1991) 'Burke and Keynes', in B.W. Bateman and J.B. Davis (eds) *Keynes and Philosophy: essays on the origins of Keynes's thought*, Aldershot: Elgar.

Helczmanovski, H. (1973) 'Die Entwicklung der Bevölkerung Österreichs in den letzen hundert Jahren nach den wichtigsten demographischen Komponenten', in H. Helczmanovski (ed.) *Beitrage zur Bevölkerungs und Sozialgeschichte Österreichs*, Vienna: Verlag für Geschichte und Politik.

Henderson, W.O. (1983) *Friedrich List: Economist and Visionary 1789–1846*, London: Cass.

Hendrickson, J.E. (1981) 'The Constitutional Development of Colonial Vancouver Island and British Columbia', in W.P. Ward and R.A.J. McDonald (eds) *British Columbia: historical readings*, Vancouver: Douglas and McIntyre.

Hettne, B. (1990) *Development Theory and the Three Worlds*, Harlow: Longman.

Heyck, T.W. (1974) *The Dimensions of British Radicalism: the case of Ireland 1874–95*, Urbana, Illinois: University of Illinois Press.

Heyer, J., Ireri, D. and Moris, J. (1971) *Rural Development in Kenya*, Nairobi: East African Publishing House.

Heyer, P. (1981) 'Innis and the History of Communication: antecedents, parallels, and unsuspected biases', in W.H. Melody, L. Salter and P. Heyer (eds) *Culture, Communication, and Dependency: the tradition of H.A. Innis*, Norwood, NJ: Ablex Publishing.

Hicks, J.R. (1983) 'From Classical to Post-Classical: the work of J.S. Mill', in J. Hicks, *Classics and Moderns: collected essays on economic theory*, vol. 3, Oxford: Basil Blackwell.

——— and Weber, W. (eds) (1973) *Carl Menger and the Austrian School of Economics*, Oxford: Clarendon Press.

Hicks, U. (1961) *Development from Below: local government finance in developing countries of the Commonwealth*, Oxford: Clarendon Press.

Hilferding, R. (1985) [1910] *Finance Capital: a study of the latest phase of capitalist development*, London: Routledge and Kegan Paul.

Hill, M.J.D. (1991) *The Harambee Movement in Kenya: self-help, development and education among the Kamba of Kitui District*, London: Athlone Press.

Himbara, D. (1992) 'The Role of Indigenous Entrepreneurs and the State in Kenya Industrial Development', unpublished PhD thesis, Queen's University, Ontario.

——— (1994) *Kenyan Capitalists, the State and Development*, Boulder, CO: Lynne Rienner.

Hincks, F. (1849) *Canada: its financial position and resources*, London: James Ridgeway.

——— (1884) *Reminiscences of his Public Life*, Montreal: William Drysdale.

Hinga, S.N. and Heyer, J. (1971) 'The Development of the Large Farms', in J. Heyer, D. Ireri and J. Moris, *Rural Development in Kenya*, Nairobi: East African Publishing House.

Hirschman, A.O. (1958) *The Strategy of Economic Development*, New Haven: Yale University Press.

——— (1977) *The Passions and the Interests: political arguments for capitalism before its triumph*, Princeton, NJ: Princeton University Press.

Hirst, P.Q. (1989) 'Introduction', in P.Q. Hirst (ed.) *The Pluralist Theory of the State: Selected Writings of G.D.H. Cole, J.N. Figgis, and H.J. Laski*, London: Routledge.

Hobhouse, L.T. (1901) *Mind in Evolution*, London: Macmillan and Co.

——— (1927) [1913] *Development and Purpose: an essay towards a philosophy of evolution*, London: Macmillan.

——— (1929) [1906] *Morals in Evolution*, London: Chapman and Hall.

Hobsbawm, E.J. (1968) *Industry and Empire*, London: Wiedenfeld and Nicolson.

——— (1988) *The Age of Revolution: Europe*

1789–1848, London: Cardinal.

——— (1990) *Nations and Nationalism since 1780: programme, myth, reality*, Cambridge: Cambridge University Press.

Hobson, J.A. (1910) 'The General Election: a sociological interpretation', *Sociological Review* 3(2).

Hoddinott, J. (1989) 'Migration, Accumulation and Old Age Security in Western Kenya', unpublished PhD thesis, University of Oxford.

——— (1994) 'A Model of Migration and Remittances Applied to Western Kenya', *Oxford Economic Papers* 46(2).

Hollander, J.B. (ed.) (1931) *Letters of James Ramsay McCulloch to David Ricardo: 1818–1823*, Baltimore: Johns Hopkins University Press.

Hollis, M. (1991) 'Comment', in R.M. O'Donnell (ed.) *Keynes as Philosopher-Economist: the ninth Keynes seminar, 1989*, Basingstoke: Macmillan.

Holmes, J.D. (1967) 'A Note on Newman's Historical Method', in J. Coulson and A.M. Allchin (eds) *The Rediscovery of Newman: an Oxford Symposium*, London: Sheed and Ward.

Holmquist, F.W. (1975) *Peasant Organisation, Clientelism and Dependency: a case study of an agricultural producers' cooperative in Kenya*, Bloomington: Indiana University Press.

——— (1984) 'Self-Help: the state and peasant leverage in Kenya', *Africa* 54(3).

Hont, I. (1983) 'The "Rich Country–Poor Country" Debate in Scottish Classical Political Economy', in Istvan Hont and Michael Ignatieff (eds) *Wealth and Virtue: the shaping of political economy in the Scottish Enlightenment*, Cambridge: Cambridge University Press.

Hooker, J.R. (1967) *Black Revolutionary: George Padmore's path from communism to pan-africanism*, London: Pall Mall Press.

Hooper, A. (1986) 'Utilitarianism and Radicalism', mimeo.

Hopkins, A.G. (1973) *An Economic History of West Africa*, London: Longman.

House, W.J. (1981) 'Nairobi's Informal Sector: an exploratory study', in T. Killick (ed.) *Papers on the Kenyan Economy: performance, problems and policies*, Nairobi: Heinemann.

Hunt, D. (1984) *The Impending Crisis in Kenya: the case for land reform*, Aldershot: Gower.

Hutchful, E. (1991) 'Eastern Europe: consequences for Africa', *Review of African Political Economy* 50.

Hutchinson, T.W. (1973) 'Some Themes from Investigations into Method', in J.R. Hicks and W. Weber (eds) *Carl Menger and the Austrian School of Economics*, Oxford: Clarendon Press.

Hylton, P. (1984) 'The Nature of the Proposition and the Revolt against Idealism', in R. Rorty, J.B. Schneewind and Q. Skinner (eds) *Philosophy in History: essays in the historiography of philosophy*, Cambridge: Cambridge University Press.

Iggers, G. (trans. and ed.) (1958) [1829] *The Doctrine of Saint-Simon: An Exposition, First year 1828–1829*, Boston: Beacon Books.

Ikiara, G.K., Jama, M.A. and Amadi, J.O. (1993) 'Agricultural Decline, Politics and Structural Adjustment in Kenya', in P. Gibbon (ed.) *Social Change and Economic Reform in Africa*, Uppsala: Scandinavian Institute of African Studies.

ILO (1950) *Action against Unemployment*, Studies and Reports, ns 20, Geneva: ILO.

——— (1972) *Employment, Incomes and Equality: a strategy for increasing productive employment in Kenya*, Geneva: ILO.

Ilting, K.-H. (1984) 'The Dialectic of Civil Society', in Z.A. Pelczyinski (ed.) *The State and Civil Society: studies in Hegel's political philosophy*, Cambridge: Cambridge University Press.

Ingham, B. (1995) *Economics and Development*, London: McGraw-Hill.

Innis, H.A. (1927–34) 'Economic Destiny of

Canada', unpublished ms.

——— (1929) 'A Bibliography of Thorstein Veblen', *Southwestern Political and Social Science Quarterly* 10(1).

——— (1937) 'Significant Factors in Canadian Economic Development', *Canadian Historical Review*, 18(4), republished in *H.A. Innis, Essays in Canadian Economic History*, Toronto: University of Toronto Press, 1956.

——— (1952) *Changing Concepts of Time*, Toronto: University of Toronto Press.

——— (1956) *Essays in Canadian Economic History*, Toronto: University of Toronto Press.

——— (1962) [1930] *The Fur Trade in Canada: an introduction to Canadian economic history*, Toronto: University of Toronto Press.

——— (1964) *The Bias of Communication*, Toronto: University of Toronto Press.

——— (1972) 'Snarkov Island' in R. Neill, *New Theory of Value: The Canadian Economics of H.A. Innis*, Toronto: University of Toronto Press.

——— (1980) *The Idea File of Harold Adams Innis*, edited by W. Christian, Toronto: University of Toronto Press.

Irving, T.H. (1974) '1850–70', in F.K. Crowley (ed.) *A New History of Australia*, Melbourne: Heinemann.

Janeway, W.H. (1986) 'Doing Capitalism: notes on the practice of venture capitalism', *Journal of Economic Issues* 20(2).

Jebb, R. (1905) *Studies in Colonial Nationalism*, London: Edward Arnold.

Jensen, K.M. (1978) Beyond *Marx and Mach: Aleksandr Bogdanov's philosophy of living experience*, Dordrecht: Holland.

John, M. and Lichtblau, A. (1993) *Schmelztiegel Wien einst und jetzt: zur Geschichte und Gegenwart von Zuwanderung und Minderheiten*, Vienna: Bohlau Verlag.

Johnson, C. (1982) *MITI and the Japanese Miracle: the growth of industrial policy, 1925–1975*, Stanford: Stanford University Press.

Johnson, R.G. (1990) 'Adam Smith's Radical Views on Property, Distributive Justice and the Market', *Review of Social Economy* 48(3).

Johnson, T. (1987) 'In a Manner of Speaking: towards a reconstitution of property in mid-nineteenth century Quebec', *McGill Law Journal* 32(3).

Johnston, W.M. (1972) *The Austrian Mind: an intellectual and social history 1848–1938*, Berkeley: University of California Press.

Jones, G. (1980) *Social Darwinism and English Thought: the interaction between biological and social theory*, Brighton: Harvester.

Jones, P. d'A. (1968) *The Christian Socialist Revival: religion, class and social conscience in late-Victorian England*, Princeton: Princeton University Press.

Judd, D. (1977) *Radical Joe: a life of Joseph Chamberlain*, London: Hamilton.

Kain, P.J. (1982) *Schiller, Marx and Hegel: state, society and the aesthetic ideal of ancient Greece*, Kingston, Ontario: McGill-Queen's University Press.

Kanogo, T. (1987) *Squatters and the Roots of Mau Mau 1905–63*, London: James Currey.

Karanja, E. (1974) 'The Development of the Cooperative Movement in Kenya', unpublished PhD dissertation, University of Pittsburgh.

Kay, G. (1975) *Development and Underdevelopment: a Marxist analysis*, London: Macmillan.

——— and Mott, J. (1982) *Political Order and the Law of Labour*, London: Macmillan.

Keat, R. (1981) *The Politics of Social Theory: Habermas, Freud and the critique of positivism*, Oxford: Basil Blackwell.

Kendrick, H.R. (1967) 'Amor De Cosmos and Confederation', in W.G. Shelton (ed.) *British Columbia and Confederation*, Victoria, BC: Morriss.

Kenny, A. (1990) 'Newman as a Philosopher of Religion', in D. Brown (ed.) *Newman: A Man for our Time*, London: SPCK.

Kenny, T. (1957) *The Political Thought of John Henry Newman*, London: Longman.

Kenya Colony (1960) *Sessional Paper No. 10 of 1959/60: Unemployment* (Dalgleish Report), Nairobi: Government Printer.

Kenya Republic (1965) *Sessional Paper No. 10 of 1963/65: African socialism and its application to planning in Kenya*, Nairobi: Government Printer.

——— (1967–1991) *Statistical Abstract*, Nairobi: Central Bureau of Statistics.

——— (1970) *National Assembly: report of the select committee on unemployment* (Mwicigi Report), Nairobi: Government Printer.

——— (1973) *Sessional Paper No. 10 of 1973: Employment*, Nairobi: Government Printer.

——— (1978) *Lands, Settlement and Adjudication: policies and programmes for the fourth five-year plan 1979–83*, Nairobi: Ministry of Agriculture (mimeo).

——— (1982) *The Integrated Rural Surveys 1976–9: basic report*, Nairobi: Central Bureau of Statistics.

——— (1983) *Report of the Presidential Committee on Unemployment 1982/3* (Wanjigi Report), Nairobi: Government Printer.

——— (1989) *Survey of Rural Non-agricultural Enterprises 1985*, Nairobi: Central Bureau of Statistics.

——— (1991a) *Report of the Presidential Committee on Employment. Development and Unemployment in Kenya: A strategy for the transformation of the economy* (Ndegwa Report), Nairobi: Government Printer.

——— (1991b) *Rural Labour Force Survey 1988/9*, Nairobi: Ministry of Planning and National Development.

——— (1992) *Sessional Paper No.2 of 1992 on Small Enterprise and Jua Kali development in Kenya*, Nairobi: Government Printer.

Ker, I. (1990) 'Introduction', in I. Ker (ed.) *Newman the Theologian: a reader*, London: Collins.

——— and Hill, A.G. (eds) *Newman after a Hundred Years*, Oxford: Clarendon Press.

Keynes, J.M. (1972a) *Collected Works IX, Essays in Persuasion*, Cambridge: Cambridge University Press.

——— (1972b) [1933] 'My Early Beliefs', in *The Collected Writings of John Maynard Keynes, X: Essays in Biography*, London: Macmillan.

——— (1973a) [1921] *The Collected Writings of John Maynard Keynes. VIII: Treatise on Probability*, London: Macmillan.

——— (1973b) The *Collected Writings of John Maynard Keynes. XIV: the General Theory and after. Part II: defence and development*, London: Macmillan.

——— (1973c) [1936] *The Collected Writings of John Maynard Keynes. VII: The General Theory of Employment, Interest and Money*, London: Macmillan.

——— (1983) *Collected Works XI, Economic Articles and Correspondence: Academic*, Cambridge: Cambridge University Press.

Killick, T. (ed.) (1981) *Papers on the Kenyan Economy: performance, problems and policies*, Nairobi: Heinemann.

Kindleberger, C.P. (1984) *A Financial History of Western Europe*, London: George Allen and Unwin.

King, J.R. (1979) *Stabilisation Policy in an African Setting: Kenya 1963–1973*, London: Heinemann.

King, K. (1977) *The African Artisan: education and the informal sector*. London: Heinemann.

——— and Abuodha, C. (1991) *The Building of an Industrial Society: change and development in Kenya's informal (jua kali) sector 1972 to 1991*, Edinburgh: Centre of African Studies, University of Edinburgh.

Kitching, G.(1980) *Class and Economic Change in Kenya: the making of an African petite-bourgeoisie*, London: Yale.

——— (1989) [1982] *Development and Underdevelopment in Historical Perspective: populism, nationalism, industrialism*, London: Routledge.

Knight, N. (ed.) (1990) *Mao Ze dong on Dialectical Materialism*, Armonk, NY: M.E. Sharpe.

Kociumbas, J. (1992) *The Oxford History of Australia. Volume 2. 1770–1860: possessions*, Melbourne: Oxford University Press.

Koren, S. (1961) 'Struktur und Nutzung der Energiequellen Österreichs, in Weber, W. (ed.) *Österreichs Wirtschaftsstruktur gestern – heute – morgen*, Berlin: Duncker and Humblot.

Koss, S. (1973) *The Pro-Boers*, Chicago: University of Chicago Press.

Kubicek, R.V. (1969) *The Administration of Imperialism: Joseph Chamberlain at the Colonial Office*, Durham, NC: Duke University Press.

Kuntz, P.G. (1984) *Alfred North Whitehead*, Boston: Twayne.

Lall, S., Khanna A. and Alikhani, I. (1987) 'Determinants of Manufactured Export Performance in Low-Income Africa: Kenya and Tanzania', *World Development* 15(9).

La Nauze, J.A. (1967) 'The Gold Rushes and Australian Politics', *Australian Journal of Politics and History* 13(1).

Langdon, S. (1973) 'The Emergence of the Canadian Working-Class Movement, 1845–1867', *Journal of Canadian Studies* 8 (Reprinted in J.M. Bumstead (ed.) *Interpreting Canada's Past: Volume 1, Before confederation*, Toronto: Oxford University Press).

Larrain, J. (1989) *Theories of Development: capitalism, colonialism and dependency*, Oxford: Polity.

Lasch, C. (1965) *The New Radicalism in America [1889-1963]: the intellectual as a social type*, New York: Knopf.

Lash, N. (1975) *Newman on Development: the search for an explanation in history*, London: Sheed and Ward.

——— (1990) 'Tides and Twilight: Newman since Vatican II', in Ian Ker and Alan G. Hill (eds) *Newman After a Hundred Years*, Oxford: Clarendon Press.

Laski, H.J. (1917) *Studies in the Problem of Sovereignty*. New Haven: Yale University Press.

——— (1921) *The Foundations of Sovereignty and Other Essays*, London: Allen and Unwin.

Laszlo, P. and Pynsent, R.B. (eds) (1988) *Intellectuals and the Future in the Habsburg Monarchy 1890–1914*, Basingstoke: Macmillan.

Lavoie, Y. (1981) *L'émigration des Québecois aux Etats-Unis de 1840 à 1930*, Quebec: Documentation du Conseil de la Langue Française.

Lavrijsen, J.S.G (1984) *Rural Poverty and Impoverishment in Western Kenya*, Utrecht: Geographical Studies 33, Rijksuniversiteit, Utrecht.

Lawson, T. (1985) 'Uncertainty and Economic Analysis', *Economic Journal* 95(4).

——— (1987) 'The Relative/Absolute Nature of Knowledge and Economic Analysis', *Economic Journal* 97(4).

——— (1991) 'Keynes and the Analysis of Rational Behaviour', in R.M. O'Donnell (ed.) *Keynes as Philosopher-Economist: the ninth Keynes seminar, 1989*, Basingstoke: Macmillan.

——— (1993) 'Keynes and Conventions', *Review of Social Economy* 51(2).

Laxer, G. (1991) *Perspectives on Canadian Economic Development: class, staples, gender and elites*, Toronto: Oxford University Press.

Leclerc, I. (1984) 'Process and Order in Nature', in H. Holz and E. Wolf-Gazo (eds) *Whitehead and the Idea of Process*, Freiburg: Verlag Karl Alber.

Lee, F. (1988) *Fabianism and Colonialism: the life and political thought of Lord Sydney Olivier*, London: Defiant.

Le Goff, T.J.A. (1974) 'The Agricultural Crisis in Lower Canada, 1802–12: a review of a controversy', *Canadian Historical Review* 55(1).

Legum, C. (1965) [1962] *Pan Africanism: a short political guide*, London: Pall Mall.

Lenin, V.I. (1961a) [1915] 'On the Question of Dialectics', in *Collected Works Volume 38: philosophical notebooks*, Moscow: Foreign Languages Publishing House.

——— (1961b) 'A. Deborin. Dialectical Materialism', in *Collected Works Volume 38: philosophical notebooks*, Moscow: Foreign Languages Publishing House.

Lenzer, G. (ed.) (1983) *Auguste Comte and Positivism: the Essential Writings*, Chicago: University of Chicago Press.

Leo, C. (1984) *Land and Class in Kenya*, Toronto: University of Toronto Press.

Leonard, D.K. (1991) *African Successes: four public managers of Kenyan rural development*, Berkeley: California University Press.

Levitt, K. (1970) *Silent Surrender: the multinational corporation in Canada*, Toronto: Macmillan.

Lévy-Bruhl, L. (1903) *The Philosophy of Auguste Comte*, trans. F. Harrison, London: Swan Sonnenschein.

Lewis, B.D. and Thorbecke, E. (1992) 'District-level Economic Linkages in Kenya: evidence based on a small regional social accounting matrix', *World Development* 20(6).

Lewis, D.L. (1993) *W.E.B. Du Bois: biography of a race 1868–1919*, New York: Henry Holt.

Lewis, F. and McInnis, R.M. (1980) 'The efficiency of the French-Canadian farmer in the nineteenth century', *Journal of Economic History* 40(3).

Lewis, W.A. (1963) *The Theory of Economic Growth*, London: Allen and Unwin.

Leys, C. (1975) *Underdevelopment in Kenya: the political economy of neo-colonialism*, London: Heinemann.

——— (1978) 'Capital Accumulation, Class Formation and Dependency – the significance of the Kenya case', in R. Miliband and J. Saville (eds) *Socialist Register 1978*, London: Merlin.

Lipow, A. (1982) *Authoritarian Socialism in America: Edward Bellamy and the Nationalist Movement*, Berkeley: University of California Press.

Lippman, W. 1961 [1914] *Drift and Mastery*, reprt. Engelwood Cliffs, NJ: Prentice Hall.

Lipton, M. (1986) 'Farmer's Science', *Times Literary Supplement*, 4 July.

——— (1993) 'Discussion', in D. Crabtree and A.P. Thirlwall (eds) *Keynes and the Role of the State: the tenth Keynes seminar*, Basingstoke: Macmillan.

List, F. (1856) *The National System of Political Economy*, Philadelphia: J.B. Lippincott and Co.

——— (1991) [1885] *The National System of Political Economy*, New York: Augustus M. Kelly.

Little, I.M.D. (1982) *Economic Development: theory, policy and international relations*, New York: Basic Books.

Little, J.I. (1978) 'The Social and Economic Development of Settlers in two Quebec Townships, 1851–1870', in D. Akenson (ed.) *Canadian Papers in Rural History: volume 1*, Gananoque, Ont.: Langdale Press.

——— (1985) 'Imperialism and Colonization in Lower Canada: the role of William Bowman Felton', *Canadian Historical Review* 64(4).

——— (1989) *Nationalism, Capitalism and Colonization in Nineteenth-Century Quebec: The Upper St Francis District*, Kingston: McGill-Queens University Press.

Littlechild, S. (ed.) (1990) *Austrian Economics. Volume 1*, Aldershot: Edward Elgar.

Livingstone, I. (1986) *Rural Development, Employment and Incomes in Kenya*, Aldershot: Gower.

——— (1991) 'A Reassessment of Kenya's Rural and Urban Informal Sector', *World Development* 19(6).

Lockwood, W.W. (ed.) (1965) *The State and Economic Enterprise in Japan: essays in the political economy of growth*, Princeton: Princeton University Press.

Lofchie, M. (1989) *The Policy Factor: agricultural performance in Kenya and Tanzania*, Boulder: Lynne Rienner.

Logan, H.A. (1937) 'Labor Costs and Labour Standards', in H.A. Innis (ed.) *Labor in Canadian–American Relations*, Toronto: Ryerson Press.

Logan, R.W. (1962) 'The Historical Aspects of Pan-Africanism, 1900–1945', in J.A. Davis (ed.) *Pan-Africanism Reconsidered*, Berkeley: University of California Press.

Lonsdale, J. (1992) 'The Moral Economy of Mau Mau: wealth, poverty and civic virtue in Kikuyu political thought', in B. Berman and J. Lonsdale, *Unhappy Valley: conflict in Kenya and Africa*, London: James Currey.

Louis, W.R. and Stengers, J. (1968) *E.D. Morel's History of the Congo Reform Movement*, Oxford: Clarendon Press.

Lowe, V. (1985) *Alfred North Whitehead: the man and his work, volume 1: 1861–1910*, Baltimore: Johns Hopkins University Press.

Löwith, K. (1965) [1941] *From Hegel to Nietzsche*, London: Constable.

Lucas, C.P. (ed.) (1912) *Lord Durham's Report on the Affairs of British North America, vol. 2*, Oxford: Oxford University Press.

Lucas, G.R. (1989) *The Rehabilitation of Whitehead: an analytic and historical assessment of process philosophy*, Albany, NY: State University of New York Press.

Maas, M. (1986) *Women's Groups in Kiambu, Kenya: it is always a good thing to have land*, Leiden: African Studies Centre, research report 26.

Maccoby, S. (1938) *English Radicalism 1853–1886*, vol.4, London: George Allen and Unwin.

——— (1961) *English Radicalism: the end?*, London: George Allen and Unwin.

Macdonald, L.R. (1975) 'Merchants against Industry', *Canadian Historical Review* 56(3).

Macdonald, N. (1939) *Canada, 1763–1841, Immigration and Settlement: The administration of the Imperial Land Regulations*, London: Longman.

MacDonell, U.N. (1924) 'Gibbon Wakefield and Canada subsequent to the Durham Mission, 1839–42', *Bulletin 49*, Kingston, Ontario: Departments of History and Political and Economic Science, Queen's University.

MacGregor, D. (1984) *The Communist Ideal in Hegel and Marx*, London: George Allen and Unwin.

Macintosh, W.A. (1923) 'Economic Factors in Canadian History', *Canadian Historical Review* 4(1). (Reprinted in G. Laxer (ed.) (1991) *Perspectives on Canadian Economic Development: class, staples, gender and elites*, Toronto: Oxford University Press.)

MacIntyre, A. (1972) *Secularisation and Moral Change*. Oxford: Oxford University Press.

Mackenzie, N. and Mackenzie, J. (1979) [1977] *The First Fabians*, London: Quartet.

MacWilliam, S., Desaubin, F. and Timms, W. (1995) *Domestic Food Production and Political Conflict in Kenya*, Perth: University of Western Australia, Indian Ocean Centre for Peace Studies, Monograph no. 10.

Madarasz, A. (1980) 'Schumpeter's Theory of Economic Development', *Acta Oeconomica* 25(3–4).

Maitha, J.K. and Manundu, M. (1981) 'Production Techniques, Factor Proportions and Elasticities of Substitution', in T. Killick (ed.) *Papers on the Kenyan Economy: performance, problems and policies*, Nairobi: Heinemann.

Makonnen, R. (1973) *Pan-Africanism From Within*, Nairobi: Oxford University Press.

Malcolm, N. (1984) *Ludwig Wittgenstein: a memoir*, Oxford: Oxford University Press.

Malthus, T. (1970) [1815] 'An Inquiry into the Nature and Progress of Rent and the Principles by which it is Regulated' in *The Pamphlets of Thomas Robert Malthus*, New York: A.M. Kelly.

——— (1986a) [1798] *An Essay on the Principle of Population*, Harmondsworth: Penguin.

——— (1986b) [1830] *A Summary View of the Principle of Population*, Harmondsworth: Penguin.

Mao Tse-tung (1955a) 'On Contradiction', in *Selected Works of Mao Tse-tung, Volume I*, London: Lawrence and Wishart.

——— (1955b) 'On Practice', in *Selected Works of Mao Tse-tung, Volume I*, London: Lawrence and Wishart.

——— (1965) 'On the Mass Line', in *Selected Works of Mao Tse-tung, Volume III*, Peking: Foreign Publishing Press.

——— (1990) 'On Contradiction', in N. Knight (ed.) *Mao Ze dong on Dialectical Materialism*, Armonk, NY: M.E. Sharpe.

Marcuse, H. (1986) [1941] *Reason and Revolution*, London: Routledge and Kegan Paul.

Marglin, S.A. (1990a) 'Towards the Decolonisation of the Mind', in F.A. Marglin and S.A. Marglin, *Dominating Knowledge: development, culture and resistance*, Oxford: Clarendon Press.

——— (1990b) 'Losing Touch: the cultural conditions of worker accommodation and resistance', in F.A. Marglin and S.A. Marglin, *Dominating Knowledge: development, culture and resistance*, Oxford: Clarendon Press.

Marr, W.L. and Patterson, D.G. (1980) *Canada: an economic history*, Toronto: Gage.

Marsh, J. (1982) *Back to the Land: the pastoral impulse in Victorian England from 1880 to 1914*, London: Quartet.

Marshall, A. (1961) [1890] *Principles of Economics*, London: Macmillan.

Martin, D.A. (1990) 'Economics as Ideology: on making "The Invisible Hand" Invisible', *Review of Social Economy* 48(3).

Marx, K. (1965) *Capital, vol. 1*, Moscow: Progress Publishers.

——— (1967) *Writings of the Young Marx on Philosophy and Society*, Garden City, NY: Anchor.

——— (1973) [1939] *Grundrisse: foundations of the critique of political economy*, London: Penguin.

——— (1975a) 'Letter to father, 10 November 1837', in K. Marx and F. Engels, *Collected Works Volume 1: Marx and Engels: 1835–43*, Moscow: Progress Publishers.

——— (1975b) 'Marginal Notes, 12 February 1843', in K. Marx and F. Engels, *Collected Works Volume 1: Marx and Engels: 1835–43*, Moscow: Progress Publishers.

——— (1975c) [1845] 'Draft of an Article on Friedrich List's Book Das Nationale System der Politischen Oekonomie', in K. Marx and F. Engels, *Collected Works Volume 4: Marx and Engels 1844–1845*, Moscow: Progress Press.

——— (1976) [1867] *Capital, vol. 1*, Harmondsworth: Penguin.

——— (1979a) 'The British Rule in India', *New York Daily Tribune*, 25 June 1853, in K. Marx and F. Engels, *Collected Works Volume 12: Marx and Engels: 1853–54*, New York: International Publishers.

——— (1979b) 'The Future Results of British Rule in India', *New York Daily Tribune*, 8 August 1853, in K. Marx and F. Engels, *Collected Works Volume 12: Marx and Engels: 1853–54*, New York: International Publishers.

——— (1981) [1894] *Capital, vol. 3*, Harmondsworth, Penguin.

——— and Engels, F. (1947) [1845–6] *The German Ideology*, New York International Publishers.

——— and ——— (1968) [1848] *The Communist Manifesto*, New York: The Washington Square Press.

——— and ——— (1975) *The Holy Family, or, Critique of Critical Criticism: Against Bruno Bauer and company*, Moscow: Progress Publishers.

——— and ——— (1976) *The German Ideology* in *Collected Works Volume 5: Marx and*

Engels: 1845–47, Moscow: Progress Publishers.

Marz, E. (1983) 'The Austrian Credit Mobilier in a Time of Transition', in J. Komlos (ed.) *Economic Development in the Habsburg Monarchy in the Nineteenth Century: essays*, Boulder: East European Monographs.

——— (1984) [1981] *Austrian Banking and Financial Power: Creditanstalt at a turning point, 1913–1923*, London: Weidenfeld and Nicolson.

——— (1991) *Joseph Schumpeter: scholar, teacher, politician*, New Haven: Yale University Press.

Matthews, R. (1991) 'Appraising Efficiency in Kenya's Machinery Manufacturing Sector', *African Affairs* 90 (358).

Mayes, C. (1859) *The Victorian Contractors and Builders Price Book*, Melbourne: printed for C. Mayes.

——— (1861) 'Essay on the manufactures more immediately required for the economical development of the resources of the colony: With special reference to those manufactures the raw materials of which are the produce of Victoria', *The Victorian Government Prize Essays 1860*, Melbourne: Government Printer.

Mayhew, A. (1980) 'Schumpeterian Capitalism versus the "Schumpeterian Thesis"', *Journal of Economic Issues* 14(2).

Mbithi, P.M. and Rasmusson, R. (1977) *Self-reliance in Kenya: the case of harambee*, Uppsala, Scandinavian Institute of African Studies.

Mboya, T. (1963a) 'African Socialism', *Transition* (Kampala), no. 7. Reprt. in W.H. Friedland and C.G. Rosberg (eds) *African Socialism*, Stanford: Stanford University Press.

——— (1963b) *Freedom and After*, London: Andre Deutsch.

McBriar, A.M. (1987) *An Edwardian Mixed Doubles*, Oxford: Clarendon.

McCalla, D. (1987) *Perspectives on Canadian Economic History*, Toronto: Copp Clark Pitman.

McCallum, J. (1980) *Unequal Beginnings: agriculture and unequal development in Quebec and Ontario until 1870*, Toronto: University of Toronto Press.

McCraw, T.K. (1991) 'Schumpeter Ascending', *The American Scholar* 60(3).

McCulloch, J.R. (1843) *Principles of Political Economy*, Edinburgh: W. Tait.

McGregor Ross, W. (1927) *Kenya from Within: a short political history*, London: Allen and Unwin.

McInnis, R.M. (1982) 'A Reconsideration of the State of Agriculture in Lower Canada in the First Half of the Nineteenth Century', in D.H. Akenson (ed.) *Canadian Papers in Rural History Volume 3*, Gananoque, Ontario: Langdale Press.

McLuhan, M. (1953) 'The Later Innis', *Queen's Quarterly* 60(3).

McMichael, P. (1984) *Settlers and the Agrarian Question: foundations of capitalism in colonial Australia*, Cambridge: Cambridge University Press.

Mcwilliam, M. (1976) 'The Managed Economy: agricultural change, development, and finance in Kenya', in D.A. Low and A. Smith (eds) *History of East Africa, vol. 3*, Oxford: Clarendon Press.

Meek, R.L. (1976) *Social Science and the Ignoble Savage*, Cambridge: Cambridge University Press.

Meikle, S. (1985) *Essentialism in the Thought of Karl Marx*, London: Duckworth.

Meisner, M. (1982) *Marxism, Maoism and Utopianism: eight essays*, Madison: University of Wisconsin Press.

Menger, C. (1963) [1883] *Problems of Economics and Sociology*, Urbana, IL: University of Illinois Press.

Mill, J. (1975) [1858] *The History of British India*, Chicago: University of Chicago Press.

Mill, J.S. (1942) [1831] *The Spirit of the Age*, Chicago: University of Chicago Press.

——— (1965) [1849] *On Liberty*, in Max Lerner (ed.) *Essential Works of John Stuart Mill*, New York: Bantam.

——— (1968) [1858] *Memorandum of the Improvements in the Administration of India*, Farnborough: Gregg International Publishers.

——— (1974) [1843] *A System of Logic, Ratiocinative and Inductive*, in J.M. Robson (ed.) *Collected Works of John Stuart Mill, Volume 8*, Toronto: University of Toronto Press.

——— (1985) [1848] *Principles of Political Economy*, Harmondsworth: Penguin.

——— (1989) [1873] *Autobiography*, Harmondsworth: Penguin.

——— (1990) *Writings on India*, in J.M. Robson, M. Moir and Z. Moir (eds) *Collected Works of John Stuart Mill, Volume 30*, Toronto: University of Toronto Press.

Miller, E.J. (1992) 'Newman on Conscience and on Conversion', in E. Block (ed.) *Critical Essays on John Henry Newman*, English Critical Studies, University of Victoria, BC, Canada: 1992.

Miller, J.C. (1959) *Alexander Hamilton: portrait in paradox*, Greenwood, CT: Greenwood Press.

Miller, J.D.B. (1956) *Richard Jebb and the Problem of Empire*, London: Athlone Press

Minsky, H.P. (1990) 'Schumpeter: finance and evolution', in A. Heertje and M. Perlman (eds) *Evolving Technology and Market Structure: studies in Schumpeterian economics*, Ann Arbor: University of Michigan Press.

Miquelon, D. (ed.) (1977) *Society and Conquest: the debate on the bourgeosie and social change in French Canada, 1700–1850*, Toronto: Copp Clark.

Misner, P. (1973) 'Note on the Critique of Dogmas', *Theological Studies* 34.

——— (1985) 'The "Liberal" Legacy of Newman', in M.J. Weaver (ed.) *Newman and the Modernists*, Lanham, MD: University Press of America.

Mitchell, B. (1990) 'Newman as Philosopher', in Ian Ker and Alan G. Hill (eds) *Newman After a Hundred Years*, Oxford: Clarendon Press.

Mitchell, B.R. (1983) *International Historical Statistics: The Americas and Australasia*, London: Macmillan.

Moggeridge, D.E. (1992) *Maynard Keynes: an economist's biography*, London: Routledge.

Mohiddin, A. (1973) 'The Formulation and Manifestation of Two Socialist Ideologies: democratic African socialism of Kenya and the Arusha declaration of Tanzania, PhD dissertation, McGill University.

Monet, J. (1966) 'French Canada and the Annexation Crisis, 1848–50', *Canadian Historical Review* 47(3).

——— (1969) *The Last Cannon Shot: a study of French-Canadian Nationalism, 1837–1850*, Toronto: University of Toronto Press.

——— (1971) 'The 1840s', in J.M.S. Careless (ed.) *Colonists and Canadiens 1760–1867*, Toronto: Macmillan.

Monière, D. (1981) [1977] *Ideologies in Quebec: the historical development*, Toronto: University of Toronto Press.

Morel, E.D. (1902) *Affairs of West Africa*, London: Heinemann.

Morishima, M. (1982) *Why Japan has Succeeded: western technology and the Japanese ethos*, Cambridge, Cambridge University Press.

——— (1992) *Capital and Credit: a new formulation of general equilibrium theory*, Cambridge: Cambridge University Press.

Morisonneau, C. (1978) *La Terre Promise: le mythe du nord québecois*, Montreal: Hurtubise HMH.

Morrell, W.P. (1966) [1930] *British Colonial Policy in the Age of Peel and Russell*, London: Cass.

Morris, A.J.A. (ed.) (1974) *Edwardian Radicalism 1900-14: some aspects of British radicalism*, London: Routledge and Kegan Paul.

Mun, T. (1621) *A Discourse of Trade from England to the East Indies*, London.

——— (1628) *The Petition and Remonstrance of the Governor and Company of Merchants of London, Trading to the East Indies*, London.

Muriuki, G. (1974) *A History of the Kikuyu: 1500–1900*, Nairobi: Oxford University Press.

Murray, P. (1988) 'Karl Marx as a Historical Materialist Historian of Political Economy', *History of Political Economy* 20(1).

Murray-Brown, J. (1972) *Kenyatta*, London: Allen and Unwin.

Musson, A.E. (1978) *The Growth of British Industry*, London: B.T. Batsford.

Mutiso, G.C.M. (1975) *Kenya: politics, policy and society*, Nairobi: EALB.

Mwicigi Report, see Kenya Republic, 1970.

Myers, G. (1972) [1914] *A History of Canadian Wealth*, Toronto: James Lorimer.

Nadel, G. (1957) *Australia's Colonial Culture: ideas, men and institutions in mid-nineteenth century eastern Australia*, Melbourne: Cheshire.

Nandy, A. and Visvanathan, S. (1990) 'Modern Medicine and its Non-modern Critics: a study in discourse', in F.A. Marglin and S.A. Marglin, *Dominating Knowledge: development, culture and resistance*, Oxford: Clarendon Press.

Naoroji, D. (1901) [1871] *Poverty and Un-British Rule In India*, London: Swan Sonnenschein.

Naylor, R.T. (1972) 'The Rise and Fall of the Third Commercial Empire of the St Lawrence', in G. Teeple (ed.) *Capitalism and the National Question in Canada*, Toronto: University of Toronto Press.

——— (1975) *The History of Canadian Business 1867–1914*, Toronto: Lorimer.

Ndegwa Report, see Kenya Republic, 1991a.

Ndele, S.M. (1991) *The Effects of Non-Bank Financial Intermediaries on Demand for Money in Kenya*, Nairobi: AERC Research Paper 5, Initiative Publishers.

Neill, R. (1972) *A New Theory of Value: the Canadian economics of H.A. Innis*, Toronto: University of Toronto Press.

——— (1991) *A History of Canadian Economic Thought*, London: Routledge.

Nelkin, D. (1964) 'Socialist Sources of Pan-African Ideology', in W.H. Friedland and C.G. Rosberg (eds) *African Socialism*, Stanford: Stanford University Press.

Nelson, R.R. and Winter, S.G. (1982) 'The Schumpeterian Tradeoff Revisited', *American Economics Review* 72(1).

Newman, B. (1927) *Edmund Burke*, London: G. Bell and Sons Ltd.

Newman, F.W. (1851) *Lectures on Political Economy*, London: John Chapman.

Newman, J. (1986) *The Mental Philosophy of John Henry Newman* Waterloo, Ontario: Wilfred Laurier Press.

Newman, J.H. (1845) *An Essay on the Development of Christian Doctrine*, London: James Toovey.

——— (1870) *An Essay in Aid of A Grammar of Assent*, London: Burns and Oates.

——— (1878) *An Essay on the Development of Christian Doctrine*, London: Basil Montague Pickering.

——— (1918) [1850] *Certain Difficulties Felt by Anglicans in Catholic Teaching*, London.

——— (1969) *The Philosophical Notebook. Volume 1: general introduction to the study of Newman's philosophy*, edited by E.J. Sillem, Louvain: Nauwelaerts.

Newman, M. (1993) *Harold Laski: a political biography*, Basingstoke: Macmillan.

Ngau, P.M. (1987) 'Tensions in Empowerment: the experience of the harambee (self-help) movement in Kenya', *Economic Development and Cultural Change* 35(3).

Ngugi wa'Thiongo (1986) *De-colonizing the Mind: the politics of language in African literature*, London: Currey.

Nicholls, D. (1975) *The Pluralist State*, London: Macmillan.

Nicolaus, M. (1973) 'Foreword', to Karl Marx [1953] *Grundrisse: Foundations of the critique of political economy*, London: Penguin.

Nisbet, R.A. (1969) *Social Change and History: aspects of the western theory of development*, New York: Oxford University Press.

——— (1980) *History of the Idea of Progress*, New York: Basic Books.

Njonjo, A.L. (1977) 'The Africanization of the "White Highlands": a study in agrarian class struggle in Kenya 1950–1975, unpublished PhD thesis, Princeton University.

——— (1981) 'The Kenya peasantry: a reassessment', *Review of African Political Economy* 20.

Norcliffe, G.B. (1983) 'Operating Characteristics of Rural Non-farm Enterprises in Central Province, Kenya', *World Development* 11(1).

Norman, E. (1990a) 'Newman's Social and Political Thinking', in I. Ker and A.G. Hill (eds) *Newman after a Hundred Years*, Oxford: Clarendon Press.

——— (1990b) 'Tides and Twilights: Newman Since Vatican II', in I. Ker and A.G. Hill (eds) *Newman after a Hundred Years*, Oxford: Clarendon Press.

Norman, J. (1963) *Edward Gibbon Wakefield: a political reappraisal*, Fairfield, CT: University Press.

Nurkse, R. (1953) *Problems of Capital Formation in Underdeveloped Countries*, Oxford: Blackwell.

Nworah, K.K.D. (1966) 'Humanitarian Pressure Groups and British Attitudes to West Africa', unpublished PhD dissertation, University of London.

Nyong'o, Anyang' P., (1989) 'State and Society in Kenya: the disintegration of the nationalist coalitions and the rise of presidential authoritarianism 1963–78', *African Affairs* 88(351).

Oakley, A. (1990) *Schumpeter's Theory of Capitalist Motion: a critical exposition and reassessment*, Aldershot: Edward Elgar.

O'Donnell, R.M. (1989) *Keynes: Philosophy, Economics and Politics: the philosophical foundations of Keynes's thought and their influence on his economics and politics*, London: Macmillan.

——— (1991a) 'Keynes on Probability, Expectations and Uncertainty' in R.M. O'Donnell (ed.) *Keynes as Philosopher-Economist: the ninth Keynes seminar, 1989*, Basingstoke: Macmillan.

——— (1991b) 'Keynes's Weight of Argument and its Bearing on Rationality and Uncertainty', in B.W. Bateman and J.B. Davis, *Keynes and Philosophy: essays on the origins of Keynes's thought*, Aldershot: Elgar.

O'Gorman, F. (1973) *Edmund Burke: his political philosophy*, London: Allen and Unwin.

Okoth-Ogendo, H.W.O. (1991) *Tenant of the Crown: evolution of agrarian law and institutions in Kenya*, Nairobi: ACTS Press.

Oldroyd, D. (1986) *The Arch of Knowledge: an introductory study of the philosophy and methodology of science*. New York: Methuen.

Omosini, O. (1968–9) 'Origins of British Methods of Tropical Development in the West African Dependencies 1886–1906', unpublished PhD dissertation, University of Cambridge.

Orgaz, R.A. (1934) *Echeverria y el saintsimonismo*, Buenos Aires.

Ormsby, M.A. (1958) *British Columbia: a history*, Vancouver: Macmillan.

Orora, J.O. and Spiegel, H.B.C. (1979) 'Harambee: self-help development projects in Kenya', *International Journal of Comparative Sociology* 21(3–4).

Orvis, S.W. (1989) 'The Political Economy of Agriculture in Kisii, Kenya: social reproduction and household response to development policy', unpublished PhD thesis, University of Wisconsin-Madison.

Ostrander, G.M. (1983) 'Frederick Turner's Canadian Frontier Thesis', *Canadian Historical Review* 64(4).

Ouellet, F. (1980a) [1966] *Economic and Social History of Quebec, 1760–1850: structures and conjunctures*, Toronto: Macmillan.

——— (1980b) *Lower Canada 1791–1840: social change and nationalism*, Toronto: McClelland and Stewart.

——— (1991) *Economy, Class and Nation in Quebec: interpretive essays*, Toronto: Copp Clark Pitman.

Ouma, S.J. (1980) *A History of the Cooperative Movement in Kenya*, Nairobi: Bookwise.

Ozawa, T. (1974) *Japan's Technological Challenge to the West, 1950–1974: motivation and accomplishment*, Cambridge, MA: MIT Press.

Pack, H. (1976) 'The Substitution of Labour for Capital in Kenyan Manufacturing', *Economic Journal* 86(1).

——— (1977) 'Unemployment and Income Distribution in Kenya', *Economic Development and Cultural Change* 26(1).

——— (1987) *Productivity, Technology, and Industrial Development*, Oxford: Oxford University Press.

Padmore, G. (1956) *Pan-Africanism or Communism: the coming struggle for Africa*, London: Dennis Dobson.

Pal, S. (1955) 'Schumpeter and his Ideas on Economic Development', *Indian Journal of Economics* 36(1).

Pala, A.O. (1978) 'Women Power in Kenya: raising funds and awareness', *Ceres* 11(2).

Palma, G. (1981) 'Dependency and Development: a critical review', in D. Seers (ed.) *Dependency Theory: a critical reassessment*, London: Francis Pinter.

Palmer, B. (1983) *Working-class Experience: the rise and reconstruction of Canadian labour, 1800–1980*, Toronto: Butterworth.

——— (1985) 'Labour in Nineteenth-century Canada', in W.J.C. Cherwinski and G.S. Kealey, *Lectures in Canadian Labour and Working-class History*, Toronto: New Hogtown Press.

Panitch, L. (1981) 'Dependency and Class in Canadian Political Economy', *Studies in Political Economy* 6 (reprinted in G. Laxer (ed.), *Perspectives on Canadian Economic Development: class, staples, gender and elites*, Toronto: Oxford University Press, 1991).

Pankhurst, Richard (1957) *The Saint-Simonians, Mill, and Carlyle*, London: Lalibela Press.

Pannekoek, A. (1975) [1938] *Lenin as Philosopher: a critical examination of the philosophical basis of Leninism*, London: Merlin Press.

Paquet, G. and Smith, W.R. (1983) 'L'émigration des Canadiens Français vers les Etats Unis, 1790–1940: problématique et coups de sonde', *L'Actualité Economique* 59(3).

——— and Wallot, J.-P. (1972) 'Crise agricole et tensions socio-ethniques dans le bas-canada, 1802–1821', *Revue d'Histoire de l'Amérique Française* 26(2).

Parent, E. (1846) 'Industry as a means of survival for the French-Canadian nationality', in R. Cook (trans. and ed.) (1969) *French-Canadian Nationalism: an anthology*, Toronto: Macmillan.

Parizeau, G. (1975) *La Société Canadienne-Française au XIXe siècle: essais sur le milieu*, Montreal: Fides.

Parker, I. (1981) 'Innis, Marx and the Economics of Communication: a theoretical aspect of Canadian political economy', in W.H. Melody, L. Salter and P. Heyer (eds) *Culture, Communication, and Dependency: the tradition of H.A. Innis*, Norwood, NJ: Ablex Publishing.

——— (1985) 'Staples, Communications, and the Economics of Capacity, Overhead Costs, Rigidity, and Bias', in D. Cameron (ed.) *Explorations in Canadian Economic History: essays in honour of Irene M. Spry*, Ottawa: University of Ottawa Press.

Pasvolsky, L. (1928) *Economic Nationalism of the Danubian States*, London: Allen and Unwin. (Reprinted in E. Marz (1984) [1981] *Austrian Banking and Financial Power: Creditanstalt at a turning point, 1913–1923*, London:

Weidenfeld and Nicolson.)

Paterson, D.B. (1984) 'Kinship, Land and Community: the moral foundations of the Abaluyha of East Bunyore', unpublished PhD thesis, University of Washington.

Patterson, G. (1990) *History and Communications: Harold Innis, Marshall McLuhan, the interpretation of history*, Toronto: Toronto University Press.

Pattison, R. (1991) *The Great Dissent: John Henry Newman and the liberal heresy*, New York: Oxford University Press.

Paulsen, F. (1899) [1888] *A System of Ethics*, London: Kegan, Paul, French and Trubner.

——— (1938) *An Autobiography*, New York: Columbia University Press.

Pelczyinski, Z.A. (ed.) (1984a) *The State and Civil Society: studies in Hegel's Political Philosophy*, Cambridge: Cambridge University Press.

——— (1984b) 'Political Community and Individual Freedom in Hegel's philosophy of the state', in Z.A. Pelczynski (ed.) (1984) *The State and Civil Society: studies in Hegel's political philosophy*, Cambridge: Cambridge University Press.

Pentland, H.C. (1981) *Labour and Capital in Canada, 1650–1860*, Toronto: Lorimer.

Phillips, A. (1989) *The Enigma of Colonialism: British policy in West Africa*, London: Currey.

Phillips, P.A. (1967) 'Confederation and the Economy of British Columbia', in W.G. Shelton (ed.) *British Columbia and Confederation*, Victoria, BC: Morriss.

Pippin, R.B. (1991) *Modernism as a Philosophical Problem: on the dissatisfactions of European high culture*, Oxford: Blackwell.

Piva, M.J. (1985) 'Continuity and Crisis: Francis Hincks and Canadian economic policy', *Canadian Historical Review* 46(2).

——— (1992a) *The Borrowing Process: public finance in the Province of Canada, 1840–1867*, Ottawa: University of Ottawa Press.

——— (1992b) 'Government Finance and the Development of the Canadian State', in A. Greer and I. Radforth (eds) *Colonial Leviathan: state formation in mid-nineteenth-century Canada*, Toronto: University of Toronto Press.

Plant, R. (1977) 'Hegel's Social Theory – I and II', *New Left Review* 103 and 104.

Platt, D.C.M. and Adelman, J. (1990) 'London Merchant Bankers in the First Phase of Heavy Borrowing: the Grand Trunk railway of Canada', *Journal of Imperial and Commonwealth History* 18(2).

Pocock, J.G.A. (1985) *Virtue, Commonwealth and History: essays on political thought and history, chiefly in the eighteenth century*, Cambridge: Cambridge University Press.

——— (1992) 'Historical Introduction', to J. Harrington, *The Commonwealth of Oceana and a System of Politics*, Cambridge: Cambridge University Press.

Pollard, S. (1971) [1968] *The Idea of Progress: history and society*, Harmondsworth: Penguin.

Pomfret, R. (1981a) 'The Staple Theory as an Approach to Canadian and Australian Economic Development', *Australian Economic History Review* 21(2).

——— (1981b) *The Economic Development of Canada*, Toronto: Methuen.

Porter, B. (1968) *Critics of Empire*, London: Macmillan.

Porter, D., Allen, B., and Thompson, G. (1991) *Development in Practice: paved with good intentions*, London: Routledge.

Powell, E. (1977) *Joseph Chamberlain*, London: Thames and Hudson.

Prebisch, R. (1980) 'The Dynamics of Peripheral Capitalism', in L. Lefeber and L.L. North (eds) *Democracy and Development in Latin America*, Toronto: Latin American Research Unit.

Prime, M.G. and Henderson, D.R. (1975) 'Schumpeter on Preserving Private Enterprise', *History of Political Economy* 7(3).

Province of Canada (1849) (Chauveau Report) 'Report of the Select Committee appointed to inquire into the causes and importance of the emigration which takes place annually from Lower Canada to the United States; etc.', Appendix (A.A.A.A.A.) to Volume 8, *Journals of the Legislative Assembly*, Montreal: Province of Canada.

Quaife, G.R. (1967) 'The Diggers: democratic sentiment and political apathy', *Australian Journal of Politics and History*, 13(2).

Qualter, T. (1980) *Graham Wallas and the Great Society*, London: Macmillan.

Quinton, A.M. (1972) 'Absolute Idealism', *Proceedings of the British Academy* 57, London: Oxford University Press.

Raines, J.P. and Leathers, C.G. (1992) 'The Post-War Japanese Economy and Schumpeter's Corporatist Principle', *Journal of Economic Studies* 19(6).

Ramirez, B. (1991) *On the Move: French-Canadian and Italian Migrants in the North Atlantic Economy, 1860–1914*, Toronto: McClelland and Stewart.

Rao, V.K.R.V. (1952) 'Investment, Income and the Multiplier in an Underdeveloped Economy', *Indian Economic Review* 1(1).

Rempel, H. and House, W.J. (1978) *The Kenya Employment Problem: an analysis of the modern sector labour market*, Nairobi: Oxford University Press.

Reynolds, J.E. and Wallis, M.A.H. (1976) 'Self-help and Rural Development in Kenya', Nairobi: University of Nairobi Institute of Development Studies, Discussion Paper 241.

Ricci, D.M. (1969) 'Fabian Socialism: a theory of rent as exploitation', *Journal of British Studies* 9(1).

Richards, J. (1985) 'The Staple Debates', in D. Cameron (ed.) *Explorations in Canadian Economic History: essays in honour of Irene M. Spry*, Ottawa: University of Ottawa Press.

Richter, M. (1964) *The Politics of Conscience: T.H. Green and his age*, London: Weidenfeld and Nicolson.

Riddell, R.G. (1937) 'A Study in the Land Policy of the Colonial Office, 1763–1855', *Canadian Historical Review* 18(4).

——— (1939) 'The Policy of Creating Land Reserves in Canada', in R. Flenley (ed.) *Essays in Canadian History: presented to G.M. Wrong*, Toronto: Macmillan.

Riedel, M. (1984) *Between Tradition and Revolution: The Hegelian transformation of political philosophy*, Cambridge: Cambridge University Press.

Riesser, J. (1911) *The Great German Banks and their concentration in connection with the Economic Development of Germany*, Washington: Government Printing Office.

Rimmer, D. (1961) 'Schumpeter and the Underdeveloped Countries', *Quarterly Journal of Economics* 75.

Robbins, W. (1966) *The Newman Brothers: an essay in comparative intellectual biography*, London: Heinemann.

Roberts, M. (1964) 'A Socialist Looks at African Socialism', in W.H. Friedland and C.G. Rosberg (eds) *African Socialism*, Stanford: Stanford University Press.

Rodman, J. (1964) *The Political Theory of T.H. Green: selected writings*, New York: Appleton-Century-Crofts.

Roll, E. (1961) [1938] *A History of Economic Thought*, London: Faber.

Roman, D.W. (1991) 'Railway Imperialism in Canada, 1847–1865', in C.B.Davis and K.E. Wilburn with R.E. Robinson, *Railway Imperialism*, New York: Greenwood Press.

Rosdolsky, R. (1977) [1968] *The Making of Marx's Capital*, London: Pluto.

Rothschild, K.W. (1961) 'Wurzeln und Triebskrafte der österreichischen Wirtschaftsstruktur', in W. Weber (ed.) *Österreichs Wirtschaftsstruktur gestern-heute-morgen Bd. I.*, Berlin: Duncker and Humblot.

Rowland, P. (1968) *The Last Liberal Governments:*

the promised land 1905–1910, London: Barrie and Rockliff.

Roxborough, I. (1979) *Theories of Underdevelopment*, London: Macmillan.

Rudolph, R.L. (1972) 'Austria 1800–1914', in R. Cameron (ed.) *Banking and Economic Development: some lessons of history*, New York: Oxford University Press.

——— (1976) *Banking and Industrialization in Austria-Hungary: the role of banks in the industrialization of the Czech Crownlands 1873–1914*, Cambridge: Cambridge University Press.

Runde, J. (1991) 'Keynesian Uncertainty and the Instability of Beliefs', *Review of Political Economy* 3(2).

Ruotsi, J. (1992) 'Ownership, Management and Economic Performance: two agro-industries in Kenya', unpublished PhD thesis, University of Sussex.

Ryan, A. (1974) *J.S. Mill*, London: Routledge and Kegan Paul.

Sachs, W. (ed.) (1992) *The Development Dictionary: A Guide to Knowledge as Power*, London: Zed.

Samuels, W.J. (1990) 'The Reformation of German Economic Discourse, 1750–1840: a review article', *Review of Social Economy* 48(3).

Sarmiento, Domingo, F. (1972) [1868] *Life in the Argentine Republic in the Days of the Tyrants: or, civilization and barbarism*, New York: Hafner Press. (Excerpted in E. Bradford Burns, *Latin America: Conflict and Creation*, Englewood Cliffs, NJ: Prentice Hall, 1993.)

Saul, S.B. (1979) *The Myth of the Great Depression*, New York: St Martin's Press.

Saville, J. (1960) 'Henry George and the British labour movement', *Science and Society* 24(4).

Schedvin, C.B. (1990) 'Staples and Regions of Pax Britannica', *Economic History Review* 43(4).

Scherer, F.M. (1984) *Innovation and Growth: Schumpeterean Perspectives*, Cambridge, MA: MIT Press.

——— (1992) 'Schumpeter and Plausible Capitalism', *Journal of Economic Literature* 30(3).

Schorske, C.E. (1980) *Fin-de-Siècle Vienna: politics and culture*, London: Weidenfeld and Nicolson.

Schram, S. (1969) [1963] *The Political Thought of Mao Tse-tung*, Harmondsworth: Penguin.

——— (1988) *The Thought of Mao Tse-tung*, Cambridge: Cambridge University Press.

Schreuder, D. (1988) 'The Making of the Idea of Colonial Nationalism: the early travels and writings of Richard Jebb', in J. Eddy and D. Schreuder (eds) *The Rise of Colonial Nationalism: Australia, New Zealand, Canada and South Africa first assert their nationalities, 1880–1914*, Sydney: Allen and Unwin.

Schumpeter, J.A. (1908) *Das Wesen und der Hauptinhalt der theoretischen Nationalökonomie*, Leipzig: Duncker and Humblot.

——— (1912) *Theorie der wirtschaftlichen Entwicklung*, Leipzig: Duncker and Humblot.

——— (1934) [1912] *The Theory of Economic Development: an inquiry into profits, capital, credit, interest and the business cycle*, Cambridge, MA: Harvard University Press.

——— (1946) 'The Communist Manifesto in Sociology and Economics', *Journal of Political Economy* 57(3).

——— (1948) 'Keynes, the Economist', in S.E. Harris (ed.) *The New Economics: Keynes' influence on theory and public policy*, London: Dennis Dobson.

——— (1949) 'Economic Theory and Entrepreneurial History', in Research Center for Entrepreneurial History (ed.) *Change and the Entrepreneur: postulates and patterns for entrepreneurial history*, Cambridge, MA: Harvard University Press.

——— (1951) [1928] 'The Instability of Capitalism' in R.V. Clemence (ed.) *Essays of J.A. Schumpeter*, Cambridge, MA: Addison-Wesley.

——— (1954) *History of Economic Analysis*, London: Allen and Unwin.

——— (1956) [1917–18] 'Money and the Social Product', in A.T. Peacock, W.F. Stolper, R. Turvey, and E. Henderson (eds) *International Economic Papers* No. 6, London: Macmillan.

——— (1964) [1939] *Business Cycles: a theoretical, historical and statistical analysis of the capitalist process*, abridged edition, Philadelphia: Porcupine Press.

——— (1985) *Aufsätze zur Wirtschaftspolitik*, Tubingen: J.C.B. Mohr.

——— (1991a) [1918] 'The Crisis of the Tax State', in R. Swedberg (ed.) *Joseph A. Schumpeter: the economics and sociology of capitalism*, Princeton: Princeton University Press.

——— (1991b) [1918–19] 'The Sociology of Imperialisms', in R. Swedberg (ed.) *Joseph A. Schumpeter: the economics and sociology of capitalism*, Princeton: Princeton University Press.

——— (1991c) [1927] 'Social Classes in an Ethnically Homogeneous Environment', in R. Swedberg (ed.) *Joseph A. Schumpeter: the economics and sociology of capitalism*, Princeton: Princeton University Press.

——— (1991d) [1941] 'An Economic Interpretation of Our Time: the Lowell Lectures', in R. Swedberg (ed.) *Joseph A. Schumpeter: the economics and sociology of capitalism*, Princeton: Princeton University Press.

——— (1994) [1942] *Capitalism, Socialism and Democracy*, London: Routledge.

Seaman, L.C.B. (1973) *Victorian England: Aspects of English and Imperial History 1837–1901*, London: Methuen.

Seccareccia, M. (1992) 'Immigration from the British Isles and Canada's Labour Surplus Economy during the Mid-nineteenth Century', Department of Economics, University of Ottawa Research Paper 9209E.

Seers, D. (1979) 'The New Meaning of Development', in D. Lehmann, *Development Theory; four critical studies*, London: Cass.

——— (1983) *The Political Economy of Nationalism*, Oxford: Oxford University Press.

Selby, R.C. (1975) *The Principle of Reserve in the Writings of John Henry Newman*, London: Oxford University Press.

Semmel, B. (1961) 'The Philosophic Radicals and Colonialism', *Journal of Economic History* 21(4).

Sen, A.K. (1983a) 'Poor, relatively speaking', *Oxford Economic Papers* 35. Reprinted in A.K. Sen (1984) *Resources, Values and Development*, Oxford: Blackwell.

——— (1983b) 'Development: which way now?', *Economic Journal* 93(4).

——— (1984a) 'Goods and People', in A.K. Sen, *Resources, Values and Development*, Oxford: Blackwell.

——— (1984b) *Resources, Values and Development*, Oxford: Blackwell.

——— (1985) *Commodities and Capabilities*, Amsterdam: North-Holland.

——— (1987) 'The Standard of Living, Lectures I and II', in G. Hawthorn (ed.) *The Standard of Living: the Tanner Lectures 1985*, Cambridge: Cambridge University Press.

——— (1988a) 'The Concept of Development', in H. Chenery and T.N. Srinivasan, *Handbook of Development Economics, Volume 1*, Amsterdam: North-Holland.

——— (1988b) 'Freedom of Choice: concept and content', *European Economic Review* 32 (2/3).

Sender, J. and Smith, S. (1986) *The Development of Capitalism in Africa*, London: Methuen.

Serle, G. (1963) *The Golden Age: a history of the colony of Victoria 1851–1861*, Melbourne: University of Melbourne Press.

Sewell, W.H. (1980) *Work and Revolution in France*, Cambridge: Cambridge University Press.

Shamsavari, A. (1991) *Dialectics and Social Theory: the logic of Capital*, Braunton, Devon: Merlin Books.

Sharpley, J. (1986) *Economic Policies and Agricultural Performance: the case of Kenya*, Paris: OECD Development Centre.

——— and Lewis, S.R. (1988) *Kenya's Industrialisation, 1964–84*, Institute of Development Studies, University of Sussex, Discussion Paper 242.

Shelton, W.G. (eds) (1967) *British Columbia and Confederation*, Victoria, BC: Morriss.

Shenton, R.W. (1985) *The Development of Capitalism in Northern Nigeria*, London: James Currey.

Sheridan, A. (1980) *Michel Foucault: The Will to Truth*, London: Tavistock.

Shionoya, Y. (1986) 'The Science and Ideology of Schumpeter', *Rivista Internazionale di Scienze Economiche e Commerciali* 33(8).

——— (1990a) 'The Origin of the Schumpeterian Research Programme: a chapter omitted from Schumpeter's Theory of Economic Development', *Journal of Institutional and Theoretical Economics* 146.

——— (1990b) 'Instrumentalism in Schumpeter's Economic Methodology', *History of Political Economy* 22(2).

——— (1991) 'Sidgwick, Moore and Keynes: a philosophical analysis of Keynes's "My Early Beliefs"', in B.W. Bateman and J.B. Davis (eds) *Keynes and Philosophy: essays on the origin of Keynes's thought*, Aldershot: Edward Elgar.

Siggel, E. (1992) 'Productivity Measurement from a Deficient Data Base: an empirical study of Kenya's manufacturing sector', *Journal of Productivity Analysis* 3(4).

Simon, W.M. (1964) 'Auguste Comte's English Disciples', *Victorian Studies* 8(2).

Sinclair, W.A. (1971) 'The Tariff and Economic Growth in Pre-federation Victoria', *Economic Record* 47(1).

——— (1976) *The Process of Economic Development in Australia*, Melbourne: Longman Cheshire.

Singer, H.W. (1953) 'Obstacles to Economic Development', *Social Research* 20.

Skidelsky, R. (1983) *John Maynard Keynes: hopes betrayed 1883–1920*, London: Macmillan.

Skinner, A. (1982) 'A Scottish Contribution to Marxist Sociology', in I. Bradley and M. Howard (eds) *Classical and Marxian Political Economy: essays in honour of R.L. Meek*, London: Macmillan.

Smith, A. (1872) *An Inquiry into the Nature and Causes of the Wealth of Nations with a Life of the Author, an Introductory Discourse, and Supplemental Dissertations by J.R. McCulloch*, Edinburgh: A. and C. Black.

——— (1937) [1776] *An Inquiry into the Nature and Causes of the Wealth of Nations*, New York: Modern Library.

——— (1976) [1759] *The Theory of Moral Sentiments*, Oxford: Clarendon.

Smith, K. (1986) [1984] *The British Economic Crisis: its past and future*, Harmondsworth: Penguin.

Smith, L.D. (1976) 'An Overview of Agricultural Development Policy', in J.Heyer, J.K. Maitha and W.M. Senga, *Agricultural Development in Kenya: an economic assessment*, Nairobi: Oxford University Press.

Snow, E. (1968) [1938] *Red Star Over China*, London: Gollancz.

Soltau, R.H. (1959) [1931] *French Political Thought in the 19th Century*, New York: Russell and Russell.

Soo, F.Y.K. (1981) *Mao Tse-tung's Theory of Dialectic*, Dordrecht, Holland: D. Reidel.

Sorrenson, M.P.K. (1967) *Land Reform in Kikuyu Country: a study in government policy*, London: Oxford University Press.

Spadafora, D. (1990) *The Idea of Progress in Eighteenth-century Britain*, New Haven: Yale University Press.

Sraffa, P. and Dobb, M. (eds) (1966) 'Notes on Malthus's Principles of Political Economy', in *The Works and Correspondence of David Ricardo, Vol. II*, Cambridge: Cambridge University Press.

Staudt, K. (1991) *Managing Development: state, society, and international contexts*, Newbury Park, CA: Sage.

Stephen, L. (1877) 'Dy. Newman's Theory of Belief', *Fortnightly Review* 22 (n.s.).

Steuart, James (1967a) [1772] *The Principles of Money Applied to the Current State of Coin in Bengal*, Volume 5 of *The Works, Political, Metaphisical & Chronological of Sir James Steuart*, New York: A.M. Kelley.

——— (1967b) [1805] *The Works, Political, Metaphisical & Chronological of Sir James Steuart*, New York: A.M. Kelley.

Stigler, G. (1965) [1959] 'Bernard Shaw, Sidney Webb, and the theory of Fabian Socialism', in *Essays in the History of Economics*, Chicago: Chicago University Press.

Stokes, E. (1979) [1959] *The English Utilitarians and India*, Oxford: Clarendon Press.

Strachey, B. (1985) *The Strachey Line: an English family in America, in India, and at home 1570 to 1902*, London: Gollancz.

Strachey, J. and Strachey, R. (1882) *The Finances and Public Works of India from 1869 to 1881*, London: K. Paul, Trench.

Streissler, E. (1981) 'Schumpeter's Vienna and the Role of Credit in Innovation', in H. Frisch (ed.) *Schumpeterian Economics*, New York: Praeger.

Strong, M.K. (1930) *Public Welfare Administration in Canada*, Chicago: University of Chicago Press.

Stuart, R.C. (1988) *United States Expansionism and British North America, 1775–1871*, Chapel Hill: University of North Carolina Press.

Swainson, N. (1980) *The Development of Corporate Capitalism in Kenya 1918–1977*, London: Heinemann.

Swedberg, R. (1991a) 'Introduction: the man and his work', in R. Swedberg (ed.) *Joseph A. Schumpeter: the economics and sociology of capitalism*, Princeton: Princeton University Press.

——— (1991b) *Schumpeter: a biography*, Princeton: Princeton University Press.

Sweeny, A. (1976) *George-Etienne Cartier: a biography*, Toronto: McClelland and Stewart.

Szporluk, R. (1988) *Communism and Nationalism: Karl Marx versus Friedrich List*, Oxford: Oxford University Press.

Taylor, M. (1987) *The Possibility of Cooperation*, Cambridge: Cambridge University Press.

Teeple, G. (1972) 'Land, Labour and Capital in pre-Confederation Canada', in G. Teeple (ed.) *Capitalism and the National Question in Canada*, Toronto: University of Toronto Press.

Temple, W. (1972) [1673] *Observations upon the United Provinces of the Netherlands*, Oxford, Clarendon Press.

Theobald, R. (1990) *Corruption, Development and Underdevelopment*, Basingstoke: Macmillan.

Thirlwall, A.P. (1989) [1972] *Growth and Development: with special reference to developing economies*, Basingstoke: Macmillan.

Thomas, B.P. (1985) *Politics, Participation and Poverty: development through self-help in Kenya*, Boulder, CO: Westview Press.

——— (1987) 'Development through Harambee: who wins and who loses? Rural self-help projects in Kenya', *World Development* 15(4).

Thomas, G. (1987) *The Moral Philosophy of T.H. Green*, Oxford: Clarendon.

Thompson, A. (1970) 'The Enigma of Australian Manufacturing', *Australian Economic Papers* 9(1).

Thompson, H.F. (1965) 'Adam Smith's Philosophy of Science', *Quarterly Journal of Economics* 79(2).

Thompson, N. (1988) *The Market and its Critics: socialist political economy in nineteenth-century Britain*, London: Routledge.

Thurston, A. (1987) 'Smallholder Agriculture

in Colonial Kenya: The official mind and the Swynnerton plan', *Cambridge African Monographs* 8, Cambridge: African Studies Centre.

Tignor, R.L. (1993) 'Race, Nationality, and Industrialization in Decolonizing Kenya, 1945–1963', *International Journal of African Historical Studies* 26(6).

Tilley, R. (1967) 'Germany, 1815–1870', in R. Cameron (ed.) *Banking in Early Stages of the Industrial Revolution*, New York: Oxford University Press.

——— (1986) 'German Banking 1850–1914: development assistance for the strong', *Journal of European Economic History* 15(1). (Reprinted in R. Cameron (ed.) *Financing Industrialisation, Volume 1*, Aldershot: Edward Elgar, 1992.)

Todaro, M.P. (1989) [1977] *Economics for a Developing World: an introduction to principles, problems and policies for development* [first edn 1977, *Economic Development in the Third World*], London: Longman.

——— (1992) *Economics for a Developing World: an introduction to principles, problems and policies for development* [1st edn 1977, *Economic Development in the Third World*], London: Longman.

Toye, J. (1980) 'Does Development Studies Have a Core?' *IDS Bulletin* 11(3).

——— (1983) 'Interdependence from Kant to Brandt', in *Third World Studies, Block 4: The International Setting*, Milton Keynes: Open University Press.

——— (1987) *Dilemmas of Development: reflections on the counter-revolution in development theory and policy*, Oxford: Blackwell.

——— (1993) 'Keynes, Russia and the State in Developing Countries', and 'Discussion' in D. Crabtree and A.P. Thirlwall (eds) *Keynes and the Role of the State: the tenth Keynes seminar*, Basingstoke: Macmillan.

Trainor, L. (1994) *British Imperialism and Australian Nationalism: manipulation, conflict and compromise in the late nineteenth century*, Melbourne: Cambridge University Press.

Tribe, K. (1988a) 'Friedrich List and the Critique of Cosmopolitical Economy', *The Manchester School* 56(1).

——— (1988b) *Governing Economy: the reformation of German Economic Discourse 1750–1840*, Cambridge: Cambridge University Press.

Tucker, G.N. (1964) [1936] *The Canadian Commercial Revolution 1845–1851*, Toronto: McClelland and Stewart.

Tulchinsky, G.J.J. (1977) *The River Barons: Montreal businessmen and the growth of industry and transportation 1837–53*, Toronto: University of Toronto Press.

Uchida, H. (1988) *Marx's 'Grundrisse' and Hegel's 'Logic'*, London: Routledge.

Udvardy, M. (1988) 'Women's Groups near the Kenyan Coast: patron clientship in the development arena', in D. Brokensha (ed.) *Anthropology of Development and Change in East Africa*, Boulder: Westview Press.

United Kingdom (1934) *The Kenya Land Commission: Report*, Cmd 4556 (Carter Commission Report), Nairobi: Government Printer.

United Nations (1955) *Models and Techniques of Community Development in the United Kingdom Dependent and Trust Territories*, New York: United Nations.

Veblen, T. (1919) *The Place of Science in Modern Civilisation and Other Essays*, New York: Viking Press.

Vincent, A. (1986) *The Philosophy of T.H. Green*, Aldershot: Gower.

Vincent, A. and Plant, R. (1984) *Philosophy, Politics and Citizenship: The life and thought of the British Idealists*, Oxford: Blackwell.

Wager, W.W. (1967) 'Modern Views of the Origins of the Idea of Progress', *Journal of the History of Ideas* 28(1).

Wakefield, E.G. (1833) *England and America: a comparison of the social and political state of both nations, vol. 1*, London: Richard Bentley.

Wallace, A.R. (1892) *How Land Nationalisation*

will Benefit Householders, Labourers and Mechanics, London: Land Nationalisation Society (LNS Tract 3).

——— (1897) 'Reoccupation of the Land: the only solution to the problem of the unemployed', in E. Carpenter (ed.) *Forecasts of the Coming Century*, Manchester: The Labour Press.

Wallich, H.C. (1958) [1952] 'Some Notes Towards a Theory of Derived Development', in A.N. Agarwala and S.P. Singh (eds) *The Economics of Underdevelopment*, Bombay: Oxford University Press.

Walters, W.R. (1961) 'Schumpeter's Contributions and Catholic Social Thought', *Review of Social Economy* 19(2).

Wanjigi Report, see Kenya Republic, 1983.

Ward, J.M. (1976) *Colonial Self-government: the British experience 1759–1856*, London: Macmillan.

Ward, P.M. (ed.) (1989) *Corruption, Development and Inequality: soft touch or hard graft?*, London: Routledge.

Ward, W.G. (1912) *Life of John Henry Newman, vol. 1*, London: Longman.

Ward, W.P. and McDonald, R.A.J. (eds) (1981) *British Columbia: historical readings*, Vancouver: Douglas and McIntyre.

Warren, B. (1980) *Imperialism: pioneer of capitalism*, London: Verso.

Wasserman, G. (1976) *Politics of Decolonization: Kenya Europeans and the land issue 1960–1965*, Cambridge: Cambridge University Press.

Watkins, M.H. (1963) 'A Staple Theory of Economic Growth', *Canadian Journal of Economics and Political Science* 29(2). (Reprinted in G. Laxer (1981) *Perspectives on Canadian Economic Development: class, staples, gender and elites*, Toronto: Oxford University Press.)

——— (1977) 'The Staple Theory Revisited', *Journal of Canadian Studies* 12(5).

Weaver, M.J. (1985) *Newman and the Modernists*, Lanham, MD: University Press of America.

Weinberg, A. (1982) *The Influence of Auguste Comte on the Economics of John Stuart Mill*, London: E.G. Weinberg.

Weiner, M.J. (1971) *Between Two Worlds: the political thought of Graham Wallas*, Oxford: Clarendon Press.

Weinroth, H. (1974) 'Radicalism and Nationalism: an increasingly unstable equation', in A.J.A. Morris (ed.) *Edwardian Radicalism: 1900–1914*, London: Routledge and Kegan Paul.

West, C. (1989) *The American Evasion of Philosophy*, London: Macmillan.

Whitehead, A.N. (1929) *Process and Reality: an essay in cosmology*, Cambridge: Cambridge University Press.

——— (1953) [1925] *Science and the Modern World*, New York: Free Press.

——— (1966) 'Religious Assertions and Doctrinal Development', *Theological Studies* 27(4).

Wickwire, F. and Wickwire, W. (1980) *Cornwallis: the imperial years*, Chapel Hill: University of North Carolina Press.

Widner, J.A. (1992) *The Rise of a Part-state in Kenya: from 'harambee!' to nyayo!*, Berkeley: University of California Press.

Willey, B. (1973) [1949] *Nineteenth-century Studies: Coleridge to Matthew Arnold*, Harmondsworth: Penguin.

Williamson, E. (1992) *The Penguin History of Latin America*, Harmondsworth: Penguin.

Wilson, D.W. (1986) 'Equilibrium Theory and the Law of Value', London Guildhall University Department of Economics Working Paper 6.

——— (1989) 'On the Foundations of Marx's Value Theory: a mathematical model of social labour', London Guildhall University Department of Economics Working Paper 14.

——— (1991) 'Abstract General Labour',

London Guildhall University Department of Economics Working Paper 16.

Wilson, L.S. (1992) 'The Harambee Movement and Efficient Public Good Provision in Kenya', *Journal of Public Economics* 48(1).

Winfrey, R. (1907) 'A Plea for Smallholdings', *Progress* 2(1).

Winslow, E.G. (1986) '"Human Logic' and Keynes's Economics', *Eastern Economic Journal* 12(4).

——— (1989a) 'Organic Interdependence, Uncertainty and Economic Analysis', *Economic Journal* 99(4).

——— (1989b) '"Human Logic" and Keynes's Economics: A Reply', *Eastern Economic Journal* 15(1).

Wittfogel, K.A. (1963) 'Some Remarks on Mao's Handling of Concepts', *Studies in Soviet Thought* 3(4).

Wolfe, W. (1975) *From Radicalism to Socialism: Men and Ideas in the Formation of Fabian Socialist doctrines, 1881–1889*, New Haven: Yale University Press.

Wood, J.C. (1983) *British Economists and the Empire*, London: Croom Helm.

——— (ed.) (1991) *J.A. Schumpeter: critical assessments, Volume 2*, London: Routledge.

Woodcock, G. (1975) *Amor de Cosmos: journalist and reformer*, Toronto: Oxford University Press.

——— (1990) *British Columbia: a history of the province*, Vancouver: Douglas and McIntyre.

Woolf, R.D. (1974) *The Economics of Colonialism: Britain and Kenya, 1870–1930*, New Haven: Yale University Press.

World Bank (1983) *Kenya: Growth and Structural Change. Volume II*, Washington, DC: World Bank.

——— (1989) *Sub-Saharan Africa. From crisis to sustainable growth: a long-term perspective*, Washington, DC: World Bank.

——— (1992) *Kenya: reinvesting in stabilisation and growth through public sector management*, Washington, DC: World Bank.

Wright, R.W. (1993) *Economics, Enlightenment and Canadian Nationalism*, Kingston: McGill-Queen's University Press.

Wright, T.R. (1986) *The Religion of Humanity: the impact of Comtean positivism on Victorian Britain*, Cambridge: Cambridge University Press.

Wrigley, C.C. (1965) 'Kenya: the patterns of economic life 1902–45', in V. Harlow and E.M. Chilver (eds) *History of East Africa, Volume 2*, Oxford: Clarendon Press.

Wrigley, E.A. and Souden, D. (eds) (1986) *The Works of Thomas Robert Malthus, Volume 1*, London: William Pickering.

Yearley, L.H. (1978) *The Ideas of Newman: Christianity and Human Religiosity*, University Park: Penn State University Press.

Young, B.J. (1978) *Promoters and Politicians: the North Shore Railways in the history of Quebec, 1854–85*, Toronto: University of Toronto Press.

——— (1981) *George-Etienne Cartier: Montreal Bourgeois*, Kingston: McGill-Queen's University Press.

——— (1986) *In its Corporate Capacity: the seminary of Montreal as a business institution, 1816–1876*, Kingston, Ontario: McGill-Queens University Press.

——— and Dickinson, A. (1988) *A Short History of Quebec: a socio-economic perspective*, Toronto: Copp Clark Pitman.

——— (1992) 'Positive Law, Positive State: class realignment and the transformation of Lower Canada, 1815–1866', in A. Greer and I. Radforth (eds) *Colonial Leviathan: state formation in mid-nineteenth-century Canada*, Toronto: University of Toronto Press.

Young, R.M. (1985) *Darwin's Metaphor: nature's place in Victorian culture*, Cambridge: Cambridge University Press.

Yu-ning, Li (1971) *The Introduction of Socialism into China*, New York: Columbia University Press.

Zea, L. (1963) *The Latin American Mind*, trans.

J.H. Abbot and L. Durham, Norman: University of Oklahoma Press.

——— (1974) *Positivism in Mexico*, trans. J. Schulte, Austin: University of Texas Press.

Zeller, S. (1987) *Inventing Canada: early Victorian science and the idea of a transcontinental nation*, Toronto: University of Toronto Press.

Zolberg, A.R. (1964) 'The Dakar Colloquium: the search for a doctrine', in W.H. Friedland and C.G. Rosberg (eds) *African Socialism*, Stanford: Stanford University Press.

van Zwanenberg, R.M.A. (1975) *Colonial Capitalism and Labour in Kenya 1919–39*, Nairobi: East African Literature Bureau.

——— with King, A. (1975) *An Economic History of Kenya and Uganda, 1800–1970*, London: Macmillan.

Zylstra, B. (1970) [1968] *From Pluralism to Collectivism: the development of Harold Laski's political thought*, Assen, Netherlands: Van Gorcum.

INDEX